Design and Analysis of Experiments

DESIGN AND ANALYSIS OF EXPERIMENTS
Third Edition

Douglas C. Montgomery

ARIZONA STATE UNIVERSITY

JOHN WILEY & SONS
New York   Chichester   Brisbane   Toronto   Singapore

Library of Congress Cataloging in Publication Data:

Montgomery, Douglas C.
   Design and analysis of experiments / Douglas C. Montgomery.—3rd ed.
      p.   cm.

   Includes bibliographical references and index.

   1. Experimental design.   I. Title.
QA279.M66   1991
001.4'34—dc20   90-12678
                                                              CIP

Printed in Singapore.

10 9 8 7 6 5 4 3 2

To Meredith, Neil, and Colin

Preface

This is an introductory textbook dealing with the design and analysis of experiments. It is based on college-level courses in experimental design that I have taught at the Georgia Institute of Technology, the University of Washington, and Arizona State University for the last 20 years. It also reflects the experimental design methods that I have found useful in my own professional practice as an engineering consultant in the general areas of product and process design and development, process troubleshooting, and quality improvement.

The book is intended for readers who have completed a first course in statistical methods and who are familiar with the basic concepts of the normal, t, chi-square, and F distributions, confidence intervals, and the common tests of hypotheses. Some familiarity with matrix algebra is required in portions of Chapters 15 and 16.

Because of the relatively modest prerequisites, this book can be used in a second statistics course for undergraduate students in engineering, physical science, mathematics, biology, and social science. I have also used this book as the basis of a seminar on experimental design for practicing professionals. I have taught this material to engineers, scientists, technicians, and managers from a broad range of industries. The material can be covered more rapidly, with more emphasis on the mathematical aspects of the subject, in an experimental design course for first-year graduate students. There are numerical examples illustrating most of the design and analysis techniques, and this makes the book useful as a reference work for experimenters in various disciplines. These examples are all based on real-world applications of experimental design.

About the Book

The third edition is a major revision of the book. While maintaining a balance between design and analysis topics, I have added much new material and several new examples. The reorganization of the material on 2^k factorial and fractional factorial designs in Chapters 9, 10, and 11 should provide more in-depth treatment of these important subjects. Chapter 12, which covers additional topics in factorial and fractional factorial designs, is new. I have included a discussion of the Taguchi approach to parameter design along with a critique of this method and some recommended alternatives in this chapter. Chapter 16 on response surfaces contains additional topics on the selection of a response surface design and an introduction to the mixture problem.

Throughout the book I have stressed the importance of experimental design as a tool for practicing engineers to use for product design and development, process development and improvement, and process troubleshooting. The use of experimental design in helping engineers develop products and processes that are robust to environmental factors and other sources of variation is illustrated. I believe that the successful use of experimental design early in the product/process development cycle can substantially reduce development time and cost and can lead to processes and products that perform better in the field and have higher reliability than those developed using other approaches.

The book contains more material than can usually be covered comfortably in one course, and I hope instructors will be able to vary the contents of each course offering and perhaps to discuss certain topics in greater depth, depending on class interest. There are problem sets at the end of each substantive chapter. These problems vary in scope from computational exercises, designed to reinforce the fundamentals, to extensions or elaborations of basic principles.

Acknowledgments

I express my appreciation to the many students and instructors who have used the two earlier editions of this book and who have made helpful suggestions for its revision. The contributions of Dr. Lloyd S. Nelson, Dr. Andre Khuri, Dr. Stephen R. Schmidt, Dr. Russell G. Heikes, Dr. Harrison M. Wadsworth, Dr. William W. Hines, Dr. Arvind Shah, Dr. Jane Ammons, and Mr. Pat Whitcomb were particularly valuable.

The contributions of the professional practitioners with whom I have worked have been invaluable. It is impossible to mention everyone, but some of the major contributors include Mr. Tom Bingham, Mr. Dick Vaughn, Dr. Julian Anderson, Mr. Richard Alkire, and Mr. Chase Neilson of the Boeing Company; Mr. Mike Goza, Mr. Don Walton, Ms. Karen Madison, Mr. Jeff Stevens, and Mr. Bod Kohm of Alcoa; Dr. Jay Gardiner, Mr. John Butora, Mr. Dana Lesher, Mr. Lolly Marwah, Dr. Paul Tobias, and Mr. Leon Mason of IBM; Ms. Elizabeth A. Peck of the Coca-Cola Company; Dr. Sadri Khalessi and Mr. Franz Wagner of Signetics; Mr. Robert V. Baxley of Monsanto Chemicals; Mr. Harry Peterson-Nedry and Dr. Russell Boyles of Precision Castparts Corporation; Mr. Bill New, Mr. Randy Schmid, and Ms. Diane Nowatney of Allied-Signal Aerospace; Mr. John M. Fluke, Jr. of the John Fluke Manufacturing Company; Mr. Larry Newton and Mr. Kip Howlett of Georgia-Pacific; Dr. William DuMouchel, Dr. Cynthia Lowry, Dr. Ernesto Ramos, and Dr. Pat Spagon of BBN Software Products Corporation.

I am also indebted to Professor E. S. Pearson and the *Biometrika* Trustees,

John Wiley & Sons, Prentice-Hall, The American Statistical Association, The Institute of Mathematical Statistics, and the editors of *Biometrics* for permission to use copyrighted material. I am grateful to the Office of Naval Research and the IBM Corporation for supporting much of my research in engineering statistics and experimental design.

Douglas C. Montgomery
Tempe, Arizona

Contents

Chapter 1
Introduction

1-1 WHAT IS EXPERIMENTAL DESIGN?

Experiments are performed by investigators in virtually all fields of inquiry, usually to discover something about a particular process or system. Literally, an experiment is a *test*. A *designed experiment* is a test or series of tests in which purposeful changes are made to the input variables of a process or system so that we may observe and identify the reasons for changes in the output response.

The process or system under study can be represented by the model shown in Figure 1-1. We can usually visualize the process as a combination of machines, methods, people, and other resources that transforms some input (often a material) into an output that has one or more observable responses. Some of the process variables $x_1, x_2, \ldots, x_p$ are controllable, whereas other variables $z_1, z_2, \ldots, z_q$ are uncontrollable (although they may be controllable for purposes of a test). The objectives of the experiment may include the following:

1. Determining which variables are most influential on the response, y.
2. Determining where to set the influential x's so that y is almost always near the desired nominal value.
3. Determining where to set the influential x's so that variability in y is small.
4. Determining where to set the influential x's so that the effects of the uncontrollable variables $z_1, z_2, \ldots, z_q$ are minimized.

Experimental design methods play an important role in *process development* and *process troubleshooting* to improve performance. The objective in many cases may be to develop a *robust* process, that is, a process affected minimally by external sources of variability (the z's).

As an example of an experiment, suppose that a metallurgical engineer is interested in studying the effect of two different hardening processes, oil quench-

Controllable factors

Figure 1-1. General model of a process or system.

ing and saltwater quenching, on an aluminum alloy. Here the objective of the experimenter is to determine the quenching solution that produces the maximum hardness for this particular alloy. The engineer decides to subject a number of alloy specimens to each quenching medium and measure the hardness of the specimens after quenching. The average hardness of the specimens treated in each quenching solution will be used to determine which solution is best.

As we consider this simple experiment, a number of important questions come to mind:

1. Are these two solutions the only quenching media of potential interest?
2. Are there any other factors that might affect hardness that should be investigated or controlled in this experiment?
3. How many specimens of alloy should be tested in each quenching solution?
4. How should the specimens be assigned to the quenching solutions, and in what order should the data be collected?
5. What method of data analysis should be used?
6. What difference in average observed hardness between the two quenching media will be considered important?

All of these questions, and perhaps many others, will have to be answered satisfactorily before the experiment is performed.

In any experiment, the results and conclusions that can be drawn depend to a large extent on the manner in which the data were collected. To illustrate this point, suppose that the metallurgical engineer in the above experiment used specimens from one heat in the oil quench and specimens from a second heat in the saltwater quench. Now, when the mean hardness is compared, the engineer is unable to say how much of the observed difference is the result of the quenching media and how much is the result of inherent differences between the heats.[1] Thus, the method of data collection has adversely affected the conclusions that can be drawn from the experiment.

1-2   APPLICATIONS OF EXPERIMENTAL DESIGN

Experimental design methods have found broad application in many disciplines. In fact, we may view experimentation as part of the scientific process and as one of the ways we learn about how systems or processes work. Generally, we learn through a series of activities in which we make conjectures about a process, perform experiments to generate data from the process, and then use the information from the experiment to establish new conjectures, which lead to new experiments, and so on.

Experimental design is a critically important tool in the engineering world for improving the performance of a manufacturing process. It also has extensive application in the development of new processes. The application of experimental design techniques early in process development can result in

1.   Improved process yields.
2.   Reduced variability and closer conformance to nominal or target requirements.
3.   Reduced development time.
4.   Reduced overall costs.

Experimental design methods also play a major role in *engineering design* activities, where new products are developed and existing ones improved. Some applications of experimental design in engineering design include

1.   Evaluation and comparison of basic design configurations.
2.   Evaluation of material alterations.
3.   Selection of design parameters so that the product will work well under a wide variety of field conditions, that is, so that the product is *robust.*
4.   Determination of key product design parameters that impact product performance.

The use of experimental design in these areas can result in products that are easier to manufacture, products that have enhanced field performance and reliability, lower product cost, and shorter product design and development time. We now present several examples that illustrate some of these ideas.

---

[1] A specialist in experimental design would say that the effects of quenching media and heat were *confounded;* that is, the effects of these two factors cannot be separated.

Example 1-1

Characterizing a Process

A flow solder machine is used in the manufacturing process for printed circuit boards. The machine cleans the boards in a flux, preheats the boards, and then moves them along a conveyor through a wave of molten solder. This solder process makes the electrical and mechanical connections for the leaded components on the board.

The process currently operates around the 1 percent defective level. That is, about 1 percent of the solder joints on a board are defective and require manual retouching. However, since the average printed circuit board contains over 2000 solder joints, even a 1 percent defective level results in far too many solder joints requiring rework. The process engineer responsible for this area would like to use a designed experiment to determine which machine parameters are influential in the occurrence of solder defects and which adjustments should be made to those variables to reduce solder defects.

The flow solder machine has several variables that can be controlled. They include:

1. Solder temperature.
2. Preheat temperature.
3. Conveyor speed.
4. Flux type.
5. Flux specific gravity.
6. Solder wave depth.
7. Conveyor angle.

In addition to these controllable factors, there are several other factors that cannot be easily controlled during routine manufacturing, although they could be controlled for the purposes of a test. They are

1. Thickness of the printed circuit board.
2. Types of components used on the board.
3. Layout of the components on the board.
4. Operator.
5. Production rate.

In this situation, the engineer is interested in *characterizing* the flow solder machine; that is, he wants to determine which factors (both controllable and uncontrollable) affect the occurrence of defects on the printed circuit boards. To accomplish this, he can design an experiment that will enable him to estimate the magnitude and direction of the factor effects; that is, how much does the response variable (defects per unit) change when each factor is changed, and does changing the factors *together* produce different results than are obtained from individual factor adjustments. Sometimes we call an experiment such as this a *screening experiment*.

The information from this screening or characterization experiment will be used to identify the critical process factors and to determine the direction of adjustment for these factors to reduce further the number of defects per unit. The experiment may also provide information about which factors should be more carefully controlled during routine manufacturing to prevent high defect levels and erratic process performance. Thus, one result of the experiment could be the application of techniques such as control charts to one or more *process* variables (such as solder temperature), in addition to control charts on process output. Over time, if the process is improved enough, it may be possible to base most of the process control plan on controlling process input variables instead of control charting the output.

■

Example 1-2

Optimizing a Process

In a characterization experiment, we are usually interested in determining which process variables affect the response. A logical next step is to optimize, that is, to determine the region in the important factors that leads to the best possible response. For example, if the response is yield, we would look for a region of maximum yield, whereas if the response is variability in a critical product dimension, we would seek a region of minimum variability.

Suppose that we are interested in improving the yield of a chemical process. We know from the results of a characterization experiment that the two most important process variables that influence the yield are operating temperature and reaction time. The process currently runs at 155°F and 1.7 hours of reaction time, producing yields of around 75 percent. Figure 1-2 shows a view of the time–temperature region from above. In this graph, the lines of constant yield are connected to form response *contours,* and we have shown the contour lines for yields of 60 percent, 70 percent, 80 percent, 90 percent, and 95 percent. These contours are projections on the time–temperature region of cross sections of the yield surface corresponding to the aforementioned percent yields. This surface is sometimes called a *response surface.* The true response surface in Figure 1-2 is unknown to the process personnel, so experimental methods will be required to optimize the yield with respect to time and temperature.

To locate the optimum, it is necessary to perform an experiment that varies time and temperature together. This type of experiment is called a *factorial experiment,* and an example of the results of a factorial experiment with both time and temperature run at two levels is shown in Figure 1-2. The responses observed at the four corners of the square indicate that we should move in the general direction of increased temperature and decreased reaction time to increase yield. A few additional runs could be performed in this direction, and this additional experimentation would be sufficient to locate the region of maximum yield.

■

Example 1-3

A Product Design Example

Experimental design methods can often be applied in the product design process. To illustrate, suppose that a group of engineers are designing a door hinge for an automobile. The quality characteristic of interest is the check effort, or the holding ability of the door latch that prevents the door from swinging closed when the vehicle is parked on a hill. The check mechanism consists of a leaf spring and a roller. When

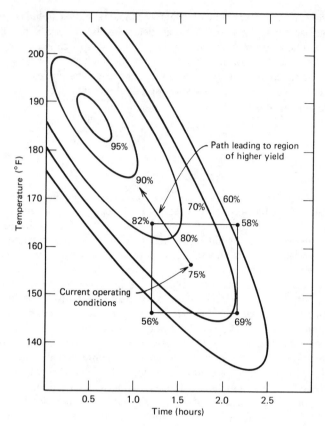

Figure 1-2. Contour plot of yield as a function of reaction time and reaction temperature, illustrating an optimization experiment.

the door is opened, the roller travels through an arc causing the leaf spring to be compressed. To close the door, the spring must be forced aside, and this creates the check effort. The engineering team thinks that check effort is a function of the following factors:

1. Roller travel distance.
2. Spring height from pivot to base.
3. Horizontal distance from pivot to spring.
4. Free height of the reinforcement spring.
5. Free height of the main spring.

The engineers can build a prototype hinge mechanism in which all of these factors can be varied over certain ranges. Once appropriate levels for these five factors have been identified, an experiment can be designed consisting of various combinations of the factor levels, and the prototype hinge can be tested at these combinations. This will produce information concerning which factors are most influential on the latch check effort, and through analysis of this information, the latch design can be improved.

■

Example 1-4

Determining System and Component Tolerances

The Wheatstone bridge shown in Figure 1-3 is a device used for measuring an unknown resistance, Y. The adjustable resistor B is manipulated until a particular current flow is obtained through the ammeter (usually $X = 0$), and then the unknown resistance is calculated as

$$Y = \frac{BD}{C} - \frac{X^2}{C^2E} \ [A(D + C) + D(B + C)][B(C + D) + F(B + C)] \quad (1\text{-}1)$$

The engineer wants to design the circuit so that overall gage capability is good; that is, he would like for the standard deviation of the measurement error to be small. He has decided that $A = 20 \ \Omega$, $C = 2 \ \Omega$, $D = 50 \ \Omega$, $E = 1.5$ V, and $F = 2 \ \Omega$ are the best choices for the design parameters so far as gage capability is concerned, but the overall measurement error is still too high. This is likely due to the tolerances that have been specified on the circuit components. These tolerances are ±1 percent for each resistor (A, B, C, D, and F) and ±5 percent for the power supply E. These tolerance bands can be used to define high and low factor levels, and an experiment can be performed to determine which circuit components have the most critical tolerances and by how much these critical tolerances must be tightened to produce adequate gage capability. This will result in a design specification that tightens only the most critical tolerances the minimum amount possible consistent with desired measurement capability; consequently, a lower-cost design that is easier to manufacture will result.

Notice that in this experiment it is unnecessary actually to build the hardware since the response from the circuit can be calculated via Equation 1-1. The actual response variable for the experiment should be the standard deviation of Y. However,

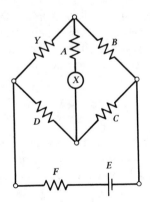

Figure 1-3. A Wheatstone bridge.

an equation for the standard deviation of Y for the circuit can be found by using a Taylor series expansion of Equation 1-1, as explained in many engineering statistics textbooks [for example, see Hines and Montgomery (1990), Chapter 5]. Therefore, the entire experiment can be performed using a computer model of the Wheatstone bridge.

■

1-3   BASIC PRINCIPLES

If an experiment such as the ones described in Examples 1-1 through 1-4 is to be performed most efficiently, then a scientific approach to planning the experiment must be employed. By the *statistical design of experiments,* we refer to the process of planning the experiment so that appropriate data that can be analyzed by statistical methods will be collected, resulting in valid and objective conclusions. The statistical approach to experimental design is necessary if we wish to draw meaningful conclusions from the data. When the problem involves data that are subject to experimental errors, statistical methodology is the only objective approach to analysis. Thus, there are two aspects to any experimental problem: the design of the experiment and the statistical analysis of the data. These two subjects are closely related since the method of analysis depends directly on the design employed. Both topics will be addressed in this book.

The three basic principles of experimental design are *replication, randomization,* and *blocking.* By replication we mean a repetition of the basic experiment. In the metallurgical experiment discussed in Section 1-1, replication would consist of treating a specimen by oil quenching and treating a specimen by saltwater quenching. Thus, if five specimens are treated in each quenching medium, we say that five *replicates* have been obtained. Replication has two important properties. First, it allows the experimenter to obtain an estimate of the experimental error. This estimate of error becomes a basic unit of measurement for determining whether observed differences in the data are really *statistically* different. Second, if the sample mean (e.g., $\bar{y}$) is used to estimate the effect of a factor in the experiment, then replication permits the experimenter to obtain a more precise estimate of this effect. For example; if σ^2 is the variance of the data, and there are n replicates, then the variance of the sample mean is

$$\sigma_{\bar{y}}^2 = \frac{\sigma^2}{n}$$

The practical implication of this is that, if we had $n = 1$ replicates and observed $y_1 = 145$ (oil quench) and $y_2 = 147$ (saltwater quench), we would probably be unable to make satisfactory inferences about the effect of the quenching medium—that is, the observed difference could be the result of experimental error. On the other hand, if n was reasonably large, and the experimental error was sufficiently small, then if we observed $\bar{y}_1 < \bar{y}_2$, we would be reasonably safe in concluding that saltwater quenching produces a higher hardness in this particular aluminum alloy than does oil quenching.

Randomization is the cornerstone underlying the use of satistical methods in experimental design. By randomization we mean that both the allocation of the experimental material and the order in which the individual runs or trials of the experiment are to be performed are randomly determined. Statistical methods require that the observations (or errors) be independently distributed random variables. Randomization usually makes this assumption valid. By properly randomizing the experiment, we also assist in "averaging out" the effects of extraneous factors that may be present. For example, suppose that the specimens in the above experiment are of slightly different thicknesses and that the effectiveness of the quenching medium may be affected by specimen thickness. If all the specimens subjected to the oil quench are thicker than those subjected to the saltwater quench, then we may be continually handicapping one quenching medium over the other. Randomly assigning the specimens to the quenching media alleviates this problem.

Blocking is a technique used to increase the precision of an experiment. A block is a portion of the experimental material that should be more homogeneous than the entire set of material. Blocking involves making comparisons among the conditions of interest in the experiment within each block. A simple example of the blocking principle is given in Section 2-5.1 of Chapter 2.

These basic principles of experimental design are an important part of every experiment. We will illustrate and emphasize them repeatedly throughout this book.

1-4 GUIDELINES FOR DESIGNING EXPERIMENTS

To use the statistical approach in designing and analyzing an experiment, it is necessary that everyone involved in the experiment have a clear idea in advance of exactly what is to be studied, how the data are to be collected, and at least a qualitative understanding of how these data are to be analyzed. An outline of the recommended procedure is as follows:

1. **Recognition of and statement of the problem.** This may seem to be a rather obvious point, but in practice it is often not simple to realize that a problem requiring experimentation exists, nor is it simple to develop a clear and generally accepted statement of this problem. It is necessary to develop all ideas about the objectives of the experiment. Usually, it is important to solicit input from all concerned parties: engineering, quality assurance, manufacturing, marketing, management, the customer, and operating personnel (who usually have much insight and who are too often ignored). A clear statement of the problem often contributes substantially to a better understanding of the phenomena and the final solution of the problem.

2. **Choice of factors and levels.** The experimenter must choose the factors to be varied in the experiment, the ranges over which these factors will be varied, and the specific levels at which runs will be made. Thought must also be given to how these factors are to be controlled at the desired values and how they are to be measured. For instance, in the flow solder experiment, the engineer has defined 12 variables that may affect the occurrence of solder defects. The engineer will also have to decide on a region of interest for each variable (that is, on the range over which each factor will be varied) and on how many levels of each variable to use. Process knowledge is required to do this. This process knowledge is usually a combination of practical experience and theoretical understanding. It is important to investigate all factors that may be of importance, and to not be overly-influenced by past experience; particularly when we are in the early stages of experimentation or when the process is not very mature. When the objective is factor screening or process characterization it is usually best to keep the number of factor levels low (most often two levels are used).

3. **Selection of the response variable.** In selecting the response variable, the experimenter should be certain that this variable really provides useful information about the process under study. Most often, the average or standard deviation (or both) of the measured characteristic will be the response variable. Multiple responses are not unusual. Gage capability (or measurement error) is also an important factor. If gage capability is poor, then only relatively large factor effects will be detected by the experiment or additional replication will be required.

4. **Choice of experimental design.** If the first three steps are done correctly, this step is relatively easy. Choice of design involves the consideration of sample size (number of replicates), the selection of a suitable run order for the experimental trials, and the determination of whether or not blocking or other randomization restrictions are involved. This book discusses some of the more important types of experimental designs, and it can ultimately be used as a catalog for selecting an appropriate experimental design for a wide variety of problems.

    In selecting the design, it is important to keep the experimental objectives in mind. In many engineering experiments, we already know at the outset that some of the factor levels will result in different values for the response. Consequently, we are interested in identifying *which* factors causes this difference and in estimating the *magnitude* of the response change. In other situations, we may be more interested in verifying uniformity. For example, two production conditions A and B may be compared, A being the standard and B being a more cost-effective alternative. The experimenter will then be interested in demonstrating that, say, there is no difference in yield between the two conditions.

5. **Performing the experiment.** When running the experiment, it is vital to monitor the process carefully to ensure that everything is being done according to plan. Errors in experimental procedure at this stage will usually

destroy experimental validity. Up-front planning is crucial to success. It is easy to underestimate the logistical and planning aspects of running a designed experiment in a complex manufacturing or research and development environment.

6. **Data analysis.** Statistical methods should be used to analyze the data so that results and conclusions are objective rather than judgmental in nature. If the experiment has been designed correctly and if it has been performed according to the design, then the statistical methods required are not elaborate. There are many excellent software packages designed to assist in data analysis, and simple graphical methods play an important role in data interpretation. Residual analysis and model adequacy checking are also important analysis techniques.

   Remember that statistical methods cannot prove that a factor (or factors) has a particular effect. They only provide guidelines as to the reliability and validity of results. Properly applied, statistical methods do not allow anything to be proved, experimentally, but they do allow us to measure the likely error in a conclusion or to attach a level of confidence to a statement. The primary advantage of statistical methods is that they add objectivity to the decision-making process. Statistical techniques coupled with good engineering or process knowledge and common sense will usually lead to sound conclusions.

7. **Conclusions and recommendations.** Once the data have been analyzed, the experimenter must draw *practical* conclusions about the results and recommend a course of action. Graphical methods are often useful in this stage, particularily in presenting the results to others. *Follow-up runs* and *confirmation testing* should also be performed to validate the conclusions from the experiment.

   Throughout this entire process, it is important to keep in mind that experimentation is an important part of the learning process, where we tentatively formulate hypotheses about a system, perform experiments to investigate these hypotheses, and on the basis of the results formulate new hypotheses, and so on. This suggests that experimentation is *iterative*. It is usually a major mistake to design a single, large, comprehensive experiment at the start of a study. A successful experiment requires knowledge of the important factors, the ranges over which these factors should be varied, the appropriate number of levels to use, and the proper units of measurement for these variables. Generally, we do not perfectly know the answers to these questions, but we learn about them as we go along. As an experimental program progresses, we often drop some imput variables, add others, change the region of exploration for some factors, or add new response variables. Consequently, we usually experiment *sequentially,* and as a general rule, no more than about 25 percent of the available resources should be invested in the first experiment. This will insure that sufficient resources are available to perform confirmation runs and ultimately accomplish the final objective of the experiment.

1-5 HISTORICAL PERSPECTIVE

The late Sir Ronald A. Fisher was the innovator in the use of statistical methods in experimental design. For several years he was responsible for statistics and data analysis at the Rothamsted Agricultural Experiment Station in London, England. Fisher developed and first used the analysis of variance as the primary method of statistical analysis in experimental design. In 1933, Fisher took a professorship at the University of London. He later was on the faculty of Cambridge University and held visiting professorships at several universities throughout the world. For an excellent biography of Fisher, see J. F. Box (1978). While Fisher was clearly the pioneer, there have been many other significant contributors to the literature of experimental design, including F. Yates, R. C. Bose, O. Kempthorne, W. G. Cochran, and G. E. P. Box. The Bibliography at the end of the book contains several works by these authors.

Many of the early applications of experimental design methods were in the agricultural and biological sciences, and as a result, much of the terminology of the field is derived from this heritage. However, the first industrial applications of experimental design began to appear in the 1930s, initially in the British textile and woolen industry. After World War II, experimental design methods were introduced to the chemical and process industries in the United States and Western Europe. These industry groups are still very fertile areas for using experimental design in product and process development. The semiconductor and electronics industry has also used experimental design methods for many years with considerable success.

In recent years, there has been a revival of interest in experimental design in the United States because many industries discovered that their off-shore competitors have been using designed experiments for many years and that this has been an important factor in their competitive success. The day is approaching (hopefully rapidly) when all engineers will receive formal training in experimental design as part of their undergraduate education. The successful integration of experimental design into the engineering profession is a key factor in the future competitiveness of the industrial base of the United States.

1-6 USING STATISTICAL TECHNIQUES IN EXPERIMENTATION

Much of the research in engineering, science, and industry is empirical and makes extensive use of experimentation. Statistical methods can greatly increase the efficiency of these experiments and often strengthens the conclusions so obtained. The intelligent use of statistical techniques in experimentation requires that the experimenter keep the following points in mind:

1. **Use your nonstatistical knowledge of the problem.** Experimenters are usually highly knowledgeable in their fields. For example, a civil engineer working

on a problem in hydrology typically has considerable practical experience and formal academic training in this area. In some fields there is a large body of physical theory on which to draw in explaining relationships between factors and responses. This type of nonstatistical knowledge is invaluable in choosing factors, determining factor levels, deciding how many replicates to run, interpreting the results of the analysis, and so forth. Using statistics is no substitute for thinking about the problem.

2. **Keep the design and analysis as simple as possible.** Don't be overzealous in the use of complex, sophisticated statistical techniques. Relatively simple design and analysis methods are almost always best. This is a good place to reemphasize step 4 of the procedure recommended in Section 1-4. If you do the design carefully and correctly, the analysis will almost always be relatively straightforward. However, if you botch the design badly, it is unlikely that even the most complex and elegant statistics can save the situation.

3. **Recognize the difference between practical and statistical significance.** Just because two experimental conditions produce mean responses that are statistically different, there is no assurance that this difference is large enough to have any practical value. For example, an engineer may determine that a modification to an automobile fuel injection system may produce a true mean improvement in gasoline mileage of 0.1 mi/gal. This is a statistically significant result. However, if the cost of the modification is $1000, then the 0.1 mi/gal difference is probably too small to be of any practical value.

4. **Experiments are usually iterative.** Remember that, in most situations, it is unwise to design too comprehensive an experiment at the start of a study. Successful design requires knowledge of the important factors, the ranges over which these factors are varied, the appropriate number of levels for each factor, and the proper units of measurement for each factor and response. Generally, we are not well-equipped to answer these questions at the beginning of the experiment, but we learn the answers as we go along. This argues in favor of the *iterative* or *sequential* approach discussed previously. Of course, there are situations where comprehensive experiments are entirely appropriate, but as a general rule, most experiments should be iterative. Consequently, we usually should not invest more than about 25 percent of the resources of experimentation (runs, budget, time, etc.) in the initial design. Often these first efforts are just learning experiences, and some resources must be available to accomplish the final objectives of the experiment.

Chapter 2
Simple Comparative Experiments

In this chapter, we consider experiments to compare two conditions (sometimes called *treatments*). These are often called *simple comparative experiments*. We begin with an example of an experiment performed to determine whether two different formulations of a product give equivalent results. The discussion leads to a review of several basic statistical concepts, such as random variables, probability distributions, random samples, sampling distributions, and tests of hypotheses.

2-1 INTRODUCTION

The tension bond strength of portland cement mortar is an important characteristic of the product. An engineer is interested in comparing the strength of a modified formulation in which polymer latex emulsions have been added during mixing to the strength of the unmodified mortar. The experimenter has collected 10 observations of strength for the modified formulation and another 10 observations for the unmodified formulation. The data are shown in Table 2-1. We could refer to the two different formulations as two *treatments* or as two levels of the *factor* formulations.

The data from this experiment are plotted in Figure 2-1. This display is called a *dot diagram*. Visual examination of these data give the immediate impression that the strength of the unmodified mortar is greater than the strength of the modified mortar. This impression is supported by comparing the *average* tension bond strengths, $\bar{y}_1 = 16.76$ kgf/cm^2 for the modified mortar and $\bar{y}_2 = 17.92$ kgf/cm^2 for the unmodified mortar. The average tension bond strengths in these two samples differ by what seems to be a nontrivial amount. However, it is not

Table 2-1 Tension Bond Strength Data for the
Portland Cement Formulation Experiment

j	Modified Mortar y_{1j}	Unmodified Mortar y_{2j}
1	16.85	17.50
2	16.40	17.63
3	17.21	18.25
4	16.35	18.00
5	16.52	17.86
6	17.04	17.75
7	16.96	18.22
8	17.15	17.90
9	16.59	17.96
10	16.57	18.15

obvious that this difference is large enough to imply that the two formulations really *are* different. Perhaps this observed difference in average strengths is the result of sampling fluctuation and the two formulations are really identical. Possibly another two samples would give opposite results, with the strength of the modified mortar exceeding that of the unmodified formulation.

A technique of statistical inference called *hypothesis testing* (some prefer *significance testing*) can be used to assist the experimenter in comparing these two formulations. Hypothesis testing allows the comparison of the two formulations to be made on *objective* terms, with a knowledge of the risks associated with reaching the wrong conclusion. To present procedures for hypothesis testing in simple comparative experiments, however, it is first necessary to develop and review some elementary statistical concepts.

2-2 BASIC STATISTICAL CONCEPTS

Each of the observations in the portland cement experiment described above would be called a *run*. Notice that the individual runs differ, so there is fluctuation

Figure 2-1. Dot diagram for the tension bond strength data in Table 2-1.

or *noise* in the results. This noise is usually called *experimental error* or simply *error*. It is a *statistical* error, meaning that it arises from variation that is uncontrolled and generally unavoidable. The presence of error or noise implies that the response variable, tension bond strength, is a *random variable*. A random variable may be either *discrete* or *continuous*. If the set of all possible values of the random variable is either finite or countably infinite, then the random variable is discrete, whereas if the set of all possible values of the random variable is an interval, then the random variable is continuous.

Graphical Description of Variability ▪ We often use simple graphical methods to assist in analyzing the data from an experiment. The *dot diagram,* illustrated in Figure 2-1, is a very useful device for displaying a small body of data (say up to about 20 observations). The dot diagram enables the experimenter to see quickly the general *location* or central tendency of the observations and their *spread*. For exmaple, in the portland cement tension bond experiment, the dot diagram reveals that the two formulations probably differ in mean strength but that both formulations produce about the same variation in strength.

If the data are fairly numerous, then the dots in a dot diagram become difficult to distinguish, and a histogram may be preferable. Figure 2-2 presents a histogram for 200 observations on the metal recovery or yield from a smelting process. The histogram shows the central tendency, spread, and general shape of the distribution of the data. Recall that a histogram is constructed by dividing the horizontal axis into intervals (usually of equal length) and drawing a rectangle over the *j*th interval with area of the rectangle proportional to n_j, the number of observations that fall in that interval.

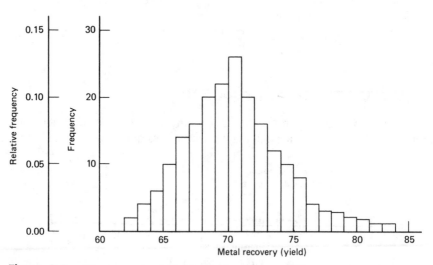

Figure 2-2. Histogram for 200 observations on metal recovery (yield) from a smelting process.

The *box plot* (or *box and whisker plot*) is a very useful way to display data. A box plot displays the minimum, the maximum, the lower and upper quartiles (the 25th percentile and the 75th percentiles, respectively), and the median (the 50th percentile) on a rectangular box aligned either horizontally or vertically. The box extends from the lower quartile to the upper quartile, and a line is drawn through the box at the median. Lines (or whiskers) extend from the ends of the box to the minimum and maximum values.

Figure 2-3 presents the box plots for the two samples of tension bond strength in the portland cement mortar experiment. This display clearly reveals the difference in mean strength between the two formulations. It also indicates that both formulations produce reasonably symmetric distributions of strength with similar variability or spread.

Dot diagrams, histograms, and box plots are useful for summarizing the information in a *sample* of data. To describe the observations that might occur in a sample more completely; we use the concept of the probability distribution.

Probability Distributions ▪ The probability structure of a random variable, say y, is described by its *probability distribution*. If y is discrete, we often call the probability distribution of y, say $p(y)$, the probability function of y. If y is continuous, the probability distribution of y, say $f(y)$, is often called the probability density function for y.

Figure 2-4 illustrates hypothetical discrete and continuous probability distributions. Notice that in the discrete probability distribution, it is the height of the

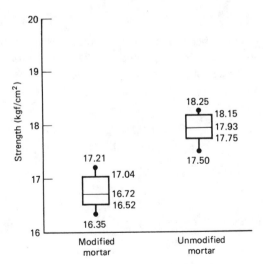

Figure 2-3. Box plots for the portland cement tension bond strength experiment.

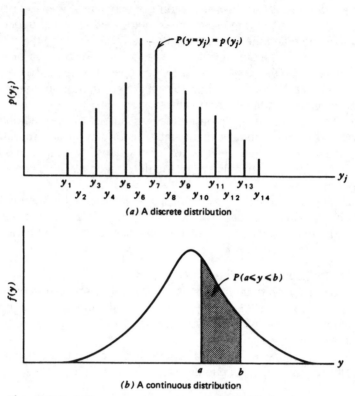

Figure 2-4.   Discrete and continuous probability distributions.

function $p(y)$ that represents probability, whereas in the continuous case, it is the area under the curve $f(y)$ associated with a given interval that represents probability. The properties of probability distributions may be summarized quantitatively as follows:

y discrete:
$$0 \le p(y_j) \le 1 \qquad \text{all values of } y_j$$
$$P(y = y_j) = p(y_j) \qquad \text{all values of } y_j$$
$$\sum_{\substack{\text{all values} \\ \text{of } y_j}} p(y_j) = 1$$

y continuous:
$$0 \le f(y)$$
$$P(a \le y \le b) = \int_a^b f(y)\, dy$$
$$\int_{-\infty}^{\infty} f(y)\, dy = 1$$

Mean, Variance, and Expected Values ▪ The mean of a probability distribution is a measure of its central tendency or location. Mathematically, we define the mean (e.g., μ) as

$$\mu = \begin{cases} \int_{-\infty}^{\infty} y f(y) \, dy & y \text{ continuous} \\ \sum_{\text{all } y} y p(y) & y \text{ discrete} \end{cases} \quad (2\text{-}1)$$

We may also express the mean in terms of the *expected value* or the long-run average value of the random variable y as

$$\mu = E(y) = \begin{cases} \int_{-\infty}^{\infty} y f(y) \, dy & y \text{ continuous} \\ \sum_{\text{all } y} y p(y) & y \text{ discrete} \end{cases} \quad (2\text{-}2)$$

where E denotes the expected value *operator*.

The spread or dispersion of a probability distribution can be measured by the *variance*, defined as

$$\sigma^2 = \begin{cases} \int_{-\infty}^{\infty} (y - \mu)^2 f(y) \, dy & y \text{ continuous} \\ \sum_{\text{all } y} (y - \mu)^2 p(y) & y \text{ discrete} \end{cases} \quad (2\text{-}3)$$

Note that the variance can be expressed entirely in terms of expectation since

$$\sigma^2 = E[(y - \mu)^2] \quad (2\text{-}4)$$

Finally, the variance is used so extensively that it is convenient to define a variance operator V such that

$$V(y) \equiv E[(y - \mu)^2] = \sigma^2 \quad (2\text{-}5)$$

The concepts of expected value and variance are used extensively throughout this book, and it may be helpful to review several elementary results concerning these operators. If y is a random variable with mean μ and variance σ^2 and c is a constant, then

1. $E(c) = c$
2. $E(y) = \mu$
3. $E(cy) = cE(y) = c\mu$
4. $V(c) = 0$
5. $V(y) = \sigma^2$
6. $V(cy) = c^2 V(y) = c^2 \sigma^2$

If there are two random variables, for example, y_1 with $E(y_1) = \mu_1$ and $V(y_1) = \sigma_1^2$ and y_2 with $E(y_2) = \mu_2$ and $V(y_2) = \sigma_2^2$, then we have

7. $E(y_1 + y_2) = E(y_1) + E(y_2) = \mu_1 + \mu_2$

It is possible to show that

8. $V(y_1 + y_2) = V(y_1) + V(y_2) + 2 \, \text{Cov}(y_1, y_2)$

where

$$\text{Cov}(y_1, y_2) = E[(y_1 - \mu_1)(y_2 - \mu_2)] \qquad (2\text{-}6)$$

is the *covariance* of the random variables y_1 and y_2. The covariance is a measure of the linear association between y_1 and y_2. More specifically, we may show that if y_1 and y_2 are independent,[1] then $\text{Cov}(y_1, y_2) = 0$. We may also show that

9. $V(y_1 - y_2) = V(y_1) + V(y_2) - 2 \, \text{Cov}(y_1, y_2)$

If y_1 and y_2 are *independent,* then we have

10. $V(y_1 \pm y_2) = V(y_1) + V(y_2) = \sigma_1^2 + \sigma_2^2$

and

11. $E(y_1 \cdot y_2) = E(y_1) \cdot E(y_2) = \mu_1 \cdot \mu_2$

However, note that, in general,

12. $E\left(\dfrac{y_1}{y_2}\right) \neq \dfrac{E(y_1)}{E(y_2)}$

regardless of whether or not y_1 and y_2 are independent.

2-3 SAMPLING AND SAMPLING DISTRIBUTIONS

Random Samples, Sample Mean, and Sample Variance ▪ The objective of satistical inference is to draw conclusions about a population using a sample from that population. Most of the methods that we will study assume that random samples are used. That is, if the population contains N elements and a sample

---

[1] Note that the converse of this is not necessarily so; that is, we may have $\text{Cov}(y_1, y_2) = 0$ and yet this does not imply independence. For an example, see Hines and Montgomery (1990, pp. 128–129).

of n of them is to be selected, then if each of the $N!/(N - n)!n!$ possible samples has an equal probability of being chosen, the procedure employed is called *random sampling*. In practice, it is sometimes difficult to obtain random samples, and tables of random numbers, such as Table XI in the Appendix, may be helpful.

Statistical inference makes considerable use of quantities computed from the observations in the sample. We define a *statistic* as any function of the observations in a sample that does not contain unknown parameters. For example, suppose that $y_1, y_2, \ldots, y_n$ represents a sample. Then the *sample mean*

$$\bar{y} = \frac{\sum\limits_{i=1}^{n} y_i}{n} \tag{2-7}$$

and the *sample variance*

$$S^2 = \frac{\sum\limits_{i=1}^{n} (y_i - \bar{y})^2}{n - 1} \tag{2-8}$$

are both statistics. These quantities are measures of the central tendency and dispersion of the sample, respectively. Sometimes $S = \sqrt{S^2}$, called the *sample standard deviation*, is used as a measure of dispersion. Engineers often prefer to use the standard deviation to measure dispersion because its units are the same as those for the variable of interest y.

Properties of the Sample Mean and Variance ▪ The sample mean $\bar{y}$ is a point estimator of the population mean μ, and the same variance S^2 is a point estimator of the population variance σ^2. In general, an *estimator* of an unknown parameter is a statistic that corresponds to that parameter. Note that a point estimator is a random variable. A particular numerical value of an estimator, computed from sample data, is called an *estimate*. For example, suppose we wish to estimate the mean and variance of the breaking strength of a particular type of textile fiber. A random sample of $n = 25$ fiber specimens is tested and the breaking strength is recorded for each. The sample mean and variance are computed according to Equations 2-7 and 2-8, respectively, and are $\bar{y} = 18.6$ and $S^2 = 1.20$. Therefore, the estimate of μ is $\bar{y} = 18.6$ and the estimate of σ^2 is $S^2 = 1.20$.

There are several properties required of good point estimators. Two of the most important are the following:

1. The point estimator should be *unbiased*. That is, the long-run average or expected value of the point estimator should be the parameter that is being estimated. Although unbiasedness is desirable, this property alone does not always make an estimator a good one.

2. An unbiased estimator should have minimum variance. This property states that the minimum variance point estimator has a variance that is smaller than the variance of any other estimator of that parameter.

We may easily show that $\bar{y}$ and S^2 are unbiased estimators of μ and σ^2 respectively. First consider $\bar{y}$. Using the properties of expectation, we have

$$E(\bar{y}) = E\left(\frac{\sum\limits_{i=1}^{n} y_i}{n}\right)$$

$$= \frac{1}{n} E\left(\sum\limits_{i=1}^{n} y_i\right)$$

$$= \frac{1}{n} \sum\limits_{i=1}^{n} E(y_i)$$

$$= \frac{1}{n} \sum\limits_{i=1}^{n} \mu$$

$$= \mu$$

since the expected value of each observation y_i is μ. Thus, $\bar{y}$ is an unbiased estimator of μ.

Now consider the sample variance S^2. We have

$$E(S^2) = E\left[\frac{\sum\limits_{i=1}^{n} (y_i - \bar{y})^2}{n - 1}\right]$$

$$= \frac{1}{n - 1} E\left[\sum\limits_{i=1}^{n} (y_i - \bar{y})^2\right]$$

$$= \frac{1}{n - 1} E(SS)$$

where $SS = \sum_{i=1}^{n}(y_i - \bar{y})^2$ is the corrected sum of squares of the observations y_i. Now

$$E(SS) = E\left[\sum\limits_{i=1}^{n} (y_i - \bar{y})^2\right] \qquad (2\text{-}9)$$

$$= E\left[\sum\limits_{i=1}^{n} y_i^2 - n\bar{y}^2\right]$$

$$= \sum\limits_{i=1}^{n} (\mu^2 + \sigma^2) - n(\mu^2 + \sigma^2/n)$$

$$= (n - 1)\sigma^2 \qquad (2\text{-}10)$$

Therefore,

$$E(S^2) = \frac{1}{n - 1} E(SS)$$

$$= \sigma^2$$

and we see that S^2 is an unbiased estimator of σ^2.

Degrees of Freedom ▪ The quantity $n - 1$ in Equation 2-10 is called the *number of degrees of freedom* of the sum of squares SS. This is a very general result; that is, if y is a random variable with variance σ^2 and $SS = \Sigma(y_i - \bar{y})^2$ has v degrees of freedom, then

$$E\left(\frac{SS}{v}\right) = \sigma^2 \tag{2-11}$$

The number of degrees of freedom of a sum of squares is equal to the number of independent elements in that sum of squares. For example, $SS = \Sigma_{i=1}^{n}(y_i - \bar{y})^2$ in Equation 2-9 consists of the sum of squares of the n elements $y_1 - \bar{y}$, $y_2 - \bar{y}, \ldots, y_n - \bar{y}$. These elements are not all independent since $\Sigma_{i=1}^{n}(y_i - \bar{y}) = 0$; in fact, only $n - 1$ of them are independent, implying that SS has $n - 1$ degrees of freedom.

The Normal and Other Sampling Distributions ▪ Often we are able to determine the probability distribution of a particular statistic if we know the probability distribution of the population from which the sample was drawn. The probability distribution of a statistic is called a *sampling distribution*. We will now briefly discuss several useful sampling distributions.

One of the most important sampling distributions is the *normal* distribution. If y is a normal random variable, then the probability distribution of y is

$$f(y) = \frac{1}{\sigma\sqrt{2\pi}} e^{-(1/2)[(y-\mu)/\sigma]^2} \qquad -\infty < y < \infty \tag{2-12}$$

where $-\infty < \mu < \infty$ is the mean of the distribution and $\sigma^2 > 0$ is the variance. The normal distribution is shown in Figure 2-5.

Because sample runs that differ as a result of experimental error often are well described by the normal distribution, the normal plays a central role in the analysis of data from designed experiments. Many important sampling distributions may also be defined in terms of normal random variables. We often use the notation $y \sim N(\mu, \sigma^2)$ to denote that y is distributed normally with mean μ and variance σ^2.

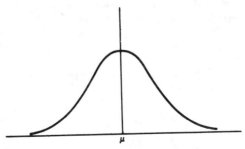

Figure 2-5.   The normal distribution.

An important special case of the normal distribution is the *standard normal distribution;* that is, $\mu = 0$ and $\sigma^2 = 1$. We see that if $y \sim N(\mu, \sigma^2)$, then the random variable

$$z = \frac{y - \mu}{\sigma} \tag{2-13}$$

follows the standard normal distribution, denoted $z \sim N(0, 1)$. The operation demonstrated in Equation 2-13 is often called *standardizing* the normal random variable y. A table of the cumulative standard normal distribution is given in Appendix Table I.

Many statistical techniques assume that the random variable is normally distributed. The central limit theorem is often a justification of approximate normality.

THEOREM 2-1   THE CENTRAL LIMIT THEOREM   If $y_1, y_2, \ldots, y_n$ is a sequence of n independent and identically distributed random variables with $E(y_i) = \mu$ and $V(y_i) = \sigma^2$ (both finite) and $x = y_1 + y_2 + \cdots + y_n$, then

$$z_n = \frac{x - n\mu}{\sqrt{n\sigma^2}}$$

has an approximate $N(0, 1)$ distribution in the sense that, if $F_n(z)$ is the distribution function of z_n and $\Phi(z)$ is the distribution function of the $N(0, 1)$ random variable, then $\lim_{n \to \infty} [F_n(z)/\Phi(z)] = 1$.

This result states essentially that the sum of n independent and identically distributed random variables is approximately normally distributed. In many cases this approximation is good for very small n, say $n < 10$, whereas in other cases large n is required, say $n > 100$. Frequently, we think of the error in an experiment as arising in an additive manner from several independent sources; consequently, the normal distribution becomes a plausible model for the combined experimental error.

An important sampling distribution that can be defined in terms of normal random variables is the chi-square or χ^2 distribution. If $z_1, z_2, \ldots, z_k$ are normally and independently distributed random variables with mean 0 and variance 1, abbreviated NID(0, 1), then the random variable

$$\chi^2_k = z_1^2 + z_2^2 + \cdots + z_k^2$$

follows the chi-square distribution with k degrees of freedom. The density function of chi-square is

$$f(\chi^2) = \frac{1}{2^{k/2} \Gamma\left(\dfrac{k}{2}\right)} (\chi^2)^{(k/2)-1} e^{-\chi^2/2} \qquad \chi^2 > 0 \tag{2-14}$$

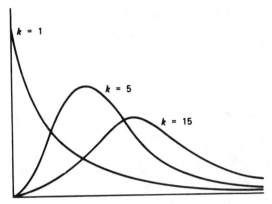

Figure 2-6. Several chi-square distributions.

Several chi-square distributions are shown in Figure 2-6. The distribution is asymmetric or *skewed*, with mean and variance

$$\mu = k$$
$$\sigma^2 = 2k$$

respectively. A table of percentage points of the chi-square distribution is given in Table III of the Appendix.

As an example of a random variable that follows the chi-square distribution, suppose that $y_1, y_2, \ldots, y_n$ is a random sample from an $N(\mu, \sigma^2)$ distribution. Then

$$\frac{SS}{\sigma^2} = \frac{\sum_{i=1}^{n} (y_i - \bar{y})^2}{\sigma^2} \sim \chi_{n-1}^2 \tag{2-15}$$

That is, SS/σ^2 is distributed as chi-square with $n - 1$ degrees of freedom.

Many of the techniques used in this book involve the computation and manipulation of sums of squares. The result given in Equation 2-15 is extremely important and occurs repeatedly; a sum of squares in normal random variables when divided by σ^2 follows the chi-square distribution.

Examining Equation 2-8, we see that the sample variance can be written as

$$S^2 = \frac{SS}{n - 1} \tag{2-16}$$

If the observations in the sample are $NID(\mu, \sigma^2)$, then the distribution of S^2 is $[\sigma^2/(n - 1)]\chi_{n-1}^2$. Thus, the sampling distribution of the sample variance is a constant times the chi-square distribution if the population is normally distributed.

If z and χ_k^2 are independent standard normal and chi-square random variables, respectively, then the random variable

$$t_k = \frac{z}{\sqrt{\chi_k^2/k}} \tag{2-17}$$

follows the t distribution with k degrees of freedom, denoted t_k. The density function of t is

$$f(t) = \frac{\Gamma[(k + 1)/2]}{\sqrt{k\pi}\Gamma(k/2)} \frac{1}{[(t^2/k) + 1]^{(k+1)/2}} \qquad -\infty < t < \infty \tag{2-18}$$

and the mean and variance of t are $\mu = 0$ and $\sigma^2 = k/(k - 2)$ for $k > 2$, respectively. Several t distributions are shown in Figure 2-7. Note that if $k = \infty$, the t distribution becomes the standard normal distribution. A table of percentage points of the t distribution is given in Table II of the Appendix. If $y_1, y_2, \ldots , y_n$ is a random sample from the $N(\mu, \sigma^2)$ distribution, then the quantity

$$t = \frac{\bar{y} - \mu}{S/\sqrt{n}} \tag{2-19}$$

is distributed as t with $n - 1$ degrees of freedom.

The final sampling distribution that we will consider is the F distribution. If χ_u^2 and χ_v^2 are two independent chi-square random variables with u and v degrees of freedom, respectively, then the ratio

$$F_{u,v} = \frac{\chi_u^2/u}{\chi_v^2/v} \tag{2-20}$$

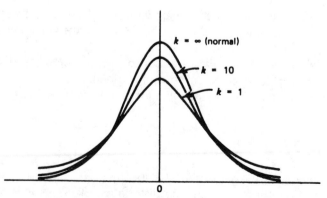

Figure 2-7.   Several t distributions.

follows the F distribution with u *numerator* degrees of freedom and v *denominator* degrees of freedom. The probability distribution of F is

$$h(F) = \frac{\Gamma\left(\dfrac{u+v}{2}\right)\left(\dfrac{u}{v}\right)^{u/2} F^{(u/2)-1}}{\Gamma\left(\dfrac{u}{2}\right)\Gamma\left(\dfrac{v}{2}\right)\left[\left(\dfrac{u}{v}\right)F + 1\right]^{(u+v)/2}} \qquad 0 < F < \infty \qquad (2\text{-}21)$$

Several F distributions are shown in Figure 2-8. This distribution is very important in the statistical analysis of designed experiments. A table of percentage points of the F distribution is given in Table IV of the Appendix.

As an example of a statistic that is distributed as F, suppose we have two independent normal populations with common variance σ^2. If $y_{11}, y_{12}, \ldots, y_{1n_1}$ is a random sample of n_1 observations from the first population, and if $y_{21}, y_{22}, \ldots, y_{2n_2}$ is a random sample of n_2 observations from the second, then

$$\frac{S_1^2}{S_2^2} \sim F_{n_1-1,\, n_2-1} \qquad (2\text{-}22)$$

where S_1^2 and S_2^2 are the two sample variances. This result follows directly from Equations 2-15 and 2-20.

2-4 INFERENCES ABOUT THE DIFFERENCES IN MEANS, RANDOMIZED DESIGNS

We are now ready to return to the portland cement mortar problem posed in Section 2-1. Recall that two different formulations of mortar were being investigated to determine if they differ in tension bond strength. In this section we discuss how the data from this simple comparative experiment can be analyzed using *hypothesis testing* and *confidence interval* procedures for comparing two treatment means.

Throughout this section we assume that a *completely randomized experi-*

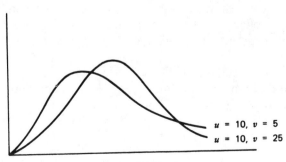

$u = 10, v = 5$
$u = 10, v = 25$

Figure 2-8. Several F distributions.

mental design is used. In such a design, the data are viewed as if they were a random sample from a normal distribution.

2-4.1   Hypothesis Testing

A statistical hypothesis is a statement about the parameters of a probability distribution. For example, in the portland cement experiment, we may think that the mean tension bond strengths of the two mortar formulations are equal. This may be stated formally as

$$H_0: \mu_1 = \mu_2$$
$$H_1: \mu_1 \neq \mu_2$$

where μ_1 is the mean tension bond strength of the modified mortar and μ_2 is the mean tension bond strength of the unmodified mortar. The statement $H_0: \mu_1 = \mu_2$ is called the *null hypothesis* and $H_1: \mu_1 \neq \mu_2$ is called the *alternative hypothesis*. The alternative hypothesis specified here is called a *two-sided* alternative hypothesis since it would be true either if $\mu_1 < \mu_2$ or if $\mu_1 > \mu_2$.

To test a hypothesis we devise a procedure for taking a random sample, computing an appropriate test statistic, and then rejecting or failing to reject the null hypothesis H_0. Part of this procedure is specifying the set of values for the test statistic that leads to rejection of H_0. This set of values is called the *critical region* or *rejection region* for the test.

Two kinds of errors may be committed when testing hypotheses. If the null hypothesis is rejected when it is true, then a type I error has occurred. If the null hypothesis is *not* rejected when it is false, then a type II error has been made. The probabilities of these two errors are given special symbols:

$$\alpha = P(\text{type I error}) = P(\text{reject } H_0 | H_0 \text{ is true})$$
$$\beta = P(\text{type II error}) = P(\text{fail to reject } H_0 | H_0 \text{ is false})$$

Sometimes it is more convenient to work with the *power* of the test, where

$$\text{Power} = 1 - \beta = P(\text{reject } H_0 | H_0 \text{ is false})$$

The general procedure in hypothesis testing is to specify a value of the probability of type I error α, often called the *significance level* of the test, and then design the test procedure so that the probability of type II error β has a suitably small value.

Suppose that we could assume that the variances of tension bond strengths were identical for both mortar formulations. Then an appropriate test statistic to use for comparing two treatment means in the completely randomized design is

$$t_0 = \frac{\bar{y}_1 - \bar{y}_2}{S_p \sqrt{\dfrac{1}{n} + \dfrac{1}{n}}} \tag{2-23}$$

where $\bar{y}_1$ and $\bar{y}_2$ are the sample means, n_1 and n_2 are the sample sizes, S_p^2 is an estimate of the common variance $\sigma_1^2 = \sigma_2^2 = \sigma^2$ computed from

$$S_p^2 = \frac{(n_1 - 1)S_1^2 + (n_2 - 1)S_2^2}{n_1 + n_2 - 2} \tag{2-24}$$

and S_1^2 and S_2^2 are the two individual sample variances. To determine whether to reject $H_0 : \mu_1 = \mu_2$, we would compare t_0 to the t distribution with $n_1 + n_2 - 2$ degrees of freedom. If $|t_0| > t_{\alpha/2,\, n_1+n_2-2}$, where $t_{\alpha/2,\, n_1+n_2-2}$, is the upper $\alpha/2$ percentage point of the t distribution with $n_1 + n_2 - 2$ degrees of freedom, we would *reject* H_0 and conclude that the mean strengths of the two formulations of portalnd cement mortar differ.

This procedure may be justified as follows. If we are sampling from independent normal distributions, then the distribution of $\bar{y}_1 - \bar{y}_2$ is $N[\mu_1 - \mu_2,\ \sigma^2(1/n_1 + 1/n_2)]$. Thus, if σ^2 were known, and if $H_0 : \mu_1 = \mu_2$ were true, the distribution of

$$Z_0 = \frac{\bar{y}_1 - \bar{y}_2}{\sigma\sqrt{\dfrac{1}{n_1} + \dfrac{1}{n_2}}} \tag{2-25}$$

would be $N(0, 1)$. However, in replacing σ in Equation 2-25 by S_p, we change the distribution of Z_0 from standard normal to t with $n_1 + n_2 - 2$ degrees of freedom. Now if H_0 is true, t_0 in Equation 2-23 is distributed as $t_{n_1+n_2-2}$ and, consequently, we would expect $100(1 - \alpha)$ percent of the values of t_0 to fall between $-t_{\alpha/2,\, n_1+n_2-2}$ and $t_{\alpha/2,\, n_1+n_2-2}$. A sample producing a value of t_0 outside these limits would be unusual if the null hypothesis were true and is evidence that H_0 should be rejected. Note that α is the probability of type I error for the test.

In some problems, one may wish to reject H_0 only if one mean is larger than the other. Thus, one would specify a *one-sided alternative* hypothesis $H_1 : \mu_1 > \mu_2$ and would reject H_0 only if $t_0 > t_{\alpha,\, n_1+n_2-2}$. If one wants to reject H_0 only if μ_1 is less than μ_2, then the alternative hypothesis is $H_1 : \mu_1 < \mu_2$, and one would reject H_0 if $t_0 < -t_{\alpha,\, n_1+n_2-2}$.

To illustrate the procedure, consider the portland cement data in Table 2-1. For these data, we find that

Modified Mortar	Unmodified Mortar
$\bar{y}_1 = 16.76$ kgf/cm^2	$\bar{y}_2 = 17.92$ kfg/cm^2
$S_1^2 = 0.100$	$S_2^2 = 0.061$
$S_1 = 0.316$	$S_2 = 0.247$
$n_1 = 10$	$n_2 = 10$

and

$$S_p^2 = \frac{(n_1 - 1)S_1^2 + (n_2 - 1)S_2^2}{n_1 + n_2 - 2}$$

$$= \frac{9(0.100) + 9(0.061)}{10 + 10 - 2}$$

$$= 0.081$$

$$S_p = 0.284$$

Notice that, since the sample variances are similar, it is probably not unreasonable to conclude that the population variances are equal. The test statistic (Equation 2-23) is

$$t_0 = \frac{\bar{y}_1 - \bar{y}_2}{S_p\sqrt{\frac{1}{n_1} + \frac{1}{n_2}}}$$

$$= \frac{16.76 - 17.92}{0.284\sqrt{\frac{1}{10} + \frac{1}{10}}}$$

$$= -9.13$$

Now the upper $2\frac{1}{2}$ percent point of the t distribution with $n_1 + n_2 - 2 = 10 + 10 - 2 = 18$ degrees of freedom is $t_{.025, 18} = 2.101$; thus, since $|t_0| = 9.13 > t_{.025, 18} = 2.101$, we would reject H_0 and conclude that the mean tension bond strengths of the two formulations of portland cement mortar are different.

An Alternate Justification to the t Test ▪ The two-sample t test we have just presented depends on the underlying assumption that the two populations from which the samples were randomly selected are normal. Although the normality assumption is required to develop the test procedure formally, moderate departures from normality will not seriously affect the results. It can be argued [e.g., see Box, Hunter, and Hunter (1978)] that the use of a randomized design enables one to test hypotheses without any assumptions regarding the form of the distribution. Briefly, the reasoning is as follows. If the formulas have no effect, then all $[20!/(10!10!)] = 184,756$ possible ways that the 20 observations could occur are equally likely. Corresponding to each of these 184,756 possible arrangements is a value of t_0. If the value of t_0 actually obtained from the data is unusually large or unusually small with reference to the set of 184,756 possible values, then it is an indication that $\mu_1 \neq \mu_2$.

This type of procedure is called a *randomization test*. It can be shown that the t test is a good approximation of the randomization test. Thus, we will use t tests (and other procedures that can be regarded as approximations of randomization tests) without extensive concern about the assumption of normality.

2-4.2   Choice of Sample Size

Selection of an appropriate sample size is one of the most important aspects of any experimental design problem. The choice of sample size and the probability of type II error β are closely connected. Suppose that we are testing the hypothesis

$$H_0:\mu_1 = \mu_2$$
$$H_1:\mu_1 \neq \mu_2$$

and that the means are *not* equal so that $\delta = \mu_1 - \mu_2$. Since $H_0:\mu_1 = \mu_2$ is not true, we are concerned about wrongly failing to reject H_0. The probability of type II error depends on the true difference in means δ. A graph of β versus δ for a particular sample size is called the *operating characteristic curve* or O.C. curve for the test. The β error is also a function of sample size. Generally, for a given value of δ, the β error decreases as the sample size increases. That is, a specified difference in means is easier to detect for larger sample sizes than for smaller ones.

A set of operating characteristic curves for the hypotheses

$$H_0:\mu_1 = \mu_2$$
$$H_1:\mu_1 \neq \mu_2$$

for the case where the two population variances σ_1^2 and σ_2^2 are unknown but equal ($\sigma_1^2 = \sigma_2^2 = \sigma^2$) and for a level of significance of $\alpha = .05$ is shown in Figure 2-9. The curves also assume that the sample sizes from the two populations are equal; that is, $n_1 = n_2 = n$. The parameter on the horizontal axis in Figure 2-9 is

$$d = \frac{|\mu_1 - \mu_2|}{2\sigma} = \frac{|\delta|}{2\sigma}$$

Dividing $|\delta|$ by 2σ allows the experimenter to use the same set of curves, regardless of the value of the variance (the difference in means is expressed in standard deviation units). Furthermore, the sample size used to construct the curves is actually $n^* = 2n - 1$.

From examining these curves, we note the following:

1.   The greater the difference in means, $\mu_1 - \mu_2$, the smaller the probability of type II error for a given sample size and α. That is, for a specified sample size and α, the test will detect large differences more easily than small ones.

2.   As the sample size gets larger, the probability of type II error gets smaller for a given difference in means and α. That is, to detect a specified difference δ, we may make the test more powerful by increasing the sample size.

Operating characteristic curves are often helpful in selecting a sample size to use in an experiment. For example, consider the portland cement mortar problem discussed previously. Suppose that if the two formulations differ in mean

Figure 2-9. Operating characteristic curves for the two-sided t test with $\alpha = .05$. (Reproduced with permission from "Operating Characteristics for the Common Statistical Tests of Significance," C. L. Ferris, F. E. Grubbs, and C. L. Weaver, *Annals of Mathematical Statistics*, June 1946.)

strength by as much as 0.5 kgf/cm² we would like to detect it with a high probability. Thus, since $\mu_1 - \mu_2 = 0.5$ kgf/cm² is the "critical" difference in means we wish to detect, we find that d, the parameter on the horizontal axis of the operating characteristic curve in Figure 2-9, is

$$d = \frac{|\mu_1 - \mu_2|}{2\sigma} = \frac{0.5}{2\sigma} = \frac{0.25}{\sigma}$$

Unfortunately, d involves the unknown parameter σ. However, suppose that we feel on the basis of prior experience that it is very unlikely that the standard deviation of any observation on strength would exceed 0.25 kgf/cm². Then using $\sigma = 0.25$ in the above expression for d yields $d = 1$. If we wish to reject the null hypothesis 95 percent of the time when $\mu_1 - \mu_2 = 0.5$, then $\beta = 0.05$, and Figure 2-9 with $\beta = 0.05$ and $d = 1$ yields $n^* = 16$, approximately. Therefore, since $n^* = 2n - 1$, the required sample size is

$$n = \frac{n^* + 1}{2} = \frac{16 + 1}{2} = 8.5 \approx 9$$

and we would use sample sizes of $n_1 = n_2 = n = 9$.

In our example, the experimenter actually used a sample size of 10. Perhaps the experimenter elected to increase the sample size slightly to guard against the possibility that the prior estimate of the common standard deviation σ was too conservative and was likely to be somewhat larger than 0.25.

Operating characteristic curves often play an important role in the choice of sample size in experimental design problems. Their use in this respect is discussed in subsequent chapters. For a discussion of the uses of operating characteristic curves for other simple comparative experiments similar to the two-sample t test see Hines and Montgomery (1990).

2-4.3 Confidence Intervals

Although hypothesis testing is a useful procedure, it sometimes does not tell the entire story. It is often preferable to provide an interval within which the value of the parameter or parameters in question would be expected to lie. These interval statements are called *confidence intervals*. In many engineering and industrial experiments, the experimenter already knows that the means μ_1 and μ_2 differ; consequently, hypothesis testing on $\mu_1 = \mu_2$ is of little interest. The experimenter would usually be more interested in a confidence interval on the difference in means $\mu_1 - \mu_2$.

To define a confidence interval, suppose that θ is an unknown parameter. To obtain an interval estimate of θ, we need to find two statistics L and U such that the probability statement

$$P(L \leq \theta \leq U) = 1 - \alpha \qquad (2\text{-}26)$$

is true. The interval

$$L \leq \theta \leq U \qquad (2\text{-}27)$$

is called a $100(1 - \alpha)$ percent *confidence interval* for the parameter θ. The interpretation of this interval is that if, in repeated random samplings, a large number of such intervals are constructed, $100(1 - \alpha)$ percent of them will contain the true value of θ. The statistics L and U are called the lower and upper *confidence limits*, respectively, and $1 - \alpha$ is called the *confidence coefficient*. If $\alpha = .05$, then Equation 2-27 is called a 95 percent confidence interval for θ. Note that confidence intervals have a frequency interpretation; that is, we do not know if the statement is true for this specific sample, but we do know that the *method* used to produce the confidence interval yields correct statements $100(1 - \alpha)$ percent of the time.

Suppose that we wish to find a $100(1 - \alpha)$ percent confidence interval on the true difference in means $\mu_1 - \mu_2$ for the portland cement problem. The interval can be derived in the following way. The statistic

$$\frac{\bar{y}_1 - \bar{y}_2 - (\mu_1 - \mu_2)}{S_p \sqrt{\dfrac{1}{n_1} + \dfrac{1}{n_2}}}$$

is distributed as $t_{n_1 + n_2 - 2}$. Thus,

$$P\left(-t_{\alpha/2,\ n_1+n_2-2} \leq \frac{\bar{y}_1 - \bar{y}_2 - (\mu_1 - \mu_2)}{S_p\sqrt{\dfrac{1}{n_1} + \dfrac{1}{n_2}}} \leq t_{\alpha/2,\ n_1+n_2-2}\right) = 1 - \alpha$$

or

$$P\left(\bar{y}_1 - \bar{y}_2 - t_{\alpha/2,\ n_1+n_2-2}\ S_p\sqrt{\frac{1}{n_1} + \frac{1}{n_2}} \leq \mu_1 - \mu_2\right.$$

$$\left. \leq \bar{y}_1 - \bar{y}_2 + t_{\alpha/2,\ n_1+n_2-2}\ S_p\sqrt{\frac{1}{n_1} + \frac{1}{n_2}}\right) = 1 - \alpha \quad (2\text{-}28)$$

Comparing Equations 2-28 and 2-26, we see that

$$\bar{y}_1 - \bar{y}_2 - t_{\alpha/2,\ n_1+n_2-2}\ S_p\sqrt{\frac{1}{n_1} + \frac{1}{n_2}} \leq \mu_1 - \mu_2$$

$$\leq \bar{y}_1 - \bar{y}_2 + t_{\alpha/2,\ n_1+n_2-2}\ S_p\sqrt{\frac{1}{n_1} + \frac{1}{n_2}} \quad (2\text{-}29)$$

is a $100(1 - \alpha)$ percent confidence interval for $\mu_1 - \mu_2$. The actual 95 percent confidence interval estimate is found by substituting in Equation 2-29 as follows:

$$16.76 - 17.92 - (2.101)0.284\sqrt{\tfrac{1}{10} + \tfrac{1}{10}} \leq \mu_1 - \mu_2 \leq 16.76 - 17.92$$

$$+ (2.101)0.284\sqrt{\tfrac{1}{10} + \tfrac{1}{10}}$$

$$-1.16 - 0.27 \leq \mu_1 - \mu_2 \leq -1.16 + 0.27$$

$$-1.43 \leq \mu_1 - \mu_2 \leq -0.89$$

Thus, the 95 percent interval estimate on the difference in means extends from -1.43 kgf/cm^2 to -0.89 kgf/cm^2. Put another way, the confidence interval is $\mu_1 - \mu_2 = -1.16$ kgf/cm$^2 \pm 0.27$ kgf/cm^2, or the difference in means strengths is -1.16 kgf/cm^2, and the accuracy of this estimate is ± 0.27 kgf/cm^2. Note that since $\mu_1 - \mu_2 = 0$ is *not* included in this interval, the data do not support the hypothesis that $\mu_1 = \mu_2$ at the 5 percent level of significance. It is likely that the mean strength of the unmodified formulation exceeds the mean strength of the modified formulation.

2-4.4   The Case Where $\sigma_1^2 \neq \sigma_2^2$

If we are testing

$$H_0: \mu_1 = \mu_2$$
$$H_1: \mu_1 \neq \mu_2$$

and cannot reasonably assume that the variances σ_1^2 and σ_2^2 are equal, then the two-sample t test must be modified slightly. The test statistic becomes

$$t_0 = \frac{\bar{y}_1 - \bar{y}_2}{\sqrt{\dfrac{S_1^2}{n_1} + \dfrac{S_2^2}{n_2}}} \qquad (2\text{-}30)$$

This statistic is not distributed exactly as t. However, the distribution of t_0 is well-approximated by t if we use

$$v = \frac{\left(\dfrac{S_1^2}{n_1} + \dfrac{S_2^2}{n_2}\right)^2}{\dfrac{(S_1^2/n_1)^2}{n_1 - 1} + \dfrac{(S_2^2/n_2)^2}{n_2 - 1}} \qquad (2\text{-}31)$$

as the degrees of freedom.

2-4.5 The Case Where σ_1^2 and σ_2^2 Are Known

If the variances of both populations are *known*, then the hypothesis

$$H_0 : \mu_1 = \mu_2$$
$$H_1 : \mu_1 \neq \mu_2$$

may be tested using the statistic

$$Z_0 = \frac{\bar{y}_1 - \bar{y}_2}{\sqrt{\dfrac{\sigma_1^2}{n_1} + \dfrac{\sigma_2^2}{n_2}}} \qquad (2\text{-}32)$$

If both populations are normal, or if the sample sizes are large enough so that the central limit theorem applies, the distribution of Z_0 is $N(0, 1)$ if the null hypothesis is true. Thus, the critical region would be found using the normal distribution rather than the t. Specifically, we would reject H_0 if $|Z_0| > Z_{\alpha/2}$, where $Z_{\alpha/2}$ is the upper $\alpha/2$ percentage point of the standard normal distribution.

Unlike the t test of the previous sections, the test on means with known variances does not require the assumption of sampling from normal populations. One can use the central limit theorem to justify an approximate normal distribution for the difference in sample means $\bar{y}_1 - \bar{y}_2$.

The $100(1 - \alpha)$ percent confidence interval on $\mu_1 - \mu_2$ where the variances are known is

$$\bar{y}_1 - \bar{y}_2 - Z_{\alpha/2}\sqrt{\frac{\sigma_1^2}{n_1} + \frac{\sigma_2^2}{n_2}} \leq \mu_1 - \mu_2 \leq \bar{y}_1 - \bar{y}_2 + Z_{\alpha/2}\sqrt{\frac{\sigma_1^2}{n_1} + \frac{\sigma_2^2}{n_2}} \qquad (2\text{-}33)$$

As noted previously, the confidence interval is often a useful supplement to the hypothesis testing procedure.

2-4.6 Comparing a Single Mean to a Specified Value

Some experiments involve comparing only one population mean μ to a specified value, say μ_0. The hypotheses are

$$H_0: \mu = \mu_0$$
$$H_1: \mu \neq \mu_0$$

If the population is normal with known variance, or if the population is nonnormal but the sample size is large enough so that the central limit theorem applies, then the hypothesis may be tested using a direct application of the normal distribution. The test statistic is

$$Z_0 = \frac{\bar{y} - \mu_0}{\sigma/\sqrt{n}} \tag{2-34}$$

If $H_0: \mu = \mu_0$ is true, then the distribution of Z_0 is $N(0, 1)$. Therefore, the decision rule for $H_0: \mu = \mu_0$ is to reject the null hypothesis if $|Z_0| > Z_{\alpha/2}$. The value of the mean μ_0 specified in the null hypothesis is usually determined in one of three ways. It may result from past evidence, knowledge, or experimentation. It may be the result of some theory or model describing the situation under study. Finally, it may be the result of contractual specifications.

The $100(1 - \alpha)$ percent confidence interval on the true population mean is

$$\bar{y} - Z_{\alpha/2}\sigma/\sqrt{n} \leq \mu \leq \bar{y} + Z_{\alpha/2}\sigma/\sqrt{n} \tag{2-35}$$

Example 2-1

A vendor submits lots of fabric to a textile manufacturer. The manufacturer wants to know if the lot average breaking strength exceeds 200 psi. If so, she wants to accept the lot. Past experience indicates that a reasonable value for the variance of breaking strength is $100(\text{psi})^2$. The hypotheses to be tested are

$$H_0: \mu = 200$$
$$H_1: \mu > 200$$

Note that this is a one-sided alternative hypothesis. Thus, we would accept the lot only if the null hypothesis $H_0: \mu = 200$ could be rejected (i.e., if $Z_0 > Z_a$).

Four specimens are randomly selected, and the average breaking strength observed is $\bar{y} = 214$ psi. The value of the test statistic is

$$Z_0 = \frac{\bar{y} - \mu_0}{\sigma/\sqrt{n}} = \frac{214 - 200}{10/\sqrt{n}} = 2.80$$

If a type I error of $\alpha = .05$ is specified, we find $Z_\alpha = Z_{.05} = 1.645$ from Appendix Table I. Thus H_0 is rejected, and we conclude that the lot average breaking strength exceeds 200 psi.

∎

If the variance of the population is unknown, we must make the additional assumption that the population is normally distributed, although moderate departures from normality will not seriously affect the results.

To test $H_0: \mu = \mu_0$ in the variance unknown case, the sample variance S^2 is used to estimate σ^2. Replacing σ with S in Equation 2-34, we have the test statistic

$$t_0 = \frac{\bar{y} - \mu_0}{S/\sqrt{n}} \tag{2-36}$$

The null hypothesis $H_0: \mu = \mu_0$ would be rejected if $|t_0| > t_{\alpha/2, n-1}$, where $t_{\alpha/2, n-1}$ denotes the upper $\alpha/2$ percentage point of the t distribution with $n - 1$ degrees of freedom. The $100(1 - \alpha)$ percent confidence interval in this case is

$$\bar{y} - t_{\alpha/2, n-1} S/\sqrt{n} \leq \mu \leq \bar{y} + t_{\alpha/2, n-1} S/\sqrt{n} \tag{2-37}$$

2-4.7 Summary

Tables 2-2 and 2-3 summarize the test procedures discussed above for sample means. Critical regions are shown for both two-sided and one-sided alternative hypotheses.

Table 2-2   Tests on Means with Variance Known

Hypothesis	Test Statistic	Criteria for Rejection
$H_0: \mu = \mu_0$ $H_1: \mu \neq \mu_0$		$|Z_0| > Z_{\alpha/2}$
$H_0: \mu = \mu_0$ $H_1: \mu < \mu_0$	$Z_0 = \dfrac{\bar{y} - \mu_0}{\sigma/\sqrt{n}}$	$Z_0 < -Z_\alpha$
$H_0: \mu = \mu_0$ $H_1: \mu > \mu_0$		$Z_0 > Z_\alpha$
$H_0: \mu_1 = \mu_2$ $H_1: \mu_1 \neq \mu_2$		$|Z_0| > Z_{\alpha/2}$
$H_0: \mu_1 = \mu_2$ $H_1: \mu_1 < \mu_2$	$Z_0 = \dfrac{\bar{y}_1 - \bar{y}_2}{\sqrt{\dfrac{\sigma_1^2}{n_1} + \dfrac{\sigma_2^2}{n_2}}}$	$Z_0 < -Z_\alpha$
$H_0: \mu_1 = \mu_2$ $H_1: \mu_1 > \mu_2$		$Z_0 > Z_\alpha$

Table 2-3   Tests on Means of Normal Distributions, Variance Unknown

Hypothesis	Test Statistic	Criteria for Rejection
$H_0: \mu = \mu_0$ $H_1: \mu \neq \mu_0$		$|t_0| > t_{\alpha/2, n-1}$
$H_0: \mu = \mu_0$ $H_1: \mu < \mu_0$	$t_0 = \dfrac{\bar{y} - \mu_0}{S/\sqrt{n}}$	$t_0 < -t_{\alpha, n-1}$
$H_0: \mu = \mu_0$ $H_1: \mu > \mu_0$		$t_0 > t_{\alpha, n-1}$
$H_0: \mu_1 = \mu_2$ $H_1: \mu_1 \neq \mu_2$	$t_0 = \dfrac{\bar{y}_1 - \bar{y}_2}{S_p \sqrt{\dfrac{1}{n_1} + \dfrac{1}{n_2}}}$ $v = n_1 + n_2 - 2$ or	$|t_0| > t_{\alpha/2, v}$
$H_0: \mu_1 = \mu_2$ $H_1: \mu_1 < \mu_2$	$t_0 = \dfrac{\bar{y}_1 - \bar{y}_2}{\sqrt{\dfrac{S_1^2}{n_1} + \dfrac{S_2^2}{n_2}}}$	$t_0 < -t_{\alpha, v}$
$H_0: \mu_1 = \mu_2$ $H_1: \mu_1 > \mu_2$	$v = \dfrac{\left(\dfrac{S_1^2}{n_1} + \dfrac{S_2^2}{n_2}\right)^2}{\dfrac{(S_1^2/n_1)^2}{n_1 - 1} + \dfrac{(S_2^2/n_2)^2}{n_2 - 1}}$	$t_0 > t_{\alpha, v}$

2-5   INFERENCES ABOUT THE DIFFERENCES IN MEANS, PAIRED COMPARISON DESIGNS

2-5.1   The Paired Comparison Problem

In some simple comparative experiments we can greatly improve the precision by making comparisons within matched pairs of experimental material. For example, consider a hardness testing machine that presses a rod with a pointed tip into a metal specimen with a known force. By measuring the depth of the depression caused by the tip, the hardness of the specimen is determined. Two different tips are available for this machine, and although the precision (variability) of the measurements made by the two tips seems to be the same, it is suspected that one tip produces different hardness readings than the other.

An experiment could be performed as follows. A number of metal specimens (e.g., 20) could be randomly selected. Half of these specimens could be tested by tip 1 and the other half by tip 2. The exact assignment of specimens to tips would be randomly determined. Since this is a completely randomized design, the average hardness of the two samples could be compared using the t test described in Section 2-4.

A little reflection will reveal a serious disadvantage in the completely randomized design for this problem. Suppose the metal specimens were cut from different bar stock that were produced in different heats or that were not exactly homogeneous in some other way that might affect the hardness. This lack of homogeneity between specimens will contribute to the variability of the hardness measurements and will tend to inflate the experimental error, thus making a true difference between tips harder to detect.

To protect against this possibility, consider an alternate experimental design. Assume that each specimen is large enough so that *two* hardness determinations may be made on it. This alternative design would consist of dividing each specimen into two parts, then randomly assigning one tip to one-half of each specimen and the other tip to the remaining half. The order in which the tips are tested for a particular specimen would also be randomly selected. The experiment, when performed according to this design with 10 specimens, produced the data shown in Table 2-4.

We may write a statistical model that describes the data from this experiment as

$$y_{ij} = \mu_i + \beta_j + \epsilon_{ij} \begin{cases} i = 1,2 \\ j = 1,2, \dots, 10 \end{cases} \qquad (2\text{-}38)$$

where y_{ij} is the observation on hardness for tip i on specimen j, μ_i is the true mean hardness of the ith tip, β_j is an effect on hardness due to the jth specimen, and ϵ_{ij} is a random experimental error with mean zero and variance σ_i^2. That is, σ_1^2 is the variance of the hardness measurements from tip 1 and σ_2^2 is the variance of the hardness measurements from tip 2. We will frequently use models such as this one to describe the results of a designed experiment.

Table 2-4 Data for the Hardness Testing Experiment

Specimen	Tip 1	Tip 2
1	7	6
2	3	3
3	3	5
4	4	3
5	8	8
6	3	2
7	2	4
8	9	9
9	5	4
10	4	5

Note that if we compute the jth paired difference

$$d_j = y_{1j} - y_{2j} \qquad j = 1, 2, \ldots, 10 \qquad (2\text{-}39)$$

the expected value of this difference is

$$
\begin{aligned}
\mu_d &= E(d_j) \\
&= E(y_{1j} - y_{2j}) \\
&= E(y_{1j}) - E(y_{2j}) \\
&= \mu_1 + \beta_j - (\mu_2 + \beta_j) \\
&= \mu_1 - \mu_2
\end{aligned}
$$

That is, we may make inferences about the difference in the mean hardness readings of the two tips $\mu_1 - \mu_2$ by making inferences about the mean of the differences μ_d. Notice that the additive effect of the specimens β_j cancels out when the observations are paired in this manner.

Testing $H_0: \mu_1 = \mu_2$ is equivalent to testing

$$H_0: \mu_d = 0$$
$$H_1: \mu_d \neq 0$$

The test statistic for this hypothesis is

$$t_0 = \frac{\bar{d}}{S_d / \sqrt{n}} \qquad (2\text{-}40)$$

where

$$\bar{d} = \frac{1}{n} \sum_{j=1}^{n} d_j \qquad (2\text{-}41)$$

is the sample mean of the differences and

$$S_d = \left[\frac{\sum_{j=1}^{n} (d_j - \bar{d})^2}{n-1} \right]^{1/2} = \left[\frac{\sum_{j=1}^{n} d_j^2 - \frac{1}{n} \left(\sum_{j=1}^{n} d_j \right)^2}{n-1} \right]^{1/2} \qquad (2\text{-}42)$$

is the sample standard deviation of the differences. $H_0: \mu_d = 0$ would be rejected if $|t_0| > t_{\alpha/2,\, n-1}$. For the data in Table 2-4, we find

$$
\begin{array}{ll}
d_1 = 7 - 6 = 1 & d_6 = 3 - 2 = 1 \\
d_2 = 3 - 3 = 0 & d_7 = 2 - 4 = -2 \\
d_3 = 3 - 5 = -2 & d_8 = 9 - 9 = 0 \\
d_4 = 4 - 3 = 1 & d_9 = 5 - 4 = 1 \\
d_5 = 8 - 8 = 0 & d_{10} = 4 - 5 = -1
\end{array}
$$

Thus,

$$\bar{d} = \frac{1}{n} \sum_{j=1}^{n} d_j = \frac{1}{10}(-1) = -0.10$$

$$S_d = \left[\frac{\sum_{j=1}^{n} d_j^2 - \frac{1}{n}\left(\sum_{j=1}^{n} d_j\right)^2}{n-1} \right]^{1/2} = \left[\frac{13 - \frac{1}{10}(-1)^2}{10-1} \right]^{1/2} = 1.20$$

The test statistic is

$$t_0 = \frac{\bar{d}}{S_d/\sqrt{n}}$$

$$= \frac{-0.10}{1.20/\sqrt{10}}$$

$$= -0.26$$

and since $t_{.025,9} = 2.262$, we cannot reject the hypothesis $H_0 : \mu_d = 0$. That is, there is no evidence to indicate that the two tips produce different hardness readings.

2-5.2 Advantages of the Paired Comparison Design

The design actually used for this experiment is called the *paired comparison design* and it illustrates the blocking principle discussed in Section 1-3. Actually, it is a special case of a more general type of design called the *randomized block design*. The term "block" refers to a relatively homogeneous experimental unit (in our case, the metal specimens are the blocks), and the block represents a restriction on complete randomization because the treatment combinations are only randomized within the block. We look at designs of this type in Chapters 5 and 6. In those chapters the mathematical model for the design, Equation 2-38, is written in a slightly different form.

Before leaving this experiment, several points should be made. Note that, although $2n = 2(10) = 20$ observations have been taken, only $n - 1 = 9$ degrees of freedom are available for the t statistic. (We know that as the degrees of freedom for t increase the test becomes more sensitive.) By blocking or pairing, we have effectively "lost" $n - 1$ degrees of freedom, but we hope we have gained a better knowledge of the situation by eliminating an additional source of variability (the difference between specimens). We may obtain an indication of the quality of information produced from the paired design by comparing the standard deviation of the differences S_d with the pooled standard deviation S_p that would have resulted had the experiment been conducted in a completely randomized manner and the data of Table 2-4 been obtained. Using the data in Table 2-4 as two independent samples, we compute the pooled standard deviation from Equation 2-24 to be $S_p = 2.32$. Comparing this value to $S_d = 1.20$, we see that blocking or pairing has reduced the estimate of variability by nearly 50 percent. We may also express this information in terms of a confidence interval on $\mu_1 - \mu_2$. Using the paired data, a 95 percent confidence interval on $\mu_1 - \mu_2$. Using the paired data, a 95 percent confidence interval on $\mu_1 - \mu_2$ is

$$\bar{d} \pm t_{.025,9} S_d / \sqrt{n}$$
$$-0.10 \pm (2.262)(1.20)/\sqrt{10}$$
$$-0.10 \pm 0.86$$

Conversely, using the pooled or independent analysis, a 95 percent confidence interval on $\mu_1 - \mu_2$ is

$$\bar{y}_1 - \bar{y}_2 \pm t_{.025,18} \, S_p \sqrt{\frac{1}{n_1} + \frac{1}{n_2}}$$
$$4.80 - 4.90 \pm (2.101)(2.32)\sqrt{\tfrac{1}{10} + \tfrac{1}{10}}$$
$$-0.10 \pm 2.18$$

The confidence interval based on the paired analysis is much narrower than the confidence interval from the independent analysis. This illustrates the *noise reduction* property of blocking.

Blocking is not always the best design strategy. If the within-block variability is the same as the between-block variability, then the variance of $\bar{y}_1 - \bar{y}_2$ will be the same regardless of which design is used. Actually, blocking in this situation would be a poor choice of design because blocking results in the loss of $n - 1$ degrees of freedom and will actually lead to a wider confidence interval on $\mu_1 - \mu_2$. A further discussion of blocking is given in Chapter 5.

2-6   INFERENCES ABOUT THE VARIANCES OF NORMAL DISTRIBUTIONS

In many experiments, we are interested in possible differences in the mean response for two treatments. However, in some experiments it is the comparison of variability in the data that is important. In the food and beverage industry, for example, it is important that the variability of filling equipment be small so that all packages have close to the nominal net weight or volume of content. In chemical laboratories, we may wish to compare the variability of two analytical methods. We now briefly examine tests of hypotheses and confidence intervals for variances of normal distributions. Unlike the tests on means, the procedures for tests on variances are rather sensitive to the normality assumption. A good discussion of the normality assumption is in Appendix 2A of Davies (1956).

Suppose we wish to test the hypothesis that the variance of a normal population equals a constant, for example, σ_0^2. Stated formally, we wish to test

$$H_0: \sigma^2 = \sigma_0^2 \tag{2-43}$$
$$H_1: \sigma^2 \neq \sigma_0^2$$

The test statistic for Equation 2-43 is

$$\chi_0^2 = \frac{SS}{\sigma_0^2} = \frac{(n-1)S^2}{\sigma_0^2} \tag{2-44}$$

where $SS = \Sigma_{i=1}(y_i - y)^2$ is the corrected sum of squares of the sample data. The null hypothesis is rejected if $\chi_0^2 > \chi_{\alpha/2,n-1}^2$ or if $\chi_0^2 < \chi_{1-(\alpha/2),n-1}^2$, where $\chi_{\alpha/2,n-1}^2$ and $\chi_{1-(\alpha/2),n-1}^2$ are the upper $\alpha/2$ and lower $1 - (\alpha/2)$ percentage points of the chi-square distribution with $n-1$ degrees of freedom, respectively. Table 2-5 gives the critical regions for the one-sided alternative hypotheses. The $100(1 - \alpha)$ percent confidence interval on σ^2 is

$$\frac{(n-1)S^2}{\chi_{\alpha/2,n-1}^2} \leq \sigma^2 \leq \frac{(n-1)S^2}{\chi_{1-(\alpha/2),n-1}^2} \tag{2-45}$$

Now consider testing the equality of the variances of two normal populations. If independent random samples of size n_1 and n_2 are taken from populations 1 and 2, respectively, then the test statistic for

$$H_0: \sigma_1^2 = \sigma_2^2 \tag{2-46}$$
$$H_1: \sigma_1^2 \neq \sigma_2^2$$

is the ratio of the sample variances

$$F_0 = \frac{S_1^2}{S_2^2} \tag{2-47}$$

The null hypothesis would be rejected if $F_0 > F_{\alpha/2,n_1-1,n_2-1}$ or if $F_0 < F_{1-(\alpha/2),n_1-1,n_2-1}$, where $F_{\alpha/2,n_1-1,n_2-1}$ and $F_{1-(\alpha/2),n_1-1,n_2-1}$ denote the upper $\alpha/2$ and lower $1 - (\alpha/2)$ percentage points of the F distribution with $n_1 - 1$ and $n_2 - 1$ degrees of freedom. Table IV of the Appendix gives only upper-tail percentage points of F; however, the upper- and lower-tail points are related by

$$F_{1-\alpha,v_1,v_2} = \frac{1}{F_{\alpha,v_2,v_1}} \tag{2-48}$$

Test procedures for more than two variances are discussed in Chapter 4, Section 4-1.3. We will also discuss the use of the variance or standard deviation as a response variable in more general experimental settings.

Example 2-2

A chemical engineer is investigating the inherent variability of two types of test equipment that can be used to monitor the output of a production process. He suspects

Table 2-5   Tests on Variances of Normal Distributions

Hypothesis	Test Statistic	Criteria for Rejection
$H_0: \sigma^2 = \sigma_0^2$ $H_1: \sigma^2 \neq \sigma_0^2$		$\chi_0^2 > \chi_{\alpha/2, n-1}^2$ or $\chi_0^2 < \chi_{1-\alpha/2, n-1}^2$
$H_0: \sigma^2 = \sigma_0^2$ $H_1: \sigma^2 < \sigma_0^2$	$\chi_0^2 = \dfrac{(n-1)S^2}{\sigma_0^2}$	$\chi_0^2 < \chi_{1-\alpha, n-1}^2$
$H_0: \sigma^2 = \sigma_0^2$ $H_1: \sigma^2 > \sigma_0^2$		$\chi_0^2 > \chi_{\alpha, n-1}^2$
$H_0: \sigma_1^2 = \sigma_2^2$ $H_1: \sigma_1^2 \neq \sigma_2^2$	$F_0 = \dfrac{S_1^2}{S_2^2}$	$F_0 > F_{\alpha/2, n_1-1, n_2-1}$ or $F_0 < F_{1-\alpha/2, n_1-1, n_2-1}$
$H_0: \sigma_1^2 = \sigma_2^2$ $H_1: \sigma_1^2 < \sigma_2^2$	$F_0 = \dfrac{S_2^2}{S_1^2}$	$F_0 > F_{\alpha, n_2-1, n_1-1}$
$H_0: \sigma_1^2 = \sigma_2^2$ $H_1: \sigma_1^2 > \sigma_2^2$	$F_0 = \dfrac{S_1^2}{S_2^2}$	$F_0 > F_{\alpha, n_1-1, n_2-1}$

that the old equipment, type 1, has a larger variance than the new one. Thus, he wishes to test the hypothesis

$$H_0: \sigma_1^2 = \sigma_2^2$$
$$H_1: \sigma_1^2 > \sigma_2^2$$

Two random samples of $n_1 = 12$ and $n_2 = 10$ observations are taken, and the sample variances are $S_1^2 = 14.5$ and $S_2^2 = 10.8$. The test statistic is

$$F_0 = \frac{S_1^2}{S_2^2} = \frac{14.5}{10.8} = 1.34$$

From Appendix Table IV we find that $F_{.05,11,9} = 3.10$, so the null hypothesis cannot be rejected. That is, we have found insufficient statistical evidence to conclude that the variance of the old equipment is greater than the variance of the new equipment.

∎

The $100(1 - \alpha)$ confidence interval for the ratio of the population variances σ_1^2/σ_2^2 is

$$\frac{S_1^2}{S_2^2} F_{1-\alpha/2, n_2-1, n_1-1} \leq \frac{\sigma_1^2}{\sigma_2^2} \leq \frac{S_1^2}{S_2^2} F_{\alpha/2, n_2-1, n_1-1} \tag{2-49}$$

To illustrate the use of Equation 2-49, the 95 percent confidence interval for the ratio of variances σ_1^2/σ_2^2 in Example 2-2 is, using $F_{.025,9,11} = 3.59$ and $F_{.975,9,11} = 1/F_{.025,11,9} = 1/3.92 = 0.255$,

$$\frac{14.5}{10.8}(0.255) \leq \frac{\sigma_1^2}{\sigma_2^2} \leq \frac{14.5}{10.8}(3.59)$$

$$0.34 \leq \frac{\sigma_1^2}{\sigma_2^2} \leq 4.81$$

2-7   PROBLEMS

2-1   The breaking strength of a fiber is required to be at least 150 psi. Past experience has indicated that the standard deviation of breaking strength is $\sigma = 3$ psi. A random sample of four specimens is tested, and the average breaking strength is observed to be 148 psi. Should the fiber be judged acceptable with $\alpha = .05$? Construct a 95 percent confidence interval on the mean breaking strength.

2-2   The diameters of steel shafts produced by a certain manufacturing process are known to have a standard deviation of $\sigma = 0.0001$ inch. A random sample of 10 shafts has an average diameter of 0.2545 inches. Test the hypothesis that the true mean diameter equals 0.255 inches using $\alpha = .05$. Construct a 90 percent confidence interval on the mean diameter.

2-3   A normally distributed random variable has an unknown mean μ and a known variance $\sigma^2 = 9$. Find the sample size required to construct a 95 percent confidence interval on the mean, which has total width of 1.0.

2-4   The shelf life of a carbonated beverage is of interest. Ten bottles are randomly selected and tested, and the following results are obtained:

Days	
108	138
124	163
124	159
106	134
115	139

Assume that the alternative hypothesis is that the mean shelf life is greater than 125 days. Can the null hypothesis $H_0: \mu = 125$ be rejected? Construct a 95 percent confidence interval on the true mean shelf life.

2-5 The time to repair an electronic instrument is a normally distributed random variable measured in hours. The repair times for 16 such instruments chosen at random are as follows:

Hours			
159	280	101	212
224	379	179	264
222	362	168	250
149	260	485	170

Does it seem reasonable that the true mean repair time is greater than 225 hours? Determine a confidence interval on the true mean repair time.

2-6 Two machines are used for filling plastic bottles with a net volume of 16.0 ounces. The filling processes can be assumed to be normal, with standard deviations of $\sigma_1 = 0.015$ and $\sigma_2 = 0.018$. The quality engineering department suspects that both machines fill to the same net volume, whether or not this volume is 16.0 ounces. A random sample is taken from the output of each machine.

Machine 1		Machine 2	
16.03	16.01	16.02	16.03
16.04	15.96	15.97	16.04
16.05	15.98	15.96	16.02
16.05	16.02	16.01	16.01
16.02	15.99	15.99	16.00

Do you think the quality engineering department is correct?

2-7 Two types of plastic are suitable for use by an electronic calculator manufacturer. The breaking strength of this plastic is important. It is known that $\sigma_1 = \sigma_2 = 1.0$ psi. From random samples of $n_1 = 10$ and $n_2 = 12$ we obtain $\bar{y}_1 = 162.5$ and $\bar{y}_2 = 155.0$. The company will not adopt plastic 1 unless its breaking strength exceeds that of plastic 2 by at least 10 psi. Based on the sample information, should they use plastic 1? Construct a 99 percent confidence interval on the true mean difference in breaking strength.

2-8 The following are the burning times of flares of two different design types. The design engineers are interested in both the mean and variance of the burning times.

Type 1		Type 2	
65	82	64	56
81	67	71	69
57	59	83	74
66	75	59	82
82	70	65	79

(a) Test the hypothesis that the two variances are equal. Use $\alpha = .05$.

(b) Using the results of (a), test the hypothesis that the mean burning times are equal.

(c) Discuss the role of the normality assumption in this problem.

2-9 An article in *Solid State Technology*, "Orthogonal Design for Process Optimization and Its Application to Plasma Etching" by G. Z. Yin and D. W. Jillie (May, 1987) describes an experiment to determine the effect of the C_2F_6 flow rate on the uniformity of the etch on a silicon wafer used in integrated circuit manufacturing. Data for two flow rates are as follows:

C_2F_6 Flow (SCCM)	Uniformity Observation					
	1	2	3	4	5	6
125	2.7	4.6	2.6	3.0	3.2	3.8
200	4.6	3.4	2.9	3.5	4.1	5.1

(a) Does the C_2F_6 flow rate affect average etch uniformity?

(b) Does the C_2F_6 flow rate affect the wafer-to-wafer variability in etch uniformity?

(c) Would box plots assist in the interpretation of the data from this experiment?

2-10 A new filtering device is installed in a chemical unit. Before its installation, a random sample yielded the following information about the percentage of impurity: $\bar{y}_1 = 12.5$, $S_1^2 = 101.17$, and $n_1 = 8$. After installation, a random sample yielded $\bar{y}_2 = 10.2$, $S_2^2 = 94.73$, $n_2 = 9$. Can you conclude that the two variances are equal? Has the filtering device reduced the percentage of impurity significantly?

2-11 Twenty observations on etch uniformity on silicon wafers are taken during a qualification experiment for a plasma etcher. The data are as follows:

5.34	6.65	4.76	5.98	7.25
6.00	7.55	5.54	5.62	6.21
5.97	7.35	5.44	4.39	4.98
5.25	6.35	4.61	6.00	5.32

(a) Compute a 95 percent confidence interval estimate of σ^2.

(b) Test the hypothesis that $\sigma^2 = 1.0$.

(c) Discuss the normality assumption and its role in this problem.

2-12 The diameter of a ball bearing was measured by 12 inspectors, each using two different kinds of calipers. The results were

Inspector	Caliper 1	Caliper 2
1	0.265	0.264
2	0.265	0.265
3	0.266	0.264
4	0.267	0.266
5	0.267	0.267
6	0.265	0.268
7	0.267	0.264
8	0.267	0.265
9	0.265	0.265
10	0.268	0.267
11	0.268	0.268
12	0.265	0.269

Is there a significant difference between the means of the population of measurements represented by the two samples? Use $\alpha = .05$.

2-13 An article in the *Journal of Strain Analysis* (vol. 18, no. 2, 1983) compares several procedures for predicting the shear strength for steel plate girders. Data for nine girders in the form of the ratio of predicted to observed load for two of these procedures, the Karlsruhe and Lehigh methods, are as follows

Girder	Karlsruhe Method	Lehigh Method
S1/1	1.186	1.061
S2/1	1.151	0.992
S3/1	1.322	1.063
S4/1	1.339	1.062
S5/1	1.200	1.065
S2/1	1.402	1.178
S2/2	1.365	1.037
S2/3	1.537	1.086
S2/4	1.559	1.052

Is there any difference between the two methods?

2-14 Two popular pain medications are being compared on the basis of the speed of absorption by the body. Specifically, tablet 1 is claimed to be absorbed twice as fast as tablet 2. Assume that σ_1^2 and σ_2^2 are known. Develop a test statistic for

$$H_0: 2\mu_1 = \mu_2$$
$$H_1: 2\mu_1 \neq \mu_2$$

2-15   Suppose we are testing

$$H_0:\mu_1 = \mu_2$$

$$H_1:\mu_1 \neq \mu_2$$

where σ_1^2 and σ_2^2 are known. Our sampling resources are constrained such that $n_1 + n_2 = N$. How should we allocate the N observations between the two populations to obtain the most powerful test?

2-16   Develop Equation 2-45 for a $100(1 - \alpha)$ percent confidence interval for the variance of a normal distribution.

2-17   Develop Equation 2-49 for a $100(1 - \alpha)$ percent confidence interval for the ratio σ_1^2/σ_2^2, where σ_1^2 and σ_2^2 are the variances of two normal distributions.

Chapter 3

Experiments with a Single Factor: The Analysis of Variance

In Chapter 2 we discussed methods for comparing two conditions or treatments. For example, the portland cement tension bond experiment involved two different mortar formulations. Another way to describe this experiment is as a single-factor experiment with two levels of the factor, where the factor is mortar formulation and the two levels are the two different formulation methods. Many experiments of this type involve more than two levels of the factor. In this chapter and the next, we present methods for the design and analysis of single-factor experiments with a levels of the factor (or a treatments). We will assume that the experiment has been completely randomized.

3-1 AN EXAMPLE

A product development engineer is interested in maximizing the tensile strength of a new synthetic fiber that will be used to make cloth for men's shirts. The engineer knows from previous experience that the strength is affected by the percentage of cotton in the fiber. Furthermore, he suspects that increasing the cotton content will increase the strength, at least initially. He also knows that cotton content should range between about 10 percent and 40 percent if the final cloth is to have other quality characteristics that are desired (such as the ability to take a permanent-press finishing treatment). The engineer decides to test specimens at five levels of cotton percentage: 15 percent, 20 percent, 25 percent, 30 percent, and 35 percent. He also decides to test five specimens at each level of cotton content.

This is an example of a single-factor experiment with $a = 5$ *levels* of the factor and $n = 5$ *replicates*. The 25 runs should be made in random order. To illustrate how the run order may be randomized, suppose that we number the runs as follows:

Cotton Percentage	Experimental Run Number				
	1	2	3	4	5
15	1	2	3	4	5
20	6	7	8	9	10
25	11	12	13	14	15
30	16	17	18	19	20
35	21	22	23	24	25

Now select a random number between 1 and 25. Suppose this number is 8. Then the number 8 observation (20 percent cotton) is run first. This process would be repeated until all 25 observations have been assigned a position in the test sequence.[1] Suppose that the test sequence obtained is

Test Sequence	Run Number	Percentage of Cotton
1	8	20
2	18	30
3	10	20
4	23	35
5	17	30
6	5	15
7	14	25
8	6	20
9	15	25
10	20	30
11	9	20
12	4	15
13	12	25
14	7	20
15	1	15
16	24	35
17	21	35
18	11	25
19	2	15
20	13	25
21	22	35
22	16	30
23	25	35
24	19	30
25	3	15

[1] The only restriction on randomization here is that if the same number (e.g., 8) is drawn again it is discarded. This is a minor restriction and is ignored.

Table 3-1   Data (in lb/in²) from the Tensile Strength Experiment

Cotton Percentage	Observations					Total	Average
	1	2	3	4	5		
15	7	7	15	11	9	49	9.8
20	12	17	12	18	18	77	15.4
25	14	18	18	19	19	88	17.6
30	19	25	22	19	23	108	21.6
35	7	10	11	15	11	54	10.8
						376	15.04

This randomized test sequence is necessary to prevent the effects of unknown nuisance variables, perhaps varying out of control during the experiment, from contaminating the results. To illustrate, suppose that we were to run the 25 test specimens in the original nonrandomized order (that is, all five 15 percent cotton specimens are tested first, all five 20 percent cotton specimens are tested next, and so on). If the tensile strength testing machine exhibits a warm-up effect such that the longer it is on, the lower the observed tensile strength readings will be, then the warm-up effect will potentially contaminate the tensile strength data and destroy the validity of the experiment.

Suppose that the engineer runs the test in the random order we have determined. The observations that he obtains on tensile strength are shown in Table 3-1.

It is always a good idea to examine experimental data graphically. Figure 3-1 presents box plots for tensile strength at each level of cotton percentage, and Figure 3-2 is a scatter diagram of tensile strength versus cotton percentage. In

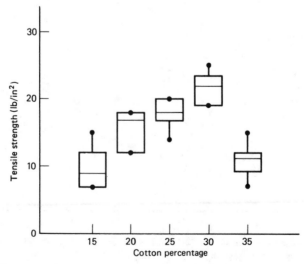

Figure 3-1.   Box plots of tensile strength versus cotton percentage experiment.

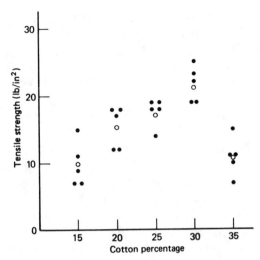

Figure 3-2.   Scatter diagram for the tensile strength versus cotton percentage experiment.

Figure 3-2, the solid dots are the individual observations and the open circles are the average observed tensile strengths. Both graphs indicate that tensile strength increases as cotton content increases, up to about 30 percent cotton. Beyond 30 percent cotton, there is a marked decrease in tensile strength. There is no strong evidence to suggest that the variability in tensile strength around the average depends on the cotton percentage. Based on this simple graphical analysis, we strongly suspect that (1) cotton content affects tensile strength and (2) around 30 percent cotton would result in maximum strength.

Suppose that we wish to be more objective in our analysis of the data. Specifically, suppose that we wish to test for differences between the mean strengths at all $a = 5$ levels of cotton percentage. Thus, we are interested in testing the equality of all five means. It might seem that this problem could be solved by performing a t test on all the possible pairs of means. However, this solution would be incorrect, since it would lead to considerable distortion in the type I error. For example, suppose we wish to test the equality of the five means using pairwise comparisons. There are 10 possible pairs, and if the probability of correctly accepting the null hypothesis for each individual test is $1 - \alpha = .95$, then the probability of correctly accepting the null hypothesis for all 10 tests is $(.95)^{10} = .60$ if the tests are independent. Thus, a substantial increase in the type I error has occurred.

The appropriate procedure for testing the equality of several means is the analysis of variance. However, the analysis of variance has a much wider application than the problem above. It is probably the most useful technique in the field of statistical inference.

3-2   THE ANALYSIS OF VARIANCE

Suppose we have a treatments or diferent levels of a single factor that we wish to compare. The observed response from each of the a treatments is a random variable. The data would appear as in Table 3-2. An entry in Table 3-2 (e.g., y_{ij}) represents the jth observation taken under treatment i. There will be, in general, n observations under the ith treatment. Notice that Table 3-2 is the general case of the data from the tensile strength experiment in Table 3-1.

We will find it useful to describe the observations with a linear statistical model:

$$y_{ij} = \mu + \tau_i + \epsilon_{ij} \begin{cases} i = 1,2, \ldots, a \\ j = 1,2, \ldots, n \end{cases} \tag{3-1}$$

where y_{ij} is the (ij)th observation, μ is a parameter common to all treatments called the *overall* mean, τ_i is a parameter unique to the ith treatment called the ith *treatment effect*, and ϵ_{ij} is a random error component. Our objectives will be to test appropriate hypotheses about the treatment effects and to estimate them. For hypothesis testing, the model errors are assumed to be normally and independently distributed random variables with mean zero and variance σ^2. The variance σ^2 is assumed to be constant for all levels of the factor.

This model is called the *one-way* or *single-factor* analysis of variance because only one factor is investigated. Furthermore, we will require that the experiment be performed in random order so that the environment in which the treatments are used (often called the *experimental units*) is as uniform as possible. Thus, the experimental design is a *completely randomized* design.

Actually, the statistical model, Equation 3-1, describes two different situations with respect to the treatment effects. First, the a treatments could have been specifically chosen by the experimenter. In this situation we wish to test hypotheses about the treatment means, and our conclusions will apply only to the factor levels considered in the analysis. The conclusions cannot be extended to similar treatments that were not explicitly considered. We may also wish to estimate the model parameters (μ, τ_i, σ^2). This is called the *fixed effects model*. Alternatively, the a treatments could be a *random sample* from a larger population of treatments. In this situation we should like to be able to extend the conclusions (which are based on the sample of treatments) to all treatments in the population, whether they were explicitly considered in the analysis or not. Here the τ_i are random variables, and knowledge about the particular ones investigated is relatively useless. Instead, we test hypotheses about the variability of the τ_i and try to estimate this variability. This is called the *random effects model* or *components of variance model*.

Table 3-2   Typical Data for a Single-Factor Experiment

Treatment (level)	Observations				Totals	Averages
1	y_{11}	y_{12}	$\cdots$	y_{1n}	$y_{1.}$	$\bar{y}_{..}$
2	y_{21}	y_{22}	$\cdots$	y_{2n}	$y_{2.}$	$\bar{y}_{2.}$
$\vdots$	$\vdots$	$\vdots$	$\cdots$	$\vdots$	$\vdots$	$\vdots$
a	y_{a1}	y_{a2}	$\cdots$	y_{an}	$y_{a.}$	$\bar{y}_{a.}$
					$y_{..}$	$\bar{y}_{..}$

3-3   ANALYSIS OF THE FIXED EFFECTS MODEL

In this section, we develop the single-factor analysis of variance for the fixed effects model. In the fixed effects model, the treatment effects τ_i are usually defined as deviations from the overall mean, so

$$\sum_{i=1}^{a} \tau_i = 0 \qquad (3\text{-}2)$$

Let $y_{i.}$ represent the total of the observations under the ith treatment and $\bar{y}_{i.}$ represent the average of the observations under the ith treatment. Similarly, let $y_{..}$ represent the grand total of all the observations and $\bar{y}_{..}$ represent the grand average of all the observations. Expressed symbolically,

$$y_{i.} = \sum_{j=1}^{n} y_{ij}, \qquad \bar{y}_{i.} = y_{i.}/n \qquad i = 1,2,\ldots,a$$

$$y_{..} = \sum_{i=1}^{a} \sum_{j=1}^{n} y_{ij} \qquad \bar{y}_{..} = y_{..}/N \qquad (3\text{-}3)$$

where $N = an$ is the total number of observations. We see that the "dot" subscript notation implies summation over the subscript that it replaces.

The mean of the ith treatment is $E(y_{ij}) \equiv \mu_i = \mu + \tau_i$, $i = 1,2,\ldots,a$. Thus, the mean of the ith treatment consists of the overall mean plus the ith treatment effect. We are interested in testing the equality of the a treatment means; that is,

$$H_0: \mu_1 = \mu_2 = \cdots = \mu_a$$
$$H_1: \mu_i \neq \mu_j \qquad \text{for at least one pair } (i,j)$$

Note that if H_0 is true, all treatments have a common mean μ. An equivalent way to write the above hypotheses is in terms of the treatment effects τ_i, say

$$H_0: \tau_1 = \tau_2 = \cdots \tau_a = 0$$
$$H_1: \tau_i \neq 0 \qquad \text{for at least one } i$$

Thus, we may speak of testing the equality of treatment means or testing that the treatment effects (the τ_i) are zero. The appropriate procedure for testing the equality of a treatment means is the analysis of variance.

3-3.1   Decomposition of the Total Sum of Squares

The name *analysis of variance* is derived from a partitioning of total variability into its component parts. The total corrected sum of squares

$$SS_T = \sum_{i=1}^{a} \sum_{j=1}^{n} (y_{ij} - \bar{y}_{..})^2$$

is used as a measure of overall variability in the data. Intuitively, this is reasonable since, if we were to divide SS_T by the appropriate number of degrees of freedom (in this case, $an - 1 = N - 1$), we would have the *sample variance* of the y's. The sample variance is, of course, a standard measure of variability.

Note that the total corrected sum of squares SS_T may be written as

$$\sum_{i=1}^{a} \sum_{j=1}^{n} (y_{ij} - \bar{y}_{..})^2 = \sum_{i=1}^{a} \sum_{j=1}^{n} [(\bar{y}_{i.} - \bar{y}_{..}) + (y_{ij} - \bar{y}_{i.})]^2 \qquad (3\text{-}4)$$

or

$$\sum_{i=1}^{a} \sum_{j=1}^{n} (y_{ij} - \bar{y}_{..})^2 = n \sum_{i=1}^{a} (\bar{y}_{i.} - \bar{y}_{..})^2 + \sum_{i=1}^{a} \sum_{j=1}^{n} (y_{ij} - \bar{y}_{i.})^2$$
$$+ 2 \sum_{i=1}^{a} \sum_{j=1}^{n} (\bar{y}_{i.} - \bar{y}_{..})(y_{ij} - \bar{y}_{i.}) \qquad (3\text{-}5)$$

However, the cross-product term in Equation 3-5 is zero, since

$$\sum_{j=1}^{n} (y_{ij} - \bar{y}_{i.}) = y_{i.} - n\bar{y}_{i.} = y_{i.} - n(y_{i.}/n) = 0$$

Therefore, we have

$$\sum_{i=1}^{a} \sum_{j=1}^{n} (y_{ij} - \bar{y}_{..})^2 = n \sum_{i=1}^{a} (\bar{y}_{i.} - \bar{y}_{..})^2 + \sum_{i=1}^{a} \sum_{j=1}^{n} (y_{ij} - \bar{y}_{i.})^2 \qquad (3\text{-}6)$$

Equation 3-6 states that the total variability in the data, as measured by the total corrected sum of squares, can be partitioned into a sum of squares of the differences *between* the treatment averages and the grand average, plus a sum of squares of the differences of observations *within* treatments from the treatment average. Now, the difference between the observed treatment averages and the grand average is a measure of the differences between treatment means, whereas the differences of observations within a treatment from the treatment average can be due only to random error. Thus, we may write Equation 3-6 symbolically as

$$SS_T = SS_{\text{Treatments}} + SS_E$$

where $SS_{\text{Treatments}}$ is called the sum of squares due to treatments (i.e., between treatments), and SS_E is called the sum of squares due to error (i.e., within treatments). There are $an = N$ total observations; thus, SS_T has $N - 1$ degrees of freedom. There are a levels of the factor (and a treatment means), so $SS_{\text{Treatments}}$ has $a - 1$ degrees of freedom. Finally, within any treatment there are n replicates providing $n - 1$ degrees of freedom with which to estimate the experimental error. Since there are a treatments, we have $a(n - 1) = an - a = N - a$ degrees of freedom for error.

It is helpful to examine explicitly the two terms on the right-hand side of the fundamental analysis of variance identity (Equation 3-6). Consider the error sum of squares

$$SS_E = \sum_{i=1}^{a} \sum_{j=1}^{n} (y_{ij} - \bar{y}_{i.})^2 = \sum_{i=1}^{a} \left[\sum_{j=1}^{n} (y_{ij} - \bar{y}_{i.})^2 \right]$$

In this form it is easy to see that the term within square brackets, if divided by $n - 1$, is the sample variance in the ith treatment, or

$$S_i^2 = \frac{\sum_{j=1}^{n} (y_{ij} - \bar{y}_{i.})^2}{n - 1} \qquad i = 1, 2, \ldots, a$$

Now a sample variances may be combined to give one estimate of the common population variance as follows:

$$\frac{(n - 1)S_1^2 + (n - 1)S_2^2 + \cdots + (n - 1)S_a^2}{(n - 1) + (n - 1) + \cdots + (n - 1)} = \frac{\sum_{i=1}^{a} \left[\sum_{j=1}^{n} (y_{ij} - \bar{y}_{i.})^2 \right]}{\sum_{i=1}^{a} (n - 1)}$$

$$= \frac{SS_E}{(N - a)}$$

Thus, $SS_E/(N - a)$ is an estimate of the common variance within each of the a treatments.

Similarly, if there were no differences between the a treatment means, we could use the variation of the treatment averages from the grand average to estimate σ^2. Specifically,

$$\frac{SS_{\text{Treatments}}}{a - 1} = \frac{n \sum_{i=1}^{a} (\bar{y}_{i.} - \bar{y}_{..})^2}{a - 1}$$

is an estimate of σ^2 if the treatment means are equal. The reason for this may be intuitively seen as follows: the quality $\sum_{i=1}^{a}(\bar{y}_{i.} - \bar{y}_{..})^2/(a - 1)$ estimates σ^2/n, the variance of the treatment averages, so $n\sum_{i=1}^{a}(\bar{y}_{i.} - \bar{y}_{..})^2/(a - 1)$ must estimate σ^2 if there are no differences in treatment means.

We see that the analysis of variance identity (Equation 3-6) provides us with two estimates of σ^2—one based on the inherent variability within treatments and one based on the variability between treatments. If there are no differences in the treatment means, these two estimates should be very similar, and if they are not, we suspect that the observed difference must be caused by differences in the treatment means. Although we have used an intuitive argument to develop this result, a somewhat more formal approach can be taken.

The quantities

$$MS_{\text{Treatments}} = \frac{SS_{\text{Treatments}}}{a - 1}$$

and

$$MS_E = \frac{SS_E}{N - a}$$

are called *mean squares*. We now examine the *expected values* of these mean squares. Consider

$$E(MS_E) = E\left(\frac{SS_E}{N - a}\right) = \frac{1}{N - a} E\left[\sum_{i=1}^{a} \sum_{j=1}^{n} (y_{ij} - \bar{y}_{i.})^2\right]$$

$$= \frac{1}{N - a} E\left[\sum_{i=1}^{a} \sum_{j=1}^{n} (y_{ij}^2 - 2y_{ij}\bar{y}_{i.} + \bar{y}_{i.}^2)\right]$$

$$= \frac{1}{N - a} E\left[\sum_{i=1}^{a} \sum_{j=1}^{n} y_{ij}^2 - 2n \sum_{i=1}^{a} \bar{y}_{i.}^2 + n \sum_{i=1}^{a} \bar{y}_{i.}^2\right]$$

$$= \frac{1}{N - a} E\left[\sum_{i=1}^{a} \sum_{j=1}^{n} y_{ij}^2 - \frac{1}{n} \sum_{i=1}^{a} y_{i.}^2\right]$$

Substituting the model (Equation 3-1) into this equation, we obtain

$$E(MS_E) = \frac{1}{N - a} E\left[\sum_{i=1}^{a} \sum_{j=1}^{n} (\mu + \tau_i + \epsilon_{ij})^2 - \frac{1}{n} \sum_{i=1}^{a} \left(\sum_{j=1}^{n} \mu + \tau_i + \epsilon_{ij}\right)^2\right]$$

Now when squaring and taking expectation of the quantity within the brackets, we see that terms involving ϵ_{ij}^2 and $\epsilon_{i.}^2$ are replaced by σ^2 and $n\sigma^2$, respectively, because $E(\epsilon_{ij}) = 0$. Furthermore, all cross-products involving ϵ_{ij} have zero expectation. Therefore, after squaring and taking expectation, the last equation becomes

$$E(MS_E) = \frac{1}{N - a}\left[N\mu^2 + n \sum_{i=1}^{a} \tau_i^2 + N\sigma^2 - N\mu^2 - n \sum_{i=1}^{a} \tau_i^2 - a\sigma^2\right]$$

or

$$E(MS_E) = \sigma^2$$

By a similar approach, we may also show that

$$E(MS_{Treatments}) = \sigma^2 + \frac{n \sum_{i=1}^{a} \tau_i^2}{a - 1}$$

Thus, as we argued heuristically, $MS_E = SS_E/(N - a)$ estimates σ^2, and, if there are no differences in treatment means (which implies that $\tau_i = 0$), $MS_{Treatments} = SS_{Treatments}/(a - 1)$ also estimates σ^2. However, note that if treatment means do differ, the expected value of the treatment mean square is greater than σ^2.

It seems clear that a test of the hypothesis of no difference in treatment means can be performed by comparing $MS_{Treatments}$ and MS_E. We now consider how this comparison may be made.

3-3.2 Statistical Analysis

We now investigate how a formal test of the hypothesis of no differences in treatment means ($H_0: \mu_1 = \mu_2 = \cdots = \mu_a$, or equivalently, $H_0: \tau_1 = \tau_2 = \cdots$ $\tau_a = 0$) can be performed. Since we have assumed that the errors ϵ_{ij} are normally and independently distributed with mean zero and variance σ^2, the observations y_{ij} are normally and independently distributed with mean $\mu + \tau_i$ and variance σ^2. Thus, SS_T is a sum of squares in normally distributed random variables; consequently, it can be shown that SS_T/σ^2 is distributed as chi-square with $N - 1$ degrees of freedom. Furthermore, we can show that SS_E/σ^2 is chi-square with $N - a$ degrees of freedom and that $SS_{\text{Treatments}}/\sigma^2$ is chi-square with $a - 1$ degrees of freedom if the null hypothesis $H_0: \tau_i = 0$ is true. However, all three sums of squares are not independent since $SS_{\text{Treatments}}$ and SS_E add to SS_T. The following theorem, which is a special form of one attributed to William Cochran, is useful in establishing the independence of SS_E and $SS_{\text{Treatments}}$.

THEOREM 3-1. COCHRAN'S THEOREM   Let Z_i be NID$(0,1)$ for $i = 1, 2, \ldots,$ v and

$$\sum_{i=1}^{v} Z_i^2 = Q_1 + Q_2 + \cdots + Q_s$$

where $s \leq v$, and Q_i has v_i degrees, of freedom $(i = 1, 2, \ldots, s)$. Then $Q_1, Q_2,$ $\ldots, Q_s$ are independent chi-square random variables with $v_1, v_2, \ldots, v_s$ degrees of freedom, respectively, if and only if

$$v = v_1 + v_2 + \cdots + v_s$$

Since the degrees of freedom for $SS_{\text{Treatments}}$ and SS_E and to $N - 1$, the total number of degrees of freedom, Cochran's theorem implies that $SS_{\text{Treatments}}/\sigma^2$ and SS_E/σ^2 are independently distributed chi-square random variables. Therefore, if the null hypothesis of no difference in treatment means is true, the ratio

$$F_0 = \frac{SS_{\text{Treatments}}/(a - 1)}{SS_E/(N - a)} = \frac{MS_{\text{Treatments}}}{MS_E} \tag{3-7}$$

is distributed as F with $a - 1$ and $N - a$ degrees of freedom. Equation 3-7 is the test statistic for the hypothesis of no differences in treatment means.

From the expected mean squares we see that, in general, MS_E is an unbiased estimator of σ^2. Also, under the null hypothesis, $MS_{\text{Treatments}}$ is an unbiased estimator of σ^2. However, if the null hypothesis is false, then the expected value of $MS_{\text{Treatments}}$ is greater than σ^2. Therefore, under the alternative hypothesis, the expected value of the numerator of the test statistic (Equation 3-7) is greater than the expected value of the denominator, and we should reject H_0 on values of the test statistic that are too large. This implies an upper-tail, one-tail critical region. We would reject H_0 if

$$F_0 > F_{\alpha, \, a-1, \, N-a}$$

where F_0 is computed from Equation 3-7.

Table 3-3 The Analysis of Variance Table for the Single-Factor, Fixed Effects Model

Source of Variation	Sum of Squares	Degrees of Freedom	Mean Square	F_0
Between treatments	$SS_{Treatments}$	$a - 1$	$MS_{Treatments}$	$F_0 = \dfrac{MS_{Treatments}}{MS_E}$
Error (within treatments)	SS_E	$N - a$	MS_E	
Total	SS_T	$N - 1$		

Computing formulas for the sums of squares may be obtained by rewriting and simplifying the definitions of $MS_{Treatments}$ and SS_T in Equation 3-6. This yields

$$SS_T = \sum_{i=1}^{a} \sum_{j=1}^{n} y_{ij}^2 - \frac{y_{..}^2}{N} \tag{3-8}$$

and

$$SS_{Treatments} = \sum_{i=1}^{a} \frac{y_{i.}^2}{n} - \frac{y_{..}^2}{N} \tag{3-9}$$

The error sum of squares is obtained by subtraction as

$$SS_E = SS_T - SS_{Treatments} \tag{3-10}$$

The test procedure is summarized in Table 3-3. This is called an analysis of variance table.

Example 3-1

The Tensile Strength Experiment

To illustrate the analysis of variance, return to the example first discussed in Section 3-1. Recall that the development engineer is interested in determining if varying the cotton content in a synthetic fiber affects the tensile strength, and he has run a completely randomized experiment with five levels of cotton percentage and five replicates. For convenience, we repeat the data from Table 3-1 here:

Percentage of Cotton	Observed tensile strength (lb/in²)					Totals y_i	Averages $\bar{y}_{i.}$
	1	2	3	4	5		
15	7	7	15	11	9	49	9.8
20	12	17	12	18	18	77	15.4
25	14	18	18	19	19	88	17.6
30	19	25	22	19	23	108	21.6
35	7	10	11	15	11	54	10.8
						$y_{..} = 376$	$\bar{y}_{..} = 15.04$

The sums of squares required for the analysis of variance are computed as follows:

$$SS_T = \sum_{i=1}^{5} \sum_{j=1}^{5} y_{ij}^2 - \frac{y_{..}^2}{N}$$

$$= (7)^2 + (7)^2 + (15)^2 + \cdots + (15)^2 + (11)^2 - \frac{(376)^2}{25} = 636.96$$

$$SS_{Treatments} = \sum_{i=1}^{5} \frac{y_{i.}^2}{n} - \frac{y_{..}^2}{N}$$

$$= \frac{(49)^2 + \cdots + (54)^2}{5} - \frac{(376)^2}{25} = 475.76$$

$$SS_E = SS_T - SS_{Treatments}$$

$$= 636.96 - 475.76 = 161.20$$

Usually, these calculations would be performed on a computer, using a software package with the capability to analyze data from designed experiments.

The analysis of variance is summarized in Table 3-4. Note that the between-treatment mean square (118.94) is many times larger than the within-treatment or error mean square (8.06). This indicates that it is unlikely that the treatment means are equal. More formally, we can compute the F ratio $F_0 = 118.94/8.06 = 14.76$ and compare this to $F_{\alpha,4,20}$. Since $F_{.01,4,20} = 4.43$, we reject H_0 and conclude that the treatment means differ; that is, the percentage of cotton in the fiber significantly affects the mean tensile strength.

■

A Remark on Computations ▪ The careful reader has likely noted that we defined sum of squares in terms of averages; that is, from Equation 3-6,

$$SS_{Treatments} = n \sum_{i=1}^{a} (\bar{y}_{i.} - \bar{y}_{..})^2$$

but we developed the computing formulas using totals. For example, to compute $SS_{Treatments}$, we would use Equation 3-9:

$$SS_{Treatments} = \sum_{i=1}^{a} \frac{y_{i.}^2}{n} - \frac{y_{..}^2}{n}$$

The reason for this is numerical accuracy; the totals $y_{i.}$ and $y_{..}$ are not subject to rounding error, whereas the averages $\bar{y}_{i.}$ and $\bar{y}_{..}$ are.

Table 3-4 Analysis of Variance for the Tensile Strength Data

Source of Variation	Sum of Squares	Degrees of Freedom	Mean Square	F_0
Cotton percentage	475.76	4	118.94	$F_0 = 14.76$
Error	161.20	20	8.06	
Total	636.96	24		

Generally, we need not be too concerned with computing, as there are many widely available computer programs for performing the calculations. These computer programs are also helpful in performing many other analyses associated with experimental design (such as *residual* analysis, discussed in Chapter 4). In many cases, these programs will also assist the experimenter in setting up the design.

When hand calculations are necessary, it is sometimes helpful to code the observations. This is illustrated in the next example.

Example 3-2

Coding the Observations

The calculations in the analysis of variance may often be made more accurate or simplified by coding the observations. For example, consider the tensile strength data in Example 3-1. Suppose we subtract 15 from each observation. The coded data are shown in Table 3-5. It is easy to verify that

$$SS_T = (-8)^2 + (-8)^2 + \cdots + (-4)^2 - \frac{(1)^2}{25} = 636.96$$

$$SS_{Treatments} = \frac{(-26)^2 + (2)^2 + \cdots + (-21)^2}{5} - \frac{(1)^2}{25} = 475.76$$

and

$$SS_E = 161.20$$

Comparing these sums of squares to those obtained in Example 3-1, we see that subtracting a constant from the original data does not change the sums of squares.

Now suppose that we multiply each observation in Example 3-1 by 2. It is easy to verify that the sums of squares for the transformed data are $SS_T = 2547.84$, $SS_{Treatments} = 1903.04$, and $SS_E = 644.80$. These sums of squares appear to differ considerably from those obtained in Example 3-1. However, if they are divided by 4 (i.e., 2^2), then the results are identical. For example, for the treatment sum of squares $1903.04/4 = 475.76$. Also, for the coded data, the F ratio is $F = (1903.04/4)/(644.80/20) = 14.76$, which is identical to the F ratio for the original data. Thus, the analyses of variance are equivalent.

■

Table 3-5 Coded Tensile Strength Data for Example 3-2

Percentage of Cotton	Observations					Totals $y_{i.}$
	1	2	3	4	5	
15	-8	-8	0	-4	-6	-26
20	-3	2	-3	3	3	2
25	-1	3	3	4	4	13
30	4	10	7	4	8	33
35	-8	-5	-4	0	-4	-21
						$1 = y_{..}$

Randomization Tests and Analysis of Variance ▪ In our development of the F test analysis of variance, we have used the assumption that the random errors ϵ_{ij} are normally and independently distributed random variables. The F test can also be justified as an approximation to a *randomization test*. To illustrate, suppose that we have five observations on each of two treatments and that we wish to test the equality of treatment means. The data would look like this:

Treatment 1	Treatment 2
y_{11}	y_{21}
y_{12}	y_{22}
y_{13}	y_{23}
y_{14}	y_{24}
y_{15}	y_{25}

We could use the F test analysis of variance to test $H_0: \mu_1 = \mu_2$. Alternatively, we could use a somewhat different approach. Suppose we consider all the possible ways of allocating the 10 numbers in the sample above to the two treatments. There are $10!/5!5! = 252$ possible arrangements of the 10 observations. If there is no difference in treatment means, all 252 arrangements are equally likely. For each of the 252 arrangements, we calculate the value of the F statistic using Equation 3-7. The distribution of these F values is called a *randomization distribution*, and a large value of F indicates that the data are not consistent with the hypothesis $H_0: \mu_1 = \mu_2$. For example, if the value of F actually observed was exceeded by only 5 of the values of the randomization distribution, then this would correspond to rejection of $H_0: \mu_1 = \mu_2$ at a significance level of $\alpha = 5/252 = .0198$ (or 1.98 percent). Notice that no normality assumption is required in this approach.

The difficulty with this approach is that, even for relatively small problems, it is computationally prohibitive to enumerate the exact randomization distribution. However, numerous studies have shown that the exact randomization distribution is well-approximated by the usual normal-theory F distribution. Thus, even without the normality assumption, the F test can be viewed as an approximation to the randomization test. For further reading on randomization tests in the analysis of variance, see Box, Hunter, and Hunter (1978).

3-3.3 Estimation of the Model Parameters

We now develop estimators for the parameters in the single-factor model

$$y_{ij} = \mu + \tau_i + \epsilon_{ij}$$

using the method of least squares. In estimating μ and τ_i by least squares, the normality assumption on the errors ϵ_{ij} is not needed. To find the least squares estimators of μ and τ_i, we form the sum of squares of the errors

$$L = \sum_{i=1}^{a} \sum_{j=1}^{n} \epsilon_{ij}^2 = \sum_{i=1}^{a} \sum_{j=1}^{n} (y_{ij} - \mu - \tau_i)^2 \qquad (3\text{-}11)$$

and choose values of μ and τ_i, say $\hat{\mu}$ and $\hat{\tau}_i$, that minimize L. The appropriate values would be the solutions to the $a + 1$ simultaneous equations

$$\left. \frac{\partial L}{\partial \mu} \right|_{\hat{\mu}, \hat{\tau}_i} = 0$$

$$\left. \frac{\partial L}{\partial \tau_i} \right|_{\hat{\mu}, \hat{\tau}_i} = 0 \qquad i = 1, 2, \ldots, a$$

Differentiating Equation 3-11 with respect to μ and τ_i and equating to zero, we obtain

$$-2 \sum_{i=1}^{a} \sum_{j=1}^{n} (y_{ij} - \hat{\mu} - \hat{\tau}_i) = 0$$

and

$$-2 \sum_{j=1}^{n} (y_{ij} - \hat{\mu} - \hat{\tau}_i) = 0 \qquad i = 1, 2, \ldots, a$$

which, after simplification, yield

$$\begin{aligned}
N\hat{\mu} + n\hat{\tau}_1 + n\hat{\tau}_2 + \cdots + n\hat{\tau}_a &= y_{..} \\
n\hat{\mu} + n\hat{\tau}_1 \qquad\qquad\qquad &= y_{1.} \\
n\hat{\mu} \qquad + n\hat{\tau}_2 \qquad\qquad &= y_{2.} \qquad\qquad (3\text{-}12) \\
&\;\;\vdots \qquad\qquad\quad \vdots \\
n\hat{\mu} \qquad\qquad\qquad\quad + n\hat{\tau}_a &= y_{a.}
\end{aligned}$$

The $a + 1$ equations (Equation 3-12) in $a + 1$ unknowns are called the *least squares normal equations*. Notice that if we add the last a normal equations, we obtain the first normal equation. Therefore, the normal equations are not linearly independent, and no unique solution for $\mu, \tau_1, \ldots, \tau_a$ exists. This difficulty can be overcome by several methods. Since we have defined the treatment effects as deviations from the overall mean, it seems reasonable to apply the constraint

$$\sum_{i=1}^{a} \hat{\tau}_i = 0 \qquad (3\text{-}13)$$

Using this constraint, we obtain as the solution to the normal equations

$$\begin{aligned}
\hat{\mu} &= \bar{y}_{..} \\
\hat{\tau}_i &= \bar{y}_{i.} - \bar{y}_{..} \qquad i = 1, 2, \ldots, a
\end{aligned} \qquad (3\text{-}14)$$

This solution has considerable intuitive appeal because we can see that the over-all mean is estimated by the grand average of the observations and that any treatment effect is just the difference between the treatment average and the grand average.

This solution is obviously not unique and depends on the constraint (Equation 3-13) that we have chosen. At first this may seem unfortunate because two different experimenters could analyze the same data and obtain different results if they apply different constraints. However, certain *functions* of the model parameter *are* uniquely estimated, regardless of the constraint. Some examples are $\tau_i - \tau_j$, which would be estimated by $\hat{\tau}_i - \hat{\tau}_j = \bar{y}_{i.} - \bar{y}_{j.}$, and the *i*th treatment mean $\mu_i = \mu + \tau_i$, which would be estimated by $\hat{\mu}_i = \hat{\mu} + \hat{\tau}_i = \bar{y}_{i.}$.

Since we are usually interested in differences among the treatment effects rather than their actual values, it causes no concern that the τ_i cannot be uniquely estimated. In general, any function of the model parameters that is a linear combination of the left-hand side of the normal equations (Equations 3-12) can be uniquely estimated. Functions that are uniquely estimated regardless of which constraint is used are called *estimable* functions.

A confidence interval estimate of the *i*th treatment mean may be easily determined. The mean of the *i*th treatment is

$$\mu_i = \mu + \tau_i$$

A point estimator of μ_i would be $\hat{\mu}_i = \hat{\mu} + \hat{\tau}_i = \bar{y}_{i.}$. Now, if we assume that the errors are normally distributed, each $\bar{y}_{i.}$ is $\text{NID}(\mu_i, \sigma^2/n)$. Thus, if σ^2 were known, we could use the normal distribution to define the confidence interval. Using the MS_E as an estimator of σ^2, we would base the confidence interval on the *t* distribution. Therefore, a $100(1 - \alpha)$ percent confidence interval on the *i*th treatment mean μ_i is

$$[\bar{y}_{i.} \pm t_{\alpha/2, N-a} \sqrt{MS_E/n}] \tag{3-15}$$

A $100(1 - \alpha)$ percent confidence interval on the difference in any two treatments means, say $\mu_i - \mu_j$, would be

$$[\bar{y}_{i.} - \bar{y}_{j.} \pm t_{\alpha/2, N-a} \sqrt{2MS_E/n}] \tag{3-16}$$

Example 3-3

Using the data in Example 3-1, we may find the estimates of the overall mean and the treatment effects as $\hat{\mu} = 376/25 = 15.04$ and

$$\hat{\tau}_1 = \bar{y}_{1.} - \bar{y}_{..} = 9.80 - 15.04 = -5.24$$
$$\hat{\tau}_2 = \bar{y}_{2.} - \bar{y}_{..} = 15.40 - 15.04 = +0.36$$
$$\hat{\tau}_3 = \bar{y}_{3.} - \bar{y}_{..} = 17.60 - 15.04 = -2.56$$
$$\hat{\tau}_4 = \bar{y}_{4.} - \bar{y}_{..} = 21.60 - 15.04 = +6.56$$
$$\hat{\tau}_5 = \bar{y}_{5.} - \bar{y}_{..} = 10.80 - 15.04 = -4.24$$

A 95 percent confidence interval on the mean of treatment 4 is computed from Equation 3-15 as

$$[21.60 \pm (2.086)\sqrt{8.06/5}]$$

or

$$[21.60 \pm 2.65]$$

Thus, the desired confidence interval is $18.95 \leqslant \mu_4 \leqslant 25.25$.

■

3-3.4 Model Adequacy Checking: Preview

The decomposition of the variability in observations through an analysis of variance identity (Equation 3-6) is a purely algebraic relationship. However, the use of the partitioning to test formally for no differences in treatment means requries that certain assumptions be satisfied. Specifically, these assumptions are that the observations are adequately described by the model

$$y_{ij} = \mu + \tau_i + \epsilon_{ij}$$

and that the errors are normally and independently distributed with mean zero and constant but unknown variance σ^2. If these assumptions are valid, then the analysis of variance procedure is an exact test of the hypothesis of no difference in treatment means.

In practice, however, these assumptions will usually not hold exactly. Consequently, it is usually unwise to rely on the analysis of variance until the validity of these assumptions has been checked. Violations of the basic assumptions and model adequacy can be easily investigated by the examination of *residuals*. We define the residual for observation j in treatment i as

$$e_{ij} = y_{ij} - \hat{y}_{ij} \tag{3-17}$$

where $\hat{y}_{ij}$ is an estimate of the corresponding observation y_{ij} obtained as follows:

$$\begin{aligned} \hat{y}_{ij} &= \hat{\mu} + \hat{\tau}_i \\ &= \bar{y}_{..} + (\bar{y}_{i.} - \bar{y}_{..}) \tag{3-18} \\ &= \bar{y}_{i.} \end{aligned}$$

Equation 3-18 gives the intuitively appealing result that the estimate of any observation in the ith treatment is just the corresponding treatment average.

Examination of the residuals should be an automatic part of any analysis of variance. If the model is adequate, the residuals should be *structureless*; that is, they should contain no obvious patterns. Through a study of residuals, many types of model inadequacies and violations of the underlying assumptions can be discovered. In Chapter 4, we take up model diagnostic checking and look at how to deal with several commonly occurring abnormalities.

3-3.5   The Unbalanced Case

In some single-factor experiments the number of observations taken within each treatment may be different. We then say that the design is *unbalanced*. The analysis of variance described above may still be used, but slight modifications must be made in the sum of squares formulas. Let n_i observations be taken under treatment $i(i = 1, 2, \ldots, a)$ and $N = \sum_{i=1}^{a} n_i$. The computational formulas for SS_T and $SS_{\text{Treatments}}$ become

$$SS_T = \sum_{i=1}^{a} \sum_{j=1}^{n_i} y_{ij}^2 - \frac{y_{..}^2}{N}$$

and

$$SS_{\text{Treatments}} = \sum_{i=1}^{a} \frac{y_{i.}^2}{n_i} - \frac{y_{..}^2}{N}$$

In solving the normal equations, the constraint $\sum_{i=1}^{a} n_i \hat{\tau}_i = 0$ is used. No other changes are required in the analysis of variance.

There are two advantages in choosing a balanced design. First, the test statistic is relatively insensitive to small departures from the assumption of equal variances for the a treatments if the sample sizes are equal. This is not the case for unequal sample sizes. Second, the power of the test is maximized if the samples are of equal size.

3-4   COMPARISON OF INDIVIDUAL TREATMENT MEANS

Suppose that in conducting an analysis of variance for the fixed effects model the null hypothesis is rejected. Thus, there are differences between the treatment means, but exactly *which* means differ is not specified. In this situation, further comparisons between groups of treatment means may be useful. The ith treatment mean is defined as $\mu_i = \mu + \tau_i$, and μ_i is estimated by $\bar{y}_{i.}$. Comparisons between treatment means are made in terms of either the treatment totals $\{y_{i.}\}$ or the treatment averages $\{\bar{y}_{i.}\}$. The procedures for making these comparisons are usually called *multiple comparison methods*.

3-4.1   Graphical Comparison of Means

It is very easy to develop a graphical procedure for the comparison of means following an analysis of variance. Suppose that the factor of interest has a levels and that $\bar{y}_{1.}, \bar{y}_{2.}, \ldots, \bar{y}_{a.}$ are the treatment averages. If we know σ, any treatment average would have a standard deviation $\sigma/\sqrt{n}$. Consequently, if all factor level means are identical, the observed sample means $\bar{y}_{i.}$ would behave as if they were a set of observations drawn at random from a normal distribution with mean $\bar{y}_{..}$

and standard deviation $\sigma/\sqrt{n}$. Visualize a normal distribution capable of being slid along an axis below which the $\bar{y}_{1.}, \bar{y}_{2.}, \ldots, \bar{y}_{a.}$ are plotted. if the treatment means are all equal, there should be some position for this distribution that makes it obvious that the $\bar{y}_{i.}$ values were drawn from the same distribution. If this is not the case, then the $\bar{y}_{i.}$ values that appear *not* to have been drawn from this distribution are associated with factor levels that produce different mean responses.

The only flaw in this logic is that σ is unknown. However, we can replace σ with $\sqrt{MS_E}$ from the analysis of variance and use a t distribution with a scale factor $\sqrt{MS_E/n}$ instead of the normal. Such an arrangement for the tensile strength data of Example 3-1 is shown in Figure 3-3.

To sketch the t distribution in Figure 3-3, simply multiply the abscissa t value by the scale factor

$$\sqrt{MS_E/n} = \sqrt{8.06/5} = 1.27$$

and plot this against the ordinate *of t* at that point. Since the t distribution looks much like the normal, except that it is a little flatter near the center and has longer tails, this sketch is usually easily constructed by eye. If you wish to be more precise, there is a table of abscissa t values and the corresponding ordinates in Box, Hunter, and Hunter (1978). The distribution can have an arbitrary origin, although it is usually best to choose one in the region of the $\bar{y}_{i.}$ values to be compared. In Figure 3-3, the origin is 15 lb/in^2.

Now visualize sliding the t distribution in Figure 3-3 along the horizontal axis, and examine the five means plotted in the figure. Notice that there is no location for the distribution such that all five averages could be thought of as typical, randomly selected observations from the distribution. This implies that all five means are not equal; thus, the figure is a graphical display of the results of the analysis of variance. The figure does indicate that 30 percent cotton produces much higher tensile strengths than would 20 or 25 percent cotton (which are about the same) or that either 15 or 35 percent cotton (which are about the same) would result in even lower tensile strengths.

This simple procedure is a rough but effective technique for many multiple comparisons problems. However, there are more formal methods. We now turn to a brief discussion of these procedures.

Figure 3-3. Tensile strength averages from the cotton percentage experiment in relation to a t distribution with a scale factor $\sqrt{MS_E/n} = \sqrt{8.06/5} = 1.27$.

3-4.2 Contrasts

Many multiple comparison methods use the idea of a contrast. Consider the synthetic fiber testing problem of Example 3-1. Since the hypothesis $H_0:\tau_i = 0$ was rejected, we know that some cotton percentages produce different tensile strengths than others, but which ones actually cause this difference? We might suspect at the outset of the experiment that cotton percentages 4 and 5 produce the same tensile strength, implying that we would like to test the hypothesis

$$H_0:\mu_4 = \mu_5$$
$$H_1:\mu_4 \neq \mu_5$$

This hypothesis could be tested by investigating an appropriate linear combination of treatment totals, say

$$y_{4.} - y_{5.} = 0$$

If we had suspected that the *average* of cotton percentages 1 and 3 did not differ from the *average* of cotton percentages 4 and 5, then the hypothesis would have been

$$H_0:\mu_1 + \mu_3 = \mu_4 + \mu_5$$
$$H_1:\mu_1 + \mu_3 \neq \mu_4 + \mu_5$$

which implies the linear combination

$$y_{1.} + y_{3.} - y_{4.} - y_{5.} = 0$$

In general, the comparison of treatment means of interest will imply a linear combination of treatment totals such as

$$C = \sum_{i=1}^{a} c_i y_{i.}$$

with the restriction that $\sum_{i=1}^{a} c_i = 0$. Such linear combinations are called *contrasts*. The sum of squares for any contrast is

$$SS_C = \frac{\left(\sum\limits_{i=1}^{a} c_i y_{i.}\right)^2}{n \sum\limits_{i=1}^{a} c_i^2} \tag{3-19}$$

and has a single degree of freedom. If the design is unbalanced, then the comparison of treatment means requires that $\sum_{i=1}^{a} n_i c_i = 0$, and Equation 3-19 becomes

$$SS_C = \frac{\left(\sum\limits_{i=1}^{a} c_i y_{i.}\right)^2}{\sum\limits_{i=1}^{a} n_i c_i^2} \tag{3-20}$$

A contrast is tested by comparing its sum of squares to the error mean square. The resulting statistic would be distributed as F with 1 and $N - a$ degrees of freedom. Many important comparisons between treatment means may be made using contrasts.

3-4.3   Orthogonal contrasts

A very important special case of the above procedure is that of *orthogonal* contrasts. Two contrasts with coefficients $\{c_i\}$ and $\{d_i\}$ are orthogonal if

$$\sum_{i=1}^{a} c_i d_i = 0$$

or, for an unbalanced design, if

$$\sum_{i=1}^{a} n_i c_i d_i = 0$$

For a treatments the set of $a - 1$ orthogonal contrasts partition the sum of squares due to treatments into $a - 1$ independent single-degree-of-freedom components. Thus, tests performed on orthogonal contrasts are independent.

There are many ways to choose the orthogonal contrast coefficients for a set of treatments. Usually, something in the nature of the experiment should suggest which comparisons will be of interest. For example, if there are $a = 3$ treatments with treatment 1 a control and treatments 2 and 3 actual levels of the factor of interest to the experimenter, then appropriate orthogonal contrasts might be as follows:

Treatment	Coefficients for Orthogonal Contrasts	
1 (control)	-2	0
2 (level 1)	1	-1
3 (level 2)	1	1

Note that contrast 1 with $c_i = -2, 1, 1$ compares the average effect of the factor with the control, whereas contrast 2 with $d_i = 0, 0, -1, 1$ compares the two levels of the factor of interest.

Contrast coefficients must be chosen prior to running the experiment and examining the data. The reason for this is that, if comparisons are selected after examining the data, most experimenters would construct tests that correspond to large observed differences in means. These large differences could be the result of the presence of real effects or they could be the result of random error. If experimenters consistently pick the largest differences to compare, they will inflate the type I error of the test since it is likely that, in an unusually high percentage of the comparisons selected, the observed differences will be the result of error.

Example 3-4

Consider the data in Example 3-1. There are five treatment means and four degrees of freedom between these treatments. The set of comparisons between these means and the associated orthogonal contrasts are

Hypothesis	**Contrast**
$H_0: \mu_4 = \mu_5$	$C_1 = \qquad\qquad\qquad -y_{4.} + y_{5.}$
$H_0: \mu_1 + \mu_3 = \mu_4 + \mu_5$	$C_2 = y_{1.} \quad + y_{3.} - y_{4.} - y_{5.}$
$H_0: \mu_1 = \mu_3$	$C_3 = y_{1.} \quad - y_{3.}$
$H_0: 4\mu_2 = \mu_1 + \mu_3 + \mu_4 + \mu_5$	$C_4 = -y_{1.} + 4y_{2.} - y_{3.} - y_{4.} - y_{5.}$

Notice that the contrast coefficients are orthogonal. Using the data in Table 3-4, we find the numerical values of the contrasts and the sums of squares to be as follows.

$$C_1 = \qquad\qquad -1(108) + 1(54) = -54 \quad SS_{C_1} = \frac{(-54)^2}{5(2)} = 291.60$$

$$C_2 = +1(49) \quad +1(88) \quad -1(108) - 1(54) = -25 \quad SS_{C_2} = \frac{(-25)^2}{5(4)} = 31.25$$

$$C_3 = +1(49) \quad -1(88) \qquad\qquad\qquad = -39 \quad SS_{C_3} = \frac{(-39)^2}{5(2)} = 152.10$$

$$C_4 = -1(49) + 4(77) - 1(88) \quad -1(108) - 1(54) = 9 \quad SS_{C_4} = \frac{(9)^2}{5(20)} = 0.81$$

These contrast sums of squares completely partition the treatment sum of squares. The tests on the contrasts are usually incorporated in the analysis of variance, as shown in Table 3-6. From this display, we conclude that there are significant differ-

Table 3-6 Analysis of Variance for the Tensile Strength Data

Source of Variation	Sum of Squares	Degrees of Freedom	Mean Square	F_0
Cotton percentage	475.76	4	118.94	14.76[a]
orthogonal contrasts				
$C_1: \mu_4 = \mu_5$	(291.60)	1	291.60	36.18[a]
$C_2: \mu_1 + \mu_3 = \mu_4 + \mu_5$	(31.25)	1	31.25	3.88
$C_3: \mu_1 = \mu_3$	(152.10)	1	152.10	18.87[a]
$C_4: 4\mu_2 = \mu_1 + \mu_3 + \mu_4 + \mu_5$	(0.81)	1	0.81	0.10
Error	161.20	20	8.06	
Total	636.96	24		

[a] Significant at 1 percent.

ences between cotton percentages 4 and 5 and 1 and 3 but that the *average* of 1 and 3 does not differ from the average of 4 and 5 and that 2 does not differ from the average of the other four percentages.

■

3-4.4 Scheffé's Method for Comparing All Contrasts

In many situations, experimenters may not know in advance which contrasts they wish to compare, or they may be interested in more than $a - 1$ possible comparisons. In many exploratory experiments, the comparisons of interest are discovered only after preliminary examination of the data. Scheffé (1953) has proposed a method for comparing any and all possible contrasts between treatment means. In the Scheffé method, the type I error is at most α for any of the possible comparisons.

Suppose that a set of m contrasts in the treatment means

$$\Gamma_u = c_{1u}\mu_1 + c_{2u}\mu_2 + \cdots + c_{au}\mu_a \qquad u = 1, 2, \ldots, m \qquad (3\text{-}21)$$

of interest have been determined. The corresponding contrast in the treatment averages $\bar{y}_{i.}$ is

$$C_u = c_{1u}\bar{y}_{1.} + c_{2u}\bar{y}_{2.} + \cdots + c_{au}\bar{y}_{a.} \qquad u = 1, 2, \ldots, m \qquad (3\text{-}22)$$

and the standard error of this contrast is

$$S_{C_u} = \sqrt{MS_E \sum_{i=1}^{a} (c_{iu}^2/n_i)} \qquad (3\text{-}23)$$

where n_i is the number of observations in the ith treatment. It can be shown that the critical value against which C_u should be compared is

$$S_{\alpha,u} = S_{C_u} \sqrt{(a-1) F_{\alpha,a-1,N-a}} \qquad (3\text{-}24)$$

To test the hypothesis that the contrast Γ_u differs significantly from zero, refer C_u to the critical value. If $|C_u| > S_{\alpha,u}$, then the hypothesis that the contrast Γ_u equals zero is rejected. The Scheffé procedure can also be used to form confidence intervals for all possible contrasts among treatment means. The resulting intervals, say $C_u - S_{\alpha,u} \leq \Gamma_u \leq C_u + S_{\alpha,u}$, are *simultaneous* confidence intervals in that the probability is at least $1 - \alpha$ that all of them are simultaneously true.

To illustrate the procedure, consider the data in Example 3-1 and suppose that the contrasts of interests are

$$\Gamma_1 = \mu_1 + \mu_3 - \mu_4 - \mu_5$$

and

$$\Gamma_2 = \mu_1 - \mu_4$$

The numerical values of these contrasts are

$$C_1 = \bar{y}_{1.} + \bar{y}_{3.} - \bar{y}_{4.} - \bar{y}_{5.}$$
$$= 9.80 + 17.60 - 21.60 - 10.80$$
$$= 5.00$$

and

$$C_2 = \bar{y}_{1.} - \bar{y}_{4.}$$
$$= 9.80 - 21.60$$
$$= -11.80$$

and the standard errors are found from Equation 3-23 as

$$S_{C_1} = \sqrt{MS_E \sum_{i=1}^{5} (c_{i1}^2/n_i)} = \sqrt{8.06(1 + 1 + 1 + 1)/5} = 2.54$$

and

$$S_{C_2} = \sqrt{MS_E \sum_{i=1}^{5} (c_{i2}^2/n_i)} = \sqrt{8.06(1 + 1)/5} = 1.80$$

From Equation 3-24, the 1 percent critical values are

$$S_{.01,1} = S_{C_1} \sqrt{(a - 1) F_{.01,a-1,N-a}} = 2.54 \sqrt{4(4.43)} = 10.69$$

and

$$S_{.01,2} = S_{C_2} \sqrt{(a - 1) F_{.01,a-1,a(n-1)}} = 1.80 \sqrt{4(4.43)} = 7.58$$

Since $|C_1| < S_{.01,1}$, we conclude that the contrast $\Gamma_1 = \mu_1 + \mu_3 - \mu_4 - \mu_5$ equals zero; that is there is no strong evidence to conclude that the means of treatments 1 and 3 as a group differ from the means of treatments 4 and 5 as a group. However, since $|C_2| > S_{.01,2}$, we conclude that the contrast $\Gamma_2 = \mu_1 - \mu_4$ does not equal zero; that is, the mean strengths of treatments 1 and 4 differ significantly.

In many practical situations, we will wish to compare only *pairs* of means. Frequently, we can determine which means differ by testing the differences between *all* pairs of treatment means. Thus, we are interested in contrasts of the form $\Gamma = \mu_i - \mu_j$ for all $i \neq j$. Although the Scheffé method could be easily applied to this problem, it is not the most sensitive procedure for such comparisons. We now turn to a consideration of methods specifically designed for pairwise comparisons between all a population means.

3-4.5 Comparing Pairs of Treatment Means

Suppose that we are interested in comparing all pairs of a treatment means and that the null hypotheses that we wish to test are $H_0: \mu_i = \mu_j$ for all $i \neq j$. We now present four methods for making such comparisons.

The Least Significant Difference (LSD) Method ▪ Suppose that, following an analysis of variance F test where the null hypothesis is rejected, we wish to test $H_0: \mu_i = \mu_j$ for all $i \neq j$. This could be done by employing the t statistic

$$t_0 = \frac{\bar{y}_{i.} - \bar{y}_{j.}}{\sqrt{MS_E \left(\dfrac{1}{n_i} + \dfrac{1}{n_j} \right)}} \tag{3-25}$$

Assuming a two-sided alternative, the pair of means μ_i and μ_j would be declared significantly different if $|\bar{y}_{i.} - \bar{y}_{j.}| > t_{\alpha/2, N-a} \sqrt{MS_E (1/n_i + 1/n_j)}$. The quantity

$$LSD = t_{\alpha/2, N-a} \sqrt{MS_E \left(\frac{1}{n_i} + \frac{1}{n_j} \right)} \tag{3-26}$$

is called the *least significant difference*. If the design is balanced, then $n_1 = n_2 = \cdots = n_a = n$, and

$$LSD = t_{\alpha/2, N-a} \sqrt{\frac{2MS_E}{n}} \tag{3-27}$$

To use the LSD procedure, simply compare the observed difference between each pair of averages to the corresponding LSD. If $|\bar{y}_{i.} - \bar{y}_{j.}| > LSD$, we conclude that the population means μ_i and μ_j differ.

Example 3-5

To illustrate the procedure, if we use the data in Example 3-1, the LSD at $\alpha = .05$ is

$$LSD = t_{.025,20} \sqrt{\frac{2MS_E}{n}} = 2.086 \sqrt{\frac{2(8.06)}{5}} = 3.75$$

Thus, any pair of treatment averages that differ in absolute value by more than 3.75 would imply that the corresponding pair of population means are significantly different. The five treatment averages are

$$\bar{y}_{1.} = 9.8 \qquad \bar{y}_{2.} = 15.4 \qquad \bar{y}_{3.} = 16.6 \qquad \bar{y}_{4.} = 21.6 \qquad \bar{y}_{5.} = 10.8$$

and the differences in averages are

$$\bar{y}_{1.} - \bar{y}_{2.} = \quad 9.8 - 15.4 = \quad -5.6^*$$
$$\bar{y}_{1.} - \bar{y}_{3.} = \quad 9.8 - 17.6 = \quad -7.8^*$$
$$\bar{y}_{1.} - \bar{y}_{4.} = \quad 9.8 - 21.6 = -11.8^*$$
$$\bar{y}_{1.} - \bar{y}_{5.} = \quad 9.8 - 10.8 = \quad -1.0$$
$$\bar{y}_{2.} - \bar{y}_{3.} = 15.4 - 17.6 = \quad -2.2$$
$$\bar{y}_{2.} - \bar{y}_{4.} = 15.4 - 21.6 = \quad -6.2^*$$

$$\bar{y}_{2.} - \bar{y}_{5.} = 15.4 - 10.8 = \quad 4.6^*$$

$$\bar{y}_{3.} - \bar{y}_{4.} = 17.6 - 21.6 = \quad -4.0^*$$

$$\bar{y}_{3.} - \bar{y}_{5.} = 17.6 - 10.8 = \quad 6.8^*$$

$$\bar{y}_{4.} - \bar{y}_{5.} = 21.6 - 10.8 = \quad 10.8^*$$

The starred values indicate pairs of means that are significantly different. It is helpful to draw a graph, such as Figure 3-4, underlining pairs of means that do not differ significantly. Clearly, the only pairs of means that do not differ significantly are 1 and 5 and 2 and 3, and treatment 4 produces a significantly greater tensile strength than the other treatments.

■

Several comments should be made relative to this procedure. Note that the α risk may be considerably inflated using this method. Specifically, as a gets larger, the type I error of the experiment (the ratio of the number of experiments in which at least 1 type I error is made to the total number of experiments) becomes large. We also occasionally find that the overall F statistic in the analysis of variance is significant, but the LSD method fails to find any significant pairwise differences. This situation occurs because the F test is simultaneously considering all possible comparisons between the treatment means, not just pairwise comparisons. In the data at hand, the significant contrasts may not be of the form $\mu_i - \mu_j$.

Duncan's Multiple Range Test ▪ A widely used procedure for comparing all pairs of means is the multiple range test developed by Duncan (1955). To apply Duncan's multiple range test for equal sample sizes, the a treatment averages are arranged in ascending order, and the standard error of each average is determined as

$$S_{\bar{y}_{i.}} = \sqrt{\frac{MS_E}{n}} \tag{3-28}$$

For unequal sample sizes, replace n in Equation 3-28 by the harmonic mean n_h of the $\{n_i\}$, where

$$n_h = \frac{a}{\displaystyle\sum_{i=1}^{a} (1/n_i)} \tag{3-29}$$

$\bar{y}_{1.}$	$\bar{y}_{5.}$	$\bar{y}_{2.}$	$\bar{y}_{3.}$	$\bar{y}_{4.}$
9.8	10.8	15.4	17.6	21.6

Figure 3-4. Results of the LSD procedure.

Note that if $n_1 = n_2 = \cdots = n_a$, then $n_h = n$. From Duncan's table of significant ranges (Appendix Table VII), obtain the values $r_\alpha(p, f)$ for $p = 2, 3, \ldots, a$, where α is the significance level and f is the number of degrees of freedom for error. Convert these ranges into a set of $a - 1$ least significant ranges (e.g., R_p) for $p = 2, 3, \ldots, a$ by calculating

$$R_p = r_\alpha(p, f)S_{\bar{y}_{i.}} \qquad \text{for } p = 2, 3, \ldots, a$$

Then, the observed differences between means are tested, beginning with largest versus smallest, which would be compared with the least significant range R_a. Next, the difference of the largest and the second smallest is computed and compared with the least significant range R_{a-1}. These comparisons are continued until all means have been compared with the largest mean. Finally, the difference of the second largest mean and the smallest is computed and compared against the least significant range R_{a-1}. This process is continued until the differences of all possible $a(a - 1)/2$ pairs of means have been considered. If an observed difference is greater than the corresponding least significant range, then we conclude that the pair of means in question is significantly different. To prevent contradictions, no differences between a pair of means are considered significant if the two means involved fall between two other means that do not differ significantly.

Example 3-6

We can apply Duncan's multiple range test to the data of Example 3-1. Recall that $MS_E = 8.06$, $N = 25$, $n = 5$, and there are 20 error degrees of freedom. Ranking the treatment averages in ascending order, we have

$$\bar{y}_{1.} = 9.8$$
$$\bar{y}_{5.} = 10.8$$
$$\bar{y}_{2.} = 15.4$$
$$\bar{y}_{3.} = 17.6$$
$$\bar{y}_{4.} = 21.6$$

The standard error of each average is $S_{\bar{y}_{i.}} = \sqrt{8.06/5} = 1.27$. From the table of significant ranges in Appendix Table VII for 20 degrees of freedom and $\alpha = .05$, we obtain $r_{.05}(2,20) = 2.95$, $r_{.05}(3,20) = 3.10$, $r_{.05}(4,20) = 3.18$, and $r_{.05}(5,20) = 3.25$. Thus, the least significant ranges are

$$R_2 = r_{.05}(2,20)S_{\bar{y}_{i.}} = (2.95)(1.27) = 3.75$$

$$R_3 = r_{.05}(3,20)S_{\bar{y}_{i.}} = (3.10)(1.27) = 3.94$$

$$R_4 = r_{.05}(4,20)S_{\bar{y}_{i.}} = (3.18)(1.27) = 4.04$$

$$R_5 = r_{.05}(5,20)S_{\bar{y}_{i.}} = (3.25)(1.27) = 4.13$$

The comparisons would yield

$$\text{4 vs. 1:} \quad 21.6 - 9.8 = 11.8 > 4.13(R_5)$$
$$\text{4 vs. 5:} \quad 21.6 - 10.8 = 10.8 > 4.04(R_4)$$

$$4 \text{ vs. } 2: \quad 21.6 - 15.4 = \quad 6.2 > 3.94(R_3)$$
$$4 \text{ vs. } 3: \quad 21.6 - 17.6 = \quad 4.0 > 3.75(R_2)$$
$$3 \text{ vs. } 1: \quad 17.6 - \;\, 9.8 = \quad 7.8 > 4.04(R_4)$$
$$3 \text{ vs. } 5: \quad 17.6 - 10.8 = \quad 6.8 > 3.95(R_3)$$
$$3 \text{ vs. } 2: \quad 17.6 - 15.4 = \quad 2.2 < 3.75(R_2)$$
$$2 \text{ vs. } 1: \quad 15.4 - \;\, 9.8 = \quad 5.6 > 3.94(R_3)$$
$$2 \text{ vs. } 5: \quad 15.4 - 10.8 = \quad 4.6 > 3.75(R_2)$$
$$5 \text{ vs. } 1: \quad 10.8 - \;\, 9.8 = \quad 1.0 < 3.75(R_2)$$

From the analysis we see that there are significant differences between all pairs of means except 3 and 2 and 5 and 1. A graph underlining those means that are not significantly different is shown in Figure 3-5. Notice that, in this example, Duncan's multiple range test and the LSD method produce identical conclusions.

■

Duncan's test requires a greater observed difference to detect significantly different pairs of means as the number of means included in the group increases. For instance, in the above example $R_2 = 3.75$ (two means) whereas $R_3 = 3.94$ (three means). For two means, the critical value R_2 will always exactly equal the LSD value from the t test. The values $r_\alpha(p, f)$ in Appendix Table VII are chosen so that a specified *protection level* is obtained. That is, when two means that are p steps apart are compared, the protection level is $(1 - \alpha)^{p-1}$, where α is the level of significance specified for two *adjacent* means. Thus, the error rate is $1 - (1 - \alpha)^{p-1}$ of reporting at least one incorrect significant difference between two means when the group size is p. For example, if $\alpha = .05$, then $1 - (1 - .05)^1 = .05$ is the significance level for comparing the adjacent pair of means, $1 - (1 - .05)^2 = .10$ is the significance level for means that are one step apart, and so forth.

Generally, if the protection level is α, then tests on means have a significance level that is greater than or equal to α. Consequently, Duncan's procedure is quite powerful; that is, it is very effective at detecting differences between means when real differences exist. For this reason, Duncan's multiple range test is very popular.

The Newman–Keuls Test ▪ This test was devised by Newman (1939). Because new interest in Newman's test was generated by Keuls (1952), the procedure is usually called the Newman–Keuls test. Operationally, the procedure is similar to Duncan's multiple range test, except that the critical differences between

$\bar{y}_{1.}$	$\bar{y}_{5.}$	$\bar{y}_{2.}$	$\bar{y}_{3.}$	$\bar{y}_{4.}$
9.8	10.8	15.4	17.6	21.6

Figure 3-5. Results of Duncan's multiple range test.

means are calculated somewhat differently. Specifically, we compute a set of critical values

$$K_p = q_\alpha(p, f)S_{\bar{y}_{i.}} \qquad p = 2,3, \ldots, a \qquad (3\text{-}30)$$

where $q_\alpha(p, f)$ is the upper α percentage point of the studentized range for groups of means of size p and f error degrees of freedom. The studentized range is defined as

$$q = \frac{\bar{y}_{max} - \bar{y}_{min}}{\sqrt{MS_E/n}}$$

where $\bar{y}_{max}$ and $\bar{y}_{min}$ are the largest and smallest sample means, respectively, out of a group of p sample means. Appendix Table VIII contains values of $q_\alpha(p, f)$, the upper α percentage point of q. Once the values K_p are computed from Equation 3-30, extreme pairs of means in groups of size p are compared with K_p exactly as in Duncan's multiple range test.

The Newman–Keuls test is more conservative than Duncan's test in that the type I error rate is smaller. Specifically, the type I error of the experiment is α for all tests involving the same number of means. Consequently, because α is generally lower, the power of the Newman–Keuls test is generally less than that of Duncan's test. To demonstrate that the Newman–Keuls procedure leads to a less powerful test than Duncan's multiple range test, we note from a comparison of Tables VII and VIII of the Appendix that for $p > 2$ we have $q_\alpha(p, f) > r_\alpha(p, f)$. That is, it is "harder" to declare a pair of means significantly different using the Newman–Keuls test than it is using Duncan's procedure. This is illustrated below for the case $\alpha = .01$, $a = 8$, and $f = 20$:

p	2	3	4	5	6	7	8
$r_{.01}(p,20)$	4.02	4.22	4.33	4.40	4.47	4.53	4.58
$q_{.01}(p,20)$	4.02	4.64	5.02	5.29	5.51	5.69	5.84

Tukey's Test ▪ Tukey (1953) proposed a multiple comparison procedure that is also based on the studentized range statistic. His procedure requires the use of $q_\alpha(a,f)$ to determine the critical value for *all* pairwise comparisons, regardless of how many means are in the group. Thus, Tukey's test declares two means significantly different if the absolute value of their sample differences exceeds

$$T_\alpha = q_\alpha(a,f)S_{\bar{y}_{i.}} \qquad (3\text{-}31)$$

where $S_{\bar{y}_{i.}}$ is as defined in Equation 3-28. Note that a single critical value is used in all comparisons. Thus, in Example 3-6, we would use $q_{.05}(5,20) = 4.23$ as the basis of determining the critical value; consequently, no pair of means would be declared significantly different unless $|\bar{y}_{i.} - \bar{y}_{j.}|$ exceeded

$$T_{.05} = q_{.05}(5,20)S_{\bar{y}_{i.}} = (4.23)(1.27) = 5.37$$

Using this critical value for the data in Example 3-6, we find now that $\bar{y}_{3.}$ and $\bar{y}_{4.}$ do not differ significantly; nor do $\bar{y}_{2.}$ and $\bar{y}_{5.}$. The graphical display of the results is

$\bar{y}_{1.}$	$\bar{y}_{5.}$	$\bar{y}_{2.}$	$\bar{y}_{3.}$	$\bar{y}_{4.}$
9.8	10.8	15.4	17.6	21.6

Note that these results are somewhat more difficult to interpret than those in either Figure 3-4 or Figure 3-5. Furthermore, Tukey's test has a type I error rate of α for all pairwise comparisons on an experimentwise basis. This is more conservative (it has a smaller type I error rate) than either the Newman–Keuls or Duncan tests. Consequently, Tukey's test has less power than either the Duncan or Newman–Keuls procedure.

Which Method Do I Use? ▪ Certainly a logical question at this point is which one of these procedures should I use? Unfortunately, there is no clear-cut answer to this question, and professional statisticians often disagree over the utility of the various procedures. Carmer and Swanson (1973) have conducted Monte Carlo simulation studies of a number of multiple comparison procedures, including others not discussed here. They report that the least significant difference method is a very effective test for detecting true differences in means if it is applied *only after* the F test in the analysis of variance is significant at 5 percent. They also report good performance in detecting true differences with Duncan's multiple range test. This is not surprising, since these two methods are the most powerful of those we have discussed. Duncan's multiple range test is also available in many computer software packages for the analysis of variance. It should be satisfactory for many general applications.

   As indicated above, there are several other multiple comparison procedures. For survey articles describing these methods, see Miller (1977) and O'Neill and Wetherill (1971). The book by Miller (1966) is also recommended.

3-4.6 Comparing Treatments with a Control

In many experiments, one of the treatments is a *control,* and the analyst is interested in comparing each of the other $a - 1$ treatment means with the control. Thus, there are only $a - 1$ comparisons to be made. A procedure for making these comparisons has been developed by Dunnett (1964). Suppose that treatment a is the control. Then we wish to test the hypotheses

$$H_0: \mu_i = \mu_a$$
$$H_1: \mu_i \neq \mu_a$$

for $i = 1, 2, \ldots, a - 1$. Dunnett's procedure is a modification of the usual t test. For each hypothesis, we compute the observed differences in the sample means

$$|\bar{y}_{i.} - \bar{y}_{a.}| \qquad i = 1, 2, \ldots, a - 1$$

The null hypothesis $H_0: \mu_i = \mu_a$ is rejected using a type I error rate α if

$$|\bar{y}_{i.} - \bar{y}_{a.}| > d_\alpha(a - 1, f) \sqrt{MS_E \left(\frac{1}{n_i} + \frac{1}{n_a} \right)} \tag{3-32}$$

where the constant $d_\alpha(a - 1, f)$ is given in Appendix Table IX (Both two-sided and one-sided tests are possible.) Note that α is the *joint* significance level associated with all $a - 1$ tests.

Example 3-7

To illustrate Dunnett's test, consider the data from Example 3-1 with treatment 5 considered the control. In this example, $a = 5$, $a - 1 = 4$, $f = 20$, and $n_i = n = 5$. At the 5 percent level, we find from Appendix Table IX that $d_{.05}(4,20) = 2.65$. Thus, the critical difference becomes

$$d_{.05}(4,20) \sqrt{\frac{2MS_E}{n}} = 2.65 \sqrt{\frac{2(8.06)}{5}} = 4.76$$

(Note that this is a simplification of Equation 3-32 resulting from a balanced design.) Thus, any treatment mean that differs from the control by more than 4.76 would be declared significantly different. The observed differences are

$$1 \text{ vs. } 5: \bar{y}_{1.} - \bar{y}_{5.} = \quad 9.8 - 10.8 = -1.0$$
$$2 \text{ vs. } 5: \bar{y}_{2.} - \bar{y}_{5.} = 15.4 - 10.8 = 4.6$$
$$3 \text{ vs. } 5: \bar{y}_{3.} - \bar{y}_{5.} = 17.6 - 10.8 = 6.8$$
$$4 \text{ vs. } 5: \bar{y}_{4.} - \bar{y}_{5.} = 21.6 - 10.8 = 10.8$$

Only the differences $\bar{y}_{3.} - \bar{y}_{5.}$ and $\bar{y}_{4.} - \bar{y}_{5.}$ indicate any significant difference when compared to the control; thus, we conclude that $\mu_3 \neq \mu_5$ and $\mu_4 \neq \mu_5$.

When comparing treatments with a control, it is a good idea to use more observations for the control treatment (say n_a) than for the other treatments (say n), assuming equal numbers of observations for the remaining $a - 1$ treatments. The ratio n_a/n should be chosen to be approximately equal to the square root of the total number of treatments. That is, choose $n_a/n = \sqrt{a}$.

■

3-5 THE RANDOM EFFECTS MODEL

An experimenter is frequently interested in a factor that has a large number of possible levels. If the experimenter randomly selects a of these levels from the population of factor levels, then we say that the factor is *random*. Because the levels of the factor actually used in the experiment were chosen randomly, inferences are made about the entire population of factor levels. We assume that the population of factor levels is either of infinite size or is large enough to be considered infinite. Situations in which the population of factor levels is small enough to employ a finite population approach are not encountered frequently. One may refer to Bennett and Franklin (1954) and Searle and Fawcett (1970) for a discussion of the finite population case.

The linear statistical model is

$$y_{ij} = \mu + \tau_i + \epsilon_{ij} \begin{cases} i = 1,2, \ldots, a \\ j = 1,2, \ldots, n \end{cases} \tag{3-33}$$

where *both* τ_i and ϵ_{ij} are random variables. If τ_i has variance a σ_τ^2 and is independent of ϵ_{ij}, the variance of any observation is

$$V(y_{ij}) = \sigma_\tau^2 + \sigma^2$$

The variances σ_τ^2 and σ^2 are called *variance components,* and the model (Equation 3-33) is called the *components of variance* or *random effects model.* To test hypotheses in this model, we require that the $\{\epsilon_{ij}\}$ are NID(0, σ^2), that the $\{\tau_i\}$ are NID(0, σ_τ^2), and that τ_i and ϵ_{ij} are independent.[2]

The sum of squares identity

$$SS_T = SS_{\text{Treatments}} + SS_E \tag{3.34}$$

is still valid. That is, we partition the total variability in the observations into a component that measures the variation between treatments ($SS_{\text{Treatments}}$) and a component that measures the variation within treatments (SS_E). Testing hypotheses about individual treatment effects is meaningless, so instead we test the hypotheses

$$H_0 : \sigma_\tau^2 = 0$$

$$H_1 : \sigma_\tau^2 > 0$$

If $\sigma_\tau^2 = 0$, all treatments are identical; but if $\sigma_\tau^2 > 0$, variability exists between treatments. As before, SS_E/σ^2 is distributed as chi-square with $N - a$ degrees of freedom, and under the null hypothesis, $SS_{\text{Treatments}}/\sigma^2$ is distributed as chi-square with $a - 1$ degrees of freedom. Both random variables are independent. Thus, under the null hypothesis $\sigma_\tau^2 = 0$, the ratio

$$F_0 = \frac{\dfrac{SS_{\text{Treatments}}}{a - 1}}{\dfrac{SS_E}{N - a}} = \frac{MS_{\text{Treatments}}}{MS_E} \tag{3-35}$$

---

[2] The assumption that the $\{\tau_i\}$ are independent random variables implies that the usual assumption of $\sum_{i=1}^{a} \tau_i = 0$ from the fixed effects model does not apply to the random effects model.

is distributed as F with $a - 1$ and $N - a$ degrees of freedom. However, we need to examine the expected mean sauares to fully describe the test procedure.

Consider

$$E(MS_{Treatments}) = \frac{1}{a-1} E(SS_{Treatments}) = \frac{1}{a-1} E\left[\sum_{i=1}^{a} \frac{y_{i.}^2}{n} - \frac{y_{..}^2}{N}\right]$$

$$= \frac{1}{a-1} E\left[\frac{1}{n}\sum_{i=1}^{a}\left(\sum_{j=1}^{n} \mu + \tau_i + \epsilon_{ij}\right)^2 - \frac{1}{N}\left(\sum_{i=1}^{a}\sum_{j=1}^{n} \mu + \tau_i + \epsilon_{ij}\right)^2\right]$$

When squaring and taking expectation of the quantities in brackets, we see that terms involving τ_i^2 are replaced by σ_τ^2 as $E(\tau_i) = 0$. Also, terms involving $\epsilon_{i.}^2$, $\epsilon_{..}^2$, and $\sum_{i=1}^{a} \sum_{j=1}^{n} \tau^2$ are replaced by $n\sigma^2$, $an\sigma^2$, and $an^2\sigma_\tau^2$, respectively. Furthermore, all cross-product terms involving τ_i and ϵ_{ij} have zero expectation. This leads to

$$E(MS_{Treatments}) = \frac{1}{a-1} [N\mu^2 + N\sigma_\tau^2 + a\sigma^2 - N\mu^2 - n\sigma_\tau^2 - \sigma^2]$$

or

$$E(MS_{Treatments}) = \sigma^2 + n\sigma_\tau^2 \qquad (3\text{-}36)$$

Similarly, we may show that

$$E(MS_E) = \sigma^2 \qquad (3\text{-}37)$$

From the expected mean squares, we see that under H_0 both the numerator and denominator of the test statistic (Equation 3-35) are unbiased estimators of σ^2, whereas under H_1 the expected value of the numerator is greater than the expected value of the denominator. Therefore, we should reject H_0 for values of F_0 that are too large. This implies an upper-tail, one-tail critical region, so we reject H_0 if $F_0 > F_{\alpha, a-1, N-a}$.

The computational procedure and analysis of the variance table for the random effects model are identical to those for the fixed effects case. The conclusions, however, are quite different because they apply to the entire population of treatments.

We are usually interested in estimating the variance components (σ^2 and σ_τ^2) in the model. The procedure that we use to estimate σ^2 and σ_τ^2 is called the "analysis of variance method" because it makes use of the lines in the analysis of variance table. The procedure consists of equating the expected mean squares to their observed values in the analysis of variance table and solving for the variance components. In equating observed and expected mean squares in the single-factor random effects model, we obtain

$$MS_{Treatments} = \sigma^2 + n\sigma_\tau^2$$

and

$$MS_E = \sigma^2$$

Therefore, the estimators of the variance components are

$$\hat{\sigma}^2 = MS_E \qquad (3\text{-}38)$$

and

$$\hat{\sigma}_\tau^2 = \frac{MS_{\text{Treatments}} - MS_E}{n} \qquad (3\text{-}39)$$

For unequal sample sizes, replace n in Equation 3-39 by

$$n_0 = \frac{1}{a-1}\left[\sum_{i=1}^{a} n_i - \frac{\displaystyle\sum_{i=1}^{a} n_i^2}{\displaystyle\sum_{i=1}^{a} n_i}\right]$$

The analysis of variance method of variance component estimation does not require the normality assumption. It does yield estimators of σ^2 and σ_τ^2 that are best quadratic unbiased (i.e., of all unbiased quadratic functions of the observations, these estimators have minimum variance).

Occasionally, the analysis of variance method produces a negative estimate of a variance component. Clearly, variance components are by definition nonnegative, so a negative estimate of a variance component is viewed with some concern. One course of action is to accept the estimate and use it as evidence that the true value of the variance component is zero, assuming that sampling variation led to the negative estimate. This has intuitive appeal, but it suffers from some theoretical difficulties. For instance, using zero in place of the negative estimate can disturb the statistical properties of other estimates. Another alternative is to reestimate the negative variance component using a method that always yields nonnegative estimates. Still another alternative is to consider the negative estimate as evidence that the assumed linear model is incorrect and reexamine the problem. A good discussion of variance component estimation is given by Searle (1971a, 1971b).

Example 3-8

A textile company weaves a fabric on a large number of looms. They would like the looms to be homogeneous so that they obtain a fabric of uniform strength. The process engineer suspects that, in addition to the usual variation in strength within samples of fabric from the same loom, there may also be significant variations in strength between looms. To investigate this, she selects four looms at random and makes four strength determinations on the fabric manufactured on each loom. This experiment is run in random order, and the data obtained are shown in Table 3-7. The analysis of variance is conducted and is shown in Table 3-8. From the analysis of variance, we conclude that the looms in the plant differ significantly.

The variance components are estimated by $\hat{\sigma}^2 = 1.90$ and

$$\hat{\sigma}_\tau^2 = \frac{29.73 - 1.90}{4} = 6.96$$

Table 3-7   Strength Data for Example 3-8

Looms	Observations				$y_{i.}$
	1	2	3	4	
1	98	97	99	96	390
2	91	90	93	92	366
3	96	95	97	95	383
4	95	96	99	98	388
					$1527 = y_{..}$

Therefore, the variance of any observation on strength is estimated by $\hat{\sigma}^2 + \hat{\sigma}_\tau^2 = 1.90 + 6.96 = 8.86$. Most of this variability is attributable to differences *between* looms.

∎

This example illustrates an important use of variance components—the isolating of different sources of variability that affect a product or system. The problem of product variability frequently arises in quality assurance, and it is often difficult to isolate the sources of variability. For example, this study may have been motivated by too much variability in the strength of the fabric, as illustrated in Figure 3-6a. Figure 3-6a displays the process output (fiber strength) modeled as a normal distribution with variance $\hat{\sigma}_\tau^2 = 8.86$. (This is the estimate of the variance of any observation on strength from Example 3-8.) Upper and lower specifications on strength are also shown in Figure 3-6a, and it is relatively easy to see that a fairly large proportion of the process output is outside the specifications (the shaded tail areas in Figure 3-6a). The process engineer has asked why so much fabric is defective and must be scrapped, reworked, or downgraded to a lower quality product. The answer is that most of the product strength variability is the result of differences between looms. Different loom performance could be the result of faulty set-up, poor maintenance, ineffective supervision, poorly trained operators, defective input fiber, and so forth. The process engineer must now try to isolate the specific causes of the difference in loom performance. If she could identify and eliminate these sources of between-loom variability, the variance of the process output could be reduced considerably, perhaps to as low

Table 3-8   Analysis of Variance for the Strength Data

Source of Variation	Sum of Squares	Degrees of Freedom	Mean Square	F_0
Looms	89.19	3	29.73	15.68[a]
Error	22.75	12	1.90	
Total	111.94	15		

[a] Significant at 5 percent.

(a) Variability of process output.

(b) Variability of process output if $\sigma_\tau^2 = 0$.

Figure 3-6. Process output in the fiber strength problem.

as $\hat{\sigma}_y^2 = 1.90$, the estimate of the within-loom (error) variance component in Example 3-8. Figure 3-6b shows a normal distribution of fiber strength with $\hat{\sigma}_y^2 = 1.90$. Note that the proportion of defective product in the output has been dramatically reduced. Although it is unlikely that *all* of the between-loom variability can be eliminated, it is clear that a significant reduction in this variance component would greatly increase the quality of the fiber produced.

We may easily find a confidence interval for the variance component σ^2. If the observations are normally and independently distributed, then $(N - a)MS_E/\sigma^2$ is distributed as χ^2_{N-a}. Thus,

$$P\left(\chi^2_{1-(\alpha/2),N-a} \leq \frac{(N - a)MS_E}{\sigma^2} \leq \chi^2_{\alpha/2,N-a}\right) = 1 - \alpha$$

and a $100(1 - \alpha)$ percent confidence interval for σ^2 is

$$\frac{(N - a)MS_E}{\chi^2_{\alpha/2,N-a}} \leq \sigma^2 \leq \frac{(N - a)MS_E}{\chi^2_{1-(\alpha/2),N-a}} \tag{3-40}$$

Now consider the variance component σ_τ^2. The point estimator of $\hat{\sigma}_\tau^2$ is

$$\sigma_\tau^2 = \frac{MS_{Treatments} - MS_E}{n}$$

The random variable $(a - 1)MS_{Treatments}/(\sigma^2 + n\sigma_\tau^2)$ is distributed as χ^2_{a-1}, and $(N - a)MS_E/\sigma^2$ is distributed as χ^2_{N-a}. Thus, the probability distribution of $\hat{\sigma}_\tau^2$ is a linear combination of two chi-square random variables, say

$$u_1\chi^2_{a-1} - u_2\chi^2_{N-a}$$

where

$$u_1 = \frac{\sigma^2 + n\sigma_\tau^2}{n(a - 1)} \quad \text{and} \quad u_2 = \frac{\sigma^2}{n(N - a)}$$

Unfortunately, a closed-form expression for the distribution of this linear combination of chi-square random variables cannot be obtained. Thus, an exact confidence interval for σ_τ^2 cannot be constructed. Approximate procedures are given in Graybill (1961) and Searle (1971a).

It is easy to find an exact expression for a confidence interval on the ratio $\sigma_\tau^2/(\sigma_\tau^2 + \sigma^2)$. This is a meaningful ratio since it reflects the *proportion* of the variance of an observation [recall that $V(y_{ij}) = \sigma_\tau^2 + \sigma^2$] that is the result of differences between treatments. To develop this confidence interval for the case of a balanced design, note that $MS_{\text{Treatments}}$ and MS_E are independent random variables and, furthermore, that it can be shown that

$$\frac{MS_{\text{Treatments}}/(n\sigma_\tau^2 + \sigma^2)}{MS_E/\sigma^2} \sim F_{a-1,N-a}$$

Thus,

$$\left(F_{1-\alpha/2,a-1,N-a} \leq \frac{MS_{\text{Treatments}}}{MS_E} \frac{\sigma^2}{n\sigma_\tau^2 + \sigma^2} \leq F_{\alpha/2,a-1,N-a} \right) = 1 - \alpha \quad (3\text{-}41)$$

By rearranging Equation 3-41, we may obtain the following:

$$P\left(L \leq \frac{\sigma_\tau^2}{\sigma^2} \leq U \right) = 1 - \alpha \quad (3\text{-}42)$$

where

$$L = \frac{1}{n}\left(\frac{MS_{\text{Treatments}}}{MS_E} \frac{1}{F_{\alpha/2,a-1,N-a}} - 1 \right) \quad (3\text{-}43a)$$

and

$$U = \frac{1}{n}\left(\frac{MS_{\text{Treatments}}}{MS_E} \frac{1}{F_{1-\alpha/2,a-1,N-a}} - 1 \right) \quad (3\text{-}43b)$$

Note that L and U are $100(1 - \alpha)$ percent lower and upper confidence limits, respectively, for the ratio σ_τ^2/σ^2. Therefore, a $100(1 - \alpha)$ percent confidence interval for $\sigma_\tau^2/(\sigma_\tau^2 + \sigma^2)$ is

$$\frac{L}{1 + L} \leq \frac{\sigma_\tau^2}{\sigma_\tau^2 + \sigma^2} \leq \frac{U}{1 + U} \quad (3\text{-}44)$$

To illustrate this procedure, we find a 95 percent confidence interval on $\sigma_\tau^2/(\sigma_\tau^2 + \sigma^2)$ for the strength data in Example 3-8. Recall that $MS_{\text{Treatments}} = 29.73$, $MS_E = 1.90$, $a = 4$, $n = 4$, $F_{.025,3,12} = 4.47$, and $F_{.975,3,12} = 1/F_{.025,12,3} = 1/14.34 = 0.070$. Therefore, from Equation 3-43,

$$L = \frac{1}{4}\left[\left(\frac{29.73}{1.90}\right)\left(\frac{1}{4.47}\right) - 1 \right] = 0.625$$

$$U = \frac{1}{4}\left[\left(\frac{29.73}{1.90}\right)\left(\frac{1}{0.070}\right) - 1 \right] = 54.883$$

and from Equation 3-44, the 95 percent confidence interval on $\sigma_\tau^2/(\sigma_\tau^2 + \sigma^2)$ is

$$\frac{0.625}{1.625} \leq \frac{\sigma_\tau^2}{\sigma_\tau^2 + \sigma^2} \leq \frac{54.883}{55.883}$$

or

$$0.39 \leq \frac{\sigma_\tau^2}{\sigma_\tau^2 + \sigma^2} \leq 0.98$$

We conclude that variability between looms accounts for between 39 and 98 percent of the variance in the observed strength of the fabric produced. This confidence interval is relatively wide because of the small sample size that was used in the experiment. Clearly, however, the variability between looms (σ_τ^2) is not negligible.

3-6 SAMPLE COMPUTER OUTPUT

Computer programs for the analysis of variance are widely available. The output from one such program, the Statistical Analysis System (SAS) General Linear Models procedure, is shown in Figure 3-7, using the data in Example 3-1. The sum of squares corresponding to the "model" is the usual $SS_{\text{Treatments}}$ for a single-factor design. That source is further identified as "cotton," and type I and type III sums of squares are displayed for that factor. These sums of squares are always identical for a balanced design, and in the case of a single-factor design, they are the same as the model sum of squares.

In addition to the basic analysis of variance, the program displays some additional useful information. The quantity "R-SQUARE" is defined as

$$R^2 = \frac{SS_{\text{Model}}}{SS_{\text{Total}}} = \frac{475.76}{636.96} = 0.746923$$

and is loosely interpreted as the proportion of the variability in the data "explained" by the analysis of variance model. Thus, in the synthetic fiber strength testing data, the factor "cotton percentage" explains about 74.69 percent of the variability in tensile strength. Clearly, we must have $0 \leq R^2 \leq 1$, with larger values being more desirable. "STD DEV" is the square root of the error mean square, $\sqrt{8.060} = 2.839$, and "C.V." is the coefficient of variation, defined as $(\sqrt{MS_E}/\bar{y})100$. The coefficient of variation measures the unexplained or residual variability in the data as a percentage of the mean of the response variable.

The computer program also calculates and displays the residuals, as defined in Equation 3-17. In the next chapter, we discuss how the residuals may be used in model adequacy checking.

S T A T I S T I C A L   A N A L Y S I S   S Y S T E M   13:16 WEDNESDAY, JANUARY 17, 1990

GENERAL LINEAR MODELS PROCEDURE

DEPENDENT VARIABLE: Y

SOURCE	DF	SUM OF SQUARES	MEAN SQUARE	F VALUE	PR > F	R-SQUARE	C.V.
MODEL	4	475.76000000	118.94000000	14.76	0.0001	0.746923	18.8764
ERROR	20	161.20000000	8.06000000			STD DEV	Y MEAN
CORRECTED TOTAL	24	636.96000000				2.83901391	15.04000000

SOURCE	DF	TYPE I SS	F VALUE	PR > F
COTTON	4	475.76000000	14.76	0.0001

SOURCE	DF	TYPE III SS	F VALUE	PR > F
COTTON	4	475.76000000	14.76	0.0001

OBSERVATION	OBSERVED VALUE	PREDICTED VALUE	RESIDUAL
1	7.00000000	9.80000000	-2.80000000
2	7.00000000	9.80000000	-2.80000000
3	15.00000000	9.80000000	5.20000000
4	11.00000000	9.80000000	1.20000000
5	9.00000000	9.80000000	-0.80000000
6	12.00000000	15.40000000	-3.40000000
7	17.00000000	15.40000000	1.60000000
8	12.00000000	15.40000000	-3.40000000
9	18.00000000	15.40000000	2.60000000
10	18.00000000	15.40000000	2.60000000
11	14.00000000	17.60000000	-3.60000000
12	18.00000000	17.60000000	0.40000000
13	18.00000000	17.60000000	0.40000000
14	19.00000000	17.60000000	1.40000000
15	19.00000000	17.60000000	1.40000000
16	19.00000000	21.60000000	-2.60000000
17	25.00000000	21.60000000	3.40000000
18	22.00000000	21.60000000	0.40000000
19	19.00000000	21.60000000	-2.60000000
20	23.00000000	21.60000000	1.40000000
21	7.00000000	10.80000000	-3.80000000
22	10.00000000	10.80000000	-0.80000000
23	11.00000000	10.80000000	0.20000000
24	15.00000000	10.80000000	4.20000000
25	11.00000000	10.80000000	0.20000000

SUM OF RESIDUALS   0.00000000
SUM OF SQUARED RESIDUALS   161.20000000
SUM OF SQUARED RESIDUALS - ERROR SS   -0.00000000
FIRST ORDER AUTOCORRELATION   -0.22555831
DURBIN-WATSON D   2.40223325

Figure 3-7. Sample computer output for Example 3-1.

3-7 PROBLEMS

3-1 The tensile strength of portland cement is being studied. Four different mixing techniques can be used economically. The following data have been collected:

Mixing Technique	Tensile Strength (lb/in.^{2})			
1	3129	3000	2865	2890
2	3200	3300	2975	3150
3	2800	2900	2985	3050
4	2600	2700	2600	2765

(a) Test the hypothesis that mixing techniques affect the strength of the cement. Use $\alpha = .05$.

(b) Use Duncan's multiple range test to make comparisons between pairs of means.

3-2 A textile mill has a large number of looms. Each loom is supposed to provide the same output of cloth per minute. To investigate this assumption, five looms are chosen at random and their output is noted at different times. The following data are obtained:

Loom	Output (lb/min)				
1	14.0	14.1	14.2	14.0	14.1
2	13.9	13.8	13.9	14.0	14.0
3	14.1	14.2	14.1	14.0	13.9
4	13.6	13.8	14.0	13.9	13.7
5	13.8	13.6	13.9	13.8	14.0

(a) Explain why this is a random effects experiment. Are the looms equal in output?

(b) Estimate the variability between looms.

(c) Estimate the experimental error variance.

(d) Find a 95 percent confidence interval for $\sigma_\tau^2/(\sigma_\tau^2 + \sigma^2)$.

3-3 An experiment was run to determine whether four specific firing temperatures affect the density of a certain type of brick. The experiment led to the following data:

Temperature	Density				
100	21.8	21.9	21.7	21.6	21.7
125	21.7	21.4	21.5	21.4	
150	21.9	21.8	21.8	21.6	21.5
175	21.9	21.7	21.8	21.4	

(a) Does the firing temperature affect the density of the bricks?

(b) Compare the means using Duncan's multiple range test.

(c) Use the graphical approach discussed in Section 3-4.1 to determine which means differ. Do the conclusions using this method agree with the conclusions from the Duncan multiple range test in part (b)?

3-4 A manufacturer of television sets is interested in the effect on tube conductivity of four different types of coating for color picture tubes. The following conductivity data are obtained:

Coating Type	Conductivity			
1	143	141	150	146
2	152	149	137	143
3	134	136	132	127
4	129	127	132	129

(a) Is there a difference in conductivity due to coating type? Use $\alpha = .05$.

(b) Estimate the overall mean and the treatment effects.

(c) Compute a 95 percent interval estimate of the mean of coating type 4. Compute a 99 percent interval estimate of the mean difference between coating types 1 and 4.

(d) Test all pairs of means using Duncan's multiple range test with $\alpha = .05$.

(e) Use the graphical method discussed in Section 3-4.1 to compare the means. Which coating type produces the highest conductivity?

(f) Assuming that coating type 4 is currently in use, what are your recommendations to the manufacturer? We wish to minimize conductivity.

3-5 The response time in milliseconds was determined for three different types of circuits that could be used in an automatic valve shutoff mechanism. The results were

Circuit Type	Response Time				
1	9	12	10	8	15
2	20	21	23	17	30
3	6	5	8	16	7

(a) Test the hypothesis that the three circuit types have the same response time. Use $\alpha = .01$.

(b) Use Tukey's test to compare pairs of treatment means.

(c) Construct a set of orthogonal contrasts, assuming that at the outset of the experiment you suspected the response time of circuit type 2 to be different from the other two.

(d) If you were the design engineer and you wished to minimize the response time, which circuit type would you select?

3-6 The effective life of insulating fluids at an accelerated load of 35 kV is being studied. Test data have been obtained for four types of fluids. The results were as follows:

Fluid Type	Life (in h) at 35 kV Load					
1	17.6	18.9	16.3	17.4	20.1	21.6
2	16.9	15.3	18.6	17.1	19.5	20.3
3	21.4	23.6	19.4	18.5	20.5	22.3
4	19.3	21.1	16.9	17.5	18.3	19.8

(a) Is there any indication that the fluids differ?

(b) Which fluid would you select, given that the objective is long life?

3-7 Four different designs for a digital computer circuit are being studied in order to compare the amount of noise present. The following data have been obtained:

Circuit Design	Noise Observed				
1	19	20	19	30	8
2	80	61	73	56	80
3	47	26	25	35	50
4	95	46	83	78	97

(a) Is the amount of noise present the same for all four designs?

(b) Estimate the components of the appropriate model for this problem. Estimate all possible *differences* between pairs of treatments. Are these treatment differences uniquely estimated? If so, why?

(c) Which circuit design would you select for use? Low noise is best.

3-8 A manufacturer suspects that the batches of raw material furnished by her supplier differ significantly in calcium content. There are a large number of batches currently in the warehouse. Five of these are randomly selected for study. A chemist makes five determinations on each batch and obtains the following data:

Batch 1	Batch 2	Batch 3	Batch 4	Batch 5
23.46	23.59	23.51	23.28	23.29
23.48	23.46	23.64	23.40	23.46
23.56	23.42	23.46	23.37	23.37
23.39	23.49	23.52	23.46	23.32
23.40	23.50	23.49	23.39	23.38

(a) Is there significant variation in calcium content from batch to batch?

(b) Estimate the components of variance.

(c) Find a 95 percent confidence interval for $\sigma_\tau^2/(\sigma_\tau^2 + \sigma^2)$.

3-9   Several ovens in a metal working shop are used to heat metal specimens. All the ovens are supposed to operate at the same temperature, although it is suspected that this may not be true. Three ovens are selected at random and their temperatures on successive heats are noted. The data collected are as follows:

Oven	Temperature					
1	491.50	498.30	498.10	493.50	493.60	
2	488.50	484.65	479.90	477.35		
3	490.10	484.80	488.25	473.00	471.85	478.65

(a) Is there significant variation in temperature between ovens?

(b) Estimate the components of variance for this model.

3-10 Four chemists are asked to determine the percentage of methyl alcohol in a certain chemical compound. Each chemist makes three determinations, and the results are the following:

Chemist	Percentage of Methyl Alcohol		
1	84.99	84.04	84.38
2	85.15	85.13	84.88
3	84.72	84.48	85.16
4	84.20	84.10	84.55

(a) Do chemists differ significantly? Use $\alpha = .05$.

(b) If chemist 2 is a new employee, construct a meaningful set of orthogonal contrasts that might have been useful at the start of the experiment.

3-11 Three brands of batteries are under study. It is suspected that the lives (in weeks) of the three brands are different. Five batteries of each brand are tested with the following results:

Weeks of Life		
Brand 1	Brand 2	Brand 3
100	76	108
96	80	100
92	75	96
96	84	98
92	82	100

(a) Are the lives of these brands of batteries different?

(b) Estimate the components of the appropriate statistical model.

(c) Construct a 95 percent interval estimate on the mean life of battery brand 2. Construct a 99 percent interval estimate on the mean difference between the lives of battery brands 2 and 3.

(d) Which brand would you select for use? If the manufacturer will replace without charge any battery that fails in less than 85 weeks, what percentage would he expect to replace?

3-12 Four catalysts that may affect the concentration of one component in a three-component liquid mixture are being investigated. The following concentrations are obtained:

	Catalyst		
1	2	3	4
58.2	56.3	50.1	52.9
57.2	54.5	54.2	49.9
58.4	57.0	55.4	50.0
55.8	55.3		51.7
54.9			

(a) Do the four catalysts have the same effect on the concentration?

(b) Estimate the components of the appropriate statistical model.

(c) Construct a 99 percent confidence interval estimate of the mean response for catalyst 1.

3-13 Consider testing the equality of the means of two normal populations, where the variances are unknown but are assumed to be equal. The appropriate test procedure is the pooled t test. Show that the pooled t test is equivalent to the single-factor analysis of variance.

3-14 Show that the variance of the linear combination $\sum_{i=1}^{a} c_i y_{i.}$ is $\sigma^2 \sum_{i=1}^{a} n_i c_i^2$.

3-15 In a fixed effects experiment, suppose that there are n observations for each of four treatments. Let Q_1^2, Q_2^2, Q_3^2 be single degree-of-freedom components for the orthogonal contrasts. Prove that $SS_{Treatments} = Q_1^2 + Q_2^2 + Q_3^2$.

3-16 Consider the data shown in Problem 3-5.

(a) Write out the least squares normal equations for this problem, and solve them for $\hat{\mu}$ and $\hat{\tau}_i$, using the usual constraint ($\sum_{i=1}^{3} \hat{\tau}_i = 0$). Estimate $\tau_1 - \tau_2$.

(b) Solve the equations in (a) using the constraint $\hat{\tau}_3 = 0$. Are the estimators $\hat{\tau}_i$ and $\hat{\mu}$ the same as you found in (a)? Why? Now estimate $\tau_1 - \tau_2$ and compare your answer with that for (a). What statement can you make about estimating contrasts in τ_i variables?

(c) Estimate $\mu + \tau_1$, $2\tau_1 - \tau_2 - \tau_3$, and $\mu + \tau_1 + \tau_2$ using the two solutions to the normal equations. Compare the results obtained in each case.

3-17   Consider the data in Problem 3-4. Use the LSD method to test for differences between pairs of means. Compare your conclusions with those for Duncan's multiple range test; see Problem 3-4(d).

3-18   Use the LSD method to find significant differences in the circuit types for the data in Problem 3-5.

3-19   Consider the one-way, balanced, random effects method. Develop a procedure for finding a $100(1 - \alpha)$ percent confidence interval for $\sigma^2/(\sigma_\tau^2 + \sigma^2)$.

3-20   Consider the data in Problem 3-5. If circuit type 1 is a control, determine if either circuit type 2 or 3 differs from circuit type 1 in terms of mean response time.

Chapter 4
More About Single Factor Experiments

This chapter discusses several additional topics concerning single-factor experiments and the analysis of variance: model adequacy checking and the consequences of departures from the underlying assumptions; methods for choosing an appropriate sample size for comparing treatment means; fitting response curves to develop interpolation equations for quantitative factors; a brief look at a more general method for developing the analysis of variance; a nonparametric alternative to the usual F test in the analysis of variance; and an introduction to experiments with repeated measures.

4-1 MODEL ADEQUACY CHECKING

We have noted previously that the assumptions underlying the analysis of variance are that the data are adequately described by the model

$$y_{ij} = \mu + \tau_i + \epsilon_{ij} \begin{cases} i = 1,2, \dots, a \\ j = 1,2, \dots, n. \end{cases} \tag{4-1}$$

and that the errors are normally and independently distributed with mean zero and constant variance σ^2. In the random effects model, we make the additional assumptions that the τ_i are normally and independently distributed with mean zero and variance σ_τ^2 and that the τ_i and ϵ_{ij} are independent. In this section, we discuss and illustrate methods for checking these assumptions and present remedial measures that are often useful when the assumptions are violated.

The primary diagnostic tools are based on the *residuals*. In Section 3-3.4, we noted that the residuals for the one-way model are

$$e_{ij} = y_{ij} - \hat{y}_{ij} \tag{4-2}$$
$$= y_{ij} - \bar{y}_{i.}$$

That is, the residuals for the ith treatment are found by subtracting the treatment average from each observation in that treatment. Model adequacy checking usually consists of plotting the residuals as described below. We recommend that such diagnostic checking be a routine part of every experimental design project.

4-1.1   The Normality Assumption

A check of the normality assumption may be made by plotting a histogram of the residuals. If the NID(0, σ^2) assumption on the errors is satisfied, then this plot should look like a sample from a normal distribution centered at zero. Unfortunately, with small samples, considerable fluctuation often occurs, so the appearance of a moderate departure from normality does not necessarily imply a serious violation of the assumptions. Gross deviations from normality are potentially serious and require further analysis.

Another useful procedure is to construct a normal probability plot of the residuals. A normal probability plot is just a graph of the cumulative distribution of the residuals on *normal probability paper*,[1] that is, graph paper with the ordinate scaled so that the cumulative normal distribution plots as a straight line. To construct a normal probability plot, arrange the residuals in increasing order and plot the kth of these ordered residuals versus the cumulative probability point $P_k = (k - 1/2)/N$ on normal probability paper. If the underlying error distribution is normal, this plot will resemble a straight line. In visualizing the straight line, place more emphasis on the central values of the plot than on the extremes.

Table 4-1 shows the original data and the residuals for the tensile strength data in Example 3-1. In Table 4-2, the residuals are ranked in ascending order and their cumulative probability points P_k are calculated. The normal probability plot is shown in Figure 4-1 with the residuals plotted versus $P_k \times 100$ on the right vertical scale. Note that the bottom of this figure also gives a dot diagram of the residuals. The general impression from examining this display is that the error distribution may be slightly skewed, with the right tail being longer than the left. The tendency of the normal probability plot to bend down slightly on the left side implies that the left tail of the error distribution is somewhat *thinner* than would be anticipated in a normal distribution; that is, the negative residuals are not quite as large (in absolute value) as expected. This plot is not grossly nonnormal, however.

In general, moderate departures from normality are of little concern in the fixed effects analysis of variance. An error distribution that has considerably thicker or thinner tails than the normal is of more concern than a skewed distri-

---

[1] Normal probability paper is available at most technical book stores. Many computer programs for the analysis of variance will also construct normal probability plots on request. When the plot is generated via computer, it is customary to plot the standardized normal score Z_j, which satisfies $P_j = \Phi(Z_j)$, on the vertical axis instead of the normal probability P_j (or $P_j \times 100$).

Table 4-1 Data and Residuals from Example 3-1[a]

Percentage of Cotton	Observations (j)										$\hat{y}_{ij} = \bar{y}_{i.}$
	1		2		3		4		5		
15	7	−2.8 (15)	7	−2.8 (19)	15	5.2 (25)	11	1.2 (12)	9	−0.8 (6)	9.8
20	12	−3.4 (8)	17	1.6 (14)	12	−3.4 (1)	18	2.6 (11)	19	2.6 (3)	15.4
25	14	−3.6 (18)	18	0.4 (13)	18	0.4 (20)	19	1.4 (7)	19	1.4 (9)	17.6
30	19	−2.6 (22)	25	3.4 (5)	22	0.4 (2)	19	−2.6 (24)	23	1.4 (10)	21.6
35	7	−3.8 (17)	10	−0.8 (21)	11	0.2 (4)	15	4.2 (16)	11	0.2 (23)	10.8

[a] The residuals are shown in the box in each cell. The numbers in parentheses indicate the order of data collection.

bution. Since the F test is only slightly affected, we say that the analysis of variance (and related procedures such as multiple comparisons) is *robust* to the normality assumption. Departures from normality usually cause both the true significance level and the power to differ slightly from the advertised values, with the power generally being lower. The random effects model is more severely impacted by nonnormality. In particular, the true confidence levels on interval estimates of variance components may differ greatly from the advertised values.

Table 4-2 Ordered Residuals and Probability Points for the Tensile Strength Data

Order k	Residual e_{ij}	$P_k = (k - \frac{1}{2})/25$	Order k	Residual e_{ij}	$P_k = (k - \frac{1}{2})/25$
1	−3.8	.0200	14	0.4	.5400
2	−3.6	.0600	15	0.4	.5800
3	−3.4	.1000	16	1.2	.6200
4	−3.4	.1400	17	1.4	.6600
5	−2.8	.1800	18	1.4	.7000
6	−2.8	.2200	19	1.4	.7400
7	−2.8	.2600	20	1.6	.7800
8	−2.6	.3000	21	2.6	.8200
9	−0.8	.3400	22	2.6	.8600
10	−0.8	.3800	23	3.4	.9000
11	0.2	.4200	24	4.2	.9400
12	0.2	.4600	25	5.2	.9800
13	0.4	.5000			

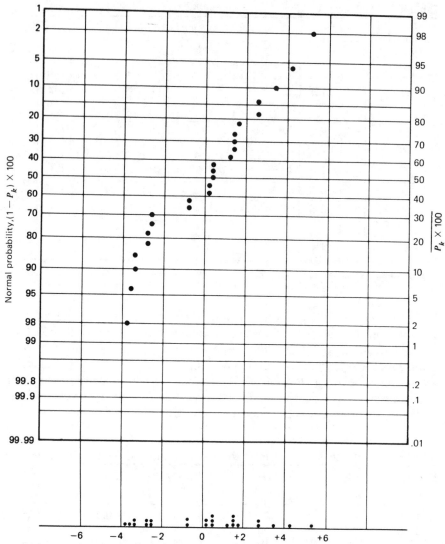

Figure 4-1.   Normal probability plot and dot diagram of residuals for Example 3-1.

A very common defect that often shows up on normal probability plots is one residual that is very much larger than any of the others. Such a residual is often called an *outlier*. The presence of one or more outliers can seriously distort the analysis of variance, so when a potential outlier is located, careful investigation is called for. Frequently, the cause of the outlier is a mistake in calculations or a data coding or copying error. If this is not the cause, then the experimental circumstances surrounding this run must be carefully studied. If the outlying response is a particularly desirable value (high strength, low cost, etc.), then the outlier may be more informative than the rest of the data. We should be careful not to reject or discard an outlying observation unless we have reasonable non-statistical grounds for doing so. At worst, you may end up with two analyses; one with the outlier, and one without.

There are several formal statistical procedures for detecting outliers [e.g., see Barnett and Lewis (1978), John and Prescott (1975), and Stefansky (1972)]. A rough check for outliers may be made by examining the standardized residuals

$$d_{ij} = \frac{e_{ij}}{\sqrt{MS_E}} \qquad (4\text{-}3)$$

If the errors ϵ_{ij} are $N(0, \sigma^2)$, then the standardized residuals should be approximately normal with mean zero and unit variance. Thus, about 68 percent of the standardized residuals should fall within the limits ±1, about 95 percent of them should fall within ±2, and virtually all of them should fall within ±3. A residual bigger than 3 or 4 standard deviations from zero is a potential outlier.

For the tensile strength data of Example 3-1, the normal probability plot and histogram give no indication of outliers. Furthermore, the largest standardized residual is

$$d_{13} = \frac{e_{13}}{\sqrt{MS_E}} = \frac{5.2}{\sqrt{8.06}} = \frac{5.2}{2.84} = 1.83$$

which should cause no concern.

4-1.2 Plot of Residuals in Time Sequence

Plotting the residuals in time order of data collection is helpful in detecting *correlation* between the residuals. A tendency to have runs of positive and negative residuals indicates positive correlation. This would imply that the *independence* assumption on the errors has been violated. This is a potentially serious problem and one that is difficult to correct, so it is important to prevent the problem if possible when the data are collected. Proper randomization of the experiment is an important step in obtaining independence.

Sometimes the skill of the experimenter (or the subjects) may change as the experiment progresses, or the process being studied may "drift" or become more erratic. This will often result in a change in the error variance over time. This condition often leads to a plot of residuals versus time that exhibits more spread at one end than at the other. Nonconstant variance is a potentially serious problem. We will have more to say on the subject in Sections 4-1.3 and 4-1.4.

Table 4-1 displays the residuals and the time sequence of data collection for the tensile strength data. A plot of these residuals versus time is shown in Figure

4-2. There is no reason to suspect any violation of the independence or constant variance assumptions.

4-1.3   Plot of Residuals Versus Fitted Values $\hat{y}_{ij}$

If the model is correct and if the assumptions are satisfied, the residuals should be structureless; in particular, they should be unrelated to any other variable including the response y_{ij}. A simple check is to plot the residuals versus the fitted values $\hat{y}_{ij}$. (For the one-way model, remember that $\hat{y}_{ij} = \bar{y}_{i.}$, the ith treatment average.) This plot should not reveal any obvious pattern. Figure 4-3 plots the residuals versus the fitted values for the tensile strength data of Example 3-1. No unusual structure is apparent.

A defect that occasionally shows up on this plot is *nonconstant variance*. Sometimes the variance of the observations increases as the magnitude of the observation increases. This would be the case if the error or background noise in the experiment was a constant percentage of the size of the observation. (This commonly happens with many measuring instruments—error is a percent of the scale reading.) If this were the case, the residuals would get larger as y_{ij} gets larger, and the plot of residuals versus $\hat{y}_{ij}$ would look like an outward-opening funnel or megaphone. Nonconstant variance also arises in cases where the data follow a nonnormal, skewed distribution because in skewed distributions the variance tends to be a function of the mean.

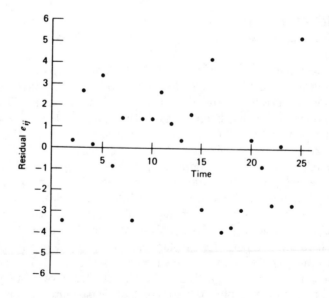

Figure 4-2.   Plot of residuals versus time.

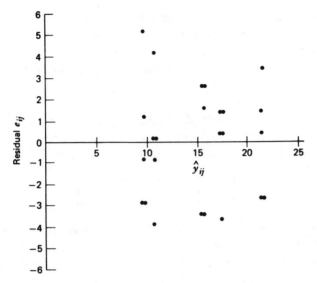

Figure 4-3.   Plot of residuals versus fitted values.

If the assumption of homogeneity of variances is violated, the F test is only slightly affected in the balanced fixed effects model. However, in unbalanced designs or in cases where one variance is very much larger than the others, the problem is more serious. For the random effects model, unequal error variances can significantly disturb inferences on variance components even if balanced designs are used.

The usual approach to dealing with nonconstant variance when it occurs for the above reasons is to apply a *variance-stabilizing transformation* and then to run the analysis of variance on the transformed data. In this approach, one should note that the conclusions of the analysis of variance apply to the *transformed populations*.

Considerable research has been devoted to the selection of an appropriate transformation. If experimenters know the theoretical distribution of the observations, they may utilize this information in choosing a transformation. For example, if the observations follow the Poisson distribution, then the square root transformation $y_{ij}^* = \sqrt{y_{ij}}$ or $y_{ij}^* = \sqrt{1 + y_{ij}}$ would be used. If the data follow the log-normal distribution, then the logarithmic transformation $y_{ij}^* = \log y_{ij}$ is appropriate. For binomial data expressed as fractions, the arcsine transformation $y_{ij}^* = \arcsin\sqrt{y_{ij}}$ is useful. When there is no obvious transformation, the experimenter usually empirically seeks a transformation that equalizes the variance regardless of the value of the mean. In factorial experiments, which we discuss in Chapter 7, another approach is to select a transformation that minimizes the interaction mean square, resulting in an experiment that is easier to interpret. For more discussion of transformations, refer to Bartlett (1947), Box and Cox (1964),

Dolby (1963), and Draper and Hunter (1969). In Section 4-1.4, we discuss in more detail methods for analytically selecting the form of the transformation. Transformations made for inequality of variance also affect the form of the error distribution. In most cases, the transformation brings the error distribution closer to normal.

Statistical Tests for Equality of Variance ▪ Although residual plots are frequently used to diagnose inequality of variance, several statistical tests have also been proposed. These tests may be viewed as formal tests of the hypotheses

$$H_0: \sigma_1^2 = \sigma_2^2 = \cdots = \sigma_a^2$$

$$H_1: \text{above not true for at least one } \sigma_i^2$$

A widely used procedure is *Bartlett's test*. The procedure involves computing a statistic whose sampling distribution is closely approximated by the chi-square distribution with $a - 1$ degrees of freedom when the a random samples are from independent normal populations. The test statistic is

$$\chi_0^2 = 2.3026 \frac{q}{c} \tag{4-4}$$

where

$$q = (N - a)\log_{10} S_p^2 - \sum_{i=1}^{a} (n_i - 1)\log_{10} S_i^2$$

$$c = 1 + \frac{1}{3(a - 1)} \left(\sum_{i=1}^{a} (n_i - 1)^{-1} - (N - a)^{-1} \right)$$

$$S_p^2 = \frac{\sum_{i=1}^{a} (n_i - 1) S_i^2}{N - a}$$

and S_i^2 is the sample variance of the ith population.

The quantity q is large when the sample variances S_i^2 differ greatly and is equal to zero when all S_i^2 are equal. Therefore, we should reject H_0 on values of χ_0^2 that are too large; that is, we reject H_0 only when

$$\chi_0^2 > \chi_{\alpha,a-1}^2$$

where $\chi_{\alpha,a-1}^2$ is the upper α percentage point of the chi-square distribution with $a - 1$ degrees of freedom. Various studies have indicated that Bartlett's test is very sensitive to the normality assumption and should not be applied when the normality assumption is doubtful. Other tests for equality of variance are reviewed by Anderson and McLean (1974).

Example 4-1

We can apply Bartlett's test to the tensile strength data from the cotton percentage experiment in Example 3-1. We first compute the sample variances in each treatment and find that $S_1^2 = 11.2$, $S_2^2 = 9.8$, $S_3^2 = 4.3$, $S_4^2 = 6.8$, and $S_5^2 = 8.2$. Then

$$S_p^2 = \frac{4(11.2) + 4(9.8) + 4(4.3) + 4(6.8) + 4(8.2)}{20} = 8.06$$

$$q = 20 \log_{10}(8.06) - 4[\log_{10}11.2 + \log_{10}9.8$$
$$+ \log_{10}4.3 + \log_{10}6.8 + \log_{10}8.2] = 0.45$$

$$c = 1 + \frac{1}{3(4)}\left(\frac{5}{4} - \frac{1}{20}\right) = 1.10$$

and the test statistic is

$$\chi_0^2 = 2.3026\frac{(0.45)}{(1.10)} = 0.93$$

Since $\chi_{.05,4}^2 = 9.49$, we cannot reject the null hypothesis and conclude that all five variances are the same. This is the same conclusion reached by analyzing the plot of residuals versus fitted values.

∎

4-1.4 Selecting a Variance-Stabilizing Transformation

In the previous section we noted that, if experimenters knew the relationship between the variance of the observations and the mean, they could use this information to guide them in selecting the form of the transformation. We now elaborate on this point and show how it is possible to estimate the form of the required transformation from the data.

Let $E(y) = \mu$ be the mean of y, and suppose that the standard deviation of y is proportional to a power of the mean of y such that

$$\sigma_y \propto \mu^\alpha$$

We want to find a transformation on y that yields a constant variance. Suppose that the transformation is a power of the original data, say

$$y^* = y^\lambda \tag{4-5}$$

Then it can be shown that

$$\sigma_{y^*} \propto \mu^{\lambda+\alpha-1} \tag{4-6}$$

Clearly, if we set $\lambda = 1 - \alpha$, then the variance of the transformed data y^* is constant.

Several of the common transformations discussed in Section 4-1.3 are summarized in Table 4-3. Note that $\lambda = 0$ implies the log transformation. These

Table 4-3 Variance Stabilizing Transformations

Relationship Between σ_y and μ	α	$\lambda = 1 - \alpha$	Transformation	Comment
$\sigma_y \propto$ Constant	0	1	No transformation	
$\sigma_y \propto \mu^{1/2}$	1/2	1/2	Square root	Poisson (count) data
$\sigma_y \propto \mu$	1	0	Log	
$\sigma_y \propto \mu^{3/2}$	3/2	−1/2	Reciprocal square root	
$\sigma_y \propto \mu^2$	2	−1	Reciprocal	

transformations are arranged in order of increasing *strength*. By the strength of a transformation we mean the amount of curvature it induces. A mild transformation applied to data spanning a narrow range has little effect on the analysis, whereas a strong transformation applied over a large range may have dramatic results. Transformations often have little effect unless the ratio y_{max}/y_{min} is larger than two or three.

Empirical Selection of α ▪ In many experimental design situations where there is replication, we can empirically estimate α from the data. Since in the *i*th treatment combination $\sigma_{y_i} \propto \mu_i^\alpha = \theta\mu_i^\alpha$, where θ is a constant of proportionality, we may take logs to obtain

$$\log \sigma_{y_i} = \log \theta + \alpha \log \mu_i \qquad (4\text{-}7)$$

Therefore, a plot of $\log \sigma_{y_i}$ versus $\log \mu_i$ would be a straight line with slope α. Since we don't know σ_{y_i} and μ_i, we may substitute reasonable estimates of them in Equation 4-7 and use the slope of the resulting straight line fit as an estimate of α. Typically, we would use the standard deviation S_i and the average $\bar{y}_{i.}$ of the *i*th treatment (or, more generally, the *i*th treatment combination or set of experimental conditions) to estimate σ_{y_i} and μ_i.

Example 4-2

A civil engineer is interested in determining whether four different methods of estimating flood flow frequency produce equivalent estimates of peak discharge when applied to the same watershed. Each procedure is used six times on the watershed, and the resulting discharge data (in cubic feet per second) are shown in Table 4-4.

Table 4-4 Peak Discharge Data

Estimation Method	Observations						$\bar{y}_{i.}$	S_i
1	0.34	0.12	1.23	0.70	1.75	0.12	0.71	0.66
2	0.91	2.94	2.14	2.36	2.86	4.55	2.63	1.09
3	6.31	8.37	9.75	6.09	9.82	7.24	7.93	1.66
4	17.15	11.82	10.95	17.20	14.35	16.82	14.72	2.77

Table 4-5   Analysis of Variance for Peak Discharge Data

Source of Variation	Sum of Squares	Degrees of Freedom	Mean Square	F_0
Methods	708.3471	3	236.1157	76.07
Error	62.0811	20	3.1041	
Total	770.4282	23		

The analysis of variance for the data, summarized in Table 4-5, implies that there is a difference in mean peak discharge estimates given by the four procedures. The plot of residuals versus fitted values, shown in Figure 4-4, is disturbing since the outward-opening funnel shape indicates that the constant variance assumption is not satisfied.

To investigate the possibility of using a variance-stabilizing transformation on the data, we plot log S_i versus log $\bar{y}_{i.}$ in Figure 4-5. The slope of a straight line passing through these four points is close to 1/2 and, from Table 4-3, this implies that the square root transformation may be appropriate. The analysis of variance for the transformed data $y^* = \sqrt{y}$ is presented in Table 4-6, and a plot of residuals is shown in Figure 4-6. This residual plot is much improved in comparison to Figure 4-4, so we conclude that the square root transformation has been helpful. Note that in Table 4-6 we have reduced the error degrees of freedom by one to account for the use of the data to estimate the transformation parameter α.

∎

Analytical Determination of the Transformation ▪ Box and Cox (1964) have shown how the transformation parameter λ in $y^* = y^{\lambda}$ may be estimated simultaneously with the other model parameters (overall mean and treatment effects) using the method of maximum likelihood. The procedure consists of performing,

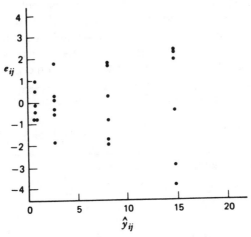

Figure 4-4.   Plot of residuals versus $\hat{y}_{ij}$ for Example 4-2.

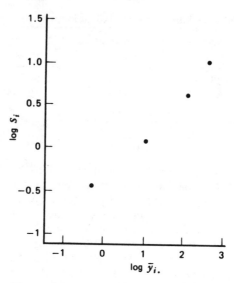

Figure 4-5.   Plot of log S_i versus log $\bar{y}_{i.}$ for Example 4-2.

for various values of λ, a standard analysis of variance on

$$
y^{(\lambda)} = \begin{cases} \dfrac{y^\lambda - 1}{\lambda \dot{y}^{\lambda-1}} & \lambda \neq 0 \\[2mm] \dot{y} \ln y & \lambda = 0 \end{cases} \tag{4-8}
$$

where $\dot{y} = \ln^{-1}[(1/n)\Sigma\ln y]$ is the geometric mean of the observations. The maximum likelihood estimate of λ is the value for which the error sum of squares, say $SS_E(\lambda)$, is a minimum. This value of λ is usually found by plotting a graph of $SS_E(\lambda)$ versus λ and then reading the value of λ that minimizes $SS_E(\lambda)$ from the graph. Usually between 10 and 20 values of λ are sufficient for estimation of the optimum value. A second iteration using a finer mesh of values could be performed if a more accurate estimate of λ is necessary. Notice that we *cannot* select the value of λ by *directly* comparing the error sums of squares from analyses of variance on y^λ because for each value of λ the error sum of squares is measured on a different scale. Equation 4-8 rescales the responses so that the error sums of squares are directly comparable. We recommend that the experimenter use

Table 4-6   Analysis of Variance for Transformed Peak Discharge Data, $y^* = \sqrt{y}$

Source of Variation	Sum of Squares	Degrees of Freedom	Mean Square	F_0
Methods	32.6842	3	10.8947	76.99
Error	2.6884	19	0.1415	
Total	35.3726	22		

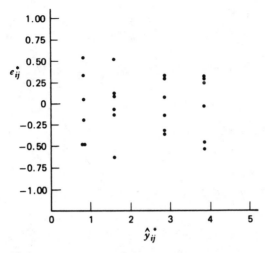

Figure 4-6. Plot of residuals from transformed data versus $\hat{y}_{ij}^*$ for Example 4-2.

simple choices for λ because the practical difference between $\lambda = 0.5$ and $\lambda = 0.55$ is likely to be small, but the former is much easier to interpret.

An approximate $100(1 - \alpha)$ percent confidence interval for λ can be found by computing

$$SS^* = SS_E(\lambda) \left(1 + \frac{t_{\alpha/2,v}^2}{v}\right) \tag{4-9}$$

where v is the number of error degrees of freedom, and then by reading the corresponding confidence limits on λ directly from the graph. If this confidence interval includes the value $\lambda = 1$, this implies that the data do not support the need for transformation.

Example 4-3

Using the peak discharge data in Table 4-4, values of $SS_E(\lambda)$ for various values of λ are determined:

λ	$SS_E(\lambda)$
-1.00	7922.11
-0.50	687.10
-0.25	232.52
0.00	91.96
0.25	46.99
0.50	35.42
0.75	40.61
1.00	62.08
1.25	109.82
1.50	208.12

A graph of values close to the minimum is shown in Figure 4-7, from which it is seen that $\lambda = 0.52$ gives a minimum value of approximately $SS_E(\lambda) = 35.00$. An approx-

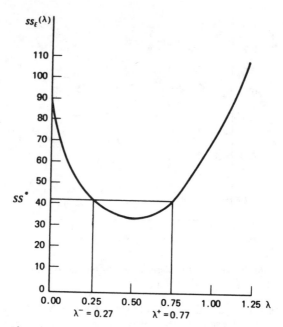

Figure 4-7. Plot of $SS_E(\lambda)$ versus λ for Example 4-3.

imate 95 percent confidence interval on λ is found by calculating the quantity SS^* from Equation 4-9 as follows:

$$SS^* = SS_E(\lambda)\left(1 + \frac{t_{.025,20}^2}{20}\right)$$

$$= 35.00\left[1 + \frac{(2.086)^2}{20}\right]$$

$$= 42.61$$

By plotting SS^* on the graph in Figure 4-7 and by reading the points on the λ scale where this line intersects the curve, we obtain lower and upper confidence limits on λ of $\lambda^- = 0.27$ and $\lambda^+ = 0.77$. Since these confidence limits do not include the value 1, the use of a transformation is indicated, and the square root transformation ($\lambda = 0.50$) actually used is easily justified.

∎

4-1.5 Plots of Residuals Versus Other Variables

If data have been collected on any other variables that might possibly affect the response, then the residuals should be plotted against these variables. For example, in the tensile strength experiment of Example 3-1, strength may be significantly affected by the thickness of the fiber, so the residuals should be plotted versus fiber thickness. If different testing machines were used to collect the data, then the residuals should be plotted against machines. Patterns in such residual plots imply that the variable affects the response. This suggests that the variable should be either controlled more carefully in future experiments or included in the analysis.

4-1.6 Discovering Dispersion Effects

In previous sections we have focused on finding which factor levels result in differences between the means. It is customary to refer to these effects as *location* effects. If there was inequality of variance at the different factor levels, we used transformations to stabilize the variance to improve our inference on the location effects. In some problems, however, we are interested in discovering whether the different factor levels affect *variability*; that is, we are interested in discovering potential *dispersion effects*. This will occur whenever the standard deviation, variance, or some other measure of variability is used as a response variable.

To illustrate these ideas, consider the data in Table 4-7, which resulted from a designed experiment in an aluminum smelter. Aluminum is produced by combining alumina with other ingredients in a reaction cell and applying heat by passing electric current through the cell. Alumina is added continuously to the cell to maintain the proper ratio of alumina to other ingredients. Four different ratio control algorithms were investigated in this experiment. The response variables studied were related to cell voltage. Specifically, a sensor scans cell voltage several times each second, producing thousands of voltage measurements during each run of the experiment. The process engineers decided to use the average voltage and the standard deviation of cell voltage (shown in parentheses) over the run as the response variables. The average voltage is important because it impacts cell temperature, and the standard deviation of voltage (called "pot noise" by the process engineers) is important because it impacts the overall cell efficiency.

An analysis of variance was performed to determine if the different ratio control algorithms affect average cell voltage. This revealed that the ratio control algorithm had no *location effect*; that is, changing the ratio control algorithms does not change the average cell voltage. (Refer to Problem 4-5.)

Table 4-7 Data For the Smelting Experiment

Ratio Control Algorithm	Observations					
	1	2	3	4	5	6
1	4.93(0.05)	4.86(0.04)	4.75(0.05)	4.95(0.06)	4.79(0.03)	4.88(0.05)
2	4.85(0.04)	4.91(0.02)	4.79(0.03)	4.85(0.05)	4.75(0.03)	4.85(0.02)
3	4.83(0.09)	4.88(0.13)	4.90(0.11)	4.75(0.15)	4.82(0.08)	4.90(0.12)
4	4.89(0.03)	4.77(0.04)	4.94(0.05)	4.86(0.05)	4.79(0.03)	4.76(0.02)

Table 4-8   Analysis of Variance for the Natural Logarithm of Pot Noise

Source of Variation	Sums of Squares	Degrees of Freedom	Mean Square	F_0
Ratio Control Algorithm	6.166	3	2.055	21.96
Error	1.872	20	0.094	
Total	8.038	23		

To investigate dispersion effects, it is usually best to use

$$\log (s)$$

as a response variable as the log transformation is effective in stabilizing variability in the distribution of the sample standard deviation. As all sample standard deviations of pot voltage are less than unity, we will use

$$y = -\ln (s)$$

as the response variable. Table 4-8 presents the analysis of variance for this response, the natural logarithm of "pot noise." Notice that the choice of a ratio control algorithm affects pot noise; that is, the ratio control algorithm has a *dispersion effect*. Standard tests of model adequacy, including normal probability plots of the residuals, indicate that there are no problems with experimental validity. (Refer to Problem 4-6.)

Figure 4-8 plots the average log pot noise for each ratio control algorithm and also presents a scaled *t* distribution for use as a *reference distribution* in discriminating between ratio control algorithms. This plot clearly reveals that ratio control algorithm 3 produces greater pot noise or greater cell voltage standard deviation than the other algorithms. There does not seem to be much difference between algorithms 1, 2, and 4.

4-2   CHOICE OF SAMPLE SIZE

In any experimental design problem, a critical decision is the choice of sample size—that is, determining the number of replicates to run. Generally, if the experimenter is interested in detecting small effects, more replicates are required than if the experimenter is interested in detecting large effects. In this section we discuss several approaches to determining sample size. Although our discussion focuses on a single-factor design, most of the methods can be used in more complex experimental situations.

4-2.1   Operating Characteristic Curves

Recall that an operating characteristic curve is a plot of the type II error probability of a statistical test for a particular sample size versus a parameter that reflects the

Figure 4-8. Average log pot noise $[-\ln(s)]$ for four ratio control algorithms relative to a scaled t distribution with scale factor $\sqrt{MS_E/n} = \sqrt{0.094/6} = 0.125$.

extent to which the null hypothesis is false. These curves can be used to guide the experimenter in selecting the number of replicates so that the design will be sensitive to important potential differences in the treatments.

We first consider the probability of type II error of the fixed effects model for the case of equal sample sizes per treatment, say

$$\beta = 1 - P\{\text{Reject } H_0 | H_0 \text{ is false}\} \tag{4-10}$$
$$= 1 - P\{F_0 > F_{\alpha,a-1,N-a} | H_0 \text{ is false}\}$$

To evaluate the probability statement in Equation 4-10, we need to know the distribution of the test statistic F_0 if the null hypothesis is false. It can be shown that, if H_0 is false, the statistic $F_0 = MS_{\text{Treatments}}/MS_E$ is distributed as a *noncentral F* random variable with $a - 1$ and $N - a$ degrees of freedom and the noncentrality parameter δ. If $\delta = 0$, then the noncentral F distribution becomes the usual (central) F distribution.

Operating characteristic curves given in Chart V of the Appendix are used to evaluate the probability statement in Equation 4-10. These curves plot the probability of type II error (β) against a parameter Φ, where

$$\Phi^2 = \frac{n \sum_{i=1}^{a} \tau_i^2}{a\sigma^2} \tag{4-11}$$

The quantity Φ^2 is related to the noncentrality parameter δ. Curves are available for $\alpha = .05$ and $\alpha = .01$ and a range of degrees of freedom for numerator and denominator.

In using the operating characteristic curves, the experimenter must specify the parameter Φ. This is often difficult to do in practice. One way to determine Φ is to choose the actual values of the treatment means for which we would like to reject the null hypothesis with high probability. Thus, if $\mu_1, \mu_2, \ldots, \mu_a$ are the specified treatment means, we find the τ_i in Equation 4-11 as $\tau_i = \mu_i - \bar{\mu}$, where $\bar{\mu} = (1/a)\sum_{i=1}^{a} \mu_i$ is the average of the individual treatment means. We also require an estimate of σ^2. Sometimes this is available from prior experience, a previous experiment, or a judgment estimate. When we are uncertain about the value of σ^2, sample sizes could be determined for a range of likely values of σ^2 to study the effect of this parameter on the required sample size before a final choice is made.

Example 4-4

Consider the tensile strength problem in Example 3-1. Suppose that the experimenter is interested in rejecting the null hypothesis with a probability of at least .90 if the five treatment means are

$$\mu_1 = 11 \qquad \mu_2 = 12 \qquad \mu_3 = 15 \qquad \mu_4 = 18 \qquad \text{and} \qquad \mu_5 = 19$$

She plans to use $\alpha = .01$. In this case, since $\sum_{i=1}^{5} \mu_i = 75$, we have $\bar{\mu} = (1/5)75 = 15$ and

$$\tau_1 = \mu_1 - \bar{\mu} = 11 - 15 = -4$$
$$\tau_2 = \mu_2 - \bar{\mu} = 12 - 15 = -3$$
$$\tau_3 = \mu_3 - \bar{\mu} = 15 - 15 = 0$$
$$\tau_4 = \mu_4 - \bar{\mu} = 18 - 15 = 3$$
$$\tau_5 = \mu_5 - \bar{\mu} = 19 - 15 = 4$$

Thus, $\sum_{i=1}^{5} \tau_i^2 = 50$. Suppose the experimenter feels that the standard deviation of tensile strength at any particular level of cotton percentage will be no larger than $\sigma = 3$ psi. Then, by using Equation 4-11, we have

$$\Phi^2 = \frac{n \sum_{i=1}^{5} \tau_i^2}{a\sigma^2} = \frac{n(50)}{5(3)^2} = 1.11n$$

We use the operating characteristic curve for $a - 1 = 5 - 1 = 4$ with $N - a = a(n - 1) = 5(n - 1)$ error degrees of freedom and $\alpha = .01$ (see Appendix Chart V). As a first guess at the required sample size, try $n = 4$ replicates. This yields $\Phi^2 = 1.11(4) = 4.44$, $\Phi = 2.11$, and $5(3) = 15$ error degrees of freedom. Consequently, from Chart V, we find that $\beta \approx .30$. Therefore, the power of the test is approximately $1 - \beta = 1 - .30 = .70$, which is less than the required .90, and so we conclude that $n = 4$ replicates are not sufficient. Proceeding in a similar manner, we can construct the following display:

n	Φ^2	Φ	$a(n - 1)$	β	Power$(1 - \beta)$
4	4.44	2.11	15	.30	.70
5	5.55	2.36	20	.15	.85
6	6.66	2.58	25	.04	.96

Thus, at least $n = 6$ replicates must be run in order to obtain a test with the required power. ∎

The only problem with this approach to using the operating characteristic curves is that it is usually difficult to select a set of treatment means on which the sample size decision should be based. An alternate approach is to select a sample size such that if the difference between any two treatment means exceeds a specified value the null hypothesis should be rejected. If the difference between any two treatment means is as large as D, then it can be shown that the minimum value of Φ^2 is

$$\Phi^2 = \frac{nD^2}{2a\sigma^2} \tag{4-12}$$

Since this is a minimum value of Φ^2, the corresponding sample size obtained from the operating characteristic curve is a conservative value; that is, it provides a power at least as great as that specified by the experimenter.

To illustrate this approach, suppose that in the tensile strength problem of Example 3-1, the experimenter wished to reject the null hypothesis with probability at least .90 if any two treatment means differed by as much as 10 psi. Then, assuming that $\sigma = 3$ psi, we find the minimum value of Φ^2 to be

$$\Phi^2 = \frac{n(10)^2}{2(5)(3^2)} = 1.11n$$

and, from the analysis in Example 4-4, we conclude that $n = 6$ replicates are required to give the desired sensitivity when $\alpha = .01$.

Now consider the random effects model. The type II error probability for the random effects model is

$$\beta = 1 - P\{\text{Reject } H_0 | H_0 \text{ is false}\} \tag{4-13}$$
$$= 1 - P\{F_0 > F_{\alpha, a-1, N-a} | \sigma_\tau^2 > 0\}$$

Once again, the distribution of the test statistic $F_0 = MS_{\text{Treatments}}/MS_E$ under the alternative hypothesis is needed. It can be shown that if H_1 is true ($\sigma_\tau^2 > 0$), the distribution of F_0 is central F with $a - 1$ and $N - a$ degrees of freedom.

Since the type II error probability of the random effects model is based on the usual central F distribution, we could use the tables of the F distribution in the Appendix to evaluate Equation 4.13. However, it is simpler to determine the sensitivity of the test through the use of operating characteristic curves. A set of these curves for various values of numerator degrees of freedom, denominator degrees of freedom, and α of .05 or .01 is provided in Chart VI of the Appendix. These curves plot the probability of type II error against the parameter λ, where

$$\lambda = \sqrt{1 + \frac{n\sigma_\tau^2}{\sigma^2}} \tag{4-14}$$

Note that λ involves two unknown parameters, σ^2 and σ_τ^2. We may be able to estimate σ_τ^2 if we have an idea about how much variability in the population of treatments it is important to detect. An estimate of σ^2 may be chosen using prior experience or judgment. Sometimes it is helpful to define the value of σ_τ^2 we are interested in detecting in terms of the ratio σ_τ^2/σ^2.

Example 4-5

Suppose we have five treatments selected at random with six observations per treatment and $\alpha = .05$, and we wish to determine the power of the test if σ_τ^2 is equal to

σ^2. Since $a = 5$, $n = 6$, and $\sigma_\tau^2 = \sigma^2$, we may compute

$$\lambda = \sqrt{1 + 6(1)} = 2.646$$

From the operating characteristic curve with $a - 1 = 4$, $N - a = 25$ degrees of freedom, and $\alpha = .05$, we find that

$$\beta \simeq .20$$

and thus the power is approximately .80.

■

4-2.2 Specifying a Standard Deviation Increase

This approach is occasionally helpful in choosing the sample size. Consider first the fixed effects model. If the treatment means do not differ, then the standard deviation of an observation chosen at random is σ. If the treatment means are different, however, then the standard deviation of a randomly chosen observation is

$$\sqrt{\sigma^2 + \left(\sum_{i=1}^{a} \tau_i^2 / a \right)}$$

If we choose a percentage P for the increase in the standard deviation of an observation beyond which we wish to reject the hypothesis that all treatment means are equal, then this is equivalent to choosing

$$\frac{\sqrt{\sigma^2 + \left(\sum_{i=1}^{a} \tau_i^2 / a \right)}}{\sigma} = 1 + 0.01P \qquad (P = \text{percent})$$

or

$$\frac{\sqrt{\sum_{i=1}^{a} \tau_i^2 / a}}{\sigma} = \sqrt{(1 + 0.01P)^2 - 1}$$

so that

$$\Phi = \frac{\sqrt{\sum_{i=1}^{a} \tau_i^2 / a}}{\sigma / \sqrt{n}} = \sqrt{(1 + 0.01P)^2 - 1}(\sqrt{n}) \qquad (4\text{-}15)$$

Thus, for a specified value of P, we may compute Φ from Equation 4-15 and then use the operating characteristic curves in Appendix Chart V to determine the required sample size.

For example, in the tensile strength problem of Example 3-1, suppose that we wish to detect a standard deviation increase of 20 percent with a probability of at least .90 and $\alpha = .05$. Then

$$\Phi = \sqrt{(1.2)^2 - 1} \; (\sqrt{n}) = 0.66\sqrt{n}$$

Reference to the operating characteristic curves shows that $n = 9$ is required to give the desired sensitivity.

A similar approach can be used for the random effects model. If the treatments are homogeneous, then the standard deviation of an observation selected at random is σ. However, if the treatments are different, then the standard deviation of a randomly chosen observation is

$$\sqrt{\sigma^2 + \sigma_\tau^2}$$

If P is the fixed percentage increase in the standard deviation of an observation beyond which rejection of the null hypothesis is desired, then

$$\frac{\sqrt{\sigma^2 + \sigma_\tau^2}}{\sigma} = 1 + 0.01P$$

or

$$\frac{\sigma_\tau^2}{\sigma^2} = (1 + 0.01P)^2 - 1$$

Therefore, using Equation 4-14, we find that

$$\lambda = \sqrt{1 + \frac{n\sigma_\tau^2}{\sigma^2}} = \sqrt{1 + n[(1 + 0.01P)^2 - 1]} \qquad (4\text{-}16)$$

For a given P, the operating characteristic curves in Appendix Chart VI can be used to find the desired sample size.

4-2.3 Confidence Interval Estimation Method

This approach assumes that the experimenter wishes to express the final results in terms of confidence intervals and is willing to specify in advance how wide he or she wants these confidence intervals to be. For example, suppose that in the tensile strength problem in Example 3-1 we wanted a 95 percent confidence interval on the difference in mean tensile strength for any two cotton percentages to be ±5 psi and a prior estimate of σ is 3. Then, using Equation 3-16, we find that the accuracy of the confidence interval is

$$\pm t_{\alpha/2, N-a} \sqrt{\frac{2MS_E}{n}}$$

Suppose that we try $n = 5$ replicates. Then, using $\sigma^2 = 3^2 = 9$ as an estimate of MS_E, the accuracy of the confidence interval becomes

$$\pm 2.086 \sqrt{\frac{2(9)}{5}} = \pm 3.96$$

which is more accurate than the requirement. Trying $n = 4$ gives

$$\pm 2.132 \sqrt{\frac{2(9)}{4}} = \pm 4.52$$

Trying $n = 3$ gives

$$\pm 2.228 \sqrt{\frac{2(9)}{3}} = \pm 5.46$$

Clearly $n = 4$ is the smallest sample size that will lead to the desired accuracy.

The quoted level of significance in the above illustration applies only to one confidence interval. However, the same general approach can be used if the experimenter wishes to prespecify a *set* of confidence intervals about which a *joint* or *simultaneous* confidence statement is made. Furthermore, the confidence intervals could be constructed about more general contrasts in the treatment means than the pairwise comparison illustrated above. For further examples of this approach to determining sample sizes, see Neter and Wasserman (1974).

4-3 FITTING RESPONSE CURVES IN THE SINGLE-FACTOR MODEL

4-3.1 General Regression Approach

The factor investigated in the single-factor model analysis of variance can be either *quantitative* or *qualitative*. A quantitative factor is one whose levels can be associated with points on a numerical scale, such as temperature, pressure, or time. Qualitative factors, on the other hand, are factors for which the levels cannot be arranged in order of magnitude. Operators, batches of raw material, and shifts are typical qualitative factors because there is no reason to rank them in any particular numerical order.

Insofar as the initial design and analysis of the experiment are concerned, both types of factors are treated identically. The experimenter is interested in determining the differences, if any, between the levels of the factors. If the factor is qualitative, such as operators, it is meaningless to consider the response for a subsequent run at an intermediate level of the factor. However, with a quantitative factor such as time, the experimenter is usually interested in the entire range of values used, particularly the response from a subsequent run at an intermediate factor level. That is, if the levels 1.0 hours, 2.0 hours, and 3.0 hours are used in the experiment, we may wish to predict the response at 2.5 hours. Thus, the experimenter is frequently interested in developing an interpolation equation from the data.

The general approach to fitting equations to data is regression analysis, which is discussed extensively in Chapter 15. This section briefly illustrates the technique using the tensile strength data of Example 3-1.

Figure 4-9 presents a scatter diagram of tensile strength y versus the cotton percentage of the fabric x for the data in Example 3-1. The open circles on the graph are the mean tensile strengths at each value of cotton percentage x. From examining the scatter diagram, it is clear that the relationship between tensile strength and cotton percentage is not linear. As a first guess, we could try fitting a quadratic equation to the data, say

$$y = \beta_0 + \beta_1 x + \beta_2 x^2 + \epsilon$$

where β_0, β_1, and β_2 are unknown parameters to be estimated and ϵ is a random error term. The least squares fit is

$$\hat{y} = -39.9886 + 4.596x - 0.0886x^2$$

(If you are unfamiliar with regression methods, see Chapter 15.)

This quadratic model is shown in Figure 4-9. It does not appear to be very satisfactory because it drastically underestimates the responses at $x = 30$ percent

Figure 4-9. Scatter diagram for the tensile strength data of Example 3-1.

cotton and overestimates the responses at $x = 25$ percent. Perhaps an improvement can be obtained by adding a cubic term in x. The resulting cubic fit is

$$\hat{y} = 62.6114 - 9.0114x + 0.4814x^2 - 0.0076x^3$$

This cubic fit is also shown in Figure 4-9. The cubic model appears to be superior to the quadratic because it provides a better fit at $x = 25$ and $x = 30$ percent cotton.

In general, we would like to fit the lowest-order polynomial that adequately describes the data. In this example, the cubic polynomial seems to fit better than the quadratic, so the extra complexity of the cubic model is justified. Selecting the order of the approximating polynomial is not always easy, however, and it is relatively easy to overfit, that is, to add high-order polynomial terms that do not really improve the fit but increase the complexity of the model and often damage its usefulness as a predictor or interpolation equation.

4-3.2 Orthogonal Polynomials

In situations where the levels of the factor are equally spaced, fitting polynomial models by the method of least squares is greatly simplified. The procedure makes use of the orthogonal contrast coefficients in Appendix Table X. In addition to the least squares fit to the polynomial, we also obtain a linear, quadratic, cubic, and so forth, effect and the sum of squares for the factor. This allows the contribution of each term to the polynomial to be tested. It is possible to extract polynomial effects up through order $a - 1$ if there are a levels of the factor involved in the experiment.

The process is illustrated in Table 4-9 using the data of Example 3-1. In this

Table 4-9 Calculation of Polynomial Effects for Example 3-1

Percentage of Cotton	Treatment Totals $y_{i.}$	Orthogonal Contrast Coefficients (c_i)			
		Linear	Quadratic	Cubic	Quartic
15	49	-2	2	-1	1
20	77	-1	-1	2	-4
25	88	0	-2	0	6
30	108	1	-1	-2	-4
35	54	2	2	1	1
Effects: $\left(\sum\limits_{i=1}^{a} c_i y_{i.} \right)$		41	-155	-57	-109
Sums of squares: $\left[\dfrac{\left(\sum\limits_{i=i}^{a} c_i y_{i.} \right)^2}{n \sum\limits_{i=1}^{a} c_i^2} \right]$		$\dfrac{(41)^2}{5(10)} =$ 33.62	$\dfrac{(-155)^2}{5(14)} =$ 343.21	$\dfrac{(-57)^2}{5(10)} =$ 64.98	$\dfrac{(-109)^2}{5(70)} =$ 33.95

problem the independent factor, percentage of cotton, is equally spaced at five levels. The sums of squares for the linear, quadratic, cubic, and quartic effects of the factor form a partition of the treatment sum of squares and can be incorporated into the analysis of variance, as shown in Table 4-10. Each effect has one degree of freedom and can be tested by comparing its sum of squares to the mean square error.

From inspection of Table 4-10, we note that the quadratic and cubic effects of cotton percentage are statistically significant when compared to $F_{.05,1,20}$. Therefore, we will fit a cubic polynomial to the data, say

$$y = \alpha_0 + \alpha_1 P_1(x) + \alpha_2 P_2(x) + \alpha_3 P_3(x) + \epsilon$$

where $P_u(x)$ is the uth order *orthogonal polynomial*, which implies that if there are a levels of x we have $\sum_{j=1}^{a} P_u(x_j)P_s(x_j) = 0$ for $u \neq s$. The first five orthogonal polynomials are

$$P_0(x) = 1$$

$$P_1(x) = \lambda_1 \left[\frac{(x - \bar{x})}{d} \right]$$

$$P_2(x) = \lambda_2 \left[\left(\frac{x - \bar{x}}{d} \right)^2 - \left(\frac{a^2 - 1}{12} \right) \right]$$

$$P_3(x) = \lambda_3 \left[\left(\frac{x - \bar{x}}{d} \right)^3 - \left(\frac{x - \bar{x}}{d} \right) \left(\frac{3a^2 - 7}{20} \right) \right]$$

$$P_4(x) = \lambda_4 \left[\left(\frac{x - \bar{x}}{d} \right)^4 - \left(\frac{x - \bar{x}}{d} \right)^2 \left(\frac{3a^2 - 13}{14} \right) + \frac{3(a^2 - 1)(a^2 - 9)}{560} \right]$$

where d is the distance between the levels of x, a is the total number of levels, and $\{\lambda_i\}$ are constants such that the polynomials have integer values. Appendix Table X lists coefficients of the orthogonal polynomials and values of the λ_i for $a \leq 10$. The least squares estimates of the parameters in the orthogonal poly-

Table 4-10   Analysis of Variance

Source of Variation	Sum of Squares	Degrees of Freedom	Mean Square	F_0
Percentage of cotton	475.76	4	118.94	14.76[a]
(Linear)	(33.62)	1	33.62	4.17
(Quadratic)	(343.21)	1	343.21	42.58[a]
(Cubic)	(64.98)	1	64.98	8.06[b]
(Quartic)	(33.95)	1	33.95	4.21
Error	161.20	20	8.06	
Total		24		

[a] Significant at 1 percent.

[b] Significant at 5 percent.

nomial model are

$$\hat{\alpha}_i = \frac{\sum y P_i(x)}{\sum [P_i(x)]^2} \qquad i = 0, 1, \dots, a - 1$$

A mathematical justification of this equation will be given in Chapter 15. For the data in Example 3-1, we may estimate the model parameters as

$$\hat{\alpha}_0 = \frac{\sum y P_0(x)}{\sum [P_0(x)]^2} = \frac{\sum y}{25} = \frac{376}{5(5)} = 15.0400$$

$$\hat{\alpha}_1 = \frac{\sum y P_1(x)}{\sum [P_1(x)]^2} = \frac{41}{5(10)} = 0.8200$$

$$\hat{\alpha}_2 = \frac{\sum y P_2(x)}{\sum [P_2(x)]^2} = \frac{-155}{5(14)} = -2.2143$$

$$\hat{\alpha}_3 = \frac{\sum y P_3(x)}{\sum [P_3(x)]^2} = \frac{-57}{5(10)} = -1.1400$$

If we wish either to add or to delete terms from the model, it is not necessary to recompute the $\{\hat{\alpha}_i\}$ already in the model because of the orthogonality property of the polynomials $\{P_i(x)\}$.

Since we have $a = 5$ levels of x and the spacing between the levels is $d = 5$, the orthogonal polynomial model becomes

$$\hat{y} = 15.0400 + 0.8200(1)\left(\frac{x - 25}{5}\right)$$

$$- 2.2143(1)\left[\left(\frac{x - 25}{5}\right)^2 - \left(\frac{5^2 - 1}{12}\right)\right]$$

$$- 1.1400(5/6)\left[\left(\frac{x - 25}{5}\right)^3 - \left(\frac{x - 25}{5}\right)\left(\frac{3(5)^2 - 7}{20}\right)\right]$$

where $\lambda_1 = \lambda_2 = 1$ and $\lambda_3 = 5/6$ have been obtained from Appendix Table X. This equation can be simplified to

$$\hat{y} = 62.6111 - 9.0100x + 0.4814x^2 - 0.00786x^3$$

which is essentially the same equation found earlier by more general regression methods. (The difference is due primarily to how the regression computer program used handles round-off error.)

We have noted that it is desirable to fit the lowest-degree polynomial that adequately describes the data. The computational simplicity of orthogonal polynomials is helpful in this type of problem because it allows the experimenter to add or delete terms easily without affecting previously computed coefficients. For further details of regression analysis, see Chapter 15.

4-4 THE REGRESSION APPROACH TO THE ANALYSIS OF VARIANCE

Thus far, we have given an intuitive or heuristic development of the analysis of variance. However, it is possible to give a more formal development. The method will be useful later in understanding the basis for the statistical analysis of more complex designs. Called the *general regression significance test,* the procedure essentially consists of finding the reduction in the total sum of squares for fitting the model with all parameters included and the reduction in sum of squares when the model is restricted to the null hypotheses. The difference between these two sums of squares is the treatment sum of squares with which a test of the null hypothesis can be conducted.

A fundamental part of the procedure is writing the normal equations for the model. These equations may always be obtained by forming the least squares function and differentiating it with respect to each unknown parameter, as we did in Section 3-3.3. However, an easier method is available. The following list of rules allow the normal equations for *any* experimental design model to be written directly:

RULE 1. There is one normal equation for each parameter in the model to be estimated.

RULE 2. The right-hand side of any normal equation is just the sum of all observations that contain the parameter associated with that particular normal equation.

To illustrate this rule consider the single-factor model. The first normal equation is for the parameter μ; therefore, the right-hand side is $y_{..}$ because *all* observations contain μ.

RULE 3. The left-hand side of any normal equation is the sum of all model parameters, where each parameter is multiplied by the number of times it appears in the total on the right-hand side. The parameters are written with a circumflex ($\hat{\ }$) to indicate that they are *estimators* and not the true parameter values.

For example, consider the first normal equation in a single-factor experiment. According to the above rules, it would be

$$N\hat{\mu} + n\hat{\tau}_1 + n\hat{\tau}_2 + \cdots + n\hat{\tau}_a = y_{..}$$

since μ appears in all N observations, τ_1 appears only in the n observations taken under the first treatment, τ_2 appears only in the n observations taken under the second treatment, and so on. From Equation 3-12, we verify that the equation shown above is correct. The second normal equation would correspond to τ_1 and is

$$n\hat{\mu} + n\hat{\tau}_1 = y_{1.}$$

since only the observations in the first treatment contain τ_1 (this gives $y_{1.}$ as the right-hand side), μ and τ_1 appear exactly n times in $y_{1.}$, and all other τ_i appear zero times. In general, the left-hand side of any normal equation is the expected value of the right-hand side.

Now, consider finding the reduction in the sum of squares by fitting a particular model to the data. By fitting a model to the data, we "explain" some of the variability; that is, we reduce the unexplained variability by some amount. The reduction in the unexplained variability is always the sum of the parameter estimates, each multiplied by the right-hand side of the normal equation that corresponds to that parameter. For example, in a single-factor experiment, the reduction due to fitting the model $y_{ij} = \mu + \tau_i + \epsilon_{ij}$ is

$$R(\mu, \tau) = \hat{\mu}y_{..} + \hat{\tau}_1 y_{1.} + \hat{\tau}_2 y_{2.} + \cdots + \hat{\tau}_a y_{a.} \tag{4-17}$$

$$= \hat{\mu}y_{..} + \sum_{i=1}^{a} \hat{\tau}_i y_{i.}$$

The notation $R(\mu, \tau)$ means that reduction in the sum of squares from fitting the model containing μ and $\{\tau_i\}$. $R(\mu, \tau)$ is also sometimes called the "regression" sum of squares for the model $y_{ij} = \mu + \tau_i + \epsilon_{ij}$. The number of degrees of freedom associated with a reduction in the sum of squares, such as $R(\mu, \tau)$, is always equal to the number of linearly independent normal equations. The remaining variability unaccounted for by the model is found from

$$SS_E = \sum_{i=1}^{a} \sum_{j=1}^{n} y_{ij}^2 - R(\mu, \tau) \tag{4-18}$$

This quantity is used in the denominator of the test statistic for $H_0: \tau_i = 0$.

We now illustrate the general regression significance test for a single-factor experiment and show that it yields the usual one-way analysis of variance. The model is $y_{ij} = \mu + \tau_i + \epsilon_{ij}$, and the normal equations are found from the above rules as

$$
\begin{aligned}
N\hat{\mu} + n\hat{\tau}_1 + n\hat{\tau}_2 + \cdots + n\hat{\tau}_a &= y_{..} \\
n\hat{\mu} + n\hat{\tau}_1 \phantom{+ n\hat{\tau}_2} &= y_{1.} \\
n\hat{\mu} \phantom{+ n\hat{\tau}_1} + n\hat{\tau}_2 &= y_{2.} \\
&\vdots \\
n\hat{\mu} \phantom{+ n\hat{\tau}_1} + n\hat{\tau}_a &= y_{a.}
\end{aligned}
$$

Compare these normal equations with those obtained in Equation 3-12.

Applying the constraint $\sum_{i=1}^{a} \hat{\tau}_i = 0$, the estimators for μ and τ_i are

$$\hat{\mu} = \bar{y}_{..} \qquad \hat{\tau}_i = \bar{y}_{i.} - \bar{y}_{..} \qquad i = 1, 2, \ldots, a$$

The reduction in the sum of squares due to fitting this model is found from

Equation 4-17 as

$$R(\mu, \tau) = \hat{\mu} y_{..} + \sum_{i=1}^{a} \hat{\tau}_i y_{i.}$$

$$= (\bar{y}_{..}) y_{..} + \sum_{i=1}^{a} (\bar{y}_{i.} - \bar{y}_{..}) y_{i.}$$

$$= \frac{y_{..}^2}{N} + \sum_{i=1}^{a} \bar{y}_{i.} y_{i.} - \bar{y}_{..} \sum_{i=1}^{a} y_{i.}$$

$$= \sum_{i=1}^{a} \frac{y_{i.}^2}{n}$$

which has a degrees of freedom because there are a linearly independent normal equations. The error sum of squares is, from Equation 4-18,

$$SS_E = \sum_{i=1}^{a} \sum_{j=1}^{n} y_{ij}^2 - R(\mu, \tau)$$

$$= \sum_{i=1}^{a} \sum_{j=1}^{n} y_{ij}^2 - \sum_{i=1}^{a} \frac{y_{i.}^2}{n}$$

and has $N - a$ degrees of freedom.

To find the sum of squares resulting from the treatment effects (the $\{\tau_i\}$), we consider the model to be restricted to the null hypothesis, that is, $\tau_i = 0$ for all i. The reduced model is $y_{ij} = \mu + \epsilon_{ij}$. There is only one normal equation for this model:

$$N\hat{\mu} = y_{..}$$

and the estimator of μ is $\hat{\mu} = \bar{y}_{..}$. Thus, the reduction in the sum of squares that results from fitting only μ is

$$R(\mu) = (\bar{y}_{..})(y_{..}) = \frac{y_{..}^2}{N}$$

Since there is only one normal equation for this reduced model, $R(\mu)$ has one degree of freedom. The sum of squares due to the $\{\tau_i\}$, given that μ is already in the model, is the difference between $R(\mu, \tau)$ and $R(\mu)$, which is

$$R(\tau|\mu) = R(\mu, \tau) - R(\mu)$$

$$= \sum_{i=1}^{a} \frac{y_{i.}^2}{n} - \frac{y_{..}^2}{N}$$

with $a - 1$ degrees of freedom, which we recognize from Equation 3-9 as $SS_{\text{Treatments}}$. Making the usual normality assumption, the appropriate statistic for testing $H_0: \tau_i = 0$ is

$$F_0 = \frac{R(\tau|\mu)/(a-1)}{\left[\sum_{i=1}^{a}\sum_{j=1}^{n} y_{ij}^2 - R(\mu, \tau)\right]/(N-a)}$$

which is distributed as $F_{a-1,N-a}$ under the null hypothesis. This is, of course, the test statistic for the single-factor analysis of variance.

By now you may suspect that there is a close connection between analysis of variance and regression. In fact, every analysis of variance model can be expressed in terms of a regression equation, and the general regression significance test described above can be used to develop a test for the hypothesis of interest. This will lead to the analysis of variance procedure we have used previously.

To illustrate this connection, suppose that we have a single-factor analysis of variance model with $a = 3$ treatments so that the model is

$$y_{ij} = \mu + \tau_i + \epsilon_{ij} \quad \begin{cases} i = 1,2,3 \\ j = 1,2, \ldots, n \end{cases}$$

The equivalent regression model is

$$y_{ij} = \beta_0 + \beta_1 x_{1j} + \beta_2 x_{2j} + \epsilon_{ij} \quad \begin{cases} i = 1,2,3 \\ j = 1,2, \ldots, n \end{cases} \tag{4-19}$$

where x_{1j} and x_{2j} are defined as follows:

$$x_{1j} = \begin{cases} 1 \text{ if observation } j \text{ is from treatment 1} \\ 0 \text{ otherwise} \end{cases}$$

$$x_{2j} = \begin{cases} 1 \text{ if observation } j \text{ is from treatment 2} \\ 0 \text{ otherwise} \end{cases}$$

The relationships among the parameters β_0, β_1, and β_2 in the regression model and the parameters μ and τ_i ($i = 1,2,3$) in the analysis of variance model are easily determined. If the observations come from treatment 1, then

$$x_{1j} = 1 \quad \text{and} \quad x_{2j} = 0$$

and Equation 4-19 becomes

$$y_{1j} = \beta_0 + \beta_1(1) + \beta_2(0) + \epsilon_{1j}$$
$$= \beta_0 + \beta_1 + \epsilon_{1j}$$

Since in the analysis of variance model an observation from treatment 1 is represented by $y_{1j} = \mu + \tau_1 + \epsilon_{1j} = \mu_1 + \epsilon_{1j}$, this implies that

$$\beta_0 + \beta_1 = \mu_1 = \mu + \tau_1$$

Similarly, if the observations are from treatment 2, then $x_{1j} = 0$ and $x_{2j} = 1$ and

$$x_{2j} = \beta_0 + \beta_1(0) + \beta_2(1) + \epsilon_{2j}$$
$$= \beta_0 + \beta_2 + \epsilon_{2j}$$

Considering the corresponding analysis of variance model, $y_{2j} = \mu + \tau_2 + \epsilon_{2j} = \mu_2 + \epsilon_{2j}$, so

$$\beta_0 + \beta_2 = \mu_2 = \mu + \tau_2$$

Finally, consider observations from treatment 3, for which $x_{1j} = x_{2j} = 0$. The regression model becomes

$$y_{3j} = \beta_0 + \beta_1(0) + \beta_2(0) + \epsilon_{3j}$$
$$= \beta_0 + \epsilon_{3j}$$

The corresponding analysis of variance model is $y_{3j} = \mu + \tau_3 + \epsilon_{3j} = \mu_3 + \epsilon_{3j}$, so

$$\beta_0 = \mu_3 = \mu + \tau_3$$

Thus, in the regression model formulation of the one-way analysis of variance model, the regression coefficients describe comparisons of the first two treatment means with the third mean; that is,

$$\beta_0 = \mu_3$$
$$\beta_1 = \mu_1 - \mu_3$$
$$\beta_2 = \mu_2 - \mu_3$$

In general, if there are a treatments, the regression model will have $a - 1$ variables, say

$$y_{ij} = \beta_0 + \beta_1 x_{1j} + \beta_2 x_{2j} + \cdots + \beta_{a-1} x_{a-1,j}$$

$$+ \epsilon_{ij} \quad \begin{cases} i = 1, 2, \ldots, a \quad (4\text{-}20) \\ j = 1, 2, \ldots, n \end{cases}$$

where

$$x_{ij} = \begin{cases} 1 & \text{if observation } j \text{ is from treatment } i \\ 0 & \text{otherwise} \end{cases}$$

The relationship between the parameters in the regression and analysis of variance models is

$$\beta_0 = \mu_a$$
$$\beta_i = \mu_i - \mu_a \quad i = 1, 2, \ldots, a - 1$$

Thus, β_0 always estimates the mean of the ath treatment and β_i estimates the difference between the means of treatment i and treatment a.

Now consider testing hypotheses. In the analysis of variance model, we want to test $H_0: \mu_1 = \mu_2 = \mu_3$ (or equivalently, $H_0: \tau_1 = \tau_2 = \tau_3 = 0$). If the hypothesis is true, then the parameters in the regression model become

$$\beta_0 = \mu$$
$$\beta_1 = 0$$
$$\beta_2 = 0$$

Therefore, testing $H_0: \beta_1 = \beta_2 = 0$ in the regression model provides a test of the equality of the three treatment means. The general regression significance test can be used to obtain the test procedure for $H_0: \beta_1 = \beta_2 = 0$. The resulting test statistic is the F test in the single-factor analysis of variance.

4-5 NONPARAMETRIC METHODS IN THE ANALYSIS OF VARIANCE

4-5.1 The Kruskal–Wallis Test

In situations where the normality assumption is unjustified, the experimenter may wish to use an alternative procedure to the F test analysis of variance that does not depend on this assumption. Such a procedure has been developed by Kruskal and Wallis (1952). The Kruskal–Wallis test is used to test the null hypothesis that the a treatments are identical against the alternative hypothesis that some of the treatments generate observations that are larger than others. Because the procedure is designed to be sensitive for testing differences in means, it is sometimes convenient to think of the Kruskal–Wallis test as a test for equality of treatment means. The Kruskal–Wallis test is a nonparametric alternative to the usual analysis of variance.

To perform a Kruskal–Wallis test, first rank the observations y_{ij} in ascending order and replace each observation by its rank, say R_{ij}, with the smallest observation having rank 1. In the case of ties (observations having the same value), assign the average rank to each of the tied observations. Let $R_{i.}$ be the sum of the ranks in the ith treatment. The test statistic is

$$H = \frac{1}{S^2}\left[\sum_{i=1}^{a} \frac{R_{i.}^2}{n_i} - \frac{N(N+1)^2}{4}\right] \tag{4-21}$$

where n_i is the number of observations in the ith treatment, N is the total number of observations, and

$$S^2 = \frac{1}{N-1}\left[\sum_{i=1}^{a}\sum_{j=1}^{n_i} R_{ij}^2 - \frac{N(N+1)^2}{4}\right] \tag{4-22}$$

Note that S^2 is just the variance of the ranks. If there are no ties, $S^2 = N(N+1)/12$, and the test statistic simplifies to

$$H = \frac{12}{N(N+1)}\sum_{i=1}^{a} \frac{R_{i.}^2}{n_i} - 3(N+1) \tag{4-23}$$

When the number of ties is moderate, there will be little difference between Equations 4-21 and 4-23, and the simpler form (Equation 4-23) may be used. If the n_i are reasonably large, say $n_i \geq 5$, then H is distributed approximately as χ^2_{a-1} under the null hypothesis. Therefore, if

$$H > \chi^2_{\alpha,a-1}$$

the null hypothesis is rejected.

Example 4-6

The data from Example 3-1 and their corresponding ranks are shown in Table 4-11. Since there is a fairly large number of ties, we use Equation 4-21 as the test statistic. From Equation 4-22 we find

$$S^2 = \frac{1}{N-1}\left[\sum_{i=1}^{a}\sum_{j=1}^{n_i} R_{ij}^2 - \frac{N(N+1)^2}{4}\right]$$

$$= \frac{1}{24}\left[5497.79 - \frac{25(26)^2}{4}\right]$$

$$= 53.03$$

and the test statistic is

$$H = \frac{1}{S^2}\left[\sum_{i=1}^{a} \frac{R_{i.}^2}{n_i} - \frac{N(N+1)^2}{4}\right]$$

$$= \frac{1}{53.03}\left[5245.0 - \frac{25(26)^2}{4}\right]$$

$$= 19.25$$

Since $H > \chi_{.01,4}^2 = 13.28$, we would reject the null hypothesis and conclude that the treatments differ. This is the same conclusion given by the usual analysis of variance F test.

■

4-5.2 General Comments on the Rank Transformation

The procedure used in the previous section of replacing the observations by their ranks is called the *rank transformation*. It is a very powerful and widely useful technique. If we were to apply the ordinary F test to the ranks rather than to the

Table 4-11   Data and Ranks for the Tensile Testing Experiment in Example 3-1

				Percentage of Cotton					
15		20		25		30		35	
y_{1j}	R_{1j}	y_{2j}	R_{2j}	y_{3j}	R_{3j}	y_{4j}	R_{4j}	y_{5j}	R_{5j}
7	2.0	12	9.5	14	11.0	19	20.5	7	2.0
7	2.0	17	14.0	18	16.5	25	25.0	10	5.0
15	12.5	12	9.5	18	16.5	22	23.0	11	7.0
11	7.0	18	16.5	19	20.5	19	20.5	15	12.5
9	4.0	18	16.5	19	20.5	23	24.0	11	7.0
$R_{i.}$	27.5		66.0		85.0		113.0		33.5

original data, we would obtain

$$F_0 = \frac{H/(a - 1)}{(N - 1 - H)/(N - a)}$$ (4-24)

as the test statistic [see Conover (1980), p. 337]. Note that, as the Kruskal–Wallis statistic H increases or decreases, F_0 also increases or decreases, so the Kruskal–Wallis test is equivalent to applying the usual analysis of variance to the ranks.

The rank transformation has wide applicability in experimental design problems for which no nonparametric alternative to the analysis of variance exists. This includes many of the designs in subsequent chapters of this book. If the data are ranked and the ordinary F test is applied, an approximate procedure that has good statistical properties results [see Conover and Iman (1976, 1981), and Conover (1980)]. When we are concerned about the normality assumption or the effect of outliers or "wild" values, we recommend that the usual analysis of variance be performed on both the original data and the ranks. When both procedures give similar results, the analysis of variance assumptions are probably satisfied reasonably well, and the standard analysis is satisfactory. When the two procedures differ, the rank transformation should be preferred since it is less likely to be distorted by nonnormality and unusual observations. In such cases, the experimenter may want to investigate the use of transformations for nonnormality and examine the data and the experimental procedure to determine if outliers are present and why they have occurred.

4-6 REPEATED MEASURES

In experimental work in the social and behavioral sciences, the experimental units are frequently people. Because of differences in experience, training, or background, the differences in the responses of different people to the same treatment may be very large in some experimental situations. Unless it is controlled, this variability between people would become part of the experimental error, and in some cases, it would significantly inflate the error mean square, making it more difficult to detect real differences between treatments.

It is possible to control this variability between people by using a design in which each of the a treatments is used on each person (or "subject"). Such a design is called a *repeated measures design*. In this section we give a brief introduction to repeated measures experiments with a single factor.

Suppose that an experiment involves a treatments and every treatment is to be used exactly once on each of n subjects. The data would appear as in Table 4-12. Note that the observation y_{ij} represents the response of subject j to treatment i and that only n subjects are used. The model that we use for this design is

$$y_{ij} = \mu + \tau_i + \beta_j + \epsilon_{ij}$$ (4-25)

where τ_i is the effect of the ith treatment and β_j is a parameter associated with the jth subject. We assume that treatments are fixed (so $\sum_{i=1}^{a} \tau_i = 0$) and that the

Table 4-12   Data for a Single-Factor Repeated Measures Design

Treatment	Subject				Treatment Total
	1	2	$\cdots$	n	
1	y_{11}	y_{12}	$\cdots$	y_{1n}	$y_{1.}$
2	y_{21}	y_{22}	$\cdots$	y_{2n}	$y_{2.}$
.	.	.		.	
.	.	.		.	
.	.	.		.	
a	y_{a1}	y_{a2}		y_{an}	$y_{a.}$
Subject totals	$y_{.1}$	$y_{.2}$	$\cdots$	$y_{.n}$	$y_{..}$

subjects employed are a random sample of subjects from some larger population of potential subjects. Thus, the subjects collectively represent a random effect, so we assume that the mean of β_j is zero and that the variance of β_j is σ_β^2. Since the term β_j is common to all a measurements on the same subject, the covariance between y_{ij} and $y_{i'j}$ is not, in general, zero. It is customary to assume that the covariance between y_{ij} and $y_{i'j}$ constant across all treatments and subjects.

Consider an analysis of the variance partitioning of the total sum of squares, say

$$\sum_{i=1}^{a} \sum_{j=1}^{n} (y_{ij} - \bar{y}_{..})^2 = a \sum_{j=1}^{n} (\bar{y}_{.j} - \bar{y}_{..})^2 + \sum_{i=1}^{a} \sum_{j=1}^{n} (y_{ij} - \bar{y}_{.j})^2 \qquad (4\text{-}26)$$

We may view the first term on the right-hand side of Equation 4-26 as a sum of squares that results from differences *between subjects* and the second term as a sum of squares of differences *within subjects*. That is,

$$SS_T = SS_{\text{Between Subjects}} + SS_{\text{Within Subjects}}$$

The sums of squares $SS_{\text{Between Subjects}}$ and $SS_{\text{Within Subjects}}$ are statistically independent, with degrees of freedom

$$an - 1 = (n - 1) + n(a - 1)$$

The differences within subjects depend on both differences in treatment effects and uncontrolled variability (noise or error). Therefore, we may decompose the sum of squares resulting from differences within subjects as follows:

$$\sum_{i=1}^{a} \sum_{j=1}^{n} (y_{ij} - \bar{y}_{.j})^2 = n \sum_{i=1}^{a} (\bar{y}_{i.} - \bar{y}_{..})^2$$

$$+ \sum_{i=1}^{a} \sum_{j=1}^{n} (y_{ij} - \bar{y}_{i.} - \bar{y}_{.j} + \bar{y}_{..})^2 \qquad (4\text{-}27)$$

The first term on the right-hand side of Equation 4-27 measures the contribution of the difference between treatment means to $SS_{\text{Within Subjects}}$, and the second term is the residual variation due to error. Both components of $SS_{\text{Within Subjects}}$ are independent. Thus,

$$SS_{\text{Within Subjects}} = SS_{\text{Treatments}} + SS_E$$

with degrees of freedom given by

$$n(a - 1) = (a - 1) + (a - 1)(n - 1)$$

respectively.

To test the hypothesis of no treatment effect, that is,

$$H_0: \quad \tau_1 = \tau_2 = \cdots = \tau_a = 0$$
$$H_1: \quad \text{At least one } \tau_i \neq 0$$

we would use the ratio

$$F_0 = \frac{SS_{\text{Treatment}}/(a - 1)}{SS_E/(a - 1)(n - 1)} = \frac{MS_{\text{Treatments}}}{MS_E} \tag{4-28}$$

If the model errors are normally distributed, then under the null hypothesis, H_0: $\tau_i = 0$, the statistic F_0 follows and $F_{a-1,(a-1)(n-1)}$ distribution. The null hypothesis would be rejected if $F_0 > F_{\alpha,a-1,(a-1)(n-1)}$.

The analysis of variance procedure is summarized in Table 4-13, which also gives convenient computing formulas for the sums of squares. Readers with prior exposure to experimental design will recognize the analysis of variance for a single-factor design with repeated measures as equivalent to the analysis for a randomized complete block design, with subjects considered to be the blocks. Randomized block designs are discussed in the next two chapters.

4-7   PROBLEMS

4-1   Consider the data and the analysis from Problem 3-1. Analyze the residuals from this experiment. Do you think that the analysis of variance assumptions are satisfied?

4-2   Consider the data and the analysis from Problem 3-2. Analyze the residuals from this experiment. Do you think that the analysis of variance assumptions are satisfied?

4-3   Consider the data and the analysis from Problem 3-3. Analyze the residuals from this experiment. Do you think that the analysis of variance assumptions are satisfied?

4-4   Consider the data and the analysis from Problem 3-8. Analyze the residuals from this experiment. Do you think that the analysis of variance assumptions are satisfied?

4-5   Refer to the aluminum smelting experiment described in Section 4-1.6. Verify that

Table 4-13 Analysis of Variance for a Single-Factor Repeated Measures Design

Source of Variation	Sums of Squares	Degrees of Freedom	Mean Square	F_0
1. Between subjects	$\sum_{i=1}^{n} \dfrac{\bar{y}_{.j}^2}{a} - \dfrac{\bar{y}_{..}^2}{an}$	$n - 1$		
2. Within subjects	$\sum_{i=1}^{a}\sum_{j=1}^{n} y_{ij}^2 - \sum_{j=1}^{n} \dfrac{\bar{y}_{.j}^2}{a}$	$n(a - 1)$		
3. (Treatments)	$\sum_{i=1}^{a} \dfrac{\bar{y}_{i.}^2}{n} - \dfrac{\bar{y}_{..}^2}{an}$	$a - 1$	$MS_{\text{Treatment}} = \dfrac{SS_{\text{Treatment}}}{a - 1}$	$\dfrac{MS_{\text{Treatment}}}{MS_E}$
4. (Error)	Subtraction: line (2) − line (3)	$(a - 1)(n - 1)$	$MS_E = \dfrac{SS_E}{(a - 1)(n - 1)}$	
5. Total	$\sum_{i=1}^{a}\sum_{j=1}^{n} y_{ij}^2 - \dfrac{\bar{y}_{..}^2}{an}$	$an - 1$		

ratio control methods do not affect average cell voltage. Construct a normal probability plot of the residuals. Plot the residuals versus the predicted values. Is there an indication that any underlying assumptions are violated?

4-6 Refer to the aluminum smelting experiment in Section 4-1.6. Verify the analysis of variance for pot noise summarized in Table 4-8. Examine the usual residual plots and comment on the experimental validity.

4-7 Four different feed rates were investigated in an experiment on a CNC machine producing a component part used in an aircraft auxiliary power unit. The manufacturing engineer in charge of the experiment knows that a critical part dimension of interest may be affected by the feed rate. However, prior experience has indicated that only dispersion effects are likely to be present. That is, changing the feed rate does not affect the *average* dimension, but it could affect dimensional variability. The engineer makes five production runs at each feed rate and obtains the standard deviation of the critical dimension (in 10^{-3} mm). The data are shown below. Assume that all runs were made in random order.

Feed Rate	Production Run				
(in/min)	1	2	3	4	5
10	0.09	0.10	0.13	0.08	0.07
12	0.06	0.09	0.12	0.07	0.12
14	0.11	0.08	0.08	0.05	0.06
16	0.19	0.13	0.15	0.20	0.11

(a) Does feed rate have any effect on the standard deviation of this critical dimension?

(b) Use the residuals from this experiment to investigate model adequacy. Are there any problems with experimental validity?

4-8 Refer to Problem 3-2.

(a) What is the probability of accepting H_0 if σ_τ^2 is four times the error variance σ^2?

(b) If the difference between looms is large enough to increase the standard deviation of an observation by 20 percent, we wish to detect this with a probability of at least .80. What sample size should be used?

4-9 Refer to Problem 3-5. If we wish to detect a maximum difference in mean response times of 10 milliseconds with a probability of at least .90, what sample size should be used? How would you obtain a preliminary estimate of σ^2?

4-10 Refer to Problem 3-8.

(a) If we wish to detect a maximum difference in mean calcium content of 0.5 percent with a probability of at least .90, what sample size should be used? Discuss how you would obtain a preliminary estimate of σ^2 for answering this question.

(b) If the difference between batches is great enough so that the standard deviation of an observation is increased by 25 percent, what sample size should be used if we wish to detect this with a probability of at least .90?

4-11 Consider the data in Problem 3-11. If we wish to construct a 95 percent confidence interval on the difference in two mean battery lives that has an accuracy of ±2 weeks, how many batteries of each branch must be tested?

4-12 Suppose that four normal populations have means of $\mu_1 = 50$, $\mu_2 = 60$, $\mu_3 = 50$, and $\mu_4 = 60$. How many observations should be taken from each population so that the probability of rejecting the null hypothesis of equal population means is at least .90? Assume that $\alpha = .05$ and that a reasonable estimate of the error variance is $\sigma^2 = 25$.

4-13 Use Bartlett's test to determine if the assumption of equal variances is satisfied in Problem 3-11.

4-14 Apply the general regression significance test to the data in Example 3-1. Show that the procedure yields the same results as the usual analysis of variance.

4-15 Use the Kruskal–Wallis test for the data in Problem 3-7. Compare the conclusions obtained with those from the usual analysis of variance.

4-16 Use the Kruskal–Wallis test for the data in Problem 3-8. Are the results comparable to those found by the usual analysis of variance?

4-17 Consider the data in Example 3-1. Suppose that the largest observation on tensile strength is incorrectly recorded as 50. What effect does this have on the usual analysis of variance? What effect does it have on the Kruskal–Wallis test?

Chapter 5
Randomized Blocks, Latin Squares, and Related Designs

5-1 THE RANDOMIZED COMPLETE BLOCK DESIGN

In many experimental problems, it is necessary to design the experiment so that variability arising from known nuisance sources can be systematically controlled. As an example, suppose we wish to determine whether or not four different tips produce different readings on a hardness testing machine. The machine operates by pressing the tip into a metal test coupon, and from the depth of the resulting depression, the hardness of the coupon can be determined. The experimenter has decided to obtain four observations for each tip. There is only one factor—tip type—and a completely randomized single-factor design would consist of randomly assigning each one of the $4 \times 4 = 16$ runs to an *experimental unit*, that is, a metal coupon, and observing the hardness reading that results. Thus, 16 different metal test coupons would be required in this experiment, one for each run in the design.

There is a potentially serious problem with a completely randomized experiment in this design situation. If the metal coupons differ slightly in their hardness, as might happen if they are taken from ingots that are produced in different heats, then the experimental units (the coupons) will contribute to the variability observed in the hardness data. As a result, the experimental error will reflect *both* random error *and* variability between coupons.

We would like to make the experimental error as small as possible; that is, we would like to remove the variability between coupons from the experimental error. A design that would accomplish this requires the experimenter to test each tip once on each of four coupons. This design, shown in Table 5-1, is called a *randomized complete block design*. The observed response is the Rockwell C scale hardness minus 40. The word "complete" indicates that each block (coupon) contains all the treatments (tips). By using this design, the blocks or coupons

Table 5-1 Randomized Complete Block Design for the Hardness Testing Experiment

Type of Tip	Test Coupon			
	1	2	3	4
1	9.3	9.4	9.6	10.0
2	9.4	9.3	9.8	9.9
3	9.2	9.4	9.5	9.7
4	9.7	9.6	10.0	10.2

form a more homogeneous experimental unit on which to compare the tips. Effectively, this design strategy improves the accuracy of the comparisons among tips by eliminating the variability among the coupons. Within a block, the order in which the four tips are tested is randomly determined. Notice the similarity of this design problem to the one of Section 2-5 in which the paired t test was discussed. The randomized complete block design is a generalization of that concept.

The randomized complete block design is one of the most widely used experimental designs. Situations for which the randomized complete block design is appropriate are numerous and easily detected with practice. Units of test equipment or machinery are often different in their operating characteristics and would be a typical blocking factor. Batches of raw material, people, and time are also common nuisance sources of variability in an experiment that can be systematically controlled through blocking.

5-1.1 Statistical Analysis

Suppose we have, in general, a treatments that are to be compared and b blocks. The randomized complete block design is shown in Figure 5-1. There is one observation per treatment in each block, and the order in which the treatments are run within each block is determined randomly. Because the only randomization of treatments is within the blocks, we often say that the blocks represent a *restriction* on randomization.

The statistical model for this design is

$$y_{ij} = \mu + \tau_i + \beta_j + \epsilon_{ij} \begin{cases} i = 1,2, \ldots, a \\ j = 1,2, \ldots, b \end{cases} \tag{5-1}$$

where μ is an overall mean, τ_i is the effect of the ith treatment, β_j is the effect in the jth block, and ϵ_{ij} is the usual NID$(0, \sigma^2)$ random error term. Treatments and blocks are considered initially as fixed factors. Furthermore, the treatment and block effects are defined as deviations from the overall mean so that

$$\sum_{i=1}^{a} \tau_i = 0$$

Figure 5-1. The randomized complete block design.

and

$$\sum_{j=1}^{b} \beta_j = 0$$

We are interested in testing the equality of the treatment means. Thus, the hypotheses of interest are

$$H_0 : \mu_1 = \mu_2 = \cdots = \mu_a$$

$$H_1 : \text{at least one } \mu_i \neq \mu_j$$

Since the ith treatment mean $\mu_i = (1/b) \sum_{j=1}^{b} (\mu + \tau_i + \beta_j) = \mu + \tau_i$, an equivalent way to write the above hypotheses is in terms of the treatment effects, say

$$H_0 : \tau_1 = \tau_2 = \cdots = \tau_a = 0$$

$$H_1 : \tau_i \neq 0 \text{ at least one } i$$

Let $y_{i.}$ be the total of all observations taken under treatment i, $y_{.j}$ be the total of all observations in block j, $y_{..}$ be the grand total of all observations, and $N = ab$ be the total number of observations. Expressed mathematically,

$$y_{i.} = \sum_{j=1}^{b} y_{ij} \qquad i = 1, 2, \ldots, a \qquad (5\text{-}2)$$

$$y_{.j} = \sum_{i=1}^{a} y_{ij} \qquad j = 1, 2, \ldots, b \qquad (5\text{-}3)$$

and

$$y_{..} = \sum_{i=1}^{a} \sum_{j=1}^{b} y_{ij} = \sum_{i=1}^{a} y_{i.} = \sum_{j=1}^{b} y_{.j} \qquad (5\text{-}4)$$

Similarly, $\bar{y}_{i.}$ is the average of the observations taken under treatment i, $\bar{y}_{.j}$ is the average of the observations in block j, and $\bar{y}_{..}$ is the grand average of all observations That is,

$$\bar{y}_{i.} = y_{i.}/b \qquad \bar{y}_{.j} = y_{.j}/a \qquad \bar{y}_{..} = y_{..}/N \qquad (5\text{-}5)$$

We may express the total corrected sum of squares as

$$\sum_{i=1}^{a} \sum_{j=1}^{b} (y_{ij} - \bar{y}_{..})^2 = \sum_{i=1}^{a} \sum_{j=1}^{b} [(\bar{y}_{i.} - \bar{y}_{..}) $$
$$ + (\bar{y}_{.j} - \bar{y}_{..}) + (y_{ij} - \bar{y}_{i.} - \bar{y}_{.j} + \bar{y}_{..})]^2 \qquad (5\text{-}6)$$

By expanding the right-hand side of Equation 5-6, we obtain

$$\sum_{i=1}^{a} \sum_{j=1}^{b} (y_{ij} - \bar{y}_{..})^2 = b \sum_{i=1}^{a} (\bar{y}_{i.} - \bar{y}_{..})^2 + a \sum_{j=1}^{b} (\bar{y}_{.j} - \bar{y}_{..})^2$$

$$+ \sum_{i=1}^{a} \sum_{j=1}^{b} (y_{ij} - \bar{y}_{i.} - \bar{y}_{.j} + \bar{y}_{..})^2 + 2 \sum_{i=1}^{a} \sum_{j=1}^{b} (\bar{y}_{i.} - \bar{y}_{..})(\bar{y}_{.j} - \bar{y}_{..})$$

$$+ 2 \sum_{i=1}^{a} \sum_{j=1}^{b} (\bar{y}_{.j} - \bar{y}_{..})(y_{ij} - \bar{y}_{i.} - \bar{y}_{.j} + \bar{y}_{..})$$

$$+ 2 \sum_{i=1}^{a} \sum_{j=1}^{b} (\bar{y}_{i.} - \bar{y}_{..})(y_{ij} - \bar{y}_{i.} - \bar{y}_{.j} + \bar{y}_{..})$$

Simple but tedious algebra proves that the three cross products are zero. Therefore,

$$\sum_{i=1}^{a} \sum_{j=1}^{b} (y_{ij} - \bar{y}_{..})^2 = b \sum_{i=1}^{a} (\bar{y}_{i.} - \bar{y}_{..})^2 + a \sum_{j=1}^{b} (\bar{y}_{.j} - \bar{y}_{..})^2$$
$$+ \sum_{i=1}^{a} \sum_{j=1}^{b} (y_{ij} - \bar{y}_{.j} - \bar{y}_{i.} + \bar{y}_{..})^2 \qquad (5\text{-}7)$$

represents a partition of the total sum of squares. Expressing the sums of squares in Equation 5-7 symbolically, we have

$$SS_T = SS_{\text{Treatments}} + SS_{\text{Blocks}} + SS_E \qquad (5\text{-}8)$$

Since there are N observations, SS_T has $N - 1$ degrees of freedom. There are a treatments and b blocks, so $SS_{\text{Treatments}}$ and SS_{Blocks} have $a - 1$ and $b - 1$ degrees of freedom, respectively. The error sum of squares is just a sum of squares between cells minus the sum of squares for treatments and blocks. There are ab cells with $ab - 1$ degrees of freedom between them, so SS_E has $ab - 1 - (a - 1) - (b - 1) = (a - 1)(b - 1)$ degrees of freedom. Furthermore, the degrees of freedom on the right-hand side of Equation 5-8 add to the total on the left; therefore, making the usual normality assumptions on the errors, one may use Theorem 3-1 to show that $SS_{\text{Treatments}}/\sigma^2$, $SS_{\text{Blocks}}/\sigma^2$, and SS_E/σ^2 are independently distributed chi-square random variables. Each sum of squares divided by its degrees of freedom is a mean square. The expected value of the mean squares, if treatments and blocks are fixed, can be shown to be

$$E(MS_{\text{Treatments}}) = \sigma^2 + \frac{b \sum\limits_{i=1}^{a} \tau_i^2}{a - 1}$$

$$E(MS_{\text{Blocks}}) = \sigma^2 + \frac{a \sum\limits_{j=1}^{b} \beta_j^2}{b - 1}$$

$$E(MS_E) = \sigma^2$$

Therefore, to test the equality of treatment means, we would use the test statistic

$$F_0 = \frac{MS_{\text{Treatments}}}{MS_E}$$

which is distributed as $F_{a-1,(a-1)(b-1)}$ if the null hypothesis is true. The critical region is the upper tail of the F distribution, and we would reject H_0 if $F_0 > F_{\alpha,a-1,(a-1)(b-1)}$.

We may also be interested in comparing block means since, if these means do not differ greatly, blocking may not be necessary in future experiments. From the expected mean squares, it seems that the hypothesis $H_0 : \beta_j = 0$ may be tested by comparing the statistic $F_0 = MS_{\text{Blocks}}/MS_E$ to $F_{\alpha,b-1,(a-1)(b-1)}$. However, recall that randomization has been applied only to treatments *within* blocks; that is, the blocks represent a *restriction on randomization*. What effect does this have on the statistic $F_0 = MS_{\text{Blocks}}/MS_E$? Some difference in treatments of this question exists. For example, Box, Hunter, and Hunter (1978) point out that the usual analysis of variance F test can be justified on the basis of randomization only,[1] without direct use of the normality assumption. They further observe that the test to compare block means cannot appeal to such a justification because of the randomization restriction; but if the errors are NID(0, σ^2), then the statistic $F_0 = MS_{\text{Blocks}}/MS_E$ can be used to compare block means. On the other hand, Anderson and McLean (1974) argue that the randomization restriction prevents this statistic from being a meaningful test for comparing block means and that this F ratio really is a test for the equality of the block means plus the randomization restriction [which they call a restriction error; see Anderson and McLean (1974) for further details.]

In practice, then, what do we do? Since the normality assumption is often questionable, to view $F_0 = MS_{\text{Blocks}}/MS_E$ as an exact F test on the equality of block means is not a good general practice. For that reason, we exclude this F test from the analysis of variance table. However, as an approximate procedure to investigate the effect of the blocking variable, examining the ratio of MS_{Blocks} to MS_E is certainly reasonable. If this ratio is large, it implies that the blocking factor has a large effect and that the noise reduction obtained by blocking was

---

[1] Actually, the normal-theory F distribution is an approximation to the randomization distribution generated by calculating F_0 from every possible assignment of the responses to the treatments.

Table 5-2   Analysis of Variance for a Randomized Complete Block Design

Source of Variation	Sum of Squares	Degrees of Freedom	Mean Square	F_0
Treatments	$\sum \dfrac{y_{i.}^2}{b} - \dfrac{y_{..}^2}{N}$	$a - 1$	$\dfrac{SS_{Treatments}}{a - 1}$	$\dfrac{MS_{Treatments}}{MS_E}$
Blocks	$\sum \dfrac{y_{.j}^2}{a} - \dfrac{y_{..}^2}{N}$	$b - 1$	$\dfrac{SS_{Blocks}}{b - 1}$	
Error	SS_E(by subtraction)	$(a - 1)(b - 1)$	$\dfrac{SS_E}{(a - 1)(b - 1)}$	
Total	$\sum \sum y_{ij}^2 - \dfrac{y_{..}^2}{N}$	$N - 1$		

probably helpful in improving the precision of the comparison of treatment means.

The procedure is usually summarized in an analysis of variance table, such as the one shown in Table 5-2. Computing formulas for the sums of squares may be obtained for the elements in Equation 5-7 by expressing them in terms of treatment and block totals. These computing formulas are

$$SS_T = \sum_{i=1}^{a} \sum_{j=1}^{b} y_{ij}^2 - \frac{y_{..}^2}{N} \tag{5-9}$$

$$SS_{Treatments} = \sum_{i=1}^{a} \frac{y_{i.}^2}{b} - \frac{y_{..}^2}{N} \tag{5-10}$$

$$SS_{Blocks} = \sum_{j=1}^{b} \frac{y_{.j}^2}{a} - \frac{y_{..}^2}{N} \tag{5-11}$$

and the error sum of squares is obtained by subtraction as

$$SS_E = SS_T - SS_{Treatments} - SS_{Blocks} \tag{5-12}$$

Example 5-1

Consider the hardness testing experiment described in Section 5-1. There are four tips and four available metal coupons. Each tip is tested once on each coupon, resulting in a randomized complete block design. The data obtained are repeated for convenience in Table 5-3. Remember that the *order* in which the tips were tested on a particular coupon was determined randomly. To simplify the calculations, we code the original data by subtracting 9.5 from each observation and multiplying the result by 10. This yields the data in Table 5-4. The sums of squares are obtained as follows:

$$SS_T = \sum_{i=1}^{4} \sum_{j=1}^{4} y_{ij}^2 - \frac{y_{..}^2}{N}$$

$$= 154.00 - \frac{(20)^2}{16} = 129.00$$

Table 5-3 Randomized Complete Block Design for the Hardness Testing Experiment

Type of Tip	Coupon (Block)			
	1	2	3	4
1	9.3	9.4	9.6	10.0
2	9.4	9.3	9.8	9.9
3	9.2	9.4	9.5	9.7
4	9.7	9.6	10.0	10.2

$$SS_{Treatments} = \sum_{i=1}^{4} \frac{y_{i.}^2}{b} - \frac{y_{..}^2}{N}$$

$$= \frac{(3)^2 + (4)^2 + (-2)^2 + (15)^2}{4} - \frac{(20)^2}{16} = 38.50$$

$$SS_{Blocks} = \sum_{j=1}^{4} \frac{y_{.j}^2}{a} - \frac{y_{..}^2}{N}$$

$$= \frac{(-4)^2 + (-3)^2 + (9)^2 + (18)^2}{4} - \frac{(20)^2}{16} = 82.50$$

$$SS_E = SS_T - SS_{Treatments} - SS_{Blocks}$$

$$= 129.00 - 38.50 - 82.50 = 8.00$$

The analysis of variance is shown in Table 5-5. Using $\alpha = .05$, the critical value of F is $F_{.05, 3, 9} = 3.86$. Since $14.44 > 3.86$, we conclude that the type of tip affects the mean hardness reading. Also, the coupons (blocks) seem to differ significantly, since the mean square for blocks is large relative to error.

It is interesting to observe the results we would have obtained had we not been aware of randomized block designs. Suppose we used four coupons, randomly assigned the tips to each, and (by chance) the same design resulted as in Table 5-3. The incorrect analysis of these data as a completely randomized single-factor design is shown in Table 5-6. Since $F_{.05, 3, 12} = 3.49$, the hypothesis of equal mean hardness measurements from the four tips cannot be rejected. Thus, the randomized block design reduces the amount of noise in the data sufficiently for differences among the four tips to be detected.

Table 5-4 Coded Data for the Hardness Testing Experiment

Type of Tip	Coupon (Block)				
	1	2	3	4	$y_{i.}$
1	−2	−1	1	5	3
2	−1	−2	3	4	4
3	−3	−1	0	2	−2
4	2	1	5	7	15
$y_{.j}$	−4	−3	9	18	20 = $y_{..}$

Table 5-5   Analysis of Variance for the Hardness Testing Experiment

Source of Variation	Sum of Squares	Degrees of Freedom	Mean Square	F_0
Treatments (type of tip)	38.50	3	12.83	14.44
Blocks (coupons)	82.50	3	27.50	
Error	8.00	9	0.89	
Total	129.00	15		

Additivity of the Randomized Block Model ▪ The linear model that we have used for the randomized block design

$$y_{ij} = \mu + \tau_i + \beta_j + \epsilon_{ij}$$

is completely *additive*. This says that, for example, if the first treatment causes the expected response to increase by five units ($\tau_1 = 5$) and if the first block increases the expected response by 2 units ($\beta_1 = 2$), then the expected increase in response of *both* treatment 1 *and* block 1 together is $E(y_{11}) = \mu + \tau_1 + \beta_1 = \mu + 5 + 2 = \mu + 7$. In general, treatment 1 *always* increases the expected response by 5 units over the sum of the overall mean and the block effect.

Although this simple additive model is often useful, there are situations where it is inadequate. Suppose, for example, that we are comparing four formulations of a chemical product using six batches of raw material; the raw material batches are considered blocks. If an impurity in batch 2 affects formulation 2 adversely, resulting in an unusually low yield, but does not affect the other formulations, then an *interaction* between formulations (or treatments) and batches (or blocks) has occurred. Similarly, interactions between treatments and blocks can occur when the response is measured on the wrong scale. Thus, a relationship that is multiplicative in the original units, say

$$E(y_{ij}) = \mu \tau_i \beta_j$$

is linear or additive in a log scale since, for example,

$$\ln E(y_{ij}) = \ln \mu + \ln \tau_i + \ln \beta_j$$

Table 5-6   Incorrect Analysis of the Hardness Testing Experiment as a Completely Randomized Design

Source of Variation	Sum of Squares	Degrees of Freedom	Mean Square	F_0
Type of tip	38.50	3	12.83	1.70
Error	90.50	12	7.54	
Total	129.00	15		

or

$$E(y_{ij}^*) = \mu^* + \tau_i^* + \beta_j^*$$

Although this type of interaction can be eliminated by a transformation, not all interactions are so easily treated. For example, transformations do not eliminate the formulation–batch interaction discussed previously. Residual analysis and other diagnostic checking procedures can be helpful in detecting nonadditivity.

If interaction is present, it can seriously affect and possibly invalidate the analysis of variance. In general, the presence of interaction inflates the error mean square and may adversely affect the comparison of treatment means. In situations where both factors, as well as their possible interaction, are of interest, *factorial designs* must be used. These designs are introduced in Chapter 7.

Random Treatments and Blocks ▪ Although we have described the test procedure considering treatments and blocks as fixed factors, the same analysis procedure is used if either the treatments or blocks (or both) are random. There are, of course, corresponding changes in the interpretation of the results. For example, if the blocks are random, then we expect the comparisons among the treatments to be the same throughout the *population* of blocks from which those used in the experiment were randomly selected. There are also corresponding changes in the expected mean squares. For example, if the blocks are independent random variables with common variance, then $E(MS_{Blocks}) = \sigma^2 + a\sigma_\beta^2$, where σ_β^2 is the variance component of the block effects. In any event, however, $E(MS_{Treatments})$ is always free of any block effect, and the test statistic for between-treatment variability is always $F_0 = MS_{Treatments}/MS_E$.

In situations where the blocks are random, if treatment–block interaction is present, then the test on treatment means is unaffected by the interaction. The reason for this is that the expected mean squares for treatments and error *both* contain the interaction effect; consequently, the test for differences in treatment means may be conducted as usual by comparing the treatment mean square to the error mean square. This procedure does not yield any information on the interaction.

Multiple Comparisons ▪ If the treatments in a randomized block design are fixed, and the analysis indicates a significant difference in treatment means, then the experimenter is usually interested in multiple comparisons to discover *which* treatments means differ. Any of the multiple comparison procedures discussed in Chapter 3 (Section 3-4) may be used for this purpose. In the formulas of Section 3-4, simply replace the number of replicates in the single-factor completely randomized design (n) by the number of blocks (b). Also, remember to use the number of error degrees of freedom for the randomized block $[(a - 1)(b - 1)]$ instead of those for the completely randomized design $[a(n - 1)]$. For example, if we wish to use Duncan's multiple range test, the standard error of a treatment mean is

$$S_{\bar{y}_{i.}} = \sqrt{\frac{MS_E}{b}}$$

and Appendix Table VII would be entered with $(a - 1)(b - 1)$ degrees of freedom to find the significant ranges.

Example 5-2

Since we concluded in Example 5-1 that the mean hardness readings differ from tip to tip, it is of interest to discover the specific nature of the differences. The coded tip type or treatment means, in ascending order, are as follows:

$$\bar{y}_{3.} = -0.50 \qquad \bar{y}_{1.} = 0.75 \qquad \bar{y}_{2.} = 1.00 \qquad \bar{y}_{4.} = 3.75$$

and the standard error of a tip type or treatment mean is

$$S_{\bar{y}_{i.}} = \sqrt{\frac{MS_E}{b}} = \sqrt{\frac{0.89}{4}} = 0.47$$

Appendix Table VII provides the following significant ranges

$$r_{.05}(2, 9) = 3.20 \qquad r_{.05}(3, 9) = 3.34 \qquad \text{and} \qquad r_{.05}(4, 9) = 3.41$$

Thus, the least significant ranges are

$$R_2 = r_{.05}(2, 9)S_{\bar{y}_{i.}} = (3.20)(0.47) = 1.50$$
$$R_3 = r_{.05}(3, 9)S_{\bar{y}_{i.}} = (3.34)(0.47) = 1.57$$
$$R_4 = r_{.05}(4, 9)S_{\bar{y}_{i.}} = (3.41)(0.47) = 1.60$$

and the pairwise comparisons are

$$4 \text{ vs. } 3: \quad 3.75 - (-0.50) = 4.25 > 1.60(R_4)$$
$$4 \text{ vs. } 1: \quad 3.75 - 0.75 = 3.00 > 1.57(R_3)$$
$$4 \text{ vs. } 2: \quad 3.75 - 1.00 = 2.75 > 1.50(R_2)$$
$$2 \text{ vs. } 3: \quad 1.00 - (-0.50) = 1.50 < 1.57(R_3)$$

Note that since the comparison of $\bar{y}_{2.}$ and $\bar{y}_{3.}$ spans the remaining means and this comparison indicates that $H_0: \mu_2 = \mu_3$ cannot be rejected, we can immediately conclude that all other pairs of means do not differ (i.e., $\mu_1 = \mu_2 = \mu_3$). Thus, tip type 4 produces a mean hardness reading that is significantly higher than the mean readings obtained from the other three types of tips.

∎

To illustrate further multiple comparisons in the randomized complete block design, Figure 5-2 plots the four tip type means from Example 5-1 relative to a scaled t distribution with a scale factor $\sqrt{MS_E/b} = \sqrt{0.89/4} = 0.47$. This plot indicates that tips 1, 2, and 3 probably produce identical average hardness measurements but that tip 4 produces a much higher mean hardness. This figure confirms the results from the Duncan multiple range test in Example 5-2.

Choice of Sample Size ▪ Choosing the sample size, or the number of *blocks* to run, is an important decision when using a randomized block design. Increasing the number of blocks increases the number of replicates and the number of error degrees of freedom, making the design more sensitive. Any of the techniques

Figure 5-2. Tip type means relative to a scaled t distribution with a scale factor $\sqrt{MS_E/b} = \sqrt{0.89/4} = 0.47$.

discussed in Chapter 4 (Section 4-2) for selecting the number of replicates to run in a completely randomized single-factor experiment may be applied directly to the randomized block design. For the case of a fixed factor, the operating characteristic curves in Appendix Chart V may be used with

$$\Phi^2 = \frac{b \sum\limits_{i=1}^{a} \tau_i^2}{a\sigma^2}$$

where there are $a - 1$ numerator degrees of freedom and $(a - 1)(b - 1)$ denominator degrees of freedom. If the factor is random, then we use Appendix Chart VI with

$$\lambda = \sqrt{1 + \frac{b\sigma_\tau^2}{\sigma^2}}$$

where there are $a - 1$ numerator degrees of freedom and $(a - 1)(b - 1)$ denominator degrees of freedom.

Example 5-3

Consider the hardness testing problem described in Example 5-1. Suppose that we wish to determine the appropriate number of blocks to run if we are interested in detecting a true maximum difference in mean hardness readings of 0.4 with a high probability, and a reasonable estimate of the standard deviation of the errors is $\sigma = 0.1$. (These values are given in the *original* units; recall that the analysis of variance was performed using *coded* data.) From Equation 4-12, the minimum value of Φ^2 is (writing b, the number of blocks, for n)

$$\Phi^2 = \frac{bD^2}{2a\sigma^2}$$

where D is the maximum difference we wish to detect. Thus,

$$\Phi^2 = \frac{b(0.4)^2}{2(4)(0.1)^2} = 2.0b$$

If we use $b = 3$ blocks, then $\Phi = \sqrt{2.0b} = \sqrt{2.0(3)} = 2.45$, and there are $(a - 1)(b - 1) = 3(2) = 6$ error degrees of freedom. Appendix Chart V with $v_1 = a - 1 = 3$ and $\alpha = .05$ indicates that the β risk for this design is approximately .10 (power $= 1 - \beta = .90$). If we use $b = 4$ blocks, then $\Phi = \sqrt{2.0b} = \sqrt{2.0(4)} = 2.83$, with $(a - 1)(b - 1) = 3(3) = 9$ error degrees of freedom, and the corresponding β risk is approximately .03 (power $= 1 - \beta = .97$). Either three or four blocks will result in a design having a high probability of detecting the difference between the mean hardness readings considered important. Since coupons (blocks) are inexpensive and readily available and the cost of testing is low, the experimenter decides to use four blocks.

■

Relative Efficiency of the Randomized Block Design ▪ In Example 5-1 we illustrated the noise-reducing property of the randomized block design. If we look at the portion of the total sum of squares not accounted for by treatments (90.50; see Table 5-6), over 90 percent (82.50) is the result of differences between blocks. Thus, if we had run a completely randomzied design, MS_E would have been much larger, and the resulting design would not have been as sensitive as the randomized block design.

It is often helpful to estimate the relative efficiency of the randomized block design compared to a completely randomized design. One way to define this relative efficiency is

$$R = \frac{(df_b + 1)(df_r + 3)}{(df_b + 3)(df_r + 1)} \cdot \frac{\sigma_r^2}{\sigma_b^2} \qquad (5\text{-}13)$$

where σ_r^2 and σ_b^2 are the experimental error variances of the completely randomized and randomized block designs, respectively, and df_r and df_b are the corresponding error degrees of freedom. This statistic may be viewed as the increase in replications that is required if a completely randomized design is used as compared to a randomized block if the two designs are to have the same sensitivity. The ratio of degrees of freedom in R is an adjustment to reflect the different number of error degrees of freedom in the two designs.

To compute the relative efficiency, we must have estimates of σ_r^2 and σ_b^2. We can use MS_E from the randomized block design to estimate σ_b^2, and it may be shown [see Cochran and Cox (1957), pp. 112–114] that

$$\hat{\sigma}_r^2 = \frac{(b - 1)MS_{\text{Blocks}} + b(a - 1)MS_E}{ab - 1} \qquad (5\text{-}14)$$

is an unbiased estimator of the error variance of a completely randomized design. To illustrate the procedure, consider the data in Example 5-1. Since $MS_E = 0.89$, we have

$$\hat{\sigma}_b^2 = 0.89$$

and, from Equation 5-14,

$$\hat{\sigma}_r^2 = \frac{(b - 1)MS_{Blocks} + b(a - 1)MS_E}{ab - 1}$$

$$= \frac{(3)27.50 + 4(3)0.89}{(4)4 - 1}$$

$$= 6.21$$

Therefore, using Equation 5-13, our estimate of the relative efficiency of the randomized block design in this example is

$$R = \frac{(df_b + 1)(df_r + 3)}{(df_b + 3)(df_r + 1)} \frac{\hat{\sigma}_r^2}{\hat{\sigma}_b^2} = \frac{(9 + 1)(12 + 3)}{(9 + 3)(12 + 1)} \frac{6.21}{0.89}$$

$$= 6.71$$

This implies that we would have to use about seven times as many replicates with a completely randomized design to obtain the same sensitivity as is obtained by blocking on the metal coupons.

Clearly, blocking has paid off handsomely in this experiment. However, suppose that blocking was not really necessary. In such cases, if experimenters choose to block, what do they stand to lose? In general, the randomized block design has $(a - 1)(b - 1)$ error degrees of freedom. If blocking was unnecessary and the experiment was run as a completely randomized design with b replicates, we would have had $a(b - 1)$ degrees of freedom for error. Thus, incorrectly blocking has cost $a(b - 1) - (a - 1)(b - 1) = b - 1$ degrees of freedom for error, and the test on treatment means has been made less sensitive needlessly. However, if block effects really are large, then the experimental error may be so inflated that significant differences in treatment means would remain undetected. (Remember the incorrect analysis of Example 5-1.) As a general rule, when the importance of block effects is in doubt, the experimenter should block and gamble that the block means differ. If the experimenter is wrong, the slight loss in error degrees of freedom will have little effect on the outcome as long as a moderate number of degrees of freedom for error are available.

5-1.2 Model Adequacy Checking

We have previously discussed the importance of checking the adequacy of the assumed model (see Chapter 4, Section 4-1). Generally, we should be alert for potential problems with the normality assumption, unequal error variance by treatment or block, and block–treatment interaction. As in the completely randomized design, residual analysis is the major tool used in this diagnostic checking. The residuals for a randomized block design are

$$e_{ij} = y_{ij} - \hat{y}_{ij}$$

and, as we find in the next section, the fitted values are $\hat{y}_{ij} = \bar{y}_{i.} + \bar{y}_{.j} - \bar{y}_{..}$, so

$$e_{ij} = y_{ij} - \bar{y}_{i.} - \bar{y}_{.j} + \bar{y}_{..} \qquad (5\text{-}15)$$

The observations, fitted values, and residuals for the coded hardness testing data in Example 5-1 are as follows:

y_{ij}	$\hat{y}_{ij}$	e_{ij}
−2.00	−1.50	−0.50
−1.00	−1.25	0.25
1.00	1.75	−0.75
5.00	4.00	1.00
−1.00	−1.25	0.25
−2.00	−1.00	−1.00
3.00	2.00	1.00
4.00	4.25	−0.25
−3.00	−2.75	−0.25
−1.00	−2.50	1.50
0.00	0.50	−0.50
2.00	2.75	−0.75
2.00	1.50	0.50
1.00	1.75	−0.75
5.00	4.75	0.25
7.00	7.00	0.00

A normal probability plot and a dot diagram of these residuals are shown in Figure 5-3. There is no severe indication of nonnormality, nor is there any evidence pointing to possible outliers. Figure 5-4 shows plots of the residuals by type of tip or treatment and by coupon or block. These plots could be, potentially, very informative. If there is more scatter in the residuals for a particular tip, that could indicate that this tip produces more erratic hardness readings than the others. More scatter in the residuals for a particular test coupon could indicate that the coupon is not of uniform hardness. However, in our example, Figure 5-4 gives no indication of inequality of variance by treatment or by block. Figure 5-5 plots the residuals versus the fitted values $\hat{y}_{ij}$. There should be no relationship between the size of the residuals and the fitted values $\hat{y}_{ij}$. This plot reveals nothing of unusual interest.

Sometimes the plot of residuals versus $\hat{y}_{ij}$ has a curvilinear shape; for example, there may be a tendency for negative residuals to occur with low $\hat{y}_{ij}$ values, positive residuals with intermediate $\hat{y}_{ij}$ values, and negative residuals with high $\hat{y}_{ij}$ values. This type of pattern is suggestive of *interaction* between blocks and treatments. If this pattern occurs, a transformation should be used in an effort to eliminate or minimize the interaction. In Chapter 7 (Section 7-3.7), we describe

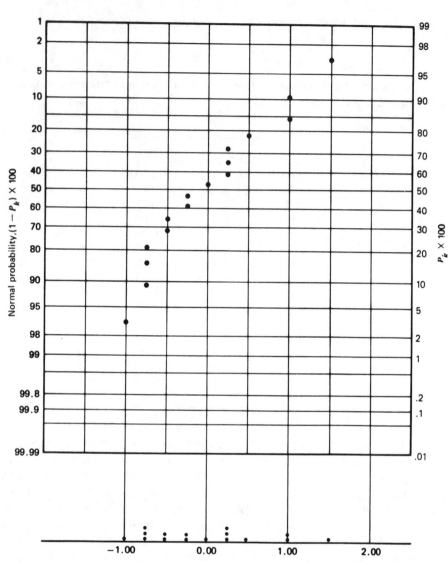

Figure 5-3.   Normal probability plot and dot diagram of residuals for Example 5-1.

a statistical test that can be used to detect the presence of interaction in a randomized block design.

5-1.3   Estimating Missing Values

When using the randomized complete block design, sometimes an observation in one of the blocks is missing. This may happen because of carelessness or error

Figure 5-4.   Plot of residuals by tip-type (treatment) and by coupon (block) for Example 5-1.

or for reasons beyond our control, such as unavoidable damage to an experimental unit. A missing observation introduces a new problem into the analysis since treatments are no longer *orthogonal* to blocks; that is, every treatment does not occur in every block. There are two general approaches to the missing value problem. The first is an approximate analysis in which the missing observation is estimated and the usual analysis of variance is performed just as if the estimated observation were real data, with the error degrees of freedom reduced by one. This approximate analysis is the subject of this section. The second is an exact analysis, which is discussed in Section 5-1.4.

Suppose the observation y_{ij} for treatment i in block j is missing. Denote the missing observation by x. As an illustration, suppose that in the hardness testing experiment of Example 5-1 specimen 3 was broken while tip 2 was being tested and that data point could not be obtained. The data might appear as in Table 5-7.

In general, we will let $y'_{..}$ represent the grand total with one missing observation, $y'_{i.}$ represent the total for the treatment with one missing observation, and $y'_{.j}$ be the total for the block with one missing observation. Suppose we wish to

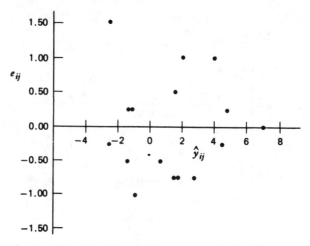

Figure 5-5.   Plot of residuals versus $\hat{y}_{ij}$ for Example 5-1.

estimate the missing observation x so that x will have a minimum contribution to the error sum of squares. Since $SS_E = \sum_{i=1}^{a}\sum_{j=1}^{b}(y_{ij} - \bar{y}_{i.} - \bar{y}_{.j} + \bar{y}_{..})^2$, this is equivalent to choosing x in order to minimize

$$SS_E = \sum_{i=1}^{a}\sum_{j=1}^{b} y_{ij}^2 - \frac{1}{b}\sum_{i=1}^{a}\left(\sum_{j=1}^{b} y_{ij}\right)^2 - \frac{1}{a}\sum_{j=1}^{b}\left(\sum_{i=1}^{a} y_{ij}\right)^2 + \frac{1}{ab}\left(\sum_{i=1}^{a}\sum_{j=1}^{b} y_{ij}\right)^2$$

or

$$SS_E = x^2 - \frac{1}{b}(y'_{i.} + x)^2 - \frac{1}{a}(y'_{.j} + x)^2 + \frac{1}{ab}(y'_{..} + x)^2 + R \quad (5\text{-}16)$$

where R includes all terms not involving x. From $dSS_E/dx = 0$, we obtain

$$x = \frac{ay'_{i.} + by'_{.j} - y'_{..}}{(a - 1)(b - 1)} \quad (5\text{-}17)$$

as the estimate of the missing observation.

For the data in Table 5-7, we find that $y'_{2.} = 1$, $y'_{.3} = 6$, and $y'_{..} = 17$. Therefore, from Equation 5-17,

$$x \equiv y_{23} = \frac{4(1) + 4(6) - 17}{(3)(3)} = 1.22$$

The usual analysis of variance may now be performed using $y_{23} = 1.22$ and reducing the error degrees of freedom by one. The analysis of variance is shown in Table 5-8. Compare the results of this approximate analysis with the results obtained for the full data set (Table 5-5).

If several observations are missing, they may be estimated by writing the error sum of squares as a function of the missing values, differentiating with

Table 5-7 Randomized Complete Block Design for the Hardness Testing Experiment with One Missing Value

Type of Tip	Specimen (Block)			
	1	2	3	4
1	−2	−1	1	5
2	−1	−2	x	4
3	−3	−1	0	2
4	2	1	5	7

respect to each missing value, equating the results to zero, and solving the resulting equations. Alternatively, we may use Equation 5-17 iteratively to estimate the missing values. To illustrate the iterative approach, suppose that two values are missing. Arbitrarily estimate the first missing value, and then use this value along with the real data and Equation 5-17 to estimate the second. Now Equation 5-17 can be used to reestimate the first missing value, and following this, the second can be reestimated. This process is continued until convergence is obtained. In any missing value problem, the error degrees of freedom are reduced by one for each missing observation.

5-1.4 Estimating Model Parameters and the General Regression Significance Test

If both treatments and blocks are fixed, we may estimate the parameters in the randomized complete block design model by least squares. Recall that the linear statistical model is

$$y_{ij} = \mu + \tau_i + \beta_j + \epsilon_{ij} \begin{cases} i = 1,2, \ldots, a \\ j = 1,2, \ldots, b \end{cases}$$

Table 5-8 Approximate Analysis of Variance for Example 5-1 with one Missing Value

Source of Variation	Sum of Squares	Degrees of Freedom	Mean Square	F_0
Type of tip	39.98	3	13.33	17.12[a]
Specimens (blocks)	79.53	3	26.51	
Error	6.22	8	0.78	
Total	125.73	14		

[a]Significant at 5 percent.

Applying the rules in Section 4-4 in writing the normal equations for an experimental design model, we obtain

$$
\begin{aligned}
\mu: &\ ab\hat{\mu} + b\hat{\tau}_1 + b\hat{\tau}_2 + \cdots + b\hat{\tau}_a + a\,\hat{\beta}_1 + a\hat{\beta}_2 + \cdots + a\hat{\beta}_b = y_{..} \\
\tau_1: &\ b\hat{\mu} + b\hat{\tau}_1 \qquad\qquad\qquad\qquad + \ \hat{\beta}_1 + \ \hat{\beta}_2 + \cdots + \ \hat{\beta}_b = y_{1.} \\
\tau_2: &\ b\hat{\mu} \qquad\quad + b\hat{\tau}_2 \qquad\qquad\qquad + \ \hat{\beta}_1 + \ \hat{\beta}_2 + \cdots + \ \hat{\beta}_b = y_{2.} \\
&\ \ \vdots \qquad\qquad\qquad\qquad\qquad\qquad\qquad\vdots \qquad\qquad\qquad\vdots \\
\tau_a: &\ b\hat{\mu} \qquad\qquad\qquad\quad + b\hat{\tau}_a + \ \hat{\beta}_1 + \ \hat{\beta}_2 + \cdots + \ \hat{\beta}_b = y_{a.} \quad (5\text{-}18) \\
\beta_1: &\ a\hat{\mu} + \ \hat{\tau}_1 + \ \hat{\tau}_2 + \cdots + \ \hat{\tau}_a + a\,\hat{\beta}_1 \qquad\qquad\qquad\quad = y_{.1} \\
\beta_2: &\ a\hat{\mu} + \ \hat{\tau}_1 + \ \hat{\tau}_2 + \cdots + \ \hat{\tau}_a \qquad\quad + a\hat{\beta}_2 \qquad\qquad\ = y_{.2} \\
&\ \ \vdots \qquad\qquad\qquad\qquad\qquad\qquad\vdots \qquad\qquad\qquad\vdots \\
\beta_b: &\ a\hat{\mu} + \ \hat{\tau}_1 + \ \hat{\tau}_2 + \cdots + \ \hat{\tau}_a \qquad\qquad\qquad\quad + a\hat{\beta}_b = y_{.b}
\end{aligned}
$$

Notice that the second through the $(a + 1)$st equations in Equation 5-18 sum to the first normal equation, as do the last b equations. Thus, there are two linear dependencies in the normal equations, implying that two constraints must be imposed to solve Equation 5-18. The usual constraints are

$$
\sum_{i=1}^{a} \hat{\tau}_i = 0 \qquad \sum_{j=1}^{b} \hat{\beta}_j = 0 \qquad\qquad (5\text{-}19)
$$

Using these constraints, the normal equations simplify considerably. In fact, they become

$$
\begin{aligned}
ab\hat{\mu} &= y_{..} \\
b\hat{\mu} + b\hat{\tau}_i &= y_{i.} \qquad i = 1,2,\ldots,a \qquad (5\text{-}20) \\
a\hat{\mu} + a\hat{\beta}_j &= y_{.j} \qquad j = 1,2,\ldots,b
\end{aligned}
$$

whose solution is

$$
\begin{aligned}
\hat{\mu} &= \bar{y}_{..} \\
\hat{\tau}_i &= \bar{y}_{i.} - \bar{y}_{..} \qquad i = 1,2,\ldots,a \qquad (5\text{-}21) \\
\hat{\beta}_j &= \bar{y}_{.j} - \bar{y}_{...} \qquad j = 1,2,\ldots,b
\end{aligned}
$$

Using the solution to the normal equation in Equation 5-21, we may find the estimated or fitted values of y_{ij} as

$$
\begin{aligned}
\hat{y}_{ij} &= \hat{\mu} + \hat{\tau}_i + \hat{\beta}_j \\
&= \bar{y}_{..} + (\bar{y}_{i.} - \bar{y}_{..}) + (\bar{y}_{.j} - \bar{y}_{..}) \\
&= \bar{y}_{i.} + \bar{y}_{.j} - \bar{y}_{..}
\end{aligned}
$$

This result was used previously in Equation 5-15 for computing the residuals from a randomized block design.

The general regression significance test can be used to develop the analysis of variance for the randomized complete block design. Using the solution to the normal equations given by Equation 5-21, the reduction in the sum of squares for fitting the complete model is

$$R(\mu, \tau, \beta) = \hat{\mu}y_{..} + \sum_{i=1}^{a} \hat{\tau}_i y_{i.} + \sum_{j=1}^{b} \hat{\beta}_j y_{.j}$$

$$= \bar{y}_{..}y_{..} + \sum_{i=1}^{a} (\bar{y}_{i.} - \bar{y}_{..})y_{i.} + \sum_{j=1}^{b} (\bar{y}_{.j} - \bar{y}_{..})y_{.j}$$

$$= \frac{y_{..}^2}{ab} + \sum_{i=1}^{a} \bar{y}_{i.}y_{i.} - \frac{y_{..}^2}{ab} + \sum_{j=1}^{b} \bar{y}_{.j}y_{.j} - \frac{y_{..}^2}{ab}$$

$$= \sum_{i=1}^{a} \frac{y_{i.}^2}{b} + \sum_{j=1}^{b} \frac{y_{.j}^2}{a} - \frac{y_{..}^2}{ab}$$

with $a + b - 1$ degrees of freedom, and the error sum of squares is

$$SS_E = \sum_{i=1}^{a}\sum_{j=1}^{b} y_{ij}^2 - R(\mu, \tau, \beta)$$

$$= \sum_{i=1}^{a}\sum_{j=1}^{b} y_{ij}^2 - \sum_{i=1}^{a} \frac{y_{i.}^2}{b} - \sum_{j=1}^{b} \frac{y_{.j}^2}{a} + \frac{y_{..}^2}{ab}$$

$$= \sum_{i=1}^{a}\sum_{j=1}^{b} (y_{ij} - \bar{y}_{i.} - \bar{y}_{.j} + \bar{y}_{..})^2$$

with $(a - 1)(b - 1)$ degrees of freedom. Compare this last equation with SS_E in Equation 5-7.

To test the hypothesis $H_0 : \tau_i = 0$, the reduced model is

$$y_{ij} = \mu + \beta_j + \epsilon_{ij}$$

which is just a single-factor analysis of variance. By analogy with Equation 4-17, the reduction in the sum of squares for fitting ths model is

$$R(\mu, \beta) = \sum_{j=1}^{b} \frac{y_{.j}^2}{a}$$

which has b degrees of freedom. Therefore, the sum of squares due to $\{\tau_i\}$ after fitting μ and $\{\beta_j\}$ is

$$R(\tau | \mu, \beta) = R(\mu, \tau, \beta) - R(\mu, \beta)$$

$$= \sum_{i=1}^{a} \frac{y_{i.}^2}{b} + \sum_{j=1}^{b} \frac{y_{.j}^2}{a} - \frac{y_{..}^2}{ab} - \sum_{j=1}^{b} \frac{y_{.j}^2}{a}$$

$$= \sum_{i=1}^{a} \frac{y_{i.}^2}{b} - \frac{y_{..}^2}{ab}$$

which we recognize as the treatment sum of squares with $a - 1$ degrees of freedom (Equation 5-10).

The block sum of squares is obtained by fitting the reduced model

$$y_{ij} = \mu + \tau_i + \epsilon_{ij}$$

which is also a single-factor analysis. Again, by analogy with Equation 4-17, the reduction in the sum of squares for fitting this model is

$$R(\mu, \tau) = \sum_{i=1}^{a} \frac{y_{i.}^2}{b}$$

with a degrees of freedom. The sum of squares for blocks (β_j) after fitting μ and τ_i is

$$
\begin{aligned}
R(\beta|\mu, \tau) &= R(\mu, \tau, \beta) - R(\mu, \tau) \\
&= \sum_{i=1}^{a} \frac{y_{i.}^2}{b} + \sum_{j=1}^{b} \frac{y_{.j}^2}{a} - \frac{y_{..}^2}{ab} - \sum_{i=1}^{a} \frac{y_{i.}^2}{b} \\
&= \sum_{j=1}^{b} \frac{y_{.j}^2}{a} - \frac{y_{..}^2}{ab}
\end{aligned}
$$

with $b - 1$ degrees of freedom, which we have found previously as Equation 5-11.

We have developed the sums of squares for treatments, blocks, and error in the randomized complete block design using the general regression significance test. Although we would not ordinarily use the general regression significance test actually to analyze data in a randomized complete block, the procedure occasionally proves useful in more general randomized block designs, such as those discussed in Chapter 6.

Exact Analysis of the Missing Value Problem ▪ In Section 5-1.3 an approximate procedure for dealing with missing observations in a randomized block design was presented. This approximate analysis consists of estimating the missing value so that the error mean square is minimized. It can be shown that the approximate analysis produces a biased mean square for treatments in the sense that $E(MS_{\text{Treatments}})$ is larger than $E(MS_E)$ if the null hypothesis is true. Consequently, too many significant results are reported.

The missing value problem may be analyzed exactly by using the general regression significance test. The missing value causes the design to be *unbalanced,* and because all the treatments do not occur in all blocks, we say that the treatments and blocks are not *orthogonal.* This method of analysis is also used in more general types of randomized block designs; it is discussed further in Chapter 6. Problem 5-26 asks the reader to perform the exact analysis for a randomized complete block design with one missing value.

5-1.5 Sample Computer Output

Computer output for the hardness testing data in Example 5-1, obtained from the GLM procedure in SAS, is shown in Figure 5-6. Recall that the original analysis

S T A T I S T I C A L   A N A L Y S I S   S Y S T E M

13:16   WEDNESDAY,   JANUARY   17,   1990

GENERAL LINEAR MODELS PROCEDURE

DEPENDENT VARIABLE: Y

SOURCE	DF	SUM OF SQUARES	MEAN SQUARE	F VALUE	PR > F	R-SQUARE	C.V.
MODEL	6	1.21000000	0.20166667	22.69	0.0001	0.937984	0.9795
ERROR	9	0.08000000	0.00888889			STD DEV	Y MEAN
CORRECTED TOTAL	15	1.29000000				0.09428090	9.62500000

SOURCE	DF	TYPE I SS	F VALUE	PR > F	DF	TYPE III SS	F VALUE	PR > F
TIPTYPE	3	0.38500000	14.44	0.0009	3	0.38500000	14.44	0.0009
BLOCK	3	0.82500000	30.94	0.0001	3	0.82500000	30.94	0.0001

OBSERVATION	OBSERVED VALUE	PREDICTED VALUE	RESIDUAL
1	9.30000000	9.35000000	-0.05000000
2	9.40000000	9.37500000	0.02500000
3	9.60000000	9.67500000	-0.07500000
4	10.00000000	9.90000000	0.10000000
5	9.40000000	9.37500000	0.02500000
6	9.30000000	9.40000000	-0.10000000
7	9.80000000	9.70000000	0.10000000
8	9.90000000	9.92500000	-0.02500000
9	9.20000000	9.22500000	-0.02500000
10	9.40000000	9.25000000	0.15000000
11	9.50000000	9.55000000	-0.50000000
12	9.70000000	9.77000000	-0.75000000
13	9.70000000	9.65000000	0.50000000
14	9.60000000	9.67500000	-0.07500000
15	10.00000000	9.97500000	0.25000000
16	10.20000000	10.20000000	0.00000000

SUM OF RESIDUALS   0.00000000
SUM OF SQUARED RESIDUALS   0.08000000
SUM OF SQUARED RESIDUALS - ERROR SS   -0.00000000
FIRST ORDER AUTOCORRELATION   -0.49218750
DURBIN-WATSON D   2.95312500

Figure 5-6.   Sample computer output for Example 5-1.

in Table 5-5 used *coded* data. (The raw responses were coded by subtracting 9.5 and multiplying the result by 10.) The computer analysis used the raw responses. Consequently, the sums of squares in Figure 5-6 are equal to those in Table 5-5 divided by 100.

The model sum of squares reported in the analysis of variance consists of the treatment sum of squares plus the block sum of squares; that is,

$$SS_{Model} = SS_{Tip\ Type} + SS_{Block}$$
$$= 0.3850 + 0.8250$$
$$= 1.21$$

The individual sources of variation resulting from types of tips and blocks are also identified in the computer output. Note that the type I and type III sums of squares for those two factors are identical, as will always be the case for balanced data.

The R^2 for this model, computed as

$$R^2 = \frac{SS_{Model}}{SS_E} = \frac{1.21}{1.29} = 0.937984$$

implies that about 94 percent of the variability in the data is explained by the model (i.e., the factors type of tip and blocks). Combined with the satisfactory residual analysis, this indicates that the model is a very good fit to the data.

5-2 THE LATIN SQUARE DESIGN

In Section 5-1 we introduced the randomized complete block design as a design to reduce the residual error in an experiment by removing variability due to a known and controlable nuisance variable. There are several other types of designs that utilize the blocking principle. For example, suppose that an experimenter is studying the effects of five different formulations of an explosive mixture used in the manufacture of dynamite on the observed explosive force. Each formulation is mixed from a batch of raw material that is only large enough for five formulations to be tested. Furthermore, the formulations are prepared by several operators, and there may be substantial differences in the skills and experience of the operators. Thus, it would seem that there are two nuisance factors to be "averaged out" in the design: batches of raw material and operators. The appropriate design for this problem consists of testing each formulation exactly once in each batch of raw material and for each formulation to be prepared exactly once by each of five operators. The resulting design, shown in Table 5-9, is called a *Latin square design*. Notice that the design is a square arrangement, and that the five formulations (or treatments) are denoted by the Latin letters A, B, C, D, and E; hence the name Latin square. We see that both batches of raw material (rows) and operators (columns) are orthogonal to treatments.

The Latin square design is used to eliminate two nuisance sources of varia-

Table 5-9   Latin Square Design for the Dynamite Formulation Problem

Batches of Raw Material	Operators				
	1	2	3	4	5
1	A = 24	B = 20	C = 19	D = 24	E = 24
2	B = 17	C = 24	D = 30	E = 27	A = 36
3	C = 18	D = 38	E = 26	A = 27	B = 21
4	D = 26	E = 31	A = 26	B = 23	C = 22
5	E = 22	A = 30	B = 20	C = 29	D = 31

bility; that is, it systematically allows blocking two directions. Thus, the rows and columns actually respresent two restrictions on randomization. In general, a Latin square for p factors, or a $p \times p$ Latin square, is a square containing p rows and p columns. Each of the resulting p^2 cells contains one of the p letters that corresponds to the treatments, and each letter occurs once and only once in each row and column. Some examples of Latin squares are

4 × 4	5 × 5	6 × 6
A B D C	A D B E C	A D C E B F
B C A D	D A C B E	B A E C F D
C D B A	C B E D A	C E D F A B
D A C B	B E A C D	D C F B E A
	E C D A B	F B A D C E
		E F B A D C

The statistical model for a Latin square is

$$y_{ijk} = \mu + \alpha_i + \tau_j + \beta_k + \epsilon_{ijk} \begin{cases} i = 1,2,\ldots,p \\ j = 1,2,\ldots,p \\ k = 1,2,\ldots,p \end{cases} \quad (5\text{-}22)$$

where y_{ijk} is the observation in the ith row and kth column for the jth treatment, μ is the overall mean, α_i is the ith row effect, τ_j is the jth treatment effect, β_k is the kth column effect, and ϵ_{ijk} is the random error. The model is completely *additive*; that is, there is no interaction between rows, columns, and treatments. Since there is only one observation in each cell, only two of the three subscripts i, j, and k are needed to denote a particular observation. For example, referring to the dynamite formulation problem in Table 5-9, if $i = 2$ and $k = 3$ we automatically find $j = 4$ (formulation D), and if $i = 1$ and $j = 3$ (formulation C) we find $k = 3$. This is a consequence of each treatment appearing exactly once in each row and column.

The analysis of variance consists of partitioning the total sum of squares of the $N = p^2$ observations into components for rows, columns, treatments, and

Table 5-10 Analysis of Variance for the Latin Square Design

Source of Variation	Sum of Squares	Degrees of Freedom	Mean Square	F_0
Treatments	$SS_{Treatments}$	$p - 1$	$\dfrac{SS_{Treatments}}{p - 1}$	$F_0 = \dfrac{MS_{Treatments}}{MS_E}$
	$= \sum\limits_{j=1}^{p} \dfrac{y_{.j.}^2}{p} - \dfrac{y_{...}^2}{N}$			
Rows	SS_{Rows}	$p - 1$	$\dfrac{SS_{Rows}}{p - 1}$	
	$= \sum\limits_{i=1}^{p} \dfrac{y_{i..}^2}{p} - \dfrac{y_{...}^2}{N}$			
Columns	$SS_{Columns}$	$p - 1$	$\dfrac{SS_{Columns}}{p - 1}$	
	$= \sum\limits_{k=1}^{p} \dfrac{y_{..k}^2}{p} - \dfrac{y_{...}^2}{N}$			
Error	SS_E (by subtraction)	$(p - 2)(p - 1)$	$\dfrac{SS_E}{(p - 2)(p - 1)}$	
Total	SS_T $= \sum\limits_{i} \sum\limits_{j} \sum\limits_{k} y_{ijk}^2 - \dfrac{y_{...}^2}{N}$	$p^2 - 1$		

error, for example,

$$SS_T = SS_{Rows} + SS_{Columns} + SS_{Treatments} + SS_E \qquad (5\text{-}23)$$

with respective degrees of freedom

$$p^2 - 1 = p - 1 + p - 1 + p - 1 + (p - 2)(p - 1)$$

Under the usual assumption that ϵ_{ijk} is NID $(0, \sigma^2)$, each sum of squares on the right-hand side of Equation 5-23 is, upon division by σ^2, an independently distributed chi-square random variable. The appropriate statistic for testing for no differences in treatment means is

$$F_0 = \frac{MS_{Treatments}}{MS_E}$$

which is distributed as $F_{p-1,(p-2)(p-1)}$ under the null hypothesis. One may also test for no row effect and no column effect by forming the ratio of MS_{Rows} or

$MS_{Columns}$ to MS_E. However, since the rows and columns represent restrictions on randomization, these tests may not be appropriate.

The computational procedure for the analysis of variance is shown in Table 5-10. From the computational formulas for the sums of squares, we see that the analysis is a simple extension of the randomized block design, with the sum of squares resulting from rows obtained from the row totals.

Example 5-4

Consider the dynamite formulation problem previously described, where both batches of raw material and operators represent randomization restrictions. The design for this experiment, shown in Table 5-9, is a 5 × 5 Latin square. After coding by subtracting 25 from each observation, we have the data in Table 5-11. The sums of squares for the total, batches (rows), and operators (columns) are computed as follows:

$$SS_T = \sum_i \sum_j \sum_k y_{ijk}^2 - \frac{y_{...}^2}{N}$$

$$= 680 - \frac{(10)^2}{25} = 676.00$$

$$SS_{Batches} = \sum_{i=1}^p \frac{y_{i..}^2}{p} - \frac{y_{...}^2}{N}$$

$$= \frac{(-14)^2 + 9^2 + 5^2 + 3^2 + 7^2}{5} - \frac{(10)^2}{25} = 68.00$$

$$SS_{Operators} = \sum_{k=1}^p \frac{y_{..k}^2}{p} - \frac{y_{...}^2}{N}$$

$$= \frac{(-18)^2 + 18^2 + (-4)^2 + 5^2 + 9^2}{5} - \frac{(10)^2}{25} = 150.00$$

The totals for the treatments (Latin letters) are

Latin Letter	Treatment Total
A	$y_{.1.} = $ 18
B	$y_{.2.} = $ −24
C	$y_{.3.} = $ −13
D	$y_{.4.} = $ 24
E	$y_{.5.} = $ 5

Table 5-11   Coded Data for the Dynamite Formulation Problem

Batches of Raw Material	Operators 1	2	3	4	5	$y_{i..}$
1	A = −1	B = −5	C = −6	D = −1	E = −1	−14
2	B = −8	C = −1	D = 5	E = 2	A = 11	9
3	C = −7	D = 13	E = 1	A = 2	B = −4	5
4	D = 1	E = 6	A = 1	B = −2	C = −3	3
5	E = −3	A = 5	B = −5	C = 4	D = 6	7
$y_{..k}$	−18	18	−4	5	9	10 = $y_{...}$

The sum of squares resulting from the formulations is computed from these totals as

$$SS_{\text{Formulations}} = \sum_{j=1}^{p} \frac{y_{.j.}^2}{p} - \frac{y_{...}^2}{N}$$

$$= \frac{18^2 + (-24)^2 + (-13)^2 + 24^2 + 5^2}{5} - \frac{(10)^2}{25} = 330.00$$

The error sum of squares is found by subtraction:

$$SS_E = SS_T - SS_{\text{Batches}} - SS_{\text{Operators}} - SS_{\text{Formulations}}$$

$$= 676.00 - 68.00 - 150.00 - 330.00 = 128.00$$

The analysis of variance is summarized in Table 5-12. We conclude that there is a significant difference in the mean explosive force generated by the different dynamite formulations. There is also an indication that there are differences between operators, so blocking on this factor was a good precaution. There is no strong evidence of a difference between batches of raw material, so it seems that in this particular experiment we were unnecessarily concerned about this source of variability. However, blocking on batches of raw material is usually a good idea.

∎

As in any design problem, the experimenter should investigate the adequacy of the model by inspecting and plotting the residuals. For a Latin square, the residuals are given by

$$e_{ijk} = y_{ijk} - \hat{y}_{ijk}$$

$$= y_{ijk} - \bar{y}_{i..} - \bar{y}_{.j.} - \bar{y}_{..k} + 2\bar{y}_{...}$$

The reader should find the residuals for Example 5-4 and construct appropriate plots.

A Latin square in which the first row and column consists of the letters written in alphabetical order is called a *standard* Latin square. The design used in Example 5-4 is a standard Latin square. A standard Latin square can always be obtained by writing the first row in alphabetical order and then writing each successive row as the row of letters just above shifted one place to the left. Table 5-13 summarizes several important facts about Latin squares and standard Latin squares.

Table 5-12 Analysis of Variance for the Dynamite Formulation Experiment

Source of Variation	Sum of Squares	Degrees of Freedom	Mean Square	F_0
Formulations	330.00	4	82.50	7.73[a]
Batches of raw material	68.00	4	17.00	
Operators	150.00	4	37.50	
Error	128.00	12	10.67	
Total	676.00	24		

[a] Significant at 1 percent.

Table 5-13   Standard Latin Squares and Number of Latin Squares of Various Sizes[a]

Size	3 × 3	4 × 4	5 × 5	6 × 6	7 × 7	$p \times p$
Examples of standard squares	A B C B C A C A B	A B C D B C D A C D A B D A B C	A B C D E B A E C D C D A E B D E B A C E C D B A	A B C D E F B C F A D E C F B E A D D E A B F C E A D F C B F D E C B A	A B C D E F G B C D E F G A C D E F G A B D E F G A B C E F G A B C D F G A B C D E G A B C D E F	A B C. . . P B C D. . . A C D E. . . B ⋮ P A B. . .$(P-1)$
Number of standard squares	1	4	56	9408	16,942,080	—
Total number of Latin squares	12	576	161,280	818,851,200	61,479,419,904,000	$p!(p-1)! \times$ (number of standard squares)

[a] Some of the information in this table is found in *Statistical Tables for Biological, Agricultural and Medical Research*, 4th edition, by R. A. Fisher and F. Yates, Oliver and Boyd, Edinburgh, 1953. Little is known about the properties of Latin squares larger than 7 × 7.

As with any experimental design, the observations in the Latin square should be taken in random order. The proper randomization procedure is to select the particular square employed at random. As we see in Table 5-13, there are a large number of Latin squares of a particular size, so it is impossible to enumerate all the squares and select one randomly. The usual procedure is to select a Latin square from a table of such designs, as in Fisher and Yates (1953), and then arrange the order of the rows, columns, and letters at random. This is discussed more completely in Fisher and Yates (1953).

Occasionally, one observation in a Latin square is missing. For a $p \times p$ Latin square, the missing value may be estimated by

$$y_{ijk} = \frac{p(y'_{i..} + y'_{.j.} + y'_{..k}) - 2y'_{...}}{(p - 2)(p - 1)} \tag{5-24}$$

where the primes indicate totals for the row, column, and treatment with the missing value, and $y'_{...}$ is the grand total with the missing value.

Latin squares can be useful in situations where the rows and columns represent factors the experimenter actually wishes to study and where there are no randomization restrictions. Thus, three factors (rows, columns, and letters), each at p levels, can be investigated in only p^2 runs. This design assumes that there is no interaction between the factors. More will be said later on the subject of interaction.

Replication of Latin Squares ▪ A disadvantage of small Latin squares is that they provide a relatively small number of error degrees of freedom. For example, a 3×3 Latin square has only 2 error degrees of freedom, a 4×4 Latin square has only 6 error degrees of freedom, and so forth. When small Latin squares are used, it is frequently desirable to replicate them to increase the error degrees of freedom.

There are several ways to replicate a Latin square. To illustrate, consider that the 5×5 Latin square used in Example 5-4 is replicated n times. This could have been done as follows:

1. Use the same batches and operators in each replicate.
2. Use the same batches but different operators in each replicate (or, equivalently, use the same operators but different batches).
3. Use different batches and different operators.

The analysis of variance depends on the method of replication.

Consider case 1, where the same levels of the row and column blocking factors are used in each replicate. Let y_{ijkl} be the observation in row i, treatment j, column k, and replicate l. There are $N = np^2$ total observations. The analysis of variance is summarized in Table 5-14.

Now consider case 2, and assume that new batches of raw material but the same operators are used in each replicate. Thus, there are now 5 new rows (in general, p new rows) within each replicate. The analysis of variance is summa-

Table 5-14 Analysis of Variance for a Replicated Latin Square, Case 1

Source of Variation	Sum of Squares	Degrees of Freedom	Mean Square	F_0
Treatments	$\sum_{j=1}^{p} \dfrac{y_{.j..}^2}{np} - \dfrac{y_{....}^2}{N}$	$p - 1$	$\dfrac{SS_{\text{Treatments}}}{p - 1}$	$\dfrac{MS_{\text{Treatments}}}{MS_E}$
Rows	$\sum_{i=1}^{p} \dfrac{y_{i...}^2}{np} - \dfrac{y_{....}^2}{N}$	$p - 1$	$\dfrac{SS_{\text{Rows}}}{p - 1}$	
Columns	$\sum_{k=1}^{p} \dfrac{y_{..k.}^2}{np} - \dfrac{y_{....}^2}{N}$	$p - 1$	$\dfrac{SS_{\text{Columns}}}{p - 1}$	
Replicates	$\sum_{l=1}^{n} \dfrac{y_{...l}^2}{p^2} - \dfrac{y_{....}^2}{N}$	$n - 1$	$\dfrac{SS_{\text{Replicates}}}{n - 1}$	
Error	Subtraction	$(p - 1)[n(p + 1) - 3]$	$\dfrac{SS_E}{(p - 1)[n(p + 1) - 3]}$	
Total	$\sum\sum\sum\sum y_{ijkl}^2 - \dfrac{y_{....}^2}{N}$	$np^2 - 1$		

rized in Table 5-15. Note that the source of variation for the rows really measures the variation between rows within the n replicates.

Finally, consider case 3, where new batches of raw material and new operators are used in each replicate. Now the variation that results from both the rows and columns measures the variation resulting from these factors within the replicates. The analysis of variance is summarized in Table 5-16.

There are other approaches to analyzing replicated Latin squares that allow some interactions between treatments and squares. Refer to Problem 5-19.

Crossover Designs and Designs Balanced for Residual Effects ▪ Occasionally, one encounters a problem in which time periods are a factor in the experiment. In general, there are p treatments to be tested in p time periods using np experimental units. For example, a human performance analyst is studying the effect of two replacement fluids on dehydration in 20 subjects. In the first period, half of the subjects (chosen at random) are given fluid A and the other half fluid B. At the end of the period, the response is measured, and a period of time is allowed to pass in which any physiological effect of the fluids is eliminated. Then the experimenter has the subjects who took fluid A take fluid B and those who took fluid B take fluid A. This design is called a *crossover design*. It is analyzed as a set of 10 Latin squares with two rows (time periods) and two treatments (fluid types). The two columns in each of the 10 squares correspond to subjects.

The layout of this design is shown in Figure 5-7. Notice that the rows in the Latin squares represent the time periods and the columns represent the subjects.

Table 5-15   Analysis of Variance for a Replicated Latin Square, Case 2

Source of Variation	Sum of Squares	Degrees of Freedom	Mean Square	F_0
Treatments	$\sum_{j=1}^{p} \dfrac{y_{.j..}^2}{np} - \dfrac{y_{....}^2}{N}$	$p - 1$	$\dfrac{SS_{\text{Treatments}}}{p - 1}$	$\dfrac{MS_{\text{Treatments}}}{MS_E}$
Rows	$\sum_{l=1}^{n} \sum_{i=1}^{p} \dfrac{y_{i..l}^2}{p} - \sum_{l=1}^{n} \dfrac{y_{....}^2}{p^2}$	$n(p - 1)$	$\dfrac{SS_{\text{Rows}}}{n(p - 1)}$	
Columns	$\sum_{k=1}^{p} \dfrac{y_{..k.}^2}{np} - \dfrac{y_{....}^2}{N}$	$p - 1$	$\dfrac{SS_{\text{Columns}}}{p - 1}$	
Replicates	$\sum_{l=1}^{n} \dfrac{y_{...l}^2}{p^2} - \dfrac{y_{....}^2}{N}$	$n - 1$	$\dfrac{SS_{\text{Replicates}}}{n - 1}$	
Error	Subtraction	$(p - 1)(np - 1)$	$\dfrac{SS_E}{(p - 1)(np - 1)}$	
Total	$\sum_i \sum_j \sum_k \sum_l y_{ijkl}^2 - \dfrac{y_{....}^2}{N}$	$np^2 - 1$		

Table 5-16 Analysis of Variance for a Replicated Latin Square, Case 3

Source of Variation	Sum of Squares	Degrees of Freedom	Mean Square	F_0
Treatments	$\sum_{j=1}^{p} \dfrac{y_{.j..}^2}{np} - \dfrac{y_{....}^2}{N}$	$p - 1$	$\dfrac{SS_{\text{Treatments}}}{p - 1}$	$\dfrac{MS_{\text{Treatments}}}{MS_E}$
Rows	$\sum_{l=1}^{n} \sum_{i=1}^{p} \dfrac{y_{i..l}^2}{p} - \sum_{l=1}^{n} \dfrac{y_{...l}^2}{p^2}$	$n(p - 1)$	$\dfrac{SS_{\text{Rows}}}{n(p - 1)}$	
Columns	$\sum_{l=1}^{n} \sum_{k=1}^{p} \dfrac{y_{..kl}^2}{p} - \sum_{l=1}^{n} \dfrac{y_{...l}^2}{p^2}$	$n(p - 1)$	$\dfrac{SS_{\text{Columns}}}{n(p - 1)}$	
Replicates	$\sum_{l=1}^{n} \dfrac{y_{...l}^2}{p^2} - \dfrac{y_{....}^2}{N}$	$n - 1$	$\dfrac{SS_{\text{Replicates}}}{n - 1}$	
Error	Subtraction	$(p - 1)[n(p - 1) - 1]$	$\dfrac{SS_E}{(p - 1)[n(p - 1) - 1]}$	
Total	$\sum_i \sum_j \sum_k \sum_l y_{ijkl}^2 - \dfrac{y_{....}^2}{N}$	$np^2 - 1$		

Latin Squares

	I		II		III		IV		V		VI		VII		VIII		IX		X	
Subject	1	2	3	4	5	6	7	8	9	10	11	12	13	14	15	16	17	18	19	20
Period 1	A	B	B	A	B	A	A	B	A	B	B	A	A	B	A	B	A	B	A	B
Period 2	B	A	A	B	A	B	B	A	B	A	A	B	B	A	B	A	B	A	B	A

Figure 5-7. A crossover design.

The 10 subjects who received fluid A first (1, 4, 6, 7, 9, 12, 13, 15, 17, and 19) are randomly determined.

An abbreviated analysis of variance is summarized in Table 5-17. The subject sum of squares is computed as the corrected sum of squares among the 20 subject totals, the period sum of squares is the corrected sum of squares among the rows, and the fluid sum of squares is computed as the corrected sum of squares among the letter totals. For further details of the statistical analysis of these designs, see Cochran and Cox (1957), John (1971), and Anderson and McLean (1974).

It is also possible to employ Latin square type designs for experiments in which the treatments have a *residual* effect—that is, for example, if the data for fluid B in period 2 still reflected some effect of fluid A taken in period 1. Designs balanced for residual effects are discussed in detail by Cochran and Cox (1957) and John (1971).

5-3   THE GRAECO–LATIN SQUARE DESIGN

Consider a $p \times p$ Latin square, and superimpose on it a second $p \times p$ Latin square in which the treatments are denoted by Greek letters. If the two squares when superimposed have the property that each Greek letter appears once and only once with each Latin letter, the two Latin squares are said to be *orthogonal*, and the design obtained is called a *Graeco-Latin square*. An example of a 4 × 4 Graeco-Latin square is shown in Table 5-18.

Table 5-17   Analysis of Variance for the Crossover Design in Figure 5-6

Source of Validation	Degrees of Freedom
Subjects (columns)	19
Periods (rows)	1
Fluids (letters)	1
Error	18
Total	39

Table 5-18   4 × 4 Graeco-Latin
Square Design

Row	Column			
	1	2	3	4
1	$A\alpha$	$B\beta$	$C\gamma$	$D\delta$
2	$B\delta$	$A\gamma$	$D\beta$	$C\alpha$
3	$C\beta$	$D\alpha$	$A\delta$	$B\gamma$
4	$D\gamma$	$C\delta$	$B\alpha$	$A\beta$

The Graeco-Latin square design can be used to control systematically three sources of extraneous variability, that is, to block in *three* directions. The design allows investigation of four factors (rows, columns, Latin letters, and Greek letters), each at p levels in only p^2 runs. Graeco-Latin squares exist for all $p \geq 3$ except $p = 6$.

The statistical model for the Graeco-Latin square design is

$$y_{ijkl} = \mu + \theta_i + \tau_j + \omega_k + \Psi_l + \epsilon_{ijkl} \quad \begin{cases} i = 1, 2, \ldots, p \\ j = 1, 2, \ldots, p \\ k = 1, 2, \ldots, p \\ l = 1, 2, \ldots, p \end{cases} \quad (5\text{-}25)$$

where y_{ijkl} is the observation in row i and column l for Latin letter j and Greek letter k, θ_i is the effect of the ith row, τ_j is the effect of Latin letter treatment j, ω_k is the effect of Greek letter treatment k, Ψ_l is the effect of column l, and ϵ_{ijkl} is an NID $(0, \sigma^2)$ random error component. Only two of the four subscripts are necessary to completely identify an observation.

The analysis of variance is very similar to that of a Latin square. Since the Greek letters appear exactly once in each row and column and exactly once with each Latin letter, the factor represented by the Greek letters is orthogonal to rows, columns, and Latin letter treatments. Therefore, a sum of squares due to the Greek letter factor may be computed from the Greek letter totals and the experimental error is further reduced by this amount. The computational details are illustrated in Table 5-19. The null hypotheses of equal row, column, Latin letter, and Greek letter treatments would be tested by dividing the corresponding mean square by mean square error. The rejection region is the upper-tail point of the $F_{p-1, (p-3)(p-1)}$ distribution.

Example 5-5

Suppose that in the dynamite formulation experiment of Example 5-4 an additional factor, test assemblies, could be of importance. Let there be five test assemblies denoted by the Greek letters α, β, γ, δ, and ϵ. The resulting 5 × 5 Graeco-Latin square design is shown in Table 5-20.

Notice that, since the totals for batches of raw material (rows), operators (col-

Table 5-19   Analysis of Variance for a Graeco-Latin Square Design

Source of Variation	Sum of Squares	Degrees of Freedom
Latin letter treatments	$SS_L = \sum_{j=1}^{p} \frac{y_{.j..}^2}{p} - \frac{y_{....}^2}{N}$	$p - 1$
Greek letter treatments	$SS_G = \sum_{k=1}^{p} \frac{y_{..k.}^2}{p} - \frac{y_{....}^2}{N}$	$p - 1$
Rows	$SS_{Rows} = \sum_{i=1}^{p} \frac{y_{i...}^2}{p} - \frac{y_{....}^2}{N}$	$p - 1$
Columns	$SS_{Columns} = \sum_{l=1}^{p} \frac{y_{...l}^2}{p} - \frac{y_{....}^2}{N}$	$p - 1$
Error	SS_E(by subtraction)	$(p - 3)(p - 1)$
Total	$SS_T = \sum_i \sum_j \sum_k \sum_l y_{ijkl}^2 - \frac{y_{....}^2}{N}$	$p^2 - 1$

umns), and formulations (Latin letters) are identical to those in Example 5-4, we have

$$SS_{Batches} = 68.00 \qquad SS_{Operators} = 150.00 \qquad \text{and} \qquad SS_{Formulations} = 330.00$$

The totals for the test assemblies (Greek letters) are

Greek Letter	Test Assembly Total
α	$y_{..1.} = 10$
β	$y_{..2.} = -6$
γ	$y_{..3.} = -3$
δ	$y_{..4.} = -4$
ϵ	$y_{..5.} = 13$

Table 5-20   Graeco-Latin Square Design for the Dynamite Formulation Problem

Batches of Raw Material	Operators					$y_{i...}$
	1	2	3	4	5	
1	$A\alpha = -1$	$B\gamma = -5$	$C\epsilon = -6$	$D\beta = -1$	$E\delta = -1$	-14
2	$B\beta = -8$	$C\delta = -1$	$D\alpha = 5$	$E\gamma = 2$	$A\epsilon = 11$	9
3	$C\gamma = -7$	$D\epsilon = 13$	$E\beta = 1$	$A\delta = 2$	$B\alpha = -4$	5
4	$D\delta = 1$	$E\alpha = 6$	$A\gamma = 1$	$B\epsilon = -2$	$C\beta = -3$	3
5	$E\epsilon = -3$	$A\beta = 5$	$B\delta = -5$	$C\alpha = 4$	$D\gamma = 6$	7
$y_{...l}$	-18	18	-4	5	9	$10 = y_{....}$

Table 5-21 Analysis of Variance for the Dynamite Formulation Problem

Source of Variation	Sum of Squares	Degrees of Freedom	Mean Square	F_0
Formulations	330.00	4	82.50	10.00[a]
Batches of raw material	68.00	4	17.00	
Operators	150.00	4	37.50	
Test assemblies	62.00	4	15.50	
Error	66.00	8	8.25	
Total	676.00	24		

[a] Significant at 1 percent.

Thus, the sum of squares due to the test assemblies is

$$SS_{Assemblies} = \sum_{k=1}^{p} \frac{y_{..k.}^2}{p} - \frac{y_{....}^2}{N}$$

$$= \frac{10^2 + (-6)^2 + (-3)^2 + (-4)^2 + 13^2}{5} - \frac{(10)^2}{25} = 62.00$$

The complete analysis of variance is summarized in Table 5-21. Formulations are significantly different at 1 percent. In comparing Tables 5-21 and 5-12, we observe that removing the variability due to test assemblies has decreased the experimental error. However, in decreasing the experimental error, we have also reduced the error degrees of freedom from 12 (in the Latin square design of Example 5-4) to 8. Thus, our estimate of error has fewer degrees of freedom, and the test may be less sensitive.

∎

The concept of orthogonal pairs of Latin squares forming a Graeco-Latin square can be extended somewhat. A $p \times p$ *hypersquare* is a design in which three or more orthogonal $p \times p$ Latin squares are superimposed. In general, up to $p + 1$ factors could be studied if a complete set of $p - 1$ orthogonal Latin squares is available. Such a design would utilize all $(p + 1)(p - 1) = p^2 - 1$ degrees of freedom, so an independent estimate of the error variance is necessary. Of course, there must be no interactions between the factors when using hypersquares.

5-4 PROBLEMS

5-1 A chemist wishes to test the effect of four chemical agents on the strength of a particular type of cloth. Because there might be variability from one bolt to another, the chemist decides to use a randomized block design, with the bolts of cloth considered as blocks. She selects five bolts and applies all four chemicals in random order to each bolt. The resulting tensile strengths follow. Analyze the data and draw appropriate conclusions.

			Bolt		
Chemical	1	2	3	4	5
1	73	68	74	71	67
2	73	67	75	72	70
3	75	68	78	73	68
4	73	71	75	75	69

5-2 Three different washing solutions are being compared to study their effectiveness in retarding bacteria growth in five-gallon milk containers. The analysis is done in a laboratory, and only three trials can be run on any day. Because days could represent a potential source of variability, the experimenter decides to use a randomized block design. Observations are taken for four days, and the data are shown here. Analyze the data and draw conclusions.

		Days		
Solution	1	2	3	4
1	13	22	18	39
2	16	24	17	44
3	5	4	1	22

5-3 Plot the mean tensile strengths observed for each chemical type in Problem 5-1 and compare them to an appropriately scaled t distribution. What conclusions would you draw from this display?

5-4 Plot the average bacteria counts for each solution in Problem 5-2 and compare them to a scaled t distribution. What conclusions can you draw?

5-5 An article in the *Fire Safety Journal* ("The Effect of Nozzle Design on the Stability and Performance of Turbulent Water Jets", Vol. 4, August 1981) describes an experiment in which a shape factor was determined for several different nozzle designs at six levels of jet efflux velocity. Interest focused on potential differences between nozzle designs, with velocity considered as a nuisance variable. The data are shown below:

Nozzle Design	Jet Efflux Velocity (m/s)					
	11.73	14.37	16.59	20.43	23.46	28.74
1	0.78	0.80	0.81	0.75	0.77	0.78
2	0.85	0.85	0.92	0.86	0.81	0.83
3	0.93	0.92	0.95	0.89	0.89	0.83
4	1.14	0.97	0.98	0.88	0.86	0.83
5	0.97	0.86	0.78	0.76	0.76	0.75

(a) Does nozzle design affect the shape factor? Compare the nozzles with box plots and with an analysis of variance.

(b) Analyze the residuals from this experiment.

(c) Which nozzle designs are different with respect to shape factor? Draw a graph of the average shape factor for each nozzle type and compare this to a scaled t distribution. Compare the conclusions that you draw from this plot to those from Duncan's multiple range test.

5-6 Consider the ratio control algorithm experiment described in Chapter 4 (p. 109). The experiment was actually conducted as a randomized block design, where six time periods were selected as the blocks, and all four ratio control algorithms were tested in each time period. The average cell voltage and the standard deviation of voltage (shown in parentheses) for each cell are as follows:

Ratio Control Algorithms	Time Period					
	1	2	3	4	5	6
1	4.93 (0.05)	4.86 (0.04)	4.75 (0.05)	4.95 (0.06)	4.79 (0.03)	4.88 (0.05)
2	4.85 (0.04)	4.91 (0.02)	4.79 (0.03)	4.85 (0.05)	4.75 (0.03)	4.85 (0.02)
3	4.83 (0.09)	4.88 (0.13)	4.90 (0.11)	4.75 (0.15)	4.82 (0.08)	4.90 (0.12)
4	4.89 (0.03)	4.77 (0.04)	4.94 (0.05)	4.86 (0.05)	4.79 (0.03)	4.76 (0.02)

(a) Analyze the average cell voltage data. Does the choice of ratio control algorithm affect the average cell voltage?

(b) Perform an appropriate analysis on the standard deviation of voltage. (Recall that this is called "pot noise.") Does the choice of ratio control algorithm affect the pot noise?

(c) Conduct any residual analyses that seem appropriate.

(d) Which ratio control algorithm would you select if your objective is to reduce both the average cell voltage *and* the pot noise?

5-7 An aluminum master alloy manufacturer produces grain refiners in ingot form. The company produces the product in four furnaces. Each furnace is known to have its own unique oeprating characteristics, so any experiment run in the foundry that involves more than one furnace will consider furnaces as a nuisance variable. The process engineers suspect that stirring rate impacts the grain size of the product. Each furnace can be run at four different stirring rates. A randomized block design is run for a particular refiner and the resulting grain size data is shown below.

Stirring Rate (rpm)	Furnace			
	1	2	3	4
5	8	4	5	6
10	14	5	6	9.
15	14	6	9	2
20	17	9	3	6

(a) Is there any evidence that stirring rate impacts grain size?

(b) Graph the residuals from this experiment on a normal probability plot. Interpret this plot.

(c) Plot the residuals versus furnace and stirring rate. Does this plot convey any useful information?

(d) What should the process engineers recommend concerning the choice of stirring rate and furnance for this particular grain refiner if small grain size is desirable?

5-8   Analyze the data in Problem 5-2 using the general regression significance test.

5-9   Assuming that chemical types and bolts are fixed, estimate the model parameters τ_i and β_j in Problem 5-1.

5-10 Draw an operating characteristic curve for the design in Problem 5-2. Does the test seem to be sensitive to small differences in the treatment effects?

5-11 Suppose that the observation for chemical type 2 and bolt 3 is missing in Problem 5-1. Analyze the problem by estimating the missing value. Perform the exact analysis and compare the results.

5-12 *Two Missing Values in a Randomized Block.* Suppose that in Problem 5-1 the observations for chemical type 2 and bolt 3 and chemical type 4 and bolt 4 are missing.

(a) Analyze the design by iteratively estimating the missing values, as described in Section 5-1.3.

(b) Differentiate SS_E with respect to the two missing values, equate the results to zero, and solve for estimates of the missing values. Analyze the design using these two estimates of the missing values.

(c) Derive general formulas for estimating two missing values when the observations are in *different* blocks.

(d) Derive general formulas for estimating two missing values when the observations are in the *same* block.

5-13 An industrial engineer is conducting an experiment on eye focus time. He is interested in the effect of the distance of the object from the eye on the focus time. Four different distances are of interest. He has five subjects available for the experiment. Because there may be differences among individuals, he decides to conduct the experiment in a randomized block design. The data obtained follow. Analyze the data and draw appropriate conclusions.

Distance (ft)	Subject				
	1	2	3	4	5
4	10	6	6	6	6
6	7	6	6	1	6
8	5	3	3	2	5
10	6	4	4	2	3

5-14 The effect of five different ingredients (A, B, C, D, E) on the reaction time of a chemical process is being studied. Each batch of new material is only large enough to permit five runs to be made. Furthermore, each run requires approximately $1\frac{1}{2}$ hours, so only five runs can be made in one day. The experimenter decides to run the experiment as a Latin square so that day and batch effects may be systematically controlled. She obtains the data that follow. Analyze these data and draw conclusions.

	Day				
Batch	1	2	3	4	5
1	A = 8	B = 7	D = 1	C = 7	E = 3
2	C = 11	E = 2	A = 7	D = 3	B = 8
3	B = 4	A = 9	C = 10	E = 1	D = 5
4	D = 6	C = 8	E = 6	B = 6	A = 10
5	E = 4	D = 2	B = 3	A = 8	C = 8

5-15 An industrial engineer is investigating the effect of four assembly methods (A, B, C, D) on the assembly time for a color television component. Four operators are selected for the study. Furthermore, the engineer knows that each assembly method produces such fatigue that the time required for the last assembly may be greater than the time required for the first, regardless of the method. That is, a trend develops in the required assembly time. To account for this source of variability, the engineer uses the Latin square design shown below. Analyze the data and draw appropriate conclusions.

Order of Assembly	Operator			
	1	2	3	4
1	C = 10	D = 14	A = 7	B = 8
2	B = 7	C = 18	D = 11	A = 8
3	A = 5	B = 10	C = 11	D = 9
4	D = 10	A = 10	B = 12	C = 14

5-16 Suppose that in Problem 5-14 the observation from batch 3 on day 4 is missing. Estimate the missing value from Equation 5-24, and perform the analysis using the value.

5-17 Consider a $p \times p$ Latin square with rows (α_i), columns (β_k), and treatments (τ_j) fixed. Obtain least squares estimates of the model parameters α_i, β_k, and τ_j.

5-18 Derive the missing value formula (Equation 5-24) for the Latin square design.

5-19 *Designs Involving Several Latin Squares.* [See Cochran and Cox (1957), John (1971).] The $p \times p$ Latin square contains only p observations for each treatment.

To obtain more replications the experimenter may use several squares, say n. It is immaterial whether the squares used are the same or different. The appropriate model is

$$y_{ijkh} = \mu + \rho_h + \alpha_{i(h)} + \tau_j + \beta_{k(h)} + (\tau\rho)_{jh} + \epsilon_{ijkh} \quad \begin{cases} i = 1,2, \ldots, p \\ j = 1,2, \ldots, p \\ k = 1,2, \ldots, p \\ h = 1,2, \ldots, n \end{cases}$$

where y_{ijkh} is the observation on treatment j in row i and column k of the hth square. Note that $\alpha_{i(h)}$ and $\beta_{k(h)}$ are the row and column effects in the hth square, ρ_h is the effect of the hth square, and $(\tau\rho)_{jh}$ is the interaction between treatments and squares.

(a) Set up the normal equations for this model, and solve for estimates of the model parameters. Assume that appropriate side conditions on the parameters are $\Sigma_h \hat{\rho}_h = 0$, $\Sigma_i \hat{\alpha}_{i(h)} = 0$, and $\Sigma_k \hat{\beta}_{k(h)} = 0$ for each h, $\Sigma_j \hat{\tau}_j = 0$, $\Sigma_j (\hat{\tau}\rho)_{jh} = 0$ for each h, and $\Sigma_h (\hat{\tau}\rho)_{jh} = 0$ for each j.

(b) Write down the analysis of variance table for this design.

5-20 Discuss how the operating characteristics curves in the Appendix may be used with the Latin square design.

5-21 Suppose that in Problem 5-14 the data taken on day 5 were incorrectly analyzed and had to be discarded. Develop an appropriate analysis for the remaining data.

5-22 The yield of a chemical process was measured using five batches of raw material, five acid concentrations, five standing times (A, B, C, D, E), and five catalyst concentrations (α, β, γ, δ, ϵ). The Graeco-Latin square that follows was used. Analyze the data and draw conclusions.

	Acid Concentration				
Batch	1	2	3	4	5
1	$A\alpha = 26$	$B\beta = 16$	$C\gamma = 19$	$D\delta = 16$	$E\epsilon = 13$
2	$B\gamma = 18$	$C\delta = 21$	$D\epsilon = 18$	$E\alpha = 11$	$A\beta = 21$
3	$C\epsilon = 20$	$D\alpha = 12$	$E\beta = 16$	$A\gamma = 25$	$B\delta = 13$
4	$D\beta = 15$	$E\gamma = 15$	$A\delta = 22$	$B\epsilon = 14$	$C\alpha = 17$
5	$E\delta = 10$	$A\epsilon = 24$	$B\alpha = 17$	$C\beta = 17$	$D\gamma = 14$

5-23 Suppose that in Problem 5-15 the engineer suspects that the workplaces used by the four operators may represent an additional source of variation. A fourth factor, workplace (α, β, γ, δ), may be introduced and another experiment conducted, yielding the Graeco-Latin square that follows. Analyze the data and draw conclusions.

Order of Assembly	Operator			
	1	2	3	4
1	$C\beta = 11$	$B\gamma = 10$	$D\delta = 14$	$A\alpha = 8$
2	$B\alpha = 8$	$C\delta = 12$	$A\gamma = 10$	$D\beta = 12$
3	$A\delta = 9$	$D\alpha = 11$	$B\beta = 7$	$C\gamma = 15$
4	$D\gamma = 9$	$A\beta = 8$	$C\alpha = 18$	$B\delta = 6$

5-24 Construct a 5 × 5 hypersquare for studying the effects of five factors. Exhibit the analysis of variance table for this design.

5-25 Consider the data in Problems 5-15 and 5-23. Suppressing the Greek letters in 5-23, analyze the data using the method developed in Problem 5-19.

5-26 Consider the randomized block design with one missing value in Table 5-7. Analyze this data by using the exact analysis of the missing value problem discussed in Section 5-1.4. Compare your results to the approximate analysis of these data given in Table 5-8.

Chapter 6
Incomplete Block Designs

6-1 INTRODUCTION

In certain experiments using randomized block designs, we may not be able to run all the treatment combinations in each block. Situations like this usually occur because of shortages of experimental apparatus or facilities or the physical size of the block. For example, in the hardness testing experiment (Example 5-1), suppose that because of their size each coupon can be used only for testing three tips. Therefore, each tip cannot be tested on each coupon. For this type of problem it is possible to use randomized block designs in which every treatment is not present in every block. These designs are known as *randomized incomplete block designs*, and they are the subject of this chapter.

6-2 BALANCED INCOMPLETE BLOCK DESIGNS

When all treatment comparisons are equally important, the treatment combinations used in each block should be selected in a balanced manner, that is, so that any pair of treatments occur together the same number of times as any other pair. Thus, a *balanced incomplete block design* is an incomplete block design in which any two treatments appear together an equal number of times. Suppose that there are a treatments and that each block can hold exactly k ($k < a$) treatments. A balanced incomplete block design may be constructed by taking $\binom{a}{k}$ blocks and assigning a different combination of treatments to each block. Frequently, however, balance can be obtained with fewer than $\binom{a}{k}$ blocks. Tables of balanced incomplete block designs are given in Fisher and Yates (1953), Davies (1956), and Cochran and Cox (1957).

As an example, suppose that a chemical engineer thinks that the time of reaction for a chemical process is a function of the type of catalyst employed. Four catalysts are currently being investigated. The experimental procedure consists of selecting a batch of raw material, loading the pilot plant, applying each catalyst in a separate run of the pilot plant, and observing the reaction time. Since variations in the batches of raw material may affect the performance of the catalysts, the engineer decides to use batches of raw material as blocks. However, each batch is only large enough to permit three catalysts to be run. Therefore, a randomized incomplete block design must be used. The balanced incomplete block design for this experiment, along with the observations recorded, is shown in Table 6-1. The order in which the catalysts are run in each block is randomized.

6-2.1 Statistical Analysis

As usual, we assume that there are a treatments and b blocks. In addition, we assume that each block contains k treatments, that each treatment occurs r times in the design (or is replicated r times), and that there are $N = ar = bk$ total observations. Furthermore, the number of times each pair of treatments appears in the same block is

$$\lambda = \frac{r(k - 1)}{a - 1}$$

If $a = b$, the design is said to be *symmetric*.

The parameter λ must be an integer. To derive the relationship for λ, consider any treatment, say treatment 1. Since treatment 1 appears in r blocks and there are $k - 1$ other treatments in each of those blocks, there are $r(k - 1)$ observations in a block containing treatment 1. These $r(k - 1)$ observations also have to represent the remaining $a - 1$ treatments λ times. Therefore, $\lambda(a - 1) = r(k - 1)$.

The statistical model is

$$y_{ij} = \mu + \tau_i + \beta_j + \epsilon_{ij} \tag{6-1}$$

where y_{ij} is the ith observation in the jth block, μ is the overall mean, τ_i is the effect of the ith treatment, β_j is the effect of the jth block, and ϵ_{ij} is the NID(0,

Table 6-1 Balanced Incomplete Block Design for Catalyst Experiment

Treatment (Catalyst)	Block (Batch of Raw Material)				$y_{i.}$
	1	2	3	4	
1	73	74	—	71	218
2	—	75	67	72	214
3	73	75	68	—	216
4	75	—	72	75	222
$y_{.j}$	221	224	207	218	$870 = y_{..}$

σ^2) random error component. The total variation in the data is expressed by the total corrected sum of squares:

$$SS_T = \sum_i \sum_j y_{ij}^2 - \frac{y_{..}^2}{N} \tag{6-2}$$

Total variability may be partitioned into

$$SS_T = SS_{\text{Treatments (adjusted)}} + SS_{\text{Blocks}} + SS_E$$

where the sum of squares for treatments is adjusted to separate the treatment and the block effects. This adjustment is necessary because each treatment is represented in a different set of r blocks. Thus, differences between unadjusted treatment totals $y_{1.}, y_{2.}, \ldots, y_{a.}$ are also affected by differences between blocks.

The block sum of squares is

$$SS_{\text{Blocks}} = \sum_{j=1}^{b} \frac{y_{.j}^2}{k} - \frac{y_{..}^2}{N} \tag{6-3}$$

where $y_{.j}$ is the total in the jth block. SS_{Blocks} has $b - 1$ degrees of freedom. The adjusted treatment sum of squares is

$$SS_{\text{Treatments (adjusted)}} = \frac{k \sum_{i=1}^{a} Q_i^2}{\lambda a} \tag{6-4}$$

where Q_i is the adjusted total for the ith treatment, which is computed as

$$Q_i = y_{i.} - \frac{1}{k} \sum_{j=1}^{b} n_{ij} y_{.j} \qquad i = 1, 2, \ldots, a \tag{6-5}$$

with $n_{ij} = 1$ if treatment i appears in block j and $n_{ij} = 0$ otherwise. Thus, $(1/k) \cdot \sum_{j=1}^{b} n_{ij} y_{.j}$ is the average of the block totals containing treatment i. The adjusted treatment totals will always sum to zero. $SS_{\text{Treatments (adjusted)}}$ has $a - 1$ degrees of freedom. The error sum of squares is computed by subtraction as

$$SS_E = SS_T - SS_{\text{Treatments (adjusted)}} - SS_{\text{Blocks}} \tag{6-6}$$

and has $N - a - b + 1$ degrees of freedom.

The appropriate statistic for testing the equality of the treatment effects is

$$F_0 = \frac{MS_{\text{Treatments (adjusted)}}}{MS_E}$$

The analysis of variance is summarized in Table 6-2.

Example 6-1

Consider the data in Table 6-1 for the catalyst experiment. This is a balanced incomplete block design with $a = 4$, $b = 4$, $k = 3$, $r = 3$, $\lambda = 2$, and $N = 12$. The

Table 6-2 Analysis of Variance for the Balanced Incomplete Block Design

Source of Variation	Sum of Squares	Degrees of Freedom	Mean Square	F_0
Treatments (adjusted)	$\dfrac{k \sum Q_i^2}{\lambda a}$	$a - 1$	$\dfrac{SS_{\text{Treatments (adjusted)}}}{a - 1}$	$F_0 = \dfrac{MS_{\text{Treatments (adjusted)}}}{MS_E}$
Blocks	$\sum \dfrac{y_{.j}^2}{k} - \dfrac{y_{..}^2}{N}$	$b - 1$	$\dfrac{SS_{\text{Blocks}}}{b - 1}$	
Error	SS_E (by subtraction)	$N - a - b + 1$	$\dfrac{SS_E}{N - a - b + 1}$	
Total	$\sum \sum y_{ij}^2 - \dfrac{y_{..}^2}{N}$	$N - 1$		

analysis of this data is as follows. The total sum of squares is

$$SS_T = \sum_i \sum_j y_{ij}^2 - \frac{y_{..}^2}{12}$$

$$= 63,156 - \frac{(870)^2}{12} = 81.00$$

The block sum of squares is found from Equation 6-3 as

$$SS_{Blocks} = \sum_{j=1}^{4} \frac{y_{.j}^2}{3} - \frac{y_{..}^2}{12}$$

$$= \frac{(221)^2 + (207)^2 + (224)^2 + (218)^2}{3} - \frac{(870)^2}{12} = 55.00$$

To compute the treatment sum of squares adjusted for blocks, we first determine the adjusted treatment totals using Equation 6-5 as

$$Q_1 = (218) - \tfrac{1}{3}(221 + 224 + 218) = -9/3$$
$$Q_2 = (214) - \tfrac{1}{3}(207 + 224 + 218) = -7/3$$
$$Q_3 = (216) - \tfrac{1}{3}(221 + 207 + 224) = -4/3$$
$$Q_4 = (222) - \tfrac{1}{3}(221 + 207 + 218) = 20/3$$

The adjusted sum of squares for treatments is computed from Equation 6-4 as

$$SS_{Treatments\ (adjusted)} = \frac{k \sum_{i=1}^{4} Q_i^2}{\lambda a}$$

$$= \frac{3[(-9/3)^2 + (-7/3)^2 + (-4/3)^2 + (20/3)^2]}{(2)(4)} = 22.75$$

The error sum of squares is obtained by subtraction as

$$SS_E = SS_T - SS_{Treatments\ (adjusted)} - SS_{Blocks}$$
$$= 81.00 - 22.75 - 55.00 = 3.25$$

The analysis of variance is shown in Table 6-3. Since $F_0 > F_{.05,3,5} = 5.41$, we conclude that the catalyst employed has a significant effect on the time of reaction.

∎

Table 6-3   Analysis of Variance for Example 6-1

Source of Variation	Sum of Squares	Degrees of Freedom	Mean Square	F_0
Treatments (adjusted for blocks)	22.75	3	7.58	11.66
Blocks	55.00	3	—	
Error	3.25	5	0.65	
Total	81.00	11		

If the factor under study is fixed, then tests on individual treatment means may be of interest. If orthogonal contrasts are employed, the contrasts must be made on the *adjusted* treatment totals, the $\{Q_i\}$ rather than the $\{y_{i.}\}$. The contrast sum of squares is

$$SS_C = \frac{k\left(\sum_{i=1}^{a} c_i Q_i\right)^2}{\lambda a \sum_{i=1}^{a} c_i^2}$$

where $\{c_i\}$ are the contrast coefficients. Other multiple comparison methods may be used to compare all the pairs of adjusted treatment means, which we will find in Section 6-2.2 are estimated by $\hat{\tau}_i = kQ_i/(\lambda a)$. The standard error of an adjusted treatment mean is

$$S = \sqrt{\frac{kMS_E}{\lambda a}} \qquad\qquad (6\text{-}7)$$

As an example, suppose we apply Duncan's multiple range test to the data in Example 6-1. The adjusted treatment means, in ascending order are

$$\hat{\tau}_1 = kQ_1/(\lambda a) = (3)(-9/3)/(2)(4) = -9/8$$

$$\hat{\tau}_2 = kQ_2/(\lambda a) = (3)(-7/3)/(2)(4) = -7/8$$

$$\hat{\tau}_3 = kQ_3/(\lambda a) = (3)(-4/3)/(2)(4) = -4/8$$

$$\hat{\tau}_4 = kQ_4/(\lambda a) = (3)(20/3)/(2)(4) = 20/8$$

From Appendix Table VII with $\alpha = .05$ and five error degrees of freedom, we obtain $r_{.05}(2,5) = 3.64$, $r_{.05}(3,5) = 3.74$, and $r_{.05}(4,5) = 3.79$. Substituting $MS_E = 0.65$ from Table 6-3 into Equation 6-7 yields

$$S = \sqrt{\frac{kMS_E}{\lambda a}} = \sqrt{\frac{3(0.65)}{2(4)}} = 0.49$$

Thus,

$$R_2 = r_{.05}(2,5)S = (3.64)(0.49) = 1.80$$

$$R_3 = r_{.05}(3,5)S = (3.74)(0.49) = 1.83$$

$$R_4 = r_{.05}(4,5)S = (3.79)(0.49) = 1.86$$

and the comparisons yield

$$4 \text{ vs. } 1 = 20/8 - (-9/8) = 29/8 > 1.86(R_4)$$

$$4 \text{ vs. } 2 = 20/8 - (-7/8) = 27/8 > 1.83(R_3)$$

$$4 \text{ vs. } 3 = 20/8 - (-4/8) = 24/8 > 1.80(R_2)$$

$$3 \text{ vs. } 1 = -4/8 - (-9/8) = 5/8 < 1.83(R_3)$$

$$3 \text{ vs. } 2 = -4/8 - (-7/8) = 3/8 < 1.80(R_2)$$

$$2 \text{ vs. } 1 = -7/8 - (-9/8) = 2/8 < 1.80(R_2)$$

The test results are summarized in Figure 6-1. Catalyst 4 differs from the other three, whereas all other pairs of means do not differ.

In the analysis we have elected to partition the total sum of squares into an adjusted sum of squares for treatments, an unadjusted sum of squares for blocks, and an error sum of squares. Sometimes we would like to assess the block effects. To do this, we require an alternate partitioning of SS_T, that is,

$$SS_T = SS_{\text{Treatments}} + SS_{\text{Blocks (adjusted)}} + SS_E$$

Here $SS_{\text{Treatments}}$ is unadjusted. If the design is symmetric, that is, if $a = b$, then a simple formula may be obtained for $SS_{\text{Blocks (adjusted)}}$. The adjusted block totals are

$$Q_j' = y_{.j} - \frac{1}{r} \sum_{i=1}^{a} n_{ij} y_{i.} \qquad j = 1, 2, \ldots, b \qquad (6\text{-}8)$$

and

$$SS_{\text{Blocks (adjusted)}} = \frac{r \sum_{j=1}^{b} (Q_j')^2}{\lambda b} \qquad (6\text{-}9)$$

The balanced incomplete block design in Example 6-1 is symmetric since $a = b = 4$. Therefore,

$$Q_1' = (221) - \tfrac{1}{3}(218 + 216 + 222) = 7/3$$
$$Q_2' = (224) - \tfrac{1}{3}(218 + 214 + 216) = 24/3$$
$$Q_3' = (207) - \tfrac{1}{3}(214 + 216 + 222) = -31/3$$
$$Q_4' = (218) - \tfrac{1}{3}(218 + 214 + 222) = 0$$

and

$$SS_{\text{Blocks (adjusted)}} = \frac{3[(7/3)^2 + (24/3)^2 + (-31/3)^2 + (0)^2]}{(2)(4)} = 66.08$$

Also,

$$SS_{\text{Treatments}} = \frac{(218)^2 + (214)^2 + (216)^2 + (222)^2}{3} - \frac{(870)^2}{12} = 11.67$$

$\hat{t}_1$	$\hat{t}_2$	$\hat{t}_3$	$\hat{t}_4$
$-9/8$	$-7/8$	$-4/8$	$20/8$

Figure 6-1. Results of Duncan's multiple range test.

A summary of the analysis of variance for the symmetric balanced incomplete block design is given in Table 6-4. Notice that the sums of squares associated with the mean squares in Table 6-4 do not add to the total sum of squares, that is,

$$SS_T \neq SS_{\text{Treatments (adjusted)}} + SS_{\text{Blocks (adjusted)}} + SS_E$$

This is a consequence of the nonorthogonality of treatments and blocks.

6-2.2   Least Squares Estimation of the Parameters

Consider estimating the treatment effects for the balanced incomplete block model. The least squares normal equations are

$$\mu: N\hat{\mu} + r \sum_{i=1}^{a} \hat{\tau}_i + k \sum_{j=1}^{b} \hat{\beta}_j = y_{..}$$

$$\tau_i: r\hat{\mu} + r\hat{\tau}_i + \sum_{j=1}^{b} n_{ij}\hat{\beta}_j = y_{i.} \qquad i = 1, 2, \ldots, a \qquad (6\text{-}10)$$

$$\beta_j: k\hat{\mu} + \sum_{i=1}^{a} n_{ij}\hat{\tau}_i + k\hat{\beta}_j = y_{.j} \qquad j = 1, 2, \ldots, b$$

Imposing $\Sigma \hat{\tau}_i = \Sigma \hat{\beta}_j = 0$, we find that $\hat{\mu} = \bar{y}_{..}$. Furthermore, using the equations for $\{\beta_j\}$ to eliminate the block effects from the equations for $\{\tau_i\}$, we obtain

$$rk\hat{\tau}_i - r\hat{\tau}_i - \sum_{j=1}^{b}\sum_{\substack{p=1 \\ p \neq i}}^{a} n_{ij}n_{pj}\hat{\tau}_p = ky_{i.} - \sum_{j=1}^{b} n_{ij}y_{.j} \qquad (6\text{-}11)$$

Note that the right-hand side of Equation 6-11 is kQ_i, where Q_i is the ith adjusted treatment total (see Equation 6-4). Now, since $\Sigma_{j=1}^{b} n_{ij}n_{pj} = \lambda$ if $p \neq i$ and $n_{pj}^2 =$

Table 6-4   Analysis of Variance Table for Example 6-1, Including Both Treatments and Blocks

Source of Variation	Sum of Squares	Degrees of Freedom	Mean Square	F_0
Treatments (adjusted)	22.75	3	7.58	11.66
Treatments (unadjusted)	11.67	3		
Blocks (unadjusted)	55.00	3		
Blocks (adjusted)	66.08	3	22.03	33.90
Error	3.25	5	0.65	
Total	81.00	11		

n_{pj} (because $n_{pj} = 0$ or 1), we may rewrite Equation 6-11 as

$$r(k - 1)\hat{\tau}_i - \lambda \sum_{\substack{p=1 \\ p \neq i}}^{a} \hat{\tau}_p = kQ_i \qquad i = 1, 2, \ldots, a \qquad (6\text{-}12)$$

Finally, note that the constraint $\sum_{i=1}^{a}\hat{\tau}_i = 0$ implies that $\sum_{\substack{p=1 \\ p \neq i}}^{a} \hat{\tau}_p = -\hat{\tau}_i$ and recall

that $r(k - 1) = \lambda(a - 1)$ to obtain

$$\lambda a \hat{\tau}_i = kQ_i \qquad i = 1, 2, \ldots, a \qquad (6\text{-}13)$$

Therefore, the least squares estimators of the treatment effects in the balanced incomplete block model are

$$\hat{\tau}_i = \frac{kQ_i}{\lambda a} \qquad i = 1, 2, \ldots, a \qquad (6\text{-}14)$$

As an illustration, consider the balanced incomplete block design in Example 6-1. Since $Q_1 = -9/3$, $Q_2 = -7/3$, $Q_3 = -4/3$, and $Q_4 = 20/3$, we obtain

$$\hat{\tau}_1 = \frac{3(-9/3)}{(2)(4)} = -9/8 \qquad \hat{\tau}_2 = \frac{3(-7/3)}{(2)(4)} = -7/8$$

$$\hat{\tau}_3 = \frac{3(-4/3)}{(2)(4)} = -4/8 \qquad \hat{\tau}_4 = \frac{3(20/3)}{(2)(4)} = 20/8$$

as we found in Section 6-2.1.

6-3 RECOVERY OF INTERBLOCK INFORMATION IN THE BALANCED INCOMPLETE BLOCK DESIGN

The analysis of the balanced incomplete block design given in Section 6-2 is usually called the *intrablock analysis* because block differences are eliminated and all contrasts in the treatment effects can be expressed as comparisons between observations in the same block. This analysis is appropriate regardless of whether the blocks are fixed or random. Yates (1940) noted that, if the block effects are uncorrelated random variables with zero means and variance σ_β^2, then one may obtain additional information about the treatment effects τ_i. Yates called the method of obtaining this additional information the *interblock analysis*.

Consider the block totals $y_{.j}$ as a collection of b observations. The model for these observations [following John (1971)] is

$$y_{.j} = k\mu + \sum_{i=1}^{a} n_{ij}\tau_i + \left(k\beta_j + \sum_{i=1}^{a} \epsilon_{ij}\right) \qquad (6\text{-}15)$$

where the term in parentheses may be regarded as error. The interblock estimators of μ and τ_i are found by minimizing the least squares function

$$L = \sum_{j=1}^{b} \left(y_{.j} - k\mu - \sum_{i=1}^{a} n_{ij}\tau_i \right)^2$$

This yields the following least squares normal equations:

$$\mu: \qquad N\tilde{\mu} + r\sum_{i=1}^{a} \tilde{\tau}_i = y_{..} \tag{6-16}$$

$$\tau_i: kr\tilde{\mu} + r\tilde{\tau}_i + \lambda \sum_{\substack{p=1 \\ p \neq i}}^{a} \tilde{\tau}_p = \sum_{j=1}^{b} n_{ij}y_{.j} \qquad i = 1,2,\ldots,a$$

where $\tilde{\mu}$ and $\tilde{\tau}_i$ denote the *interblock* estimators. Imposing the constraint $\sum_{i=1}^{a}\tilde{\tau}_i = 0$, we obtain the solutions to Equations 6-16 as

$$\tilde{\mu} = \bar{y}_{..} \tag{6-17}$$

$$\tilde{\tau}_i = \frac{\displaystyle\sum_{j=1}^{b} n_{ij}y_{.j} - kr\bar{y}_{..}}{r - \lambda} \qquad i = 1,2,\ldots,a \tag{6-18}$$

It is possible to show that the interblock estimators $\{\tilde{\tau}_i\}$ and the intrablock estimators $\{\hat{\tau}_i\}$ are uncorrelated.

The interblock estimators $\{\tilde{\tau}_i\}$ can differ from the intrablock estimators $\{\hat{\tau}_i\}$. For example, the interblock estimators for the balanced incomplete block design in Example 6-1 are computed as follows:

$$\tilde{\tau}_1 = \frac{663 - (3)(3)(72.50)}{3 - 2} = 10.50$$

$$\tilde{\tau}_2 = \frac{649 - (3)(3)(72.50)}{3 - 2} = -3.50$$

$$\tilde{\tau}_3 = \frac{652 - (3)(3)(72.50)}{3 - 2} = -0.50$$

$$\tilde{\tau}_4 = \frac{646 - (3)(3)(72.50)}{3 - 2} = -6.50$$

Note that the values of $\sum_{j=1}^{b} n_{ij}y_{.j}$ were used previously on page in computing the adjusted treatment totals in the intrablock analysis.

Now suppose we wish to combine the interblock and intrablock estimators to obtain a single, unbiased, minimum variance estimate of each τ_i. It is possible to show that both $\hat{\tau}_i$ and $\tilde{\tau}_i$ are unbiased and also that

$$V(\hat{\tau}_i) = \frac{k(a-1)}{\lambda a^2}\sigma^2 \qquad \text{(intrablock)}$$

and

$$V(\tilde{\tau}_i) = \frac{k(a-1)}{a(r-\lambda)} (\sigma^2 + k\sigma_\beta^2) \qquad \text{(interblock)}$$

We use a linear combination of the two estimators, say

$$\tau_i^* = \alpha_1 \hat{\tau}_i + \alpha_2 \tilde{\tau}_i \qquad (6\text{-}19)$$

to estimate τ_i. For this estimation method, the minimum variance unbiased combined estimator τ_i^* should have weights $\alpha_1 = u_1/(u_1 + u_2)$ and $\alpha_2 = u_2/(u_1 + u_2)$, where $u_1 = 1/V(\hat{\tau}_i)$ and $u_2 = 1/V(\tilde{\tau}_i)$. Thus, the optimal weights are inversely proportional to the variances of $\hat{\tau}_i$ and $\tilde{\tau}_i$. This implies that the best combined estimator is

$$\tau_i^* = \frac{\hat{\tau}_i \dfrac{k(a-1)}{a(r-\lambda)} (\sigma^2 + k\sigma_\beta^2) + \tilde{\tau}_i \dfrac{k(a-1)}{\lambda a^2} \sigma^2}{\dfrac{k(a-1)}{\lambda a^2} \sigma^2 + \dfrac{k(a-1)}{a(r-\lambda)} (\sigma^2 + k\sigma_\beta^2)} \qquad i = 1, 2, \ldots, a$$

which can be simplified to

$$\tau_i^* = \frac{kQ_i(\sigma^2 + k\sigma_\beta^2) + \left(\displaystyle\sum_{j=1}^{b} n_{ij} y_{.j} - kr\bar{y}_{..}\right)\sigma^2}{(r-\lambda)\sigma^2 + \lambda a(\sigma^2 + k\sigma_\beta^2)} \qquad i = 1, 2, \ldots, a \quad (6\text{-}20)$$

Unfortunately, Equation 6-20 cannot be used to estimate the τ_i because the variances σ^2 and σ_β^2 are unknown. The usual approach is to estimate σ^2 and σ_β^2 from the data and replace these parameters in Equation 6-20 by the estimates. The estimate usually taken for σ^2 is the error mean square from the intrablock analysis of variance, or the *intrablock error*. Thus,

$$\hat{\sigma}^2 = MS_E$$

The estimate of σ_β^2 is found from the mean square for blocks adjusted for treatments. In general, for a balanced incomplete block design, this mean square is

$$MS_{\text{Blocks (adjusted)}} = \left(\frac{k \displaystyle\sum_{i=1}^{a} Q_i^2}{\lambda a} + \sum_{j=1}^{b} \frac{y_{.j}^2}{k} - \sum_{i=1}^{a} \frac{y_{i.}^2}{r}\right) \Big/ (b-1) \quad (6\text{-}21)$$

and its expected value [which is derived in Graybill (1961)] is

$$E[MS_{\text{Blocks (adjusted)}}] = \sigma^2 + \frac{a(r-1)}{b-1} \sigma_\beta^2$$

Thus, if $MS_{\text{Blocks (adjusted)}} > MS_E$, the estimate of $\hat{\sigma}_\beta^2$ is

$$\hat{\sigma}_\beta^2 = \frac{[MS_{\text{Blocks (adjusted)}} - MS_E](b-1)}{a(r-1)} \qquad (6\text{-}22)$$

and if $MS_{\text{Blocks (adjusted)}} \leq MS_E$, we set $\hat{\sigma}_\beta^2 = 0$. This results in the combined estimator

$$
\tau_i^* = \begin{cases}
\dfrac{kQ_i(\hat{\sigma}^2 + k\hat{\sigma}_\beta^2) + \left(\displaystyle\sum_{j=1}^{b} n_{ij}y_{.j} - kr\bar{y}_{..}\right)\hat{\sigma}^2}{(r - \lambda)\hat{\sigma}^2 + \lambda a(\hat{\sigma}^2 + k\hat{\sigma}_\beta^2)}, & \hat{\sigma}_\beta^2 > 0 \quad (6\text{-}23a) \\[4ex]
\dfrac{y_{i.} - (1/a)y_{..}}{r}, & \hat{\sigma}_\beta^2 = 0 \quad (6\text{-}23b)
\end{cases}
$$

We now compute the combined estimates for the data in Example 6-1. From Table 6-4 we obtain $\hat{\sigma}^2 = MS_E = 0.65$ and $MS_{\text{Blocks (adjusted)}} = 22.03$. (Note that in computing $MS_{\text{Blocks (adjusted)}}$ we make use of the fact that this is a symmetric design. In general, we must use Equation 6-21.) Since $MS_{\text{Blocks (adjusted)}} > MS_E$, we use Equation 6-22 to estimate σ_β^2 as

$$
\hat{\sigma}_\beta^2 = \frac{(22.03 - 0.65)(3)}{4(3 - 1)} = 8.02
$$

Therefore, we may substitute $\hat{\sigma}^2 = 0.65$ and $\hat{\sigma}_\beta^2 = 8.02$ into Equation 6-23a to obtain the combined estimates listed below. For convenience, the intrablock and interblock estimates are also given. In this example, the combined estimates are close to the intrablock estimates because the variance of the interblock estimates is relatively large.

Parameter	Intrablock Estimate	Interblock Estimate	Combined Estimate
τ_1	−1.12	10.50	−1.09
τ_2	−0.88	−3.50	−0.88
τ_3	−0.50	−0.50	−0.50
τ_4	2.50	−6.50	2.47

6-4 PARTIALLY BALANCED INCOMPLETE BLOCK DESIGNS

Although we have concentrated on the balanced case, there are several other types of incomplete block designs that occasionally prove useful. Balanced incomplete block designs do not exist for all combinations of parameters that we might wish to employ because the constraint that λ be an integer can force either the number of blocks or the block size to be excessively large. For example, if there are eight treatments and each block can accommodate three treatments, then for λ to be an integer the smallest number of replications is $r = 21$. This leads to a design of 56 blocks, which is clearly too large for most practical problems. To reduce the number of blocks required in cases such as this, the experimenter can employ *partially balanced* incomplete block designs, in which some pairs of treatments appear together λ_1 times, some pairs appear together λ_2 times, . . . , and the remaining pairs appear together λ_m times. Pairs of treatments that appear together λ_i times are called *ith associates*. The design is then said to have *m associate classes*.

An example of a partially balanced incomplete block design is shown in Table 6-5. Some treatments appear together $\lambda_1 = 2$ times (such as treatments 1 and 2), whereas others appear together only $\lambda_2 = 1$ times (such as treatments 4 and 5). Thus, the design has two associate classes. We now describe the intra-block analysis for these designs.

A partially balanced incomplete block design with two associate classes is described by the following parameters:

1. There are a treatments arranged in b blocks. Each block contains k runs and each treatment appears in r blocks.

2. Two treatments which are *ith* associates appear together in λ_i blocks, $i = 1, 2$.

3. Each treatment has exactly n_i *ith* associates, $i = 1, 2$. The number n_i is independent of the treatment chosen.

4. If two treatments are *ith* associates, then the number of treatments that are *jth* associates of one treatment and *kth* associates of the other treatment is p^i_{jk}, $(i, j, k = 1, 2)$. It is convenient to write the p^i_{jk} as (2×2) matrices with p^i_{jk} the *jkth* element of the *ith* matrix.

For the design in Table 6-5 we may readily verify that $a = 6$, $b = 6$, $k = 3$, $r = 3$, $\lambda_1 = 2$, $\lambda_2 = 1$, $n_1 = 1$, $n_2 = 4$,

$$\{p^1_{jk}\} = \begin{bmatrix} 0 & 0 \\ 0 & 4 \end{bmatrix} \quad \text{and} \quad \{p^2_{jk}\} = \begin{bmatrix} 0 & 1 \\ 1 & 2 \end{bmatrix}$$

We now show how to determine the p^i_{jk}. Consider any two treatments that are first associates, say 1 and 2. For treatment 1, the only first associate is 2 and the second associates are 3, 4, 5, and 6. For treatment 2, the only first associate is

Table 6-5   A Partially Balanced Incomplete Block Design with Two Associate Classes

Block	Treatment Combinations		
1	1	2	3
2	3	4	5
3	2	5	6
4	1	2	4
5	3	4	6
6	1	5	6

1 and the second associates are 3, 4, 5, and 6. Combining this information produces Table 6-6. Counting the number of treatments in the cells of this table, we clearly have the $\{p_{jk}^1\}$ given above. The elements $\{p_{jk}^2\}$ are determined similarly.

The linear statistical model for the partially balanced incomplete block design with two associate classes is

$$y_{ij} = \mu + \tau_i + \beta_j + \epsilon_{ij} \tag{6-24}$$

where μ is the overall mean, τ_i is the ith treatment effect, β_j is the jth block effect, and ϵ_{ij} is the NID(0, σ^2) random error component. We compute a total sum of squares, a block sum of squares (unadjusted), and a treatment sum of squares (adjusted). As before, we call

$$Q_i = y_{i.} - \frac{1}{k}\sum_{j=1}^{b} n_{ij} y_{.j}$$

the adjusted total for the ith treatment. We also define

$$S_1(Q_i) = \sum_{s} Q_s \qquad s \text{ and } i \text{ are first associates}$$

$$\Delta = k^{-2}\{(rk - r + \lambda_1)(rk - r + \lambda_2) + (\lambda_1 - \lambda_2)$$
$$\times [r(k - 1)(p_{12}^1 - p_{12}^2) + \lambda_2 p_{12}^1 - \lambda_1 p_{12}^2]\}$$
$$c_1 = (k\Delta)^{-1}[\lambda_1(rk - r + \lambda_2) + (\lambda_1 - \lambda_2)(\lambda_2 p_{12}^1 - \lambda_1 p_{12}^2)]$$
$$c_2 = (k\Delta)^{-1}[\lambda_2(rk - r + \lambda_1) + (\lambda_1 - \lambda_2)(\lambda_2 p_{12}^1 - \lambda_1 p_{12}^2)]$$

The estimate of the ith treatment effect is

$$\hat{\tau}_i = \frac{1}{r(k-1)}[(k - c_2)Q_i + (c_1 - c_2)S_1(Q_i)]$$

and the adjusted treatment sum of squares is

$$SS_{\text{Treatments (adjusted)}} = \sum_{i=1}^{a} \hat{\tau}_i Q_i \tag{6-25}$$

The analysis of variance is summarized in Table 6-7. To test $H_0: \tau_i = 0$, we use $F_0 = MS_{\text{Treatment (adjusted)}}/MS_E$.

Table 6-6 Relationship of Treatments to 1 and 2

	Treatment 2	
Treatment 1	1st Associate	2nd Associate
1st associate		
2nd associate		3, 4, 5, 6

Table 6-7   Analysis of Variance for the Partially Balanced
Incomplete Block Design with Two Associate Classes

Source of Variation	Sum of Squares	Degrees of Freedom
Treatments (adjusted)	$\displaystyle\sum_{i=1}^{a} \hat{\tau}_i Q_i$	$a - 1$
Blocks	$\displaystyle\frac{1}{k}\sum_{j=1}^{b} y_{.j}^2 - \frac{y_{..}^2}{bk}$	$b - 1$
Error	Subtraction	$bk - b - a + 1$
Total	$\displaystyle\sum_i \sum_j y_{ij}^2 - \frac{y_{..}^2}{bk}$	$bk - 1$

We may show that the variance of any contrast of the form $\hat{\tau}_u - \hat{\tau}_v$ is

$$V(\hat{\tau}_u - \hat{\tau}_v) = \frac{2(k - c_i)\sigma^2}{r(k - 1)}$$

where treatments u and v are ith associates ($i = 1, 2$). This indicates that comparisons between treatments are not all estimated with the same precision. This is a consequence of the partial balance of the design.

We have given only the intrablock analysis. For details of the interblock analysis, refer to Bose and Shimamoto (1952) or John (1971). The second reference contains a good discussion of the general theory of incomplete block designs. An extensive table of partially balanced incomplete block designs with two associate classes has been given by Bose, Clatworthy, and Shrikhande (1954).

6-5   YOUDEN SQUARES

Youden squares are "incomplete" Latin square designs in which the number of columns does not equal the number of rows and treatments. For example, consider the design shown in Table 6-8. Notice that if we append the column (E, A, B, C, D) to this design, the result is a 5×5 Latin square. Most of these designs were developed by W. J. Youden, hence their name.

Although a Youden square is always a Latin square from which at least one column (or row or diagonal) is missing, it is not necessarily true that every Latin square with more than one column (or row or diagonal) missing is a Youden square. The arbitrary removal of more than one column, say, for a Latin square may destroy its balance. In general, a Youden square is a symmetric balanced incomplete block design in which rows correspond to blocks and each treatment occurs exactly once in each column or "position" of the block. Thus, it is possible to construct Youden squares from all symmetric balanced incomplete block de-

Table 6-8 Youden Square for Five Treatments (A, B, C, D, E)

Row	Column			
	1	2	3	4
1	A	B	C	D
2	B	C	D	E
3	C	D	E	A
4	D	E	A	B
5	E	A	B	C

signs, as shown by Smith and Hartley (1948). A table of Youden squares is given in Davies (1956), and other types of incomplete Latin squares are discussed by Cochran and Cox (1957, Chapter 13).

The linear model for a Youden square is

$$y_{ijh} = \mu + \alpha_i + \tau_j + \beta_h + \epsilon_{ijh}$$

where μ is the overall mean, α_i is the ith block effect τ_j is the jth treatment effect, β_h is the hth position effect, and ϵ_{ijh} is the usual NID(0, σ^2) error term. Since positions occur exactly once in each block and once with each treatment, positions are orthogonal to blocks and treatments. The analysis of the Youden square is similar to the analysis of a balanced incomplete block design, except that a sum of squares between the position totals may also be calculated.

Example 6-2

An industrial engineer is studying the effect of five illumination levels on the occurrence of defects in an assembly operation. Because time may be a factor in the experiment, she has decided to run the experiment in five blocks, where each block is a day of the week. However, the department in which the experiment is conducted has four work stations and these stations represent a potential source of variability. The engineer decided to run a Youden square with five rows (days or blocks), four columns (work stations), and five treatments (the illumination levels). The coded data are shown in Table 6-9.

Table 6-9 Coded Data for Example 6-2

Day (Block)	Work Station				$y_{i..}$	Treatment Totals
	1	2	3	4		
1	A = 3	B = 1	C = −2	D = 0	2	$y_{.1.} = 12(A)$
2	B = 0	C = 0	D = −1	E = 7	6	$y_{.2.} = 2(B)$
3	C = −1	D = 0	E = 5	A = 3	7	$y_{.3.} = -4(C)$
4	D = −1	E = 6	A = 4	B = 0	9	$y_{.4.} = -2(D)$
5	E = 5	A = 2	B = 1	C = −1	7	$y_{.5.} = 23(E)$
$y_{..h}$	6	9	7	9	31 = $y_{...}$	

Considering this design as a balanced incomplete block, we find $a = b = 5$, $r = k = 4$, and $\lambda = 3$. Also,

$$SS_T = \sum_i \sum_j \sum_h y_{ijh}^2 - \frac{y_{...}^2}{N} = 183.00 - \frac{(31)^2}{20} = 134.95$$

The adjusted sum of squares for treatments is found as follows:

$$Q_1 = 12 - \tfrac{1}{4}(2 + 7 + 9 + 7) = 23/4$$

$$Q_2 = 2 - \tfrac{1}{4}(2 + 6 + 9 + 7) = -16/4$$

$$Q_3 = -4 - \tfrac{1}{4}(2 + 6 + 7 + 7) = -38/4$$

$$Q_4 = -2 - \tfrac{1}{4}(2 + 6 + 7 + 9) = -32/4$$

$$Q_5 = 23 - \tfrac{1}{4}(6 + 7 + 9 + 7) = 63/4$$

$$SS_{\text{Treatments (adjusted)}} = \frac{k \sum\limits_{i=1}^{a} Q_i^2}{\lambda a}$$

$$= \frac{4[(23/4)^2 + (-16/4)^2 + (-38/4)^2 + (-32/4)^2 + (63/4)^2]}{(3)(5)} = 120.37$$

Also,

$$SS_{\text{Days}} = \sum_{i=1}^{b} \frac{y_{i..}^2}{k} - \frac{y_{...}^2}{N} = \frac{(2)^2 + (6)^2 + (7)^2 + (9)^2 + (7)^2}{4} - \frac{(31)^2}{20} = 6.70$$

$$SS_{\text{Stations}} = \sum_{h=1}^{k} \frac{y_{..h}^2}{b} - \frac{y_{...}^2}{N} = \frac{(6)^2 + (9)^2 + (7)^2 + (9)^2}{5} - \frac{(31)^2}{20} = 1.35$$

and

$$SS_E = SS_T - SS_{\text{Treatments (adjusted)}} - SS_{\text{Days}} - SS_{\text{Stations}}$$

$$= 134.95 - 120.37 - 6.70 - 1.35 = 6.53$$

Block or day effects may be assessed by computing the adjusted sum of squares for blocks. This yields

$$Q_1' = 2 - \tfrac{1}{4}(12 + 2 - 4 - 2) = 0/4$$

$$Q_2' = 6 - \tfrac{1}{4}(2 - 3 - 2 + 23) = 5/4$$

$$Q_3' = 7 - \tfrac{1}{4}(12 - 4 - 2 + 23) = -1/4$$

$$Q_4' = 9 - \tfrac{1}{4}(12 + 2 - 2 + 23) = 1/4$$

$$Q_5' = 7 - \tfrac{1}{4}(12 + 2 - 4 + 23) = -5/4$$

$$SS_{\text{Days (adjusted)}} = \frac{r \sum\limits_{j=1}^{b} Q_j'^2}{\lambda b}$$

$$= \frac{4[(0/4)^2 + (5/4)^2 + (-1/4)^2 + (1/4)^2 + (-5/4)^2]}{(3)(5)} = 0.87$$

The complete analysis of variance is shown in Table 6-10. Illumination levels are significant at 1 percent.

■

Table 6-10   Analysis of Variance for Example 6-2

Source of Variation	Sum of Squares	Degrees of Freedom	Mean Square	F_0
Illumination level, adjusted	120.37	4	30.09	36.87[a]
Days, unadjusted	6.70	4	—	
Days, adjusted	(0.87)	(4)	0.22	
Work station	1.35	3	0.45	
Error	6.53	8	0.82	
Total	134.95	19		

[a]Significant at 1 percent.

6-6   LATTICE DESIGNS

Consider a balanced incomplete block design with k^2 treatments arranged in $b = k(k + 1)$ blocks with k runs per block and $r = k + 1$ replicates. Such a design is called a *balanced lattice*. An example is shown in Table 6-11 for $k^2 = 9$ treatments in 12 blocks of 3 runs each. Notice that the blocks can be grouped into sets such that each set contains a complete replicate. The analysis of variance for the balanced lattice design proceeds like that for a balanced incomplete block design, except that a sum of squares for replicates is computed and removed from the sum of squares for blocks. Replicates will have k degrees of freedom and blocks will have $k^2 - 1$ degrees of freedom.

Lattice designs are frequently used in situations where there are a large number of treatment combinations. In order to reduce the size of the design, the experimenter may resort to *partially* balanced lattices. We briefly describe some of these designs here. Two replicates of a design for k^2 treatments in $2k$ blocks of k runs are called a *simple lattice*. For example, consider the first two replicates of the design in Table 6-11. The partial balance is easily seen, since, for example, treatment 2 appears in the same block with treatments 1, 3, 5, and 8, but does not appear at all with treatments 4, 6, 7, and 9. A lattice design with k^2 treatments in $3k$ blocks grouped into three replicates is called a *triple lattice*. An example would be the first three replicates in Table 6-11. A lattice design for k^2 treatments in $4k$ blocks arranged in four replicates is called a *quadruple lattice*.

Table 6-11   A 3 × 3 Balanced Lattice Design

Block	Replicate I				Block	Replicate III		
1	1	2	3		7	1	5	9
2	4	5	6		8	7	2	6
3	7	8	9		9	4	8	3

Block	Replicate II				Block	Replicate IV		
4	1	4	7		10	1	8	6
5	2	5	8		11	4	2	9
6	3	6	9		12	7	5	3

There are other types of lattice designs that occasionally prove useful. For example, the *cubic lattice* design can be used for k^3 treatments in k^2 blocks of k runs. A lattice design for $k(k + 1)$ treatments in $k + 1$ blocks of size k is called a *rectangular lattice*. Details of the analysis of lattice designs and tables of plans are given in Cochran and Cox (1957).

6-7 PROBLEMS

6-1 An engineer is studying the mileage performance characteristics of five types of gasoline additives. In the road test he wishes to use cars as blocks; however, because of a time constraint, he must use an incomplete block design. He runs the balanced design with the five blocks that follow. Analyze the data and draw conclusions.

	Car				
Additive	1	2	3	4	5
1		17	14	13	12
2	14	14		13	10
3	12		13	12	9
4	13	11	11	12	
5	11	12	10		8

6-2 Construct orthogonal contrasts for the data in Problem 6-1. Compute the sum of squares for each contrast.

6-3 Seven different hardwood concentrations are being studied to determine their effect on the strength of the paper produced. However the pilot plant can only produce three runs each day. As days may differ, the analyst uses the balanced incomplete block design that follows. Analyze the data and draw conclusions.

Hardwood Concentration(%)	Days						
	1	2	3	4	5	6	7
2	114				120		117
4	126	120				119	
6		137	117				134
8	141		129	149			
10		145		150	143		
12			120		118	123	
14				136		130	127

6-4 Analyze the data in Example 6-1 using the general regression significance test.

6-5 Prove that $k\Sigma_{i=1}^{a}Q_i^2/(\lambda a)$ is the adjusted sum of squares for treatments in a balanced incomplete block design.

6-6 An experimenter wishes to compare four treatments in blocks of two runs. Find a balanced incomplete block design for this experiment with six blocks.

6-7 An experimenter wishes to compare eight treatments in blocks of four runs. Find a balanced incomplete block design with 14 blocks and $\lambda = 3$.

6-8 Perform the interblock analysis for the design in Problem 6-1.

6-9 Perform the interblock analysis for the design in Problem 6-3.

6-10 Verify that a balanced incomplete block design with the parameters $a = 8$, $r = 8$, $k = 4$, and $b = 16$ does not exist.

6-11 Show that the variance of the intrablock estimators $\{\hat{\tau}_i\}$ is $k(a - 1)\sigma^2/(\lambda a^2)$.

6-12 Consider the partially balanced incomplete block design that follows. Perform the intrablock analysis discussed in Section 6-4.

			Block			
Treatment	1	2	3	4	5	6
1	14			10		16
2	10		12	15		
3	20	24			19	
4		16		11	10	
5		13	17			12
6			9		10	8

6-13 Perform the intrablock analysis for the following partially balanced incomplete block design:

				Block					
Treatment	1	2	3	4	5	6	7	8	9
1	164	153	160						
2	118							104	112
3	150			135			141		
4		137				125			121
5			103		98			109	
6		120		108	100				
7			100			92	94		
8				123		130		115	109
9					119		111		

6-14   Using the design in Table 6-5, verify the following relationships between the parameters of a partially balanced incomplete block design with two associate classes:

$$n_1 + n_2 = a - 1 \qquad n_1\lambda_1 + n_2\lambda_2 = r(k - 1)$$
$$p_{11}^1 + p_{12}^2 = n_1 \qquad p_{21}^1 + p_{22}^1 = n_2$$
$$p_{11}^1 + p_{12}^1 = n_1 - 1 \qquad p_{21}^2 + p_{22}^2 = n_2 - 1$$
$$n_1 p_{12}^1 = n_2 p_{11}^2 \qquad n_1 p_{22}^1 = n_2 p_{12}^2$$

6-15   Analyze the following Youden square design:

		Position		
Blocks	1	2	3	4
1	A = 2	B = 9	C = 0	D = 14
2	B = 6	A = 5	E = 5	C = 3
3	C = 1	D = 9	A = 0	E = 7
4	D = 8	E = 8	B = 10	A = 4
5	E = 7	C = 6	D = 11	B = 10

6-16   Consider the array

1	2	3	4
5	6	7	8
9	10	11	12
13	14	15	16

Show that the rows and columns of this arrangement form a partially balanced incomplete block design for 16 treatments in 8 blocks. Verify that this is also a simple lattice design.

6-17   *Extended Incomplete Block Designs.* Occasionally the block size obeys the relationship $a < k < 2a$. An extended incomplete block design consists of a single replicate of each treatment in each block along with an incomplete block design with $k^* = k - a$. In the balanced case, the incomplete block design will have parameters $k^* = k - a$, $r^* = r - b$, and λ^*. Write out the statistical analysis. (*Hint:* In the extended incomplete block design, we have $\lambda = 2r - b + \lambda^*$.)

Chapter 7
Introduction to Factorial Designs

7-1 BASIC DEFINITIONS AND PRINCIPLES

Many experiments involve the study of the effects of two or more factors. In general, *factorial designs* are most efficient for this type of experiment. By a factorial design, we mean that in each complete trial or replication of the experiment all possible combinations of the levels of the factors are investigated. For example, if there are a levels of factor A and b levels of factor B, then each replicate contains all ab treatment combinations. When factors are arranged in a factorial design, they are often said to be *crossed*.

The effect of a factor is defined to be the change in response produced by a change in the level of the factor. This is frequently called a *main effect* because it refers to the primary factors of interest in the experiment. For example, consider the data in Table 7-1. The main effect of factor A could be thought of as the difference between the average response at the first level of A and the average response at the second level of A. Numerically, this is

$$A = \frac{40 + 52}{2} - \frac{20 + 30}{2} = 21$$

That is, increasing factor A from level 1 to level 2 causes an average response increase of 21 units. Similarly, the main effect of B is

$$B = \frac{30 + 52}{2} - \frac{20 + 40}{2} = 11$$

If the factors appear at more than two levels, the above procedure must be modified since there are many ways to express the differences between the average responses. This point is discussed more completely later.

In some experiments, we may find that the difference in response between the levels of one factor is not the same at all levels of the other factors. When this occurs, there is an *interaction* between the factors. For example, consider

Table 7-1 A Factorial Experiment

		Factor B	
		B_1	B_2
Factor A	A_1	20	30
	A_2	40	52

the data in Table 7-2. At the first level of factor B, the A effect is

$$A = 50 - 20 = 30$$

and at the second level of factor B, the A effect is

$$A = 12 - 40 = -28$$

Since the effect of A depends on the level chosen for factor B, we see that there is interaction between A and B.

These ideas may be illustrated graphically. Figure 7-1 plots the response data in Table 7-1 against factor A for both levels of factor B. Note that the B_1 and B_2 lines are approximately parallel, indicating a lack of interaction between factors A and B. Similarly, Figure 7-2 plots the response data in Table 7-2. Here we see that the B_1 and B_2 lines are not parallel. This indicates an interaction between factors A and B. Graphs such as these are frequently very useful in interpreting significant interactions and in reporting results to nonstatistically trained management. However, they should not be utilized as the sole technique of data analysis because their interpretation is subjective and their appearance is often misleading.

Table 7-2 A Factorial Experiment with Interaction

		Factor B	
		B_1	B_2
Factor A	A_1	20	40
	A_2	50	12

Figure 7-1. A factorial experiment without interaction.

Note that when an interaction is large, the corresponding main effects have little practical meaning. For the data of Table 7-2, we would estimate the main effect of A to be

$$A = \frac{50 + 12}{2} - \frac{20 + 40}{2} = 1$$

which is very small, and we are tempted to conclude that there is no effect due to A. However, when we examine the effects of A at *different levels of factor B*, we see that this is not the case. Factor A has an effect, but it *depends on the level of factor B*. That is, knowledge of the AB interaction is more useful than knowledge of the main effect. A significant interaction will often *mask* the significance of main effects. This is clearly indicated by the data in Table 7-2. In the presence of significant interaction, the experimenter must usually examine the levels of one factor, say A, with levels of the other factors fixed to draw conclusions about the main effect of A.

7-2 THE ADVANTAGE OF FACTORIALS

The advantage of factorial designs can be easily illustrated. Suppose we have two factors A and B, each at two levels. Denote the levels of the factors by A_1, A_2,

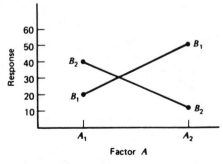

Figure 7-2. A factorial experiment with interaction.

Table 7-3   One-Factor at a Time Method

		Factor B	
		B_1	B_2
Factor A	A_1	A_1B_1	A_1B_2
	A_2	A_2B_1	

B_1, and B_2. Information on both factors could be obtained by varying the factors one at a time, as shown in Table 7-3. The effect of changing factor A is given by $A_2B_1 - A_1B_1$. Since experimental error is present, it is desirable to take two observations, say, at each treatment combination and estimate the effects of the factors using average responses. Thus, a total of six observations are required.

If a factorial experiment had been performed, an additional treatment combination, A_2B_2, would have been taken. Now, using just *four* observations, two estimates of the A effect can be made: $A_2B_1 - A_1B_1$ and $A_2B_2 - A_1B_2$. Similarly, two estimates of the B effect can be made. These two estimates of each main effect could be averaged to produce average main effects that are *just as precise* as those from the single-factor experiment, but only tour total observations are required and we would say that the relative efficiency of the factorial design to the one-factor-at-a-time experiment is $\frac{6}{4} = 1.5$. Generally, this relative efficiency will increase as the number of factors increases, as shown in Figure 7-3.

Now suppose interaction is present. If the one-factor-at-a-time design indi-

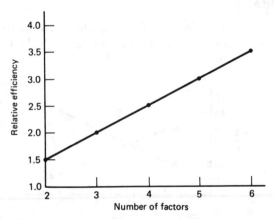

Figure 7-3.   Relative efficiency of a factorial design to a one-factor-at-a-time experiment (two-factor levels).

cated that $A_1 B_2$ and $A_2 B_1$ gave better responses than $A_1 B_1$, a logical conclusion would be that $A_2 B_2$ would be even better. However, if interaction is present, this conclusion may be *seriously in error*. For an example, refer to the data in Table 7-2.

In summary, note that factorial designs have several advantages. They are more efficient than one-factor-at-a-time experiments. Furthermore, a factorial design is necessary when interactions may be present to avoid misleading conclusions. Finally, factorial designs allow the effects of a factor to be estimated at several levels of the other factors, yielding conclusions that are valid over a range of experimental conditions.

7-3   THE TWO-FACTOR FACTORIAL DESIGN

7-3.1   An Example

The simplest types of factorial designs involve only two factors or sets of treatments. There are a levels of factor A and b levels of factor B, and these are arranged in a factorial design; that is, each replicate of the experiment contains all ab treatment combinations. In general, there are n replicates.

As an example of a factorial design involving two factors, an engineer is designing a battery for use in a device that will be subjected to some extreme variations in temperature. The only design parameter that he can select at this point is the plate material for the battery, and he has three possible choices. When the device is manufactured and is shipped to the field, the engineer has no control over the temperature extremes that the device will encounter, and he knows from experience that temperature will probably impact the effective battery life. However, temperature can be controlled in the product development laboratory for the purposes of a test.

The engineer decides to test all three plate materials at three temperature levels—15°F, 70°F, and 125°F—as these temperature levels are consistent with the product end-use environment. Four batteries are tested at each combination of plate material and temperature, and all 36 tests are run in random order. The experiment and the resulting observed battery life data are given in Table 7-4.

In this problem the engineer wants to answer the following questions:

1.   What effects do material type and temperature have on the life of the battery?

2.   Is there a choice of material that would give *uniformly long life regardless of temperature?*

This last question is particularly important. It may be possible to find a material alternative that is not greatly affected by temperature. If this is so, the engineer

Table 7-4 Life (in hours) Data for the Battery Design Example

Material Type	Temperature (°F)					
	15		70		125	
1	130	155	34	40	20	70
	74	180	80	75	82	58
2	150	188	126	122	25	70
	159	126	106	115	58	45
3	138	110	174	120	96	104
	168	160	150	139	82	60

can make the battery *robust* to temperature variation in the field. This is an example of using statistical experimental design for *robust product design,* a very important engineering problem.

This design is a specific example of the general case of a two-factor factorial. To pass to the general case, let y_{ijk} be the observed response when factor A is at the *i*th level ($i = 1, 2, \ldots, a$) and factor B is at the *j*th level ($j = 1, 2, \ldots, b$) for the *k*th replicate ($k = 1, 2, \ldots, n$). In general, the observed data will appear as in Table 7-5. The order in which the *abn* observations are taken·is selected at random so that this design is a completely randomized design.

The observations may be described by the linear statistical model

$$y_{ijk} = \mu + \tau_i + \beta_j + (\tau\beta)_{ij} + \epsilon_{ijk} \begin{cases} i = 1, 2, \ldots, a \\ j = 1, 2, \ldots, b \\ k = 1, 2, \ldots, n \end{cases} \quad (7\text{-}1)$$

where μ is the overall mean effect, τ_i is the effect of the *i*th level of the row factor A, β_j is the effect of the *j*th level of column factor B, $(\tau\beta)_{ij}$ is the effect of the

Table 7-5 General Arrangement for a Two Factorial Design

		Factor B			
		1	2	...	b
Factor A	1	$y_{111}, y_{112}, \ldots, y_{11n}$	$y_{121}, y_{122}, \ldots, y_{12n}$		$y_{1b1}, y_{1b2}, \ldots, y_{1bn}$
	2	$y_{211}, y_{212}, \ldots, y_{21n}$	$y_{221}, y_{222}, \ldots, y_{22n}$		$y_{2b1}, y_{2b2}, \ldots, y_{2bn}$
	. . .				
	a	$y_{a11}, y_{a12}, \ldots, y_{a1n}$	$y_{a21}, y_{a22}, \ldots, y_{a2n}$		$y_{ab1}, y_{ab2}, \ldots, y_{abn}$

interaction between τ_i and β_j, and ϵ_{ijk} is a random error component. Both factors are initially assumed to be fixed, and the treatment effects are defined as deviations from the overall mean, so $\Sigma_{i=1}^{a}\tau_i = 0$ and $\Sigma_{j=1}^{b}\beta_j = 0$. Similarly, the interaction effects are fixed and are defined such that $\Sigma_{i=1}^{a}(\tau\beta)_{ij} = 0$. Since there are n replicates of the experiment, there are abn total observations.

In the two-factor factorial, both row and column factors (or treatments), A and B, are of equal interest. Specifically, we are interested in testing hypotheses about the equality of row treatment effects, say

$$H_0: \quad \tau_1 = \tau_2 = \cdots = \tau_a = 0$$
$$H_1: \quad \text{at least one } \tau_i \neq 0$$

(7-2a)

and the equality of column treatment effects, say

$$H_0: \quad \beta_1 = \beta_2 = \cdots = \beta_b = 0$$
$$H_1: \quad \text{at least one } \beta_j \neq 0$$

(7-2b)

We are also interested in determining whether row and column treatments *interact*. Thus, we also wish to test

$$H_0: \quad (\tau\beta)_{ij} = 0 \qquad \text{for all } i, j$$
$$H_1: \quad \text{at least one } (\tau\beta)_{ij} \neq 0$$

(7-2c)

We now discuss how these hypotheses are tested using a two-factor analysis of variance.

7-3.2 Statistical Analysis of the Fixed Effects Model

Let $y_{i..}$ denote the total of all observations under the ith level of factor A, $y_{.j.}$ denote the total of all observations under the jth level of factor B, $y_{ij.}$ denote the total of all observations in the ijth cell, and $y_{...}$ denote the grand total of all the observations. Define $\bar{y}_{i..}$, $\bar{y}_{.j.}$, $\bar{y}_{ij.}$, and $\bar{y}_{...}$ as the corresponding row, column, cell, and grand averages. Expressed mathematically,

$$y_{i..} = \sum_{j=1}^{b}\sum_{k=1}^{n} y_{ijk} \qquad \bar{y}_{i..} = \frac{y_{i..}}{bn} \qquad i = 1, 2, \ldots, a$$

$$y_{.j.} = \sum_{i=1}^{a}\sum_{k=1}^{n} y_{ijk} \qquad \bar{y}_{.j.} = \frac{y_{.j.}}{an} \qquad j = 1, 2, \ldots, b$$

$$y_{ij.} = \sum_{k=1}^{n} y_{ijk} \qquad \bar{y}_{ij.} = \frac{y_{...}}{n} \qquad \begin{array}{l} i = 1, 2, \ldots, a \\ j = 1, 2, \ldots, b \end{array}$$

$$y_{...} = \sum_{i=1}^{a}\sum_{j=1}^{b}\sum_{k=1}^{n} y_{ijk} \qquad \bar{y}_{...} = \frac{y_{...}}{abn}$$

(7-3)

The total corrected sum of squares may be written as

$$\sum_{i=1}^{a} \sum_{j=1}^{b} \sum_{k=1}^{n} (y_{ijk} - \bar{y}_{...})^2 = \sum_{i=1}^{a} \sum_{j=1}^{b} \sum_{k=1}^{n} [(\bar{y}_{i..} - \bar{y}_{...}) + (\bar{y}_{.j.} - \bar{y}_{...})$$

$$+ (\bar{y}_{ij.} - \bar{y}_{i..} - \bar{y}_{.j.} + \bar{y}_{...}) + (y_{ijk} - \bar{y}_{ij.})]^2$$

$$= bn \sum_{i=1}^{a} (\bar{y}_{i..} - \bar{y}_{...})^2 + an \sum_{j=1}^{b} (\bar{y}_{.j.} - \bar{y}_{...})^2 \qquad (7-4)$$

$$+ n \sum_{i=1}^{a} \sum_{j=1}^{b} (\bar{y}_{ij.} - \bar{y}_{i..} - \bar{y}_{.j.} + \bar{y}_{...})^2$$

$$+ \sum_{i=1}^{a} \sum_{j=1}^{b} \sum_{k=1}^{n} (y_{ijk} - \bar{y}_{ij.})^2$$

since the six cross products on the right-hand side are zero. Notice that the total sum of squares has been partitioned into a sum of squares due to "rows" or factor A, (SS_A); a sum of squares due to "columns" or factor B, (SS_B); a sum of squares due to the interaction between A and B, (SS_{AB}); and a sum of squares due to error, (SS_E). From the last component on the right-hand side of Equation 7-4, we see that there must be at least two replicates $(n \geq 2)$ to obtain an error sum of squares.

We may write Equation 7-4 symbolically as

$$SS_T = SS_A + SS_B + SS_{AB} + SS_E \qquad (7-5)$$

The number of degrees of freedom associated with each sum of squares is

Effect	Degrees of Freedom
A	$a - 1$
B	$b - 1$
AB interaction	$(a - 1)(b - 1)$
Error	$ab(n - 1)$
Total	$abn - 1$

We may justify this allocation of the $abn - 1$ total degrees of freedom to the sums of squares as follows: The main effects A and B have a and b levels, respectively; therefore they have $a - 1$ and $b - 1$ degrees of freedom as shown. The interaction degrees of freedom are simply the number of degrees of freedom for cells (which is $ab - 1$) minus the number of degrees of freedom for the two main effects A and B; that is, $ab - 1 - (a - 1) - (b - 1) = (a - 1)(b - 1)$. Within each of the ab cells, there are $n - 1$ degrees of freedom between the n replicates; thus there are $ab(n - 1)$ degrees of freedom for error. Note that the number of degrees of freedom on the right-hand side of Equation 7-5 adds to the total number of degrees of freedom.

Each sum of squares divided by its degrees of freedom is a *mean square*. The expected values of the mean squares are

$$E(MS_A) = E\left(\frac{SS_A}{a-1}\right) = \sigma^2 + \frac{bn \sum\limits_{i=1}^{a} \tau_i^2}{a-1}$$

$$E(MS_B) = E\left(\frac{SS_B}{b-1}\right) = \sigma^2 + \frac{an \sum\limits_{j=1}^{b} \beta_j^2}{b-1}$$

$$E(MS_{AB}) = E\left(\frac{SS_{AB}}{(a-1)(b-1)}\right) = \sigma^2 + \frac{n \sum\limits_{i=1}^{a} \sum\limits_{j=1}^{b} (\tau\beta)_{ij}^2}{(a-1)(b-1)}$$

and

$$E(MS_E) = E\left(\frac{SS_E}{ab(n-1)}\right) = \sigma^2$$

Notice that if the null hypotheses of no row treatment effects, no column treatment effects, and no interaction are true, then MS_A, MS_B, MS_{AB}, and MS_E all estimate σ^2. However, if there are differences between row treatment effects, say, then MS_A will be larger than MS_E. Similalrly, if there are column treatment effects or interaction present, then the corresponding mean squares will be larger than MS_E. Therefore, to test the significance of both main effects and their interaction, simply divide the corresponding mean square by the error mean square. Large values of this ratio imply that the data do not support the null hypothesis.

If we assume that the model (Equation 7-1) is adequate and that the error terms ϵ_{ijk} are normally and independently distributed with constant variance σ^2, then each of the ratios of mean squares MS_A/MS_E, MS_B/MS_E, and MS_{AB}/MS_E are distributed as F with $a-1$, $b-1$, and $(a-1)(b-1)$ numerator degrees of freedom, respectively, and $ab(n-1)$ denominator degrees of freedom,[1] and the critical region would be the upper tail of the F distribution. The test procedure is usually summarized in an analysis of variance table, as shown in Table 7-6.

Computational formulas for the sums of squares in Equation 7-5 may be obtained easily. The total sum of squares is computed as usual by

$$SS_T = \sum_{i=1}^{a} \sum_{j=1}^{b} \sum_{k=1}^{n} y_{ijk}^2 - \frac{y_{...}^2}{abn} \tag{7-6}$$

The sums of squares for the main effects are

$$SS_A = \sum_{i=1}^{a} \frac{y_{i..}^2}{bn} - \frac{y_{...}^2}{abn} \tag{7-7}$$

[1] The F test may also be viewed as an approximation to a randomization test, as noted previously.

Table 7-6 The Analysis of Variance Table for the Two-Factor Factorial, Fixed Effects Model

Source of Variation	Sum of Squares	Degrees of Freedom	Mean Square	F_0
A treatments	SS_A	$a - 1$	$MS_A = \dfrac{SS_A}{a - 1}$	$F_0 = \dfrac{MS_A}{MS_E}$
B treatments	SS_B	$b - 1$	$MS_B = \dfrac{SS_B}{b - 1}$	$F_0 = \dfrac{MS_B}{MS_E}$
Interaction	SS_{AB}	$(a - 1)(b - 1)$	$MS_{AB} = \dfrac{SS_{AB}}{(a - 1)(b - 1)}$	$F_0 = \dfrac{MS_A}{MS_E}$
Error	SS_E	$ab(n - 1)$	$MS_E = \dfrac{SS_E}{ab(n - 1)}$	
Total	SS_T	$abn - 1$		

and

$$SS_B = \sum_{j=1}^{b} \frac{y_{.j.}^2}{an} - \frac{y_{...}^2}{abn} \tag{7-8}$$

It is convenient to obtain the SS_{AB} in two stages. First we compute the sum of squares between the ab cell totals, which is called the sum of squares due to "subtotals":

$$SS_{\text{Subtotals}} = \sum_{i=1}^{a} \sum_{j=1}^{b} \frac{y_{ij.}^2}{n} - \frac{y_{...}^2}{abn}$$

This sum of squares also contains SS_A and SS_B. Therefore, the second step is to compute SS_{AB} as

$$SS_{AB} = SS_{\text{Subtotals}} - SS_A - SS_B \tag{7-9}$$

We may compute SS_E by subtraction as

$$SS_E = SS_T - SS_{AB} - SS_A - SS_B \tag{7-10}$$

or

$$SS_E = SS_T - SS_{\text{Subtotals}}$$

Example 7-1

The Battery Design Experiment

Table 7-7 presents the effective life (in hours) observed in the battery design example described in Section 7-3.1. The row and column totals are shown in the margins of the table and the circled numbers are the cell totals.

Table 7-7 Life Data (in hours) for the Battery Design Experiment

Material Type	Temperature (°F) 15		Temperature (°F) 70		Temperature (°F) 125		$y_{i..}$
1	130 155 74 180	(539)	34 40 80 75	(229)	20 70 82 58	(230)	998
2	150 188 159 126	(623)	136 122 106 115	(479)	25 70 58 45	(198)	1300
3	138 110 168 160	(576)	174 120 150 139	(583)	96 104 82 60	(342)	1501
$y_{.j.}$	1738		1291		770		$3799 = y_{...}$

The sums of squares are computed as follows:

$$SS_T = \sum_{i=1}^{a} \sum_{j=1}^{b} \sum_{k=1}^{n} y_{ijk}^2 - \frac{y_{...}^2}{abn}$$

$$= (130)^2 + (155)^2 + (74)^2 + \cdots + (60)^2 - \frac{(3799)^2}{36} = 77,646.97$$

$$SS_{Material} = \sum_{i=1}^{a} \frac{y_{i..}^2}{bn} - \frac{y_{...}^2}{abn}$$

$$= \frac{(998)^2 + (1300)^2 + (1501)^2}{(3)(4)} - \frac{(3799)^2}{36} = 10,683.72$$

$$SS_{Temperature} = \sum_{j=1}^{b} \frac{y_{.j.}^2}{an} - \frac{y_{...}^2}{abn}$$

$$= \frac{(1738)^2 + (1291)^2 + (770)^2}{(3)(4)} - \frac{(3799)^2}{36} = 39,118.72$$

$$SS_{Interaction} = \sum_{i=1}^{a} \sum_{j=1}^{b} \frac{y_{ij.}^2}{n} - \frac{y_{...}^2}{abn} - SS_{Material} - SS_{Temperature}$$

$$= \frac{(539)^2 + (229)^2 + \cdots + (342)^2}{4} - \frac{(3799)^2}{36} - 10,683.72$$

$$- 39,118.72 = 9613.78$$

and

$$SS_E = SS_T - SS_{Material} - SS_{Temperature} - SS_{Interaction}$$

$$= 77,646.97 - 10,683.72 - 39,118.72 - 9613.78 = 18,230.75$$

The analysis of variance is shown in Table 7-8. Since $F_{.05,4,27} = 2.73$, we conclude that there is a significant interaction between material types and temperature. Furthermore, $F_{.05,2,27} = 3.35$, so the main effects of material type and temperature are also significant.

To assist in interpreting the results of this experiment, it is helpful to construct a graph of the average responses at each treatment combination. This graph is shown

Table 7-8 Analysis of Variance for Battery Life Data

Source of Variation	Sum of Squares	Degrees of Freedom	Mean Square	F_0
Material types	10,683.72	2	5,341.86	7.91
Temperature	39,118.72	2	19,558.36	28.97
Interaction	9,613.78	4	2,403.44	3.56
Error	18,230.75	27	675.21	
Total	77,646.97	35		

in Figure 7-4. The significant interaction is indicated by the lack of parallelism of the lines. In general, longer life is attained at low temperature, regardless of material type. Changing from low to intermediate temperature, battery life with material type 3 actually *increases*, whereas it decreases for types 1 and 2. From intermediate to high temperature, battery life decreases for material types 2 and 3 and is essentially unchanged for type 1. Material type 3 seems to give the best results if we want less loss of effective life as the temperature changes.

■

Multiple Comparisons ▪ When the analysis of variance indicates that row or column means differ, it is usually of interest to make comparisons between the individual row or column means to discover the specific differences. The multiple comparison methods discussed in Chapter 3 are useful in this regard.

We now illustrate the use of Duncan's multiple range test on the battery life data in Example 7-1. Note that in this experiment, interaction is significant. When interaction is significant, comparisons between the means of one factor (e.g., A) may be obscured by the AB interaction. One approach to this situation is to fix factor B at a specific level and apply Duncan's multiple range test to the means of factor A at that level. To illustrate, suppose that in Example 7-1 we are interested in detecting differences among the means of the three material types. Since

Figure 7-4. Material type–temperature plot for Example 7-1.

interaction is significant, we make this comparison at just one level of temperature, say level 2 (70 degrees). We assume that the best estimate of the error
variance is the MS_E from the analysis of variance table, utilizing the assumption
that the experimental error variance is the same over all treatment combinations.
The three material type means arranged in ascending order are

$$\bar{y}_{12.} = 57.25 \qquad \text{(material type 1)}$$
$$\bar{y}_{22.} = 119.75 \qquad \text{(material type 2)}$$
$$\bar{y}_{32.} = 145.75 \qquad \text{(material type 3)}$$

The standard error of these treatment means is

$$S_{\bar{y}_{i2.}} = \sqrt{\frac{MS_E}{n}} = \sqrt{\frac{675.21}{4}} = 12.99$$

since each mean contains $n = 4$ observations. From Appendix Table VII we
obtain the values $r_{.05}(2, 27) \simeq 2.91$ and $r_{.05}(3, 27) \simeq 3.06$. The least significant
ranges are

$$R_2 = r_{.05}(2, 27)S_{\bar{y}_{i2.}} = (2.91)(12.99) = 37.80$$
$$R_3 = r_{.05}(3, 27)S_{\bar{y}_{i2.}} = (3.06)(12.99) = 39.75$$

and the comparisons yield

$$3 \text{ vs. } 1 = 145.75 - 57.25 = 88.50 > 39.75(R_3)$$
$$3 \text{ vs. } 2 = 145.75 - 119.75 = 26.00 < 37.80(R_2)$$
$$2 \text{ vs. } 1 = 119.75 - 57.25 = 62.50 > 37.80(R_2)$$

This analysis indicates that at the temperature level 70 degrees, the mean battery
life is the same for material types 2 and 3, and that the mean battery life for
material type 1 is significantly lower in comparison to both types 2 and 3.

If interaction is significant, the experimenter could compare *all* ab cell means
to determine which ones differ significantly. In this analysis, differences between
cell means include interaction effects as well as both main effects. In Example
7-1, this would give 36 comparisons between all possible pairs of the nine cell
means.

Sample Computer Output ▪ Figure 7-5 presents computer output from the SAS
General Linear Models Procedure for the battery life data in Example 7-1. Note
that

$$SS_{\text{Model}} = SS_{\text{Material}} + SS_{\text{Temperature}} + SS_{\text{Interaction}}$$
$$= 10,683.72 + 39,118.72 + 9613.78$$
$$= 59,416.22$$

and that

$$R^2 = \frac{SS_{Model}}{SS_{Total}} = \frac{59,416.22}{77,646.97} = 0.765210$$

That is, about 77 percent of the variability in the battery life is explained by the plate material in the battery, the temperature, and the material type–temperature interaction. The program also computes the material type, temperature, and material type–temperature interaction sums of squares. (Recall that both type I and type III sums of squares are always identical for balanced data.) The residuals from the fitted model are also displayed on the computer output. In Section 7-3.3, we illustrate how to use these residuals in model adequacy checking.

7-3.3 Model Adequacy Checking

Before the conclusions from the analysis of variance are adopted, the adequacy of the underlying model should be checked. As before, the primary diagnostic tool is residual analysis. The residuals for the two-factor factorial model are

$$e_{ijk} = y_{ijk} - \hat{y}_{ijk} \qquad (7\text{-}11)$$

and since the fitted value $\hat{y}_{ijk} = \bar{y}_{ij.}$ (the average of the observations in the ijth cell), Equation 7-11 becomes

$$e_{ijk} = y_{ijk} - \bar{y}_{ij} \qquad (7\text{-}12)$$

The residuals from the battery life data in Example 7-1 are shown in Table 7-9. The normal probability plot and dot diagram of these residuals (Figure 7-6) do not reveal anything particularly troublesome, although the largest negative residual (-60.75 at 15°F for material type 1) does stand out somewhat from the others. The standardized value of this residual is $-60.75/\sqrt{675.21} = -2.34$, and this is the only residual whose absolute value is larger than 2.

Figure 7-7 plots the residuals versus the fitted values $\hat{y}_{ijk}$. This plot indicates a mild tendency for the variance of the residuals to increase as the battery life increases. Figures 7-8 and 7-9 plot the residuals versus material types and temperature, respectively. Both plots indicate mild inequality of variance, with the

Table 7-9 Residuals for Example 7-1

Material Type	Temperature (°F)					
	15		70		125	
1	−4.75	20.25	−23.25	−17.25	−37.50	12.50
	−60.75	45.25	22.75	17.75	24.50	0.50
2	−5.75	32.25	16.25	2.25	−24.50	20.50
	3.25	−29.75	−13.75	−4.75	8.50	−4.50
3	−6.00	−34.00	28.25	−25.75	10.50	18.50
	24.00	16.00	4.25	−6.75	−3.50	−25.50

S T A T I S T I C A L   A N A L Y S I S   S Y S T E M

13:16 WEDNESDAY, JANUARY 17, 1990

GENERAL LINEAR MODELS PROCEDURE

DEPENDENT VARIABLE: Y

SOURCE	DF	SUM OF SQUARES	MEAN SQUARE	F VALUE	PR > F
MODEL	8	59416.22222222	7427.02777778	11.00	0.0001
ERROR	27	18230.75000000	675.21296296		
CORRECTED TOTAL	35	77646.97222222			

R-SQUARE	C.V.
0.765210	24.6237

STD DEV	Y MEAN
25.98486026	105.52777778

SOURCE	DF	TYPE I SS	F VALUE	PR > F
MATERIAL	2	10683.72222222	7.91	0.0020
TEMP	2	39118.72222222	28.97	0.0001
MATERIAL*TEMP	4	9613.77777778	3.56	0.0186

SOURCE	DF	TYPE III SS	F VALUE	PR > F
MATERIAL	2	10683.72222222	7.91	0.0020
TEMP	2	39118.72222222	28.97	0.0001
MATERIAL*TEMP	4	9613.77777778	3.56	0.0186

OBSERVATION	OBSERVED VALUE	PREDICTED VALUE	RESIDUAL
1	130.00000000	134.75000000	-4.75000000
2	155.00000000	134.75000000	20.25000000
3	74.00000000	134.75000000	-60.75000000
4	180.00000000	134.75000000	45.25000000
5	34.00000000	57.25000000	-23.25000000
6	40.00000000	57.25000000	-17.25000000
7	80.00000000	57.25000000	22.75000000
8	75.00000000	57.25000000	17.75000000
9	20.00000000	57.50000000	-37.50000000
10	70.00000000	57.50000000	12.50000000
11	82.00000000	57.50000000	24.50000000
12	58.00000000	57.50000000	0.50000000
13	150.00000000	155.75000000	-5.75000000
14	188.00000000	155.75000000	32.25000000
15	159.00000000	155.75000000	3.25000000
16	126.00000000	155.75000000	-29.75000000
17	136.00000000	119.75000000	16.25000000
18	122.00000000	119.75000000	2.25000000
19	106.00000000	119.75000000	-13.75000000
20	115.00000000	119.75000000	-4.75000000
21	25.00000000	49.50000000	-24.50000000
22	70.00000000	49.50000000	20.50000000
23	58.00000000	49.50000000	8.50000000
24	45.00000000	49.50000000	-4.50000000
25	138.00000000	144.00000000	-6.00000000
26	110.00000000	144.00000000	-34.00000000
27	168.00000000	144.00000000	24.00000000
28	160.00000000	144.00000000	16.00000000
29	174.00000000	145.75000000	28.25000000
30	120.00000000	145.75000000	-25.75000000
31	150.00000000	145.75000000	4.25000000
32	139.00000000	145.75000000	-6.75000000
33	96.00000000	85.50000000	10.50000000
34	104.00000000	85.50000000	18.50000000
35	82.00000000	85.50000000	-3.50000000
36	60.00000000	85.50000000	-25.50000000

SUM OF RESIDUALS	0.00000000
SUM OF SQUARED RESIDUALS	18230.75000000
SUM OF SQUARED RESIDUALS - ERROR SS	-0.00000000
FIRST ORDER AUTOCORRELATION	-0.37519370
DURBIN-WATSON D	2.71348203

Figure 7-5.   Sample computer output for Example 7-1.

Figure 7-6. Normal probability plot and dot diagram of residuals for Example 7-1.

treatment combination of 15°F and material type 1 possibly having larger variance than the others.

The 70°F–material type 1 cell contains both extreme residuals (-60.75 and 45.25). These two residuals are primarily responsible for the inequality of variance detected in Figures 7-7, 7-8, and 7-9. Reexamination of the data does not reveal any obvious problem, such as an error in recording, so we accept these responses as legitimate. It is possible that this particular treatment combination

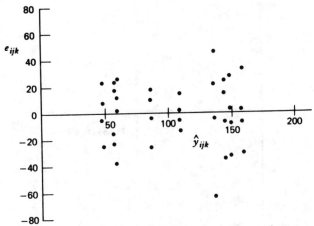

Figure 7-7. Plot of residuals versus $\hat{y}_{ijk}$ for Example 7-1.

produces slightly more erratic battery life than the others. The problem, however, is not severe enough to have a dramatic impact on the analysis and conclusions.

7-3.4 Estimating the Model Parameters

The parameters in the two-factor analysis of variance model

$$y_{ijk} = \mu + \tau_i + \beta_j + (\tau\beta)_{ij} + \epsilon_{ijk} \tag{7-13}$$

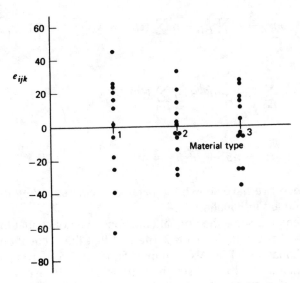

Figure 7-8. Plot of residuals versus material type for Example 7-1.

Figure 7-9.   Plot of residuals versus temperature for
Example 7-1.

may be estimated by least squares. Since the model has $1 + a + b + ab$
parameters to be estimated, there are $1 + a + b + ab$ normal equations. Using
the method of Section 4-4, it is not difficult to show that the normal equations
are

$$\mu: \quad abn\hat{\mu} + bn \sum_{i=1}^{a} \hat{\tau}_i + an \sum_{j=1}^{b} \hat{\beta}_j + n \sum_{i=1}^{a} \sum_{j=1}^{b} (\tau\beta)_{ij} = y_{...} \qquad (7\text{-}14a)$$

$$\tau_i: \quad bn\hat{\mu} + bn\hat{\tau}_i + n \sum_{j=1}^{b} \hat{\beta}_j + n \sum_{j=1}^{b} (\tau\beta)_{ij} = y_{i..} \qquad i = 1, 2, \ldots, a$$

$$(7\text{-}14b)$$

$$\beta_j: \quad an\hat{\mu} + n \sum_{i=1}^{a} \hat{\tau}_i + an\hat{\beta}_j + n \sum_{i=1}^{a} (\tau\beta)_{ij} = y_{.j.} \qquad j = 1, 2, \ldots, b$$

$$(7\text{-}14c)$$

$$(\tau\beta)_{ij}: \quad n\hat{\mu} + n\hat{\tau}_i + n\hat{\beta}_j + n(\tau\beta)_{ij} = y_{ij.} \qquad \begin{cases} i = 1, 2, \ldots, a \\ j = 1, 2, \ldots, b \end{cases} \qquad (7\text{-}14d)$$

For convenience, we have shown the parameter corresponding to each normal
equation on the left in Equation 7-14.

As we attempt to solve the normal equations, we notice that the a equations
in Equation 7-14b sum to Equation 7-14a and that the b equations of Equation
7-14c sum to Equation 7-14a. Also summing Equation 7-14d over j for a particular
i will give Equation 7-14b, and summing Equation 7-14d over i for a particular
j will give Equation 7-14c. Therefore, there are $a + b + 1$ linear dependencies
in this system of equations and no unique solution will exist. In order to obtain

a solution, we impose the constraints

$$\sum_{i=1}^{a} \hat{\tau}_i = 0 \qquad\qquad (7\text{-}15a)$$

$$\sum_{j=1}^{b} \hat{\beta}_j = 0 \qquad\qquad (7\text{-}15b)$$

$$\sum_{i=1}^{a} \widehat{(\tau\beta)}_{ij} = 0 \qquad j = 1, 2, \ldots, b \qquad (7\text{-}15c)$$

and

$$\sum_{j=1}^{b} \widehat{(\tau\beta)}_{ij} = 0 \qquad i = 1, 2, \ldots, a \qquad (7\text{-}15d)$$

Equations 7-15a and 7-15b constitute two constraints, whereas Equations 7-15c and 7-15d form $a + b - 1$ independent constraints. Therefore, we have $a + b + 1$ total constraints, the number needed.

Applying these constraints, the normal equations (Equations 7-14) simplify considerably, and we obtain the solution

$$\hat{\mu} = \bar{y}_{...}$$
$$\hat{\tau}_i = \bar{y}_{i..} - \bar{y}_{...} \qquad i = 1, 2, \ldots, a \qquad (7\text{-}16)$$
$$\hat{\beta}_j = \bar{y}_{.j.} - \bar{y}_{...} \qquad j = 1, 2, \ldots, b$$
$$\widehat{(\tau\beta)}_{ij} = \bar{y}_{ij.} - \bar{y}_{i..} - \bar{y}_{.j.} + \bar{y}_{...} \quad \begin{cases} i = 1, 2, \ldots, a \\ j = 1, 2, \ldots, b \end{cases}$$

Notice the considerable intuitive appeal of this solution to the normal equations. Row treatment effects are estimated by the row average minus the grand average; column treatments are estimated by the column average minus the grand average; and the ijth interaction is estimated by the ijth cell average minus the grand average, the ith row effect, and the jth column effect.

Using Equation 7-16, we may find the fitted value y_{ijk} as

$$\hat{y}_{ijk} = \hat{\mu} + \hat{\tau}_i + \hat{\beta}_j + \widehat{(\tau\beta)}_{ij}$$
$$= \bar{y}_{...} + (\bar{y}_{i..} - \bar{y}_{...}) + (\bar{y}_{.j.} - \bar{y}_{...})$$
$$(\bar{y}_{ij.} - \bar{y}_{i..} - \bar{y}_{.j.} + \bar{y}_{...})$$
$$= \bar{y}_{ij.}$$

That is, the kth observation in the ijth cell is estimated by the average of the n observations in that cell. This result was used in Equation 7-12 to obtain the residuals for the two-factor factorial model.

Because constraints (Equations 7-15) have been used to solve the normal equations, the model parameters are not uniquely estimated. However, certain important *functions* of the model parameters *are* estimable, that is, uniquely estimated regardless of the constraint chosen. An example is $\tau_i - \tau_u + (\tau\beta)_{i.} -$

$(\overline{\tau\beta})_{u.}$, which might be thought of as the "true" difference between the ith and uth levels of factor A. Notice that the true difference between the levels of any main effect includes an "average" interaction effect. It is this result that disturbs the tests on main effects in the presence of interaction, as noted earlier. In general, any function of the model parameters that is a linear combination of the left-hand side of the normal equations is estimable. This property was also noted in Chapter 3 when we were discussing the single-factor model.

7-3.5 Choice of Sample Size

The operating characteristic curves in Appendix Chart V can be used to assist the experimenter in determining an appropriate sample size (number of replicates, n) for a two-factor factorial design. The appropriate value of the parameter Φ^2 and the numerator and denominator degrees of freedom are shown in Table 7-10.

A very effective way to use these curves is to find the smallest value of Φ^2 corresponding to a specified difference between any two treatment means. For example, if the difference in any two row means is D, then the minimum value of Φ^2 is

$$\Phi^2 = \frac{nbD^2}{2a\sigma^2} \tag{7-17}$$

whereas if the difference in any two column means is D, then the minimum value of Φ^2 is

$$\Phi^2 = \frac{naD^2}{2b\sigma^2} \tag{7-18}$$

Table 7-10 Operating Characteristic Curve Parameters for Chart V of the Appendix for the Two-Factor Factorial, Fixed Effects Model

Factor	Φ^2	Numerator Degrees of Freedom	Denominator Degrees of Freedom
A	$\dfrac{bn\sum\limits_{i=1}^{a}\tau_i^2}{a\sigma^2}$	$a-1$	$ab(n-1)$
B	$\dfrac{an\sum\limits_{j=1}^{b}\beta_j^2}{b\sigma^2}$	$b-1$	$ab(n-1)$
AB	$\dfrac{n\sum\limits_{i=1}^{a}\sum\limits_{j=1}^{b}(\tau\beta)_{ij}^2}{\sigma^2[(a-1)(b-1)+1]}$	$(a-1)(b-1)$	$ab(n-1)$

Finally, the minimum value of Φ^2 corresponding to a difference of D between any two interaction effects is

$$\Phi^2 = \frac{nD^2}{2\sigma^2[(a-1)(b-1)+1]} \tag{7-19}$$

To illustrate the use of these equations, consider the battery life data in Example 7-1. Suppose that before running the experiment we decide that the null hypothesis should be rejected with a high probability if the difference in mean battery life between any two temperatures is as great as 40 hours. Thus $D = 40$, and if we assume that the standard deviation of battery life is approximately 25, then Equation 7-18 gives

$$\Phi^2 = \frac{naD^2}{2b\sigma^2}$$

$$= \frac{n(3)(40)^2}{2(3)(25)^2}$$

$$= 1.28n$$

as the minimum value of Φ^2. Assuming that $\alpha = .05$, we can now use Appendix Table V to construct the following display:

n	Φ^2	Φ	ω_1 = Numerator Degrees of Freedom	ω_2 = Error Degrees of Freedom	β
2	2.56	1.60	2	9	.45
3	3.84	1.96	2	18	.18
4	5.12	2.26	2	27	.06

Note that $n = 4$ replicates give a β risk of about .06 or approximately a 94 percent change of rejecting the null hypothesis if the difference in mean battery life at any two temperature levels is as large as 40 hours. Thus, we conclude that four replicates are enough to provide the desired sensitivity as long as our estimate of the standard deviation of battery life is not seriously in error. If in doubt, the experimenter could repeat the above procedure with other values of σ to determine the effect of misestimating this parameter on the sensitivity of the design.

7-3.6 The Assumption of No Interaction in a Two-Factor Model

Occasionally, an experimenter feels that a two-factor model without interaction is appropriate, say

$$y_{ijk} = \mu + \tau_i + \beta_j + \epsilon_{ijk} \begin{cases} i = 1,2,\ldots,a \\ j = 1,2,\ldots,b \\ k = 1,2,\ldots,n \end{cases} \tag{7-20}$$

Table 7-11   Analysis of Variance for Battery Life Data Assuming No Interaction

Source of Variation	Sum of Squares	Degrees of Freedom	Mean Square	F_0
Material Types	10,683.72	2	5,341.86	5.95
Temperature	39,118.72	2	19,558.36	21.78
Error	27,844.52	31	898.21	
Total	77,646.96	35		

We should be very careful in dispensing with the interaction terms, however, since the presence of significant interaction can have a dramatic impact on the interpretation of the data.

The statistical analysis of a two-factor factorial model without interaction is straightforward. Table 7-11 presents the analysis of the battery life data from Example 7-1, assuming that the no-interaction model (Equation 7-20) applies. As noted previously, both main effects are significant. However, as soon as a residual analysis is performed for these data, it becomes clear that the no-interaction model is inadequate. For the two-factor model without interaction, the fitted values are $\hat{y}_{ijk} = \bar{y}_{i..} + \bar{y}_{.j.} - \bar{y}_{...}$. A plot of $\bar{y}_{ij.} - \hat{y}_{ijk}$ (the cell averages minus the fitted value for that cell) versus the fitted value $\hat{y}_{ijk}$ is shown in Figure 7-10. Now the quantities $\bar{y}_{ij.} - \hat{y}_{ijk}$ may be viewed as the differences between the observed cell means and the estimated cell means assuming no interaction. Any pattern in these quantities is suggestive of the presence of interaction. Figure 7-10 shows a distinct pattern as the quantities $\bar{y}_{ij.} - \hat{y}_{ijk}$ move from positive to negative to positive to negative again. This structure is the result of interaction between material types and temperature.

7-3.7   One Observation per Cell

Occasionally, one encounters a two-factor experiment with only a single replicate, that is, only one observation per cell. If there are two factors and only one

Figure 7-10.   Plot of $\bar{y}_{ij.} - \hat{y}_{ijk}$ versus $\hat{y}_{ijk}$, battery life data.

observation per cell, the linear statistical model is

$$y_{ij} = \mu + \tau_i + \beta_j + (\tau\beta)_{ij} + \epsilon_{ij} \begin{cases} i = 1,2, \ldots, a \\ j = 1,2, \ldots, b \end{cases} \qquad (7\text{-}21)$$

The analysis of variance for this design is shown in Table 7-12, assuming that both factors are fixed.

From examining the expected mean squares, we see that the error variance σ^2 is not estimable; that is, the two-factor interaction effect $(\tau\beta)_{ij}$ and the experimental error cannot be separated in any obvious manner. Consequently, there are no tests on main effects unless the interaction effect is zero. If there is no interaction present, then $(\tau\beta)_{ij} = 0$ for all i and j, and a plausible model is

$$y_{ij} = \mu + \tau_i + \beta_j + \epsilon_{ij} \begin{cases} i = 1, 2, \ldots, a \\ j = 1, 2, \ldots, b \end{cases} \qquad (7\text{-}22)$$

If the model (Equation 7-22) is appropriate, then the residual mean square in Table 7-12 is an unbiased estimator of σ^2, and the main effects may be tested by comparing MS_A and MS_B to $MS_{Residual}$.

A test developed by Tukey (1949a) is helpful in determining if interaction is present. The procedure assumes that the interaction term is of a particularly simple form; namely,

$$(\tau\beta)_{ij} = \gamma\tau_i\beta_j$$

where γ is an unknown constant. By defining the interaction term this way, we may use a regression approach to test the significance of the interaction term. The test partitions the residual sum of squares into a single-degree-of-freedom component due to nonadditivity (interaction) and a component for error with

Table 7-12   Analysis of Variance for a Two-Factor Model, One Observation per Cell

Source of Variation	Sum of Squares	Degrees of Freedom	Mean Square	Expected Mean Square
Rows (A)	$\sum\limits_{i=1}^{a} \dfrac{y_{i.}^2}{b} - \dfrac{y_{..}^2}{ab}$	$a - 1$	MS_A	$\sigma^2 + \dfrac{b\sum \tau_i^2}{a - 1}$
Columns (B)	$\sum\limits_{j=1}^{b} \dfrac{y_{.j}^2}{b} - \dfrac{y_{..}^2}{ab}$	$b - 1$	MS_B	$\sigma^2 + \dfrac{a\sum \beta_j^2}{b - 1}$
Residual or AB	Subtraction	$(a - 1)(b - 1)$	$MS_{Residual}$	$\sigma^2 + \dfrac{\sum\sum (\tau\beta)_{ij}^2}{(a - 1)(b - 1)}$
Total	$\sum\limits_{i=1}^{a}\sum\limits_{j=1}^{b} y_{ij}^2 - \dfrac{y_{..}^2}{ab}$	$ab - 1$		

$(a - 1)(b - 1) - 1$ degrees of freedom. Computationally, we have

$$SS_N = \frac{\left[\displaystyle\sum_{i=1}^{a}\sum_{j=1}^{b} y_{ij}y_{i.}y_{.j} - y_{..}\left(SS_A + SS_B + \frac{y_{..}^2}{ab}\right)\right]^2}{ab\,SS_A\,SS_B} \tag{7-23}$$

with one degree of freedom, and

$$SS_{Error} = SS_{Residual} - SS_N \tag{7-24}$$

with $(a - 1)(b - 1) - 1$ degrees of freedom. To test for the presence of interaction, we compute

$$F_0 = \frac{SS_N}{SS_{Error}/[(a - 1)(b - 1) - 1]} \tag{7-25}$$

If $F_0 > F_{\alpha,1,(a-1)(b-1)-1}$, the hypothesis of no interaction must be rejected.

Example 7-2

The impurity present in a chemical product is affected by two factors—pressure and temperature. The data from a single replicate of a factorial experiment are shown in Table 7-13. The sums of squares are

$$SS_A = \sum_{i=1}^{a} \frac{y_{i.}^2}{b} - \frac{y_{..}^2}{ab}$$

$$= \frac{23^2 + 13^2 + 8^2}{5} - \frac{44^2}{(3)(5)} = 23.33$$

$$SS_B = \sum_{j=1}^{b} \frac{y_{.j}^2}{a} - \frac{y_{..}^2}{ab}$$

$$= \frac{9^2 + 6^2 + 13^2 + 6^2 + 10^2}{3} - \frac{44^2}{(3)(5)} = 11.60$$

$$SS_T = \sum_{i=1}^{a}\sum_{j=1}^{b} y_{ij}^2 - \frac{y_{..}^2}{ab}$$

$$= 166 - 129.07 = 36.93$$

Table 7-13   Impurity Data for Example 7-2

Temperature (°F)	25	30	35	40	45	$y_{i.}$
100	5	4	6	3	5	23
125	3	1	4	2	3	13
150	1	1	3	1	2	8
$y_{.j}$	9	6	13	6	10	$44 = y_{..}$

Pressure

and

$$SS_{Residual} = SS_T - SS_A - SS_B$$
$$= 36.93 - 23.33 - 11.60 = 2.00$$

The sum of squares for nonadditivity is computed from Equation 7-23 as follows:

$$\sum_{i=1}^{a} \sum_{j=1}^{b} y_{ij} y_{i.} y_{.j} = (5)(23)(9) + (4)(23)(6) + \cdots + (2)(8)(10) = 7236$$

$$SS_N = \frac{\left[\sum_{i=1}^{a} \sum_{j=1}^{b} y_{ij} y_{i.} y_{.j} - y_{..} \left(SS_A + SS_B + \frac{y_{..}^2}{ab} \right) \right]^2}{ab SS_A SS_B}$$

$$= \frac{[7236 - (44)(23.33 + 11.60 + 129.07)]^2}{(3)(5)(23.33)(11.60)}$$

$$= \frac{[20.00]^2}{4059.42} = 0.0985$$

and the error sum of squares is, from Equation 7-24,

$$SS_{Error} = SS_{Residual} - SS_N$$
$$= 2.00 - 0.0985 = 1.9015$$

The complete analysis of variance is summarized in Table 7-14. The test statistic for nonadditivity is $F_0 = 0.0985/0.2716 = 0.36$, so we conclude that there is no evidence of interaction in these data. The main effects of temperature and pressure are significant.

■

In concluding this section, we note that the two-factor factorial model with one observation per cell (Equation 7-22) looks exactly like the randomized complete block model (Equation 5-1). In fact, the Tukey single-degree-of-freedom test for nonadditivity can be directly applied to test for interaction in the randomized block model. However, remember that the *experimental situations* that lead to the randomized block and factorial models are very different. In the factorial model, *all ab* runs have been made in random order, whereas in the randomized block model, randomization occurs only *within* the block. The blocks are a

Table 7-14   Analysis of Variance for Example 7-2

Source of Variation	Sum of Squares	Degrees of Freedom	Mean Square	F_0
Temperature	23.33	2	11.67	42.97[a]
Pressure	11.60	4	2.90	10.68[a]
Nonadditivity	0.0985	1	0.0985	0.36
Error	1.9015	7	0.2716	
Total	36.93	14		

[a]Significant at 1 percent.

randomization restriction. Hence, the manner in which the data are collected and the interpretation of the two models are quite different.

7-4 RANDOM AND MIXED MODELS

In previous sections we have assumed that both factors A and B are fixed. That is, the experimenter has specifically chosen the a levels of factor A and the b levels of factor B used in the design. Consequently, the inferences drawn from the analysis of variance are applicable only to the levels of A and B actually used. In this section we consider two alternate models: the random effects model, in which the levels of A and B are randomly selected, and the mixed model, in which one factor is fixed and the other is random.

7-4.1 The Random Effects Model

Consider the situation in which the levels of factors A and B are chosen at random from larger populations. This is the *random effects* or *components of variance* model. Because the a levels of A and the b levels of B are randomly chosen, our inferences will be valid about *all* levels in the populations under study. We may represent the observations by the linear model

$$y_{ijk} = \mu + \tau_i + \beta_j + (\tau\beta)_{ij} + \epsilon_{ijk} \begin{cases} i = 1,2, \ldots, a \\ j = 1,2, \ldots, b \\ k = 1,2, \ldots, n \end{cases} \quad (7\text{-}26)$$

where the model parameters τ_i, β_j, $(\tau\beta)_{ij}$, and ϵ_{ijk} are random variables. Specifically, we assume that τ_i is NID$(0, \sigma_\tau^2)$, β_j is NID$(0, \sigma_\beta^2)$, $(\tau\beta)_{ij}$ is NID$(0, \sigma_{\tau\beta}^2)$, and ϵ_{ijk} is NID$(0, \sigma^2)$. The variance of any observation is

$$V(y_{ijk}) = \sigma_\tau^2 + \sigma_\beta^2 + \sigma_{\tau\beta}^2 + \sigma^2$$

and σ_τ^2, σ_β^2, $\sigma_{\tau\beta}^2$, and σ^2 are called *variance components*. The hypotheses that we are interested in testing are $H_0: \sigma_\beta^2 = 0$, and $H_0: \sigma_\beta^2 = 0$, and $H_0: \sigma_{\tau\beta}^2 = 0$. Notice the similarity to the single-factor random effects model.

The basic analysis of variance remains unchanged; that is, SS_A, SS_B, SS_{AB}, SS_T, and SS_E are all calculated as in the fixed effects case. However, to form the test statistics, we must examine the expected mean squares. It may be shown that

$$E(MS_A) = \sigma^2 + n\sigma_{\tau\beta}^2 + bn\sigma_\tau^2$$
$$E(MS_B) = \sigma^2 + n\sigma_{\tau\beta}^2 + an\sigma_\beta^2 \quad (7\text{-}27)$$
$$E(MS_{AB}) = \sigma^2 + n\sigma_{\tau\beta}^2$$

and

$$E(MS_E) = \sigma^2$$

From the expected mean squares, we see that the appropriate statistic for testing the no-interaction hypothesis $H_0: \sigma_{\tau\beta}^2 = 0$ is

$$F_0 = \frac{MS_{AB}}{MS_E} \tag{7-28}$$

because under H_0 both numerator and denominator of F_0 have expectation σ^2, and only if H_0 is false is $E(MS_{AB})$ greater than $E(MS_E)$. The ratio F_0 is distributed as $F_{(a-1)(b-1),ab(n-1,)}$. Similarly, for testing $H_0: \sigma_\tau^2 = 0$ we would use

$$F_0 = \frac{MS_A}{MS_{AB}} \tag{7-29}$$

which is distributed as $F_{a-1,(a-1)(b-1,)}$, and for testing $H_0: \sigma_\beta^2 = 0$ the statistic is

$$F_0 = \frac{MS_B}{MS_{AB}} \tag{7-30}$$

which is distributed as $F_{b-1,(a-1)(b-1)}$. These are all upper-tail, one-tail tests. Notice that these test statistics are not the same as those used if both factors A and B are fixed. The expected mean squares are always used as a guide to test statistic construction.

Example 7-3

Suppose that in Example 7-1 a large number of material types and many different temperatures could have been chosen, and that the ones given in Table 7-7 were selected at random. The model is now a random effects model, and the analysis of variance is shown in Table 7-15. We see that the first four columns of the analysis of variance table are identical to that for Example 7-1. Now, however, the F ratios are computed according to Equations 7-28, 7-29, and 7-30. Interaction is still significant at the 5 percent level, and since $F_{.05,2,4} = 6.94$, we conclude that the effect of temperature is significant, whereas material types do not significantly affect the battery life. However, material type is still important because it is involved in the interaction.

■

The variance components may be estimated by the *analysis of variance method,* that is, by equating the observed mean squares in the lines of the analysis

Table 7-15 Analysis of Variance for Example 7-3, Random Effects Model

Source of Variation	Sum of Squares	Degrees of Freedom	Mean Square	F_0
Material types	10,683.72	2	5,341.86	2.22
Temperature	39,118.72	2	19,558.36	8.13[a]
Interaction	9,613.78	4	2,403.44	3.56[a]
Error	18,230.75	27	675.21	
Total	77,646.97	35		

[a]Significant at 5 percent.

of variance table to their expected values and solving for the variance components. This yields

$$\hat{\sigma}^2 = MS_E$$

$$\hat{\sigma}^2_{\tau\beta} = \frac{MS_{AB} - MS_E}{n}$$

$$\hat{\sigma}^2_{\beta} = \frac{MS_B - MS_{AB}}{an} \qquad (7\text{-}31)$$

$$\hat{\sigma}^2_{\tau} = \frac{MS_A - MS_{AB}}{bn}$$

To illustrate the method, we estimate the variance components for the experimental situation described in Example 7-3. From Equations 7-31, we obtain

$$\hat{\sigma}^2_{\tau} = \frac{5341.86 - 2403.44}{(3)(4)} = 244.87$$

$$\hat{\sigma}^2_{\beta} = \frac{19,558.36 - 2403.44}{(3)(4)} = 1429.58$$

$$\hat{\sigma}^2_{\tau\beta} = \frac{2403.44 - 675.21}{4} = 432.06$$

and

$$\hat{\sigma}^2 = 675.21$$

It is possible to obtain a confidence interval estimate of σ^2 using the chi-square distribution since $ab(n - 1)MS_E/\sigma^2$ is distributed as $\chi^2_{ab(n-1)}$. However, exact confidence intervals on the other variance components cannot be obtained for the reasons discussed in Section 3-5. Approximate procedures for obtaining confidence intervals for variance components are given in Searle (1971a, 1971b).

7-4.2 Mixed Models

We now consider the situation where one of the factors A is fixed and the other B is random. This is called the *mixed model* analysis of variance. The linear statistical model is

$$y_{ijk} = \mu + \tau_i + \beta_j + (\tau\beta)_{ij} + \epsilon_{ijk} \begin{cases} i = 1,2, \ldots, a \\ j = 1,2, \ldots, b \\ k = 1,2, \ldots, n \end{cases} \qquad (7\text{-}32)$$

Here τ_i is a fixed effect, β_j is a random effect, the interaction $(\tau\beta)_{ij}$ is assumed to be a random effect, and ϵ_{ijk} is a random error. We also assume that the $\{\tau_i\}$ are fixed effects such that $\sum_{i=1}^{a} \tau_i = 0$ and β_j is a NID($0, \sigma^2_\beta$) random variable. The interaction effect, $(\tau\beta)_{ij}$, is a normal random variable with mean 0 and variance $[(a - 1)/a]\sigma^2_{\tau\beta}$; however, summing the interaction component over the fixed factor equals zero. That is,

$$\sum_{i=1}^{a} (\tau\beta)_{ij} = (\tau\beta)_{.j} = 0 \qquad j = 1,2, \ldots, b$$

This implies that certain interaction elements at different levels of the fixed factor are not independent. In fact, we may show (see Problem 7-24) that

$$\text{Cov}[(\tau\beta)_{ij}, (\tau\beta)_{i'j}] = -\frac{1}{a}\sigma_{\tau\beta}^2 \qquad i \neq i'$$

The covariance between $(\tau\beta)_{ij}$ and $(\tau\beta)_{ij'}$ for $j \neq j'$ is zero, and the random error ϵ_{ijk} is NID$(0, \sigma^2)$.

In this model the variance of $(\tau\beta)_{ij}$ is defined as $[(a-1)/a]\sigma_{\tau\beta}^2$ rather than $\sigma_{\tau\beta}^2$ in order to simplify the expected mean squares. The assumption $(\tau\beta)_{.j} = 0$ also has an effect on the expected mean squares, which we may show are

$$E(MS_A) = \sigma^2 + n\sigma_{\tau\beta}^2 + \frac{bn\sum_{i=1}^{a}\tau_i^2}{a-1}$$

$$E(MS_B) = \sigma^2 + an\sigma_{\beta}^2 \qquad (7\text{-}33)$$

$$E(MS_{AB}) = \sigma^2 + n\sigma_{\tau\beta}^2$$

and

$$E(MS_E) = \sigma^2$$

Therefore, the appropriate test statistic for testing $H_0: \tau_i = 0$ is

$$F_0 = \frac{MS_A}{MS_{AB}}$$

which is distributed as $F_{a-1,(a-1)(b-1)}$. For testing $H_0: \sigma_{\beta}^2 = 0$, the test statistic is

$$F_0 = \frac{MS_B}{MS_E}$$

which is distributed as $F_{b-1, ab(n-1)}$. Finally, for testing $H_0: \sigma_{\tau\beta}^2 = 0$, we would use

$$F_0 = \frac{MS_{AB}}{MS_E}$$

which is distributed as $F_{(a-1)(b-1), ab(n-1)}$.

In the mixed model, it is possible to estimate the fixed factor effects as

$$\hat{\mu} = \bar{y}_{...}$$

$$\hat{\tau}_i = \bar{y}_{i..} - \bar{y}_{...} \qquad i = 1,2, \ldots, a \qquad (7\text{-}34)$$

The variance components σ_{β}^2, $\sigma_{\tau\beta}^2$, and σ^2 may be estimated using the analysis of variance method. Eliminating the first equation from Equations 7-33 leaves three equations in three unknown, whose solutions are

$$\hat{\sigma}_\beta^2 = \frac{MS_B - MS_E}{an}$$

$$\hat{\sigma}_{\tau\beta}^2 = \frac{MS_{AB} - MS_E}{n} \tag{7-35}$$

and

$$\hat{\sigma}^2 = MS_E$$

This general approach can be used to estimate the variance components in any mixed model. After eliminating the mean squares containing fixed factors, there will always be a set of equations remaining that can be solved for the variance components.

Table 7-16 summarizes the analysis of variance for the two-factor mixed model. In mixed models it is possible to test hypotheses about individual treatment means for the fixed factor. In using procedures such as Duncan's multiple range test, care must be exercised to use the proper standard error of the treatment mean. Anderson and McLean (1974) show that the standard error of the fixed effect treatment mean is

$$\left[\frac{\text{Mean square for testing the fixed effect}}{\text{Number of observations in each treatment mean}}\right]^{1/2}$$

Alternate Mixed Models ▪ Several different versions of the mixed model have been proposed. These models differ from the "standard" mixed model discussed above in the assumptions made about the random components. One of these alternate models is now briefly discussed.

Consider the model

$$y_{ijk} = \mu + \alpha_i + \gamma_j + (\alpha\gamma)_{ij} + \epsilon_{ijk}$$

where the α_i ($i = 1, 2, \ldots, a$) are fixed effects such that $\sum_{i=1}^{a} \alpha_i = 0$, and γ_j, $(\alpha\gamma)_{ij}$, and ϵ_{ijk} are uncorrelated random variables having zero means and variances $V(\gamma_j) = \sigma_\gamma^2$, $V[(\alpha\gamma)_{ij}] = \sigma_{\alpha\gamma}^2$, and $V(\epsilon_{ijk}) = \sigma^2$. Note that the primary difference between this mixed model and the previous model is the assumption that the interaction effects are uncorrelated.

Table 7-16   Analysis of Variance for the Two-Factor Mixed Model

Source of Variation	Sum of Squares	Degrees of Freedom	Expected Mean Square
Rows (A)	SS_A	$a - 1$	$\sigma^2 + n\sigma_{\tau\beta}^2 + bn\Sigma\tau_i^2/(a - 1)$
Columns (B)	SS_B	$b - 1$	$\sigma^2 + an\sigma_\beta^2$
Interaction	SS_{AB}	$(a - 1)(b - 1)$	$\sigma^2 + n\sigma_{\tau\beta}^2$
Error	SS_E	$ab(n - 1)$	σ^2
Total	SS_T	$abn - 1$	

The expected mean squares for this model are

$$E(MS_A) = \sigma^2 + n\sigma_{\alpha\gamma}^2 + \frac{bn\sum_{i=1}^{a}\alpha_i^2}{a-1}$$

$$E(MS_B) = \sigma^2 + n\sigma_{\alpha\gamma}^2 + an\sigma_{\gamma}^2$$

$$E(MS_{AB}) = \sigma^2 + n\sigma_{\alpha\gamma}^2$$

and

$$E(MS_E) = \sigma^2$$

Comparing these expected mean squares with those in Table 7-16, we note that the only obvious difference is the presence of the variance component $\sigma_{\alpha\gamma}^2$ in the expected mean square for the random effect. (Actually, there are other differences because of the different definitions of the variance of the interaction effect in the two models.) Consequently, we would test the hypothesis that the variance component for the random effect equals zero ($H_0: \sigma_{\gamma}^2 = 0$) using the statistic

$$F_0 = \frac{MS_B}{MS_{AB}}$$

as contrasted with testing $H_0: \sigma_{\beta}^2 = 0$ with $F_0 = MS_B/MS_E$ in the standard model. The test should be more conservative using this model since MS_{AB} will generally be larger than MS_E.

The parameters in the two models are closely related. In fact, we may show that

$$\beta_j = \gamma_j + \overline{(\alpha\gamma)}_{.j}$$

$$(\tau\beta)_{ij} = (\alpha\gamma)_{ij} + \overline{(\alpha\gamma)}_{.j}$$

$$\sigma_{\beta}^2 = \sigma_{\beta}^2 + \frac{1}{a}\sigma_{\alpha\gamma}^2$$

and

$$\sigma_{\tau\beta}^2 = \sigma_{\alpha\gamma}^2$$

The analysis of variance method may be used to estimate the variance components. Referring to the expected mean squares, we find that the only change from Equations 7-35 is that

$$\hat{\sigma}_{\beta}^2 = \frac{MS_B - MS_{AB}}{an}$$

Both of these models are special cases of the mixed model proposed by Scheffé (1956a, 1959). This model assumes that the observations may be represented by

$$y_{ijk} = m_{ij} + \epsilon_{ijk} \begin{cases} i = 1,2,\ldots,a \\ j = 1,2,\ldots,b \\ k = 1,2,\ldots,n \end{cases}$$

where m_{ij} and ϵ_{ijk} are independent random variables. The structure of m_{ij} is

$$m_{ij} = \mu + \tau_i + b_j + c_{ij}$$

$$E(m_{ij}) = \mu + \tau_i$$

$$\sum_{i=1}^{a} \tau_i = 0$$

and

$$c_{.j} = 0 \qquad j = 1, 2, \ldots, b$$

The variances and covariances of b_j and c_{ij} are expressed through the covariances of the m_{ij}. Furthermore, the random effects parameters in other formulations of the mixed model can be related to b_j and c_{ij}. The statistical analysis of Scheffé's model is identical to that for our standard model, except that in general the statistic MS_A/MS_{AB} is not always distributed as F when $H_0\colon \tau_i = 0$ is true.

In light of this multiplicity of mixed models, a logical question is which model should one use? Most statisticians tend to prefer the standard model, and it is the most widely encountered in the literature. If the correlative structure of the random components is not large, then either mixed model is appropriate, and there are only minor differences between these models. When we subsequently refer to mixed models, we assume the standard model structure. However, if there are large correlations in the data, then Scheffé's model may have to be employed. The choice of model should always be dictated by the data. The article by Hocking (1973) is a clear summary of various mixed models.

7-4.3 Choice of Sample Size

The operating characteristic curves in the Appendix may be used for the analysis of variance of the two-factor random effects model and the mixed model. Appendix Chart VI is used for the random effects model. The parameter λ, numerator degrees of freedom, and denominator degrees of freedom are shown in the top half of Table 7-17. For the mixed model, both Charts V and VI in the Appendix must be used. The appropriate values for Φ^2 and λ are shown in the bottom half of Table 7-17.

7-5 THE GENERAL FACTORIAL DESIGN

The results for the two-factor factorial design may be extended to the general case where there are a levels of factor A, b levels of factor B, c levels of factor C, and so on, arranged in a factorial experiment. In general, there will be $abc \ldots n$ total observations if there are n replicates of the complete experiment. Once again, note that we must have at least two replicates ($n \geq 2$) in order to

ble 7-17 Operating Characteristic Curve Parameters for Tables V and VI of the Appendix for
Two-Factor Random Effects and Mixed Models

The Random Effects Model

Factor	λ	Numerator Degrees of Freedom	Denominator Degrees of Freedom
A	$\sqrt{1 + \dfrac{bn\sigma_\tau^2}{\sigma^2 + n\sigma_{\tau\beta}^2}}$	$a - 1$	$(a - 1)(b - 1)$
B	$\sqrt{1 + \dfrac{an\sigma_\beta^2}{\sigma^2 + n\sigma_{\tau\beta}^2}}$	$b - 1$	$(a - 1)(b - 1)$
AB	$\sqrt{1 + \dfrac{n\sigma_{\tau\beta}^2}{\sigma^2}}$	$(a - 1)(b - 1)$	$ab(n - 1)$

The Mixed Model

Factor	Parameter	Numerator Degrees of Freedom	Denominator Degrees of Freedom	Appendix Chart
(Fixed)	$\Phi^2 = \dfrac{bn \sum_{i=1}^{a} \tau_i^2}{a[\sigma^2 + n\sigma_{\tau\beta}^2]}$	$a - 1$	$(a - 1)(b - 1)$	V
(Random)	$\lambda = \sqrt{1 + \dfrac{an\sigma_\beta^2}{\sigma^2}}$	$b - 1$	$ab(n - 1)$	VI
	$\lambda = \sqrt{1 + \dfrac{n\sigma_{\tau\beta}^2}{\sigma^2}}$	$(a - 1)(b - 1)$	$ab(n - 1)$	VI

determine a sum of squares due to error if all possible interactions are included
in the model.

If all factors in the experiment are fixed, we may easily formulate and test
hypotheses about the main effects and interactions. For a fixed effects model,
test statistics for each main effect and interaction may be constructed by dividing
the corresponding mean square for the effect or interaction by the mean square
error. All of these F tests will be upper-tail, one-tail tests. The number of degrees
of freedom for any main effect is the number of levels of the factor minus one,
and the number of degrees of freedom for an interaction is the product of the
number of degrees of freedom associated with the individual components of the
interaction.

For example, consider the three-factor analysis of variance model:

$$y_{ijkl} = \mu + \tau_i + \beta_j + \gamma_k + (\tau\beta)_{ij} + (\tau\gamma)_{ik} + (\beta\gamma)_{jk}$$

$$+ (\tau\beta\gamma)_{ijk} + \epsilon_{ijkl} \begin{cases} i = 1,2, \ldots, a \\ j = 1,2, \ldots, b \\ k = 1,2, \ldots, c \\ l = 1,2, \ldots, n \end{cases} \tag{7-36}$$

Assuming that A, B, and C are fixed, the analysis of variance table is shown in Table 7-18. The F tests on main effects and interactions follow directly from the expected mean squares.

We will give the computing formulas for the sums of squares in Table 7-18. The total sum of squares is found in the usual way as

$$SS_T = \sum_{i=1}^{a} \sum_{j=1}^{b} \sum_{k=1}^{c} \sum_{l=1}^{n} y_{ijkl}^2 - \frac{y_{....}^2}{abcn} \tag{7-37}$$

The sums of squares for the main effects are found from the totals for factors $A(y_{i...})$, $B(y_{.j..})$, and $C(y_{..k.})$ as follows.

$$SS_A = \sum_{i=1}^{a} \frac{y_{i...}^2}{bcn} - \frac{y_{....}^2}{abcn} \tag{7-38}$$

$$SS_B = \sum_{j=1}^{b} \frac{y_{.j..}^2}{acn} - \frac{y_{....}^2}{abcn} \tag{7-39}$$

$$SS_C = \sum_{k=1}^{c} \frac{y_{..k.}^2}{abn} - \frac{y_{....}^2}{abcn} \tag{7-40}$$

To compute the two-factor interaction sums of squares, the totals for the $A \times B$, $A \times C$, and $B \times C$ cells are needed. It is frequently helpful to collapse the original data table into three two-way tables in order to compute these quantities. The sums of squares are found from

$$SS_{AB} = \sum_{i=1}^{a} \sum_{j=1}^{b} \frac{y_{ij..}^2}{cn} - \frac{y_{....}^2}{abcn} - SS_A - SS_B$$

$$= SS_{\text{Subtotals}(AB)} - SS_A - SS_B \tag{7-41}$$

$$SS_{AC} = \sum_{i=1}^{a} \sum_{k=1}^{c} \frac{y_{i.k.}^2}{bn} - \frac{y_{....}^2}{abcn} - SS_A - SS_C$$

$$= SS_{\text{Subtotals}(AC)} - SS_A - SS_C \tag{7-42}$$

and

$$SS_{BC} = \sum_{j=1}^{b} \sum_{k=1}^{c} \frac{y_{.jk.}^2}{an} - \frac{y_{....}^2}{abcn} - SS_B - SS_C$$

$$= SS_{\text{Subtotals}(BC)} - SS_B - SS_C \tag{7-43}$$

Note that the sums of squares for the two-factor subtotals are found from the

Table 7-18 The Analysis of Variance Table for the Three-Factor Fixed Effects Model

Source of Variation	Sum of Squares	Degrees of Freedom	Mean Square	Expected Mean Squares	F_0
A	SS_A	$a - 1$	MS_A	$\sigma^2 + \dfrac{bcn \sum \tau_i^2}{a - 1}$	$F_0 = \dfrac{MS_A}{MS_E}$
B	SS_B	$b - 1$	MS_B	$\sigma^2 + \dfrac{acn \sum \beta_j^2}{b - 1}$	$F_0 = \dfrac{MS_B}{MS_E}$
C	SS_C	$c - 1$	MS_C	$\sigma^2 + \dfrac{abn \sum \gamma_k^2}{c - 1}$	$F_0 = \dfrac{MS_C}{MS_E}$
AB	SS_{AB}	$(a - 1)(b - 1)$	MS_{AB}	$\sigma^2 + \dfrac{cn \sum\sum (\tau\beta)_{ij}^2}{(a - 1)(b - 1)}$	$F_0 = \dfrac{MS_{AB}}{MS_E}$
AC	SS_{AC}	$(a - 1)(c - 1)$	MS_{AC}	$\sigma^2 + \dfrac{bn \sum\sum (\tau\gamma)_{ik}^2}{(a - 1)(c - 1)}$	$F_0 = \dfrac{MS_{AC}}{MS_E}$
BC	SS_{BC}	$(b - 1)(c - 1)$	MS_{BC}	$\sigma^2 + \dfrac{an \sum\sum (\beta\gamma)_{jk}^2}{(b - 1)(c - 1)}$	$F_0 = \dfrac{MS_{BC}}{MS_E}$
ABC	SS_{ABC}	$(a - 1)(b - 1)(c - 1)$	MS_{ABC}	$\sigma^2 + \dfrac{n \sum\sum\sum (\tau\beta\gamma)_{ijk}^2}{(a - 1)(b - 1)(c - 1)}$	$F_0 = \dfrac{MS_{ABC}}{MS_E}$
Error	SS_E	$abc(n - 1)$	MS_E	σ^2	
Total	SS_T	$abcn - 1$			

totals in each two-way table. The three-factor interaction sum of squares is computed from the three-way cell totals $\{y_{ijk.}\}$ as

$$SS_{ABC} = \sum_{i=1}^{a} \sum_{j=1}^{b} \sum_{k=1}^{c} \frac{y_{ijk.}^2}{n} - \frac{y_{....}^2}{abcn} - SS_A - SS_B - SS_C - SS_{AB} - SS_{AC} - SS_{BC}$$

(7-44a)

$$= SS_{\text{Subtotals}(ABC)} - SS_A - SS_B - SS_C - SS_{AB} - SS_{AC} - SS_{BC} \qquad (7\text{-}44b)$$

The error sum of squares may be found by subtracting the sum of squares for each main effect and interaction from the total sum of squares or by

$$SS_E = SS_T - SS_{\text{Subtotals}(ABC)} \qquad (7\text{-}45)$$

Example 7-4

The Soft Drink Bottling Problem

A soft drink bottler is interested in obtaining more uniform fill heights in the bottles produced by his manufacturing process. The filling machine theoretically fills each bottle to the correct target height, but in practice, there is variation around this target, and the bottler would like to understand better the sources of this variability and eventually reduce it.

The process engineer can control three variables during the filling process: the percent carbonation (A), the operating pressure in the filler (B), and the bottles produced per minute or the line speed (C). The pressure and speed are easy to control, but the percent carbonation is more difficult to control during actual manufacturing because it varies with product temperature. However, for purposes of an experiment, the engineer can control carbonation at three levels: 10, 12, and 14 percent. He chooses two levels for pressure (25 and 30 psi) and two levels for line speed (200 and 250 bpm). He decides to run two replicates of a factorial design in these three factors, with all 24 runs taken in random order. The response variable observed is the average deviation from the target fill height observed in a production run of bottles at each set of conditions. The data that resulted from this experiment are shown in Table 7-19. Positive deviations are fill heights above the target, whereas negative deviations are fill heights below the target. The circled numbers in Table 7-19 are the three-way cell totals $y_{ijk.}$.

The total corrected sum of squares is found from Equation 7-37 as

$$SS_T = \sum_{i=1}^{a} \sum_{j=1}^{b} \sum_{k=1}^{c} \sum_{l=1}^{n} y_{ijkl}^2 - \frac{y_{....}^2}{abcn}$$

$$= 571 - \frac{(75)^2}{24} = 336.625$$

and the sums of squares for the main effects are calculated from Equations 7-38, 7-39, and 7-40 as

$$SS_{\text{Carbonation}} = \sum_{i=1}^{a} \frac{y_{i...}^2}{bcn} - \frac{y_{....}^2}{abcn}$$

$$= \frac{(-4)^2 + (20)^2 + (59)^2}{8} - \frac{(75)^2}{24} = 252.750$$

$$SS_{\text{Pressure}} = \sum_{j=1}^{b} \frac{y_{.j..}^2}{acn} - \frac{y_{....}^2}{abcn}$$

$$= \frac{(21)^2 + (54)^2}{12} - \frac{(75)^2}{24} = 45.375$$

ble 7-19   Fill Height Deviation Data for Example 7-4

Percent arbonation (A)	Operating Pressure (B)				$y_{i...}$
	25 psi		30 psi		
	Line Speed (C)		Line Speed (C)		
	200	250	200	250	
10	$\begin{matrix}-3\\-1\end{matrix}$ (−4)	$\begin{matrix}-1\\0\end{matrix}$ (−1)	$\begin{matrix}-1\\0\end{matrix}$ (−1)	$\begin{matrix}1\\1\end{matrix}$ (2)	−4
12	$\begin{matrix}0\\1\end{matrix}$ (1)	$\begin{matrix}2\\1\end{matrix}$ (3)	$\begin{matrix}2\\3\end{matrix}$ (5)	$\begin{matrix}6\\5\end{matrix}$ (11)	20
14	$\begin{matrix}5\\4\end{matrix}$ (9)	$\begin{matrix}7\\6\end{matrix}$ (13)	$\begin{matrix}7\\9\end{matrix}$ (16)	$\begin{matrix}10\\11\end{matrix}$ (21)	59
B × C Totals $y_{.jk.}$	6	15	20	34	$75 = y_{....}$
$y_{.j..}$	21		54		

A × B Totals $y_{ij..}$		
	B	
A	25	30
10	−5	1
12	4	16
14	22	37

A × C Totals $y_{i.k.}$		
	C	
A	200	250
10	−5	1
12	6	14
14	25	34

and

$$SS_{Speed} = \sum_{k=1}^{c} \frac{y_{..k.}^2}{abn} - \frac{y_{....}^2}{abcn}$$

$$= \frac{(26)^2 + (49)^2}{12} - \frac{(75)^2}{24} = 22.042$$

To calculate the sums of squares for the two-factor interactions, we must find the two-way cell totals. For example, to find the carbonation–pressure or AB interaction, we need the totals for the $A \times B$ cells $\{y_{ij..}\}$ shown in Table 7-19. Using Equation 7-41, we find the sums of squares as

$$SS_{AB} = \sum_{i=1}^{a} \sum_{j=1}^{b} \frac{y_{ij..}^2}{cn} - \frac{y_{....}^2}{abcn} - SS_A - SS_B$$

$$= \frac{(-5)^2 + (1)^2 + (4)^2 + (16)^2 + (22)^2 + (37)^2}{4} - \frac{(75)^2}{24}$$

$$- 252.750 - 45.375$$

$$= 5.250$$

The carbonation–speed or AC interaction uses the $A \times C$ cell totals $\{y_{i.k.}\}$ shown in

Table 7-19 and Equation 7-42:

$$SS_{AC} = \sum_{i=1}^{a} \sum_{k=1}^{c} \frac{y_{i.k.}^2}{bn} - \frac{y_{....}^2}{abcn} - SS_A - SS_C$$

$$= \frac{(-5)^2 + (1)^2 + (6)^2 + (14)^2 + (25)^2 + (34)^2}{4} - \frac{(75)^2}{24}$$

$$- 252.750 - 22.042$$

$$= 0.583$$

The pressure–speed or BC interaction is found from the $B \times C$ cell totals $\{y_{.jk.}\}$ shown in Table 7-19 and Equation 7-43:

$$SS_{BC} = \sum_{j=1}^{b} \sum_{k=1}^{c} \frac{y_{.jk.}^2}{an} - \frac{y_{....}^2}{abcn} - SS_B - SS_C$$

$$= \frac{(6)^2 + (15)^2 + (20)^2 + (34)^2}{6} - \frac{(75)^2}{24} - 45.375 - 22.042$$

$$= 1.042$$

The three-factor interaction sum of squares is found from the $A \times B \times C$ cell totals $\{y_{ijk.}\}$, which are circled in Table 7-19. From Equation 7-44a, we find

$$SS_{ABC} = \sum_{i=1}^{a} \sum_{j=1}^{b} \sum_{k=1}^{c} \frac{y_{ijk.}^2}{n} - \frac{y_{....}^2}{abcn} - SS_A - SS_B - SS_C - SS_{AB} - SS_{AC} - SS_{BC}$$

$$= \frac{1}{2} [(-4)^2 + (-1)^2 + (-1)^2 + \cdots + (16)^2 + (21)^2] - \frac{(75)^2}{24}$$

$$- 252.750 - 45.375 - 22.042 - 5.250 - 0.583 - 1.042 = 1.083$$

Finally, noting that

$$SS_{\text{Subtotals}(ABC)} = \sum_{i=1}^{a} \sum_{j=1}^{b} \sum_{k=1}^{c} \frac{y_{ijk.}^2}{n} - \frac{y_{....}^2}{abcn} = 328.125$$

Table 7-20   Analysis of Variance for Example 7-4

Source of Variation	Sum of Squares	Degrees of Freedom	Mean Square	F_0
Percent carbonation (A)	252.750	2	126.375	178.412[a]
Operating pressure (B)	45.375	1	45.375	64.059[a]
Line speed (C)	22.042	1	22.042	31.118[a]
AB	5.250	2	2.625	3.706[b]
AC	0.583	2	0.292	0.412
BC	1.042	1	1.042	1.471
ABC	1.083	2	0.542	0.765
Error	8.500	12	0.708	
Total	336.625	23		

[a]Significant at 1 percent.
[b]Significant at 10 percent.

we have

$$SS_E = SS_T - SS_{\text{Subtotals}(ABC)}$$
$$= 336.625 - 328.125$$
$$= 8.500$$

The analysis of variance is summarized in Table 7-20. We see that the percent carbonation, operating pressure, and line speed significantly affect the fill volume. The carbonation–pressure interaction is significant at 10 percent, indicating a mild interaction between these factors.

The next step should be an analysis of the residuals from this experiment. We leave this as an exercise for the reader but point out that a normal probability plot of the residuals and the other usual diagnostics do not indicate any major concerns.

To assist in the practical interpretation of this experiment, Figure 7-11 presents plots of the three main effects and the *AB* (carbonation–pressure) interaction. The main effect plots are just graphs of the marginal response averages at the levels of the

Figure 7-11. Main effects and interaction plots for Example 7-4. (a) Percent carbonation (*A*). (b) Pressure (*B*). (c) Line speed (*C*). (d) Carbonation-pressure interaction.

three factors. Notice that all three variables have *positive* main effects; that is, increasing the variable moves the average deviation from the fill target upward. The interaction between carbonation and pressure is fairly small, as shown by the similar shape of the two curves in Figure 7-11d.

Since the company wants the average deviation from the fill target to be close to zero, the engineer decides to recommend the low level of operating pressure (25 psi) and the high level of line speed (250 bpm, which will maximize the production rate). Figure 7-12 plots the average observed deviation from the target fill height at the three different carbonation levels for this set of operating conditions. Now the carbonation level cannot presently be perfectly controlled in the manufacturing process, and the normal distribution shown with the solid curve in Figure 7-12 approximates the variability in the carbonation levels presently experienced. As the process is impacted by the values of the carbonation level drawn from this distribution, the fill heights will fluctuate considerably. This variability in the fill heights could be reduced if the distribution of the carbonation level values followed the normal distribution shown with the dashed line in Figure 7-12. Reducing the standard deviation of the carbonation level distribution was ultimately achieved by improving temperature control during the process.

■

We have indicated that if all the factors in a factorial experiment are fixed, then test statistic construction is straightforward. The statistic for testing any main effect or interaction is always formed by dividing the mean square for the main effect or interaction by the mean square error. However, if the factorial experiment involves a random or mixed model, the test statistic construction is not immediately obvious. We must examine the expected mean squares to determine the correct tests. For example, recall the two-factor random and mixed models in Section 7-4, where the appropriate test procedure followed directly from the expected mean squares. It seems, then, that we need a procedure for developing the expected value of a mean square. We defer a complete discussion of this topic until Chapter 8, where a set of rules is given for obtaining the sums of

Figure 7-12. Average fill height deviation at high speed and low pressure for different carbonation levels.

squares, the degrees of freedom, and the expected mean squares for a general class of designs.

7-6 FITTING RESPONSE CURVES AND SURFACES

In Section 4-3 we noted that it is occasionally useful to fit a response curve to the levels of a quantitative factor so that the experimenter has an equation that relates the response to the factor. This equation might be used for interpolation, that is, for predicting the response at factor levels between those actually used in the experiment. The same principles discussed in Section 4-3 can be applied to factorial designs where at least one factor is quantitative. They are illustrated in the following two examples. In these examples, we use factor levels that are equally spaced and we use the method of orthogonal polynomials. General regression methods could also be used. In fact, if the factor levels are not equally-spaced regression methods must be used. These procedures are illustrated in Chapter 15.

Example 7-5

Consider the experiment described in Example 7-1. The temperature is quantitative and material type is qualitative. Furthermore, the three levels of temperature are equally spaced. Consequently, we can compute a linear and a quadratic temperature effect to study how temperature affects the battery life. The calculation of the linear and quadratic effects of temperature, denoted T_L and T_Q, and their sums of squares are shown in Table 7-21.

Table 7-21 Calculation of the Polynomial Effects of Temperature, Example 7-5

Temperature Levels (°F)	Treatment Totals $y_{.j.}$	Orthogonal Contrast Coefficents (c_j) Linear	Quadratic
15	1738	-1	$+1$
70	1291	0	-2
125	770	$+1$	$+1$
Effects: $\left(\sum\limits_{j=1}^{3} c_j y_{.j.} \right)$		$T_L = -968$	$T_Q = -74$
Sums of squares: $\left[\dfrac{\left(\sum\limits_{j=1}^{3} c_j y_{.j.} \right)^2}{an \sum\limits_{j=1}^{3} c_j^2} \right]$		$SS_{T_L} = \dfrac{(-968)^2}{(4)(3)(2)}$	$SS_{T_Q} = \dfrac{(-74)^2}{(4)(3)(6)}$
		$= 39,042.67$	$= 76.05$

Table 7-22 Polynomial Effects of the MT Interaction, Example 7-5

Temperature Levels (°F)	Totals for Material Types			Orthogonal Contrast Coefficients (c_j)	
	$1(y_{1j.})$	$2(y_{2j.})$	$3(y_{3j.})$	Linear	Quadratic
15	539	623	576	−1	+1
70	229	479	583	0	−2
125	230	198	342	+1	+1

Effects: $M_1 \times T_L = -309$, $M_2 \times T_L = -425$, $M_3 \times T_L = -234$

$M_1 \times T_Q = 311$, $M_2 \times T_Q = -137$, $M_3 \times T_Q = -248$

Sums of squares:

$$SS_{M \times T_L} = \frac{(-309)^2 + (-425)^2 + (-234)^2}{4(2)} - \frac{(-968)^2}{12(2)} = 2315.08$$

$$SS_{M \times T_Q} = \frac{(311)^2 + (-137)^2 + (-248)^2}{4(6)} - \frac{(-74)^2}{12(6)} = 7298.70$$

The interaction sum of squares may be partitioned into an interaction component between the material type and the linear effect of temperature ($M \times T_L$) and an interaction component between the material type and the quadratic effect of temperature ($M \times T_Q$). To compute these interaction components, the linear and quadratic effects of temperature are determined for each material type, and then the sums of squares between these effects are calculated. The process is shown in Table 7-22. Note that $SS_{MT} = SS_{M \times T_L} + SS_{M \times T_Q}$. The analysis of variance is displayed in Table 7-23. The linear temperature effect is evident, and there is no indication of significant quadratic response to temperature. Furthermore, the significant interaction is the result of the $M \times T_Q$ component.

Since the effect of temperature is significant, the experimenter could fit a response curve to the data so that the output voltage may be predicted as a function of the

Table 7-23 Analysis of Variance for Example 7-5

Source of Variation	Sum of Squares	Degrees of Freedom	Mean Square	F_0
Material types (M)	10,683.72	2	5,341.86	7.91[a]
Temperature (T)	39,118.72	2	19,559.36	28.97[a]
(Linear)	(39,042.67)	1	39,042.67	57.82[a]
(Quadratic)	(76.05)	1	76.05	0.12
MT interaction	9,613.78	4	2,403.44	3.56[b]
($M \times T_L$)	(2,315.08)	2	1157.54	1.71
($M \times T_Q$)	(7,298.70)	2	3649.35	5.41[b]
Error	18,230.75	27	675.21	
Total	77,646.97	35		

[a]Significant at 1 percent.
[b]Significant at 5 percent.

Table 7-24   Data for Tool Life Experiment

Tool Angle (degrees)	Cutting Speed (in/min) 125	150	175	$y_{i..}$
15	$\begin{array}{l} -2 \\ -1 \end{array}$ (-3)	$\begin{array}{l} -3 \\ 0 \end{array}$ (-3)	$\begin{array}{l} 2 \\ 3 \end{array}$ (5)	-1
20	$\begin{array}{l} 0 \\ 2 \end{array}$ (2)	$\begin{array}{l} 1 \\ 3 \end{array}$ (4)	$\begin{array}{l} 4 \\ 6 \end{array}$ (10)	16
25	$\begin{array}{l} -1 \\ 0 \end{array}$ (-1)	$\begin{array}{l} 5 \\ 6 \end{array}$ (11)	$\begin{array}{l} 0 \\ -1 \end{array}$ (-1)	9
$y_{.j.}$	-2	12	14	$24 = y_{....}$

temperature. However, the significant *MT* interaction implies that the output voltage response to temperature depends on which material type is used. Therefore, the experimenter would have to fit *three* response curves—one for each material type.

∎

If several factors in a factorial experiment are quantitative, then these factors may be decomposed into single-degree-of-freedom polynomial effects. Furthermore, the interactions of quantitative factors can be partitioned into single-degree-of-freedom components of interaction using a similar process. This is illustrated in the following example.

Example 7-6

The effective life of a cutting tool installed in a numerically controlled machine is thought to be affected by the cutting speed and the tool angle. Three speeds and three angles are selected, and a factorial experiment with two replicates is performed. The coded data are shown in Table 7-24. The circled numbers in the cells are the cells totals $\{y_{ij.}\}$.

Table 7-25 summarizes the analysis of variance. Both factors are treated as fixed effects. Cutting speed, tool angle, and their interaction are significant at the 5 percent level.

The linear and quadratic effects of both factors may be determined using the method presented in Example 7-5. The calculations are displayed in Table 7-26.

Table 7-25   Analysis of Variance for Tool Life Data

Source of Variation	Sum of Squares	Degrees of Freedom	Mean Square	F_0
Tool angle (*T*)	24.33	2	12.17	8.45
Cuttng speed (*C*)	25.33	2	12.67	8.80
TC interaction	61.34	4	15.34	10.65
Error	13.00	9	1.44	
Total	124.00	17		

Table 7-26 Polynomial Effects of Tool Angle (T) and Cutting Speed (C) for Example 7-6

Treatment Totals		Orthogonal Contrast Coefficients	
Tool Angle ($y_{i..}$)	Cutting Speed ($y_{.j.}$)	Linear	Quadratic
-1	-2	-1	$+1$
16	12	0	-2
9	14	$+1$	$+1$
Effects:		$T_L = 10$	$T_Q = -24$
		$C_L = 16$	$C_Q = -12$

				Totals
Sums of squares:		$SS_{T_L} = 8.33$	$SS_{T_Q} = 16.00$	24.33
		$SS_{C_L} = 21.33$	$SS_{C_Q} = 4.00$	25.33

Notice that the sum of squares for a factor is equal to the total sum of squares of the linear and quadratic components for that same factor.

Since both factors in this experiment are quantitative, it is possible to partition the TC interaction into four single-degree-of-freedom components, say

$$TC_{L \times L} \qquad TC_{L \times Q} \qquad TC_{Q \times L} \qquad \text{and} \qquad TC_{Q \times Q}$$

In general, any two-factor interaction between quantitative factors can be partitioned into $(a - 1)(b - 1)$ single-degree-of-freedom components. However, it may be difficult to interpret these interaction components. Often a particular component is significant because of atypical responses in one cell. Thus, it is relatively standard practice to compute only the lower-order components. In such a case, the remaining components are considered jointly as higher-order interaction effects. The statistical significance of these higher-order interaction components could indicate a very complex relationship between the factors.

The technique for computing the single-degree-of-freedom interaction components requires the cell totals (recall that differences between cells define an interaction effect) and the orthogonal contrast coefficients. To illustrate the approach, consider the coefficients for the $TC_{L \times L}$ interaction component shown in Table 7-27. The entries in the body of the table (e.g., c_{ij}) are obtained by multiplying the coefficients of T_L by the coefficients of C_L.

To obtain the $TC_{L \times L}$ contrast, multiply the coefficient in the ijth cell in Table

Table 7-27 $TC_{L \times L}$ Interaction Coefficients

T_L	C_L		
	-1	0	$+1$
-1	$+1$	0	-1
0	0	0	0
$+1$	-1	0	$+1$

7-27 by the ijth cell total (shown in circles in Table 7-24) and add, yielding

$$TC_{L\times L} = \sum_{i=1}^{3} \sum_{j=1}^{3} c_{ij} y_{ij}.$$

$$= +1(-3) + (0)(-3) + (-1)(5) + (0)(2) + (0)(4)$$
$$+ (0)(10) + (-1)(-1) + (0)(11) + 1(-1) = -8$$

The sum of squares for the $TC_{L\times L}$ contrast is

$$SS_{TC_{L\times L}} = \frac{(TC_{L\times L})^2}{n \sum\sum c_{ij}^2} = \frac{(-8)^2}{2[(1)^2 + (0)^2 + (-1)^2 + \cdots + (1)^2]} = \frac{(-8)^2}{2(4)} = 8.00$$

The coefficients for the $TC_{L\times Q}$, $TC_{Q\times L}$, and $TC_{Q\times Q}$ interaction components are derived similarly. They are shown in Table 7-28, along with the contrast and sum of squares for each interaction component.

∎

It is easy to see that the sum of the four single-degree-of-freedom interaction components equals the interaction sum of squares; that is,

$$SS_{TC} = SS_{TC_{L\times L}} + SS_{TC_{L\times Q}} + SS_{TC_{Q\times L}} + SS_{TC_{Q\times Q}}$$

$$= 8.00 + 42.67 + 2.67 + 8.00 = 61.34$$

The expanded analysis of variance is summarized in Table 7-29. Both linear and quadratic components of tool angle are significant, whereas cutting speed has only a linear effect on tool life. The $TC_{L\times L}$, $TC_{L\times Q}$, and $TC_{Q\times Q}$ interactions are also significant, with the $TC_{L\times Q}$ component quite large relative to the others.

Figure 7-13 plots the cell averages against the tool angle for all three speeds and illustrates the quadratic effect of tool angle. Figure 7-14 plots cell averages versus cutting speed for the tool angles.

If we let x_1 represent tool angle and x_2 represent cutting speed, then Table 7-29 implies a polynomial approximation for tool life, for example,

$$y = \alpha_0 + \alpha_1 P_1(x_1) + \alpha_{11} P_2(x_1) + \alpha_2 P_1(x_2) + \alpha_{12} P_1(x_1) P_1(x_2) \quad (7\text{-}46)$$
$$+ \alpha_{122} P_1(x_1) P_2(x_2) + \alpha_{1122} P_2(x_1) P_2(x_2) + \epsilon$$

where α_1 denotes the linear coefficient of tool angle, α_{122} denotes the coefficient for linear tool angle and quadratic cutting speed, and so on, and $P_i(x_j)$ are the orthogonal polynomials. These coefficients and polynomials are evaluated exactly as in Section 4-3; for example,

$$\hat{\alpha}_{12} = \frac{\sum_{i=1}^{3} \sum_{j=1}^{3} y_{ij} P_1(x_{1i}) P_1(x_{2j})}{\sum_{i=1}^{3} \sum_{j=1}^{3} [P_1(x_{1i}) P_1(x_{2j})]^2} = \frac{-8}{8} = -1.00$$

and

$$P_1(x_2) P_1(x_2) = \lambda_1 \left(\frac{x_1 - \bar{x}_1}{d_1} \right) \lambda_1 \left(\frac{x_2 - \bar{x}_2}{d_2} \right)$$

$$= (1) \left(\frac{x_1 - 20}{5} \right) (1) \left(\frac{x_2 - 150}{25} \right)$$

Table 7-28   Computation of the $TC_{L\times Q}$, $TC_{Q\times L}$, and $TC_{Q\times Q}$ Interactions

T_L	C_Q		
	+1	-2	+1
-1	-1	+2	-1
0	0	0	0
+1	+1	-2	+1

$$TC_{L\times Q} = \sum_{i=1}^{3}\sum_{j=1}^{3} c_{ij} y_{ij.}$$
$$= -32$$
$$SS_{TC_{L\times Q}} = \frac{(TC_{L\times Q})^2}{n\sum\sum c_{ij}^2}$$
$$= \frac{(-32)^2}{2(12)} = 42.67$$

T_Q	C_L		
	-1	0	+1
+1	-1	0	+1
-2	+2	0	-2
+1	-1	0	+1

$$TC_{Q\times L} = \sum_{i=1}^{3}\sum_{j=1}^{3} c_{ij} y_{ij.}$$
$$= -8$$
$$SS_{TC_{Q\times L}} = \frac{(TC_{Q\times L})^2}{n\sum\sum c_{ij}^2}$$
$$= \frac{(-8)^2}{2(12)} = 2.67$$

T_Q	C_Q		
	+1	-2	+1
+1	+1	-2	+1
-2	-2	+4	-2
+1	+1	-2	+1

$$TC_{Q\times Q} = \sum_{i=1}^{3}\sum_{j=1}^{3} c_{ij} y_{ij.}$$
$$= -24$$
$$SS_{TC_{Q\times Q}} = \frac{(TC_{Q\times Q})^2}{n\sum\sum c_{ij}^2}$$
$$= \frac{(-24)^2}{2(36)} = 8.00$$

Table 7-29   Analysis of Variance for Example 7-6

Source of Variance	Sum of Squares	Degrees of Freedom	Mean Square	F_0
Tool angle, T	24.33	2	12.17	8.45[a]
(T_L)	(8.33)	1	(8.33)	5.78[a]
(T_Q)	(16.00)	1	(16.00)	11.11[a]
Cutting speed, C	25.33	2	12.67	8.80[a]
(C_L)	(21.33)	1	(21.33)	14.81[a]
(C_Q)	(4.00)	1	(4.00)	2.77
TC interaction	61.34	4	15.34	10.65[a]
$(TC_{L \times L})$	(8.00)	1	(8.00)	5.56[a]
$(TC_{L \times Q})$	(42.67)	1	(42.67)	29.63[a]
$(TC_{Q \times L})$	(2.67)	1	(2.67)	1.85
$(TC_{Q \times Q})$	(8.00)	1	(8.00)	5.56[a]
Error	13.00	9	1.44	
Total	124.00			

[a]Significant at 5 percent.

Evaluating each term in Equation 7-46 in a similar manner will yield the following:

$$\hat{y} = 2.667 + 0.700(x_1 - 20) - 0.0267(x_1 - 20)^2 + 0.0532(x_2 - 150)$$
$$- 0.008(x_1 - 20)(x_1 - 150) - 0.00128(x_1 - 20)(x_1 - 150)^2 \quad (7\text{-}47)$$
$$- 0.000128(x_1 - 20)^2(x_2 - 150)^2$$

This equation is called a *response surface,* and it could be used to predict tool

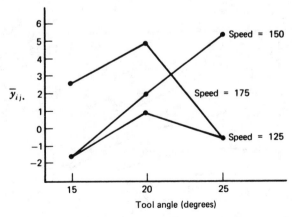

Figure 7-13. Cell averages versus tool angle for Example 7-6.

Figure 7-14. Cell averages versus cutting speed for Example 7-6.

surface generated by Equation 7-47. Examination of the response surface reveals that maximum tool life is achieved at cutting speeds around 150 rpm and tool angles of 25 degrees. The three-dimensional response surface plot in Figure 7-16 provides essentially the same information, but it provides a different and sometimes more useful perspective of the tool life response surface. Exploration of response surfaces is a very important aspect of experimental design, which we will discuss in more detail in Chapter 16.

7-7 DEALING WITH UNBALANCED DATA

The primary focus of this chapter has been the analysis of *balanced* factorial designs, that is, cases where there are an equal number of observations n in each

Figure 7-15. Two-dimensional tool life response surface for Example 7-6.

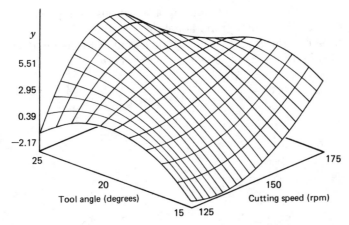

Figure 7-16.   Three-dimensional tool life response surface for Example 7-6.

cell. However, it is not unusual to encounter situations where the number of observations in the cells are unequal. These *unbalanced* factorial designs occur for various reasons. For example, the experimenter may have designed a balanced experiment initially, but because of unforeseen problems in data collection, resulting in the loss of some observations, she or he ends up with unbalanced data. On the other hand, some unbalanced experiments are deliberately designed that way. For instance, certain treatment combinations may be more expensive or more difficult to run than others, so fewer observations may be taken in those cells. Alternatively, some treatment combinations may be of greater interest to the experimenter because they represent new or unexplored conditions, so he or she may elect to obtain additional replication in those cells.

The orthogonality property of main effects and interactions present in balanced data do not carry over to the unbalanced case. This means that the usual analysis of variance techniques do not apply. Consequently, the analysis of unbalanced factorials is much more difficult than that for balanced designs.

In this section we give a brief overview of methods for dealing with unbalanced factorials, concentrating on the case of the two-factor fixed effects model. Suppose that the number of observations in the ijth cell is n_{ij}. Furthermore, let $n_{i.} = \sum_{j=1}^{b} n_{ij}$ be the number of observations in the ith row (the ith level of factor A), $n_{.j} = \sum_{i=1}^{a} n_{ij}$ be the number of observations in the jth column (the jth level of factor B), and $n_{..} = \sum_{i=1}^{a} \sum_{j=1}^{b} n_{ij}$ be the total number of observations.

7-7.1   Proportional Data: An Easy Case

One situation involving unbalanced data presents little difficulty in analysis; this is the case of *proportional data*. That is, the number of observations in the ijth cell is

$$n_{ij} = \frac{n_{i.} n_{.j}}{n_{..}} \qquad (7\text{-}48)$$

This condition implies that the number of observations in any two rows or columns are proportional. When proportional data occur, the standard analysis of variance can be employed. Only minor modifications are required in the computing formulas for the sums of squares, which become

$$SS_T = \sum_{i=1}^{a} \sum_{j=1}^{b} \sum_{k=1}^{n_{ij}} y_{ijk}^2 - \frac{y_{...}^2}{n_{..}}$$

$$SS_A = \sum_{i=1}^{a} \frac{y_{i..}^2}{n_{i.}} - \frac{y_{...}^2}{n_{..}}$$

$$SS_B = \sum_{j=1}^{b} \frac{y_{.j.}^2}{n_{.j}} - \frac{y_{...}^2}{n_{..}}$$

$$SS_{AB} = \sum_{i=1}^{a} \sum_{j=1}^{b} \frac{y_{ij.}^2}{n_{ij}} - \frac{y_{...}^2}{n_{..}} - SS_A - SS_B$$

$$SS_E = SS_T - SS_A - SS_B - SS_{AB}$$

$$= \sum_{i=1}^{a} \sum_{j=1}^{b} \sum_{k=1}^{n_{ij}} y_{ijk}^2 - \sum_{i=1}^{a} \sum_{j=1}^{b} \frac{y_{ij.}^2}{n_{ij}}$$

As an example of proportional data, consider the battery design experiment in Example 7-1. A modified version of the original data is shown in Table 7-30. Clearly, the data are proportional; for example, in cell 1, 1, we have

$$n_{11} = \frac{n_{1.}n_{.1}}{n_{..}} = \frac{10(8)}{20} = 4$$

observations. The results of applying the usual analysis of variance to these data are shown in Table 7-31. Both material type and temperature are significant,

Table 7-30   Battery Design Experiment with Proportional Data

Material Type	Temperature (°F)			
	15	70	125	
1	$n_{11} = 4$ 130 155 74 180	$n_{12} = 4$ 34 40 80 75	$n_{13} = 2$ 70 58	$n_{1.} = 10$ $y_{1..} = 896$
2	$n_{21} = 2$ 159 126	$n_{22} = 2$ 136 115	$n_{23} = 1$ 45	$n_{2.} = 5$ $y_{2..} = 581$
3	$n_{31} = 2$ 138 160	$n_{32} = 2$ 150 139	$n_{33} = 1$ 96	$n_{3.} = 5$ $y_{3..} = 683$
	$n_{.1} = 8$ $y_{.1.} = 1122$	$n_{.2} = 8$ $y_{.2.} = 769$	$n_{.3} = 4$ $y_{.3.} = 269$	$n_{..} = 20$ $y_{...} = 2160$

Table 7-31 Analysis of Variance for Battery Design Data in Table 7-30

Source of Variance	Sum of Squares	Degrees of Freedom	Mean Square	F_0
Material types	8,170.400	2	4,085.20	5.00
Temperature	16,090.875	2	8,045.44	9.85
Interaction	5,907.725	4	1,476.93	1.18
Error	8,981.000	11	816.45	
Total	39,150.000	19		

which agrees with the analysis of the full data set in Example 7-1. However, the interaction noted in Example 7-1 is not present.

7-7.2 Approximate Methods

When unbalanced data are not too far from the balanced case, it is sometimes possible to use approximate procedures that convert the unbalanced problem into a balanced one. This of course makes the analysis only approximate, but the analysis of balanced data is so easy that we are frequently tempted to use this approach. In practice, we must decide when the data are not sufficiently different from the balanced case to make the degree of approximation introduced relatively unimportant. We now briefly describe some of these approximate methods. We assume that every cell has at least one observation (i.e., $n_{ij} \geq 1$).

Estimating Missing Observations ▪ If only a few n_{ij} are different, a reasonable procedure is to estimate the missing values. For example, consider the unbalanced design in Table 7-32. Clearly, estimating the single missing value in cell 2,2 is a reasonable approach. For a model with interaction, the estimate of the missing value in the ijth cell that minimizes the error sum of squares is $\bar{y}_{ij.}$. That is, we estimate the missing value by taking the average of the observations that are available in that cell.

The estimated value is treated just like actual data. The only modification to the analysis of variance is to reduce the error degrees of freedom by the number of missing observations that have been estimated. For example, if we estimate the missing value in the 2,2 cell in Table 7-32, we would use 26 error degrees of freedom instead of 27.

Table 7-32 The n_{ij} Values for an Unbalanced Design

	Columns		
Rows	1	2	3
1	4	4	4
2	4	3	4
3	4	4	4

Setting Data Aside ▪ Consider the data in Table 7-33. Note that the 2,2 cell has only one more observation than the others. Estimating missing values for the remaining 8 cells is probably not a good idea here because this would result in estimates constituting about 18 percent of the final data. An alternative is to set aside one of the observations in cell 2,2, giving a balanced design with $n = 4$ replicates.

The observation that is set aside should be chosen randomly. Furthermore, rather than completely discarding the observation, we could return it to the design, then randomly choose another observation to set aside and repeat the analysis. And, we hope, these two analyses will not lead to conflicting interpretations of the data. If they do, then we suspect that the observation that was set aside is an outlier or a wild value and should be handled accordingly. In practice, this confusion is unlikely to occur when only small numbers of observations are set aside and the variability within the cells is small.

Method of Unweighted Means ▪ In this approach, introduced by Yates (1934), the cell averages are treated as data and are subjected to a standard balanced data analysis to obtain sums of squares for rows, columns, and interaction. The error mean square is found as

$$MS_E = \frac{\sum_{i=1}^{a}\sum_{j=1}^{b}\sum_{k=1}^{n_{ij}} (y_{ijk} - \bar{y}_{ij.})^2}{n.. - ab} \tag{7-49}$$

Now MS_E estimates σ^2, the variance of y_{ijk}, an individual observation. However, we have done an analysis of variance on the cell *averages*, and since the variance of the average in the ijth cell is σ^2/n_{ij}, the error mean square actually used in the analysis of variance should be an estimate of the average variance of the $\bar{y}_{ij.}$, say

$$\bar{V}(\bar{y}_{ij.}) = \frac{\sum_{i=1}^{a}\sum_{j=1}^{b} \sigma^2/n_{ij}}{ab} = \frac{\sigma^2}{ab}\sum_{i=1}^{a}\sum_{j=1}^{b}\frac{1}{n_{ij}} \tag{7-50}$$

Using MS_E from Equation 7-49 to estimate σ^2 in Equation 7-50, we obtain

$$MS_E' = \frac{MS_E}{ab}\sum_{i=1}^{a}\sum_{j=1}^{b}\frac{1}{n_{ij}} \tag{7-51}$$

Table 7-33   The n_{ij} Values for an Unbalanced Design

Rows	Columns		
	1	2	3
1	4	4	4
2	4	5	4
3	4	4	4

as the error mean square (with $n_{..} - ab$ degrees of freedom) to use in the analysis of variance.

The method of unweighted means is an approximate procedure because the sums of squares for rows, columns, and interaction are not distributed as chi-square random variables. The primary advantage of the method seems to be its computational simplicity. When the n_{ij} are not dramatically different, the method of unweighted means often works reasonably well.

A related technique is the *weighted squares of means method,* also proposed by Yates (1934). This technique is also based on the sums of squares of the cell means, but the terms in the sums of squares are weighted in inverse proportion to their variances. For further details of the procedure, see Searle (1971a) and Speed, Hocking, and Hackney (1978).

7-7.3 The Exact Method

In situations where approximate methods are inappropriate, such as when empty cells occur (some $n_{ij} = 0$) or when the n_{ij} are dramatically different, the experimenter must use an exact analysis. The approach used to develop sums of squares for testing main effects and interactions is to represent the analysis of variance model as a regression model, fit that model to the data, and use the general regression significance test approach. However, there are several ways that this may be done, and these methods may result in different values for the sums of squares. Furthermore, the hypotheses that are being tested are not always direct analogs of those for the balanced case, nor are they always easily interpretable. For further reading on the subject, see Searle (1971a); Speed and Hocking (1976); Hocking and Speed (1975); Hocking, Hackney, and Speed (1978); Speed, Hocking, and Hackney (1978); Searle, Speed, and Henderson (1981); and Searle (1987).

7-8 PROBLEMS

7-1 The yield of a chemical process is being studied. The two most important variables are thought to be the pressure and the temperature. Three levels of each factor are selected, and a factorial experiment with two replicates is performed. The yield data follow:

Temperature	Pressure		
	200	215	230
Low	90.4	90.7	90.2
	90.2	90.6	90.4
Medium	90.1	90.5	89.9
	90.3	90.6	90.1
High	90.5	90.8	90.4
	90.7	90.9	90.1

(a) Analyze the data and draw conclusions.

(b) Prepare appropriate residual plots and comment on the model's adequacy.

(c) Under what conditions would you operate this process?

7-2 An engineer suspects that the surface finish of a metal part is influenced by the feed rate and the depth of cut. She selects three feed rates and randomly chooses four depths of cut. She then conducts a factorial experiment and obtains the following data:

Feed Rate (in/min)	Depth of Cut (in)			
	0.15	0.18	0.20	0.25
0.20	74	79	82	99
	64	68	88	104
	60	73	92	96
0.25	92	98	99	104
	86	104	108	110
	88	88	95	99
0.30	99	104	108	114
	98	99	110	111
	102	95	99	107

(a) Analyze the data and draw conclusions.

(b) Prepare appropriate residual plots and comment on the model's adequacy.

(c) Obtain point estimates of the mean surface finish at each feed rate.

(d) Estimate the variance component for depth of cut.

7-3 For the data in Problem 7-2, compute a 95 percent interval estimate of the mean difference in response for feed rates of 0.20 and 0.25 in/min.

7-4 An article in *Industrial Quality Control*, (1956, pp. 5–8) describes an experiment to investigate the effect of the type of glass and the type of phosphor on the brightness of a television tube. The response variable is the current necessary (in microamps) to obtain a specified brightness level. The data are as follows:

Glass Type	Phosphor Type		
	1	2	3
1	280	300	290
	290	310	285
	285	295	290
2	230	260	220
	235	240	225
	240	235	230

(a) Is there any indication that either factor influences brightness?

(b) Do the two factors interact?

(c) Analyze the residuals from this experiment.

7-5 Johnson and Leone (*Statistics and Experimental Design in Engineering and the Physical Sciences*, Wiley, 1977) describe an experiment to investigate warping of copper plates The two factors studied were the temperature and the copper content of the plates. The response variable was a measure of the amount of warping. The data were as follows:

Temperature (°C)	Copper Content (%)			
	40	60	80	100
50	17, 20	16, 21	24, 22	28, 27
75	12, 9	18, 13	17, 12	27, 31
100	16, 12	18, 21	25, 23	30, 23
125	21, 17	23, 21	23, 22	29, 31

(a) Is there any indication that either factor affects the amount of warping? Is there any interaction between the factors?

(b) Analyze the residuals from this experiment.

(c) Plot the average warping at each level of copper content and compare them to an appropriately scaled t distribution. Describe the differences in the effects of the different levels of copper content on warping. If low warping is desirable, what level of copper content would you specify?

(d) Suppose that temperature cannot be easily controlled in the environment in which the copper plates are to be used. Does this change your answer for part (c)?

7-6 The factors that influence the breaking strength of a synthetic fiber are being studied. Four production machines and three operators are chosen at random and a factorial experiment is run using fiber from the same production batch. The results are as follows.

Operator	Machine			
	1	2	3	4
1	109	110	108	110
	110	115	109	108
2	110	110	111	114
	112	111	109	112
3	116	112	114	120
	114	115	119	117

(a) Analyze the data and draw conclusions.

(b) Prepare appropriate residual plots and comment on the model's adequacy.

(c) Estimate the variance components.

7-7 Suppose that in Problem 7-6 the operators were chosen at random, but the only four machines available were used for the test. Test for interaction and main effects at the 5 percent level. Does the new experimental situation influence either the analysis or your conclusions?

7-8 A mechanical engineer is studying the thrust force developed by a drill press. He suspects that the drilling speed and the feed rate of the material are the most important factors. He randomly selects four feed rates and uses a high and low drill speed chosen to represent the extreme operating conditions. He obtains the following results. Analyze the data and raw conclusions.

Drill Speed	Feed Rate			
	0.015	0.030	0.045	0.060
125	2.70	2.45	2.60	2.75
	2.78	2.49	2.72	2.86
200	2.83	2.85	2.86	2.94
	2.86	2.80	2.87	2.88

7-9 An experiment is conducted to study the influence of operating temperature and three types of face-plate glass in the light output of an oscilloscope tube. The following data are collected:

Glass Type	Temperature		
	100	125	150
1	580	1090	1392
	568	1087	1380
	570	1085	1386
2	550	1070	1328
	530	1035	1312
	579	1000	1299
3	546	1045	867
	575	1053	904
	599	1066	889

Assume that both factors are fixed. Is there a significant interaction effect? Does glass type or temperature affect the response? What conclusions can you draw? Use the method discussed in the text to partition the temperature effect into its linear and quadratic components. Break the interaction down into appropriate components.

7-10 Consider the data in Problem 7-1. Use the method described in the text to extract the linear and quadratic effects of pressure.

7-11 Use Duncan's multiple range test to determine which levels of the pressure factor are significantly different for the data in Problem 7-1.

7-12 An experiment was conducted to determine if either firing temperature or furnace position affects the baked density of a carbon anode. The data are shown below.

Position	Temperature (°C)		
	800	825	850
1	570	1063	565
	565	1080	510
	583	1043	590
2	528	988	526
	547	1026	538
	521	1004	532

Suppose we assume that no interaction exists. Write down the statistical model. Conduct the analysis of variance assuming both factors to be fixed effects. What conclusions can be drawn? Derive the expected mean squares. Comment on the model's adequacy.

7-13 Derive the expected mean squares for a two-factor analysis of variance with one observation per cell, assuming that both factors are fixed.

7-14 Consider a two-factor analysis of variance with one observation per cell. What are the expected mean squares, assuming both row and column factors are random? Under what conditions do we have exact tests for the main effects? Write down the analysis of variance table.

7-15 Consider a two-factor analysis of variance with one observation per cell. If the row factor is fixed and the column factor is random, what are the expected mean squares? Under what conditions do we have exact tests for main the effects? Write down the analysis of variance table.

7-16 Consider the following data, in which row and column factors are both fixed. Analyze the data and draw conclusions. Perform a test for nonadditivity.

Row Factor	Column Factor			
	1	2	3	4
1	36	39	36	32
2	18	20	22	20
3	30	37	33	34

7-17   The shear strength of an adhesive is thought to be affected by the application pressure and temperature. A factorial experiment is performed in which both factors are assumed to be fixed. Analyze the data and draw conclusions. Perform a test for nonadditivity.

	Temperature (°F)		
Pressure (lb/in²)	250	260	270
120	9.60	11.28	9.00
130	9.69	10.10	9.57
140	8.43	11.01	9.03
150	9.98	10.44	9.80

7-18   For the data in Problem 7-17, suppose the temperature and pressure levels are chosen at random. Assuming no interaction, conduct the analysis of variance. How do your conclusions differ, if at all, from your conclusions for Problem 7-17?

7-19   Consider the three-factor model

$$Y_{ijk} = \mu + \tau_i + \beta_j + \gamma_k + (\tau\beta)_{ij} + (\beta\gamma)_{jk} + \epsilon_{ijk} \begin{cases} i = 1,2, \ldots, a \\ j = 1,2, \ldots, b \\ k = 1,2, \ldots, c \end{cases}$$

Notice that there is only one replicate. Assuming all the factors are fixed, write down the analysis of variance table, including the expected mean squares. What would you use as the "experimental error" in order to test hypotheses?

7-20   The percentage of hardwood concentration in raw pulp, the vat pressure, and the cooking time of the pulp are being investigated for their effects on the strength of paper. Three levels of hardwood concentration, three levels of pressure, and two cooking times are selected. All the factors may be regarded as fixed effects. A factorial experiment with two replicates is conducted, and the following data are obtained:

Percentage of Hardwood Concentration	Cooking Time 3.0 hours Pressure			Cooking Time 4.0 hours Pressure		
	400	500	650	400	500	650
2	196.6	197.7	199.8	198.4	199.6	200.6
	196.0	196.0	199.4	198.6	200.4	200.9
4	198.5	196.0	198.4	197.5	198.7	199.6
	197.2	196.9	197.6	198.1	198.0	199.0
8	197.5	195.6	197.4	197.6	197.0	198.5
	196.6	196.2	198.1	198.4	197.8	199.8

(a) Analyze the data and draw conclusions.

(b) Prepare appropriate residual plots and comment on the model's adequacy.

(c) Under what set of conditions would you operate this process? Why?

7-21 The quality control department of a fabric finishing plant is studying the effect of several factors on the dyeing of cotton–synthetic cloth used to manufacture men's shirts. Three operators, three cycle times, and two temperatures were selected, and three small specimens of cloth were dyed under each set of conditions. The finished cloth was compared to a standard, and a numerical score was assigned. The results follow. Analyze the data and draw conclusions. Comment on the model's adequacy.

	Temperature					
	300°			350°		
	Operator			Operator		
Cycle Time	1	2	3	1	2	3
	23	27	31	24	38	34
40	24	28	32	23	36	36
	25	26	29	28	35	39
	36	34	33	37	34	34
50	35	38	34	39	38	36
	36	39	35	35	36	31
	28	35	26	26	36	28
60	24	35	27	29	37	26
	27	34	25	25	34	24

7-22 In Problem 7-1, suppose that we wish to reject the null hypothesis with a high probability if the difference in the true mean yield at any two pressures is as great as 0.5. If a reasonable prior estimate of the standard deviation of yield is 0.1, how many replicates should be run?

7-23 In Problem 7-6, determine the power of the test for detecting a machine effect such that $\sigma_\beta^2 = \sigma^2$, where σ_β^2 is the variance component for the machine factor. Are two replicates sufficient?

7-24 In the two-factor mixed model analysis of variance, show that $\text{Cov}[(\tau\beta)_{ij}, (\tau\beta)_{i'j}] = -(1/a)\sigma_{\tau\beta}^2$ for $i \neq i'$.

7-25 Show that the method of analysis of variance always produces unbiased point estimates of the variance components in any random or mixed model.

7-26 Invoking the usual normality assumptions, find an expression for the probability that a negative estimate of a variance component will be obtained by the analysis of variance method. Using this result, write a statement giving the probability that $\hat{\sigma}_\tau^2 < 0$ in a one-factor analysis of variance. Comment on the usefulness of this probability statement.

7-27   Consider the data in Problem 7-2, and assume that both factors are fixed. Suppose that the first observation in the 1,1 and the 2,3 cells are missing. Analyze the data by estimating the missing values. Compare this analysis with that obtained in Problem 7-2. Comment on the adequacy of this approximate analysis in this case.

7-28   Repeat Problem 7-27 using the method of unweighted means.

7-29   Consider the data in Problem 7-2, and assume that both factors are fixed. Suppose that the first observation in every cell, except the 2,2 cell is missing. Analyze this data by setting aside one observation from the 2,2 cell. Compare the analysis with that obtained in Problem 7-2. Comment on the adequacy of the approximation in this case.

7-30   Analyze the data in Problem 7-2 using both the mixed models discussed in the text. Compare the results obtained from the two models.

7-31   Consider the two-factor mixed model. Show that the standard error of the fixed factor mean (e.g., A) for Duncan's multiple range test is $[MS_{AB}/bn]^{1/2}$.

Chapter 8
Rules for Sums of Squares and Expected Mean Squares

An important part of any experimental design problem is conducting the analysis of variance. This involves determining the sum of squares for each component in the model and the number of degrees of freedom associated with each sum of squares. Then, to construct appropriate test statistics, the expected mean squares must be determined. In complex design situations, particularly those involving random or mixed models, it is frequently helpful to have a formal procedure for this process.

We will present a set of rules for writing down the sums of squares for the effects in model (the expression *effect* now includes both main effects and interactions), the number of degrees of freedom, and the expected mean squares for any balanced factorial, nested,[1] or nested factorial experiment. (Note that partially balanced arrangements, such as Latin squares and incomplete block designs, are specifically excluded.) These rules are discussed by several authors, including Scheffé (1959), Bennett and Franklin (1954), Hicks (1973), and Searle (1971a, 1971b). The two-factor factorial model is used to illustrate the application of the rules.

8-1 RULES FOR SUMS OF SQUARES

RULE 1. The error term in the model, $\epsilon_{ij\ldots m}$, is written as $\epsilon_{(ij\ldots)m}$, where the subscript m denotes the replication subscript. For the two-factor model, this rule implies that ϵ_{ijk} becomes $\epsilon_{(ij)k}$.

---

[1] Nested designs are discussed in Chapter 13.

RULE 2. In addition to an overall mean (μ) and an error term $[\epsilon_{(ij...)m}]$, the model contains all the main effects and any interactions that the experimenter assumes exist. If all possible interactions between k factors exist, then there are $\binom{k}{2}$ two-factor interactions, $\binom{k}{3}$ three-factor interactions, . . . , 1 k-factor interaction. If one of the factors in a term appears in parentheses, then there is no interaction between that factor and the other factors in that term.

RULE 3. For each term in the model, divide the subscripts into three classes: (a) live—those subscripts that are present in the term and are not in parentheses; (b) dead—those subscripts that are present in the term and are in parentheses; and (c) absent—those subscripts that are present in the model but not in that particular term.

Thus, in $(\tau\beta)_{ij}$, i and j are live and k is absent, and in $\epsilon_{(ij)k}$, k is live and i and j are dead.

RULE 4. Degrees of Freedom. The number of degrees of freedom for any term in the model is the product of the number of levels associated with each dead subscript and the number of levels minus one associated with each live subscript.

For example, the number of degrees of freedom associated with $(\tau\beta)_{ij}$ is $(a - 1)(b - 1)$, and the number of degrees of freedom associated with $\epsilon_{(ij)k}$ is $ab(n - 1)$.

RULE 5. Sums of Squares. To obtain a computing formula for the sum of squares for any effect, first expand the number of degrees of freedom for that effect. For example, for β_j the expanded degrees of freedom are simply $b - 1$. Each term in this quantity is the *symbolic form* of an uncorrected sum of squares. Then use the following procedure.

(a) Let 1 represent the correction factor; that is,

$$1 = \frac{\left(\sum_{i=1}^{a} \sum_{j=1}^{b} \cdots \sum_{m=1}^{n} y_{ij...m}\right)^2}{ab...n}$$

For each remaining symbolic form, write down the total of all observations in summation form. Thus, for b write down $\sum_{i=1}^{a} \sum_{j=1}^{b} \sum_{k=1}^{c} y_{ijk}$.

(b) Reorder the summation signs so that those relative to the letters in the symbolic form of interest (in this case b) come first, and enclose the remaining elements of the term in parentheses. Therefore, for b we obtain

$$\sum_{j=1}^{b} \left(\sum_{i=1}^{a} \sum_{k=1}^{n} y_{ijk}\right)$$

(c) Convert the quantity within parentheses to the standard "dot subscript" notation, where a dot replacing a subscript indicates summation over that subscript. Thus,

$$\sum_{j=1}^{b} \left(\sum_{i=1}^{a} \sum_{k=1}^{n} y_{ijk}\right) = \sum_{j=1}^{b} (y_{.j.})$$

(d) Square the term within parentheses and divide by the product of the number of levels of the "dot" subscripts. For our example, the b term becomes $\sum_{j=1}^{b} y_{.j.}^2/an$, which is an *uncorrected sum of squares*. The *corrected* sum of squares of the effect results from replacing each symbolic form by the corresponding uncorrected sum of squares. Thus, to obtain the sum of squares for β_j, replace the symbolic forms in $b - 1$ by the uncorrected sum of squares formed above. We then see that

$$SS_B = \sum_{j=1}^{b} \frac{y_{.j.}^2}{an} - \frac{y_{...}^2}{abn}$$

This is clearly the sum of squares for the main effect B in a two-factor analysis.

As another illustration of Rule 5, consider finding the sum of squares for $(\tau\beta)_{ij}$, that is, SS_{AB}. The expanded number of degrees of freedom is $(a - 1)(b - 1) = ab - a - b + 1$. Therefore, the sum of squares is determined as follows:

ab	$-a$	$-b$	$+1$	
$\displaystyle\sum_{i=1}^{a}\sum_{j=1}^{b}\sum_{k=1}^{n} y_{ijk}$	$\displaystyle\sum_{i=1}^{a}\sum_{j=1}^{b}\sum_{k=1}^{n} y_{ijk}$	$\displaystyle\sum_{i=1}^{a}\sum_{j=1}^{b}\sum_{k=1}^{n} y_{ijk}$	$\dfrac{y_{...}^2}{abn}$	part (a)
$\displaystyle\sum_{i=1}^{a}\sum_{j=1}^{b}\left(\sum_{k=1}^{n} y_{ijk}\right)$	$\displaystyle\sum_{i=1}^{a}\left(\sum_{j=1}^{b}\sum_{k=1}^{n} y_{ijk}\right)$	$\displaystyle\sum_{j=1}^{b}\left(\sum_{i=1}^{a}\sum_{k=1}^{n} y_{ijk}\right)$	$\dfrac{y_{...}^2}{abn}$	part (b)
$\displaystyle\sum_{i=1}^{a}\sum_{j=1}^{b} (y_{ij.})$	$\displaystyle\sum_{i=1}^{a} (y_{i..})$	$\displaystyle\sum_{j=1}^{b} (y_{.j.})$	$\dfrac{y_{...}^2}{abn}$	part (c)
$\displaystyle\sum_{i=1}^{a}\sum_{j=1}^{b} \frac{y_{ij.}^2}{n}$	$\displaystyle\sum_{i=1}^{a} \frac{y_{i..}^2}{bn}$	$\displaystyle\sum_{j=1}^{b} \frac{y_{.j.}^2}{an}$	$\dfrac{y_{...}^2}{abn}$	part (d)

Now, combining the uncorrected sums of squares in the last row according to the signs at the top of each column, we obtain

$$SS_{AB} = \sum_{i=1}^{a}\sum_{j=1}^{b} \frac{y_{ij.}^2}{n} - \sum_{i=1}^{a} \frac{y_{i..}^2}{bn} - \sum_{j=1}^{b} \frac{y_{.j.}^2}{an} + \frac{y_{...}^2}{abn}$$

This last equation is equivalent to $SS_{AB} = SS_{\text{Subtotals}} - SS_A - SS_B$, which we obtained as Equation 7-9.

8-2 RULES FOR EXPECTED MEAN SQUARES

We have seen that the expected mean squares play an important role in the analysis of variance. By examining the expected mean squares, one may develop the appropriate statistic for testing hypotheses about any model parameter. The test statistic is a ratio of mean squares that is chosen such that the expected value of the *numerator* mean square differs from the expected value of the *denominator* mean square only by the variance component or the fixed factor in which we are interested.

It is always possible to determine the expected mean squares in any model as we did in Chapter 3; that is, by the direct application of the expectation operator. This "brute force" method, as it is often called, can be very tedious. The rules that follow always produce the expected mean squares for any design without resorting to the brute force approach, and with practice, they become relatively simple to use. When applied to a mixed model, these rules produce expected mean squares that are consistent with the assumptions for the standard mixed model of Section 7-4. We illustrate the rules using the two-factor fixed effects factorial model.

RULE 1. Each effect has either a variance component (random effect) or a fixed factor (fixed effect) associated with it. If an interaction contains at least one random effect, the entire interaction is considered as random. A variance component has Greek letters as subscripts to identify the particular random effect. Thus, in a two-factor mixed model with factor A fixed and factor B random, the variance component for B is σ_β^2, and the variance component for AB is $\sigma_{\tau\beta}^2$. A fixed effect is always represented by the sum of squares of the model components associated with that factor divided by its degrees of freedom. In our example, the effect of A is

$$\frac{\sum_{i=1}^{a} \tau_i^2}{a - 1}$$

RULE 2. Expected Mean Squares. To obtain the expected mean squares, prepare the following table. There is a row for each model component (mean square) and a column for each subscript. Over each subscript, write the number of levels of the factor associated with that subscript and whether the factor is fixed (F) or random (R). Replicates are always considered to be random.

(a) In each row, write 1 if one of the dead subscripts in the row component matches the subscript in the column:

		F	F	R
		a	b	n
	Factor	i	j	k
	τ_i			
	β_j			
	$(\tau\beta)_{ij}$			
	$\epsilon_{(ij)k}$	1	1	

(b) In each row, if any of the subscripts on the row component match the subscript in the column, write 0 if the column is headed by a fixed factor and 1 if the column is headed by a random factor:

Factor	F a i	F b j	R n k
τ_i	0		
β_j		0	
$(\tau\beta)_{ij}$	0	0	
$\epsilon_{(ij)k}$	1	1	1

(c) In the remaining empty row positions, write the number of levels shown above the column heading:

Factor	F a i	F b j	R n k
τ_i	0	b	n
β_j	a	0	n
$(\tau\beta)_{ij}$	0	0	n
$\epsilon_{(ij)k}$	1	1	1

(d) To obtain the expected mean square for any model component, first cover all columns headed by live subscripts on that component. Then, in each row that contains at least the same subscripts as those on the component being considered, take the product of the visible numbers and multiply by the appropriate fixed or random factor from Rule 1. The sum of these quantities is the expected mean square of the model component being considered. To find $E(MS_A)$, for example, cover column i. The product of the visible numbers in the rows that contain at least subscript i are bn (row 1), 0 (row 3), and 1 (row 4). Note that i is missing in row 2. Therefore, the expected mean square is

$$E(MS_A) = \sigma^2 + \frac{bn \sum_{i=1}^{a} \tau_i^2}{a - 1}$$

The complete table of expected mean squares for this design is shown in Table 8-1. Tables 8-2 and 8-3 display the expected mean square derivations for the two-factor random and mixed models, respectively. A three-factor factorial design is treated in the following example.

Example 8-1

Consider a three-factor factorial experiment with a levels of factor A, b levels of factor B, c levels of factor C, and n replicates. The analysis of this design, assuming that all

Table 8-1   Expected Mean Square Derivation, Two-Factor Fixed Effects Model

Factor	F a i	F b j	R n k	Expected Mean Square
τ_i	0	b	n	$\sigma^2 + \dfrac{bn\Sigma\tau_i^2}{a-1}$
β_j	a	0	n	$\sigma^2 + \dfrac{an\Sigma\beta_j^2}{b-1}$
$(\tau\beta)_{ij}$	0	0	n	$\sigma^2 + \dfrac{n\Sigma\Sigma(\tau\beta)_{ij}^2}{(a-1)(b-1)}$
$\epsilon_{(ij)k}$	1	1	1	σ^2

the factors are fixed effects, is given in Section 7-5. We now determine the expected mean squares assuming that all the factors are *random*. The appropriate statistical model is

$$y_{ijkl} = \mu + \tau_i + \beta_j + \gamma_k + (\tau\beta)_{ij} + (\tau\gamma)_{ik} + (\beta\gamma)_{jk} + (\tau\beta\gamma)_{ijk} + \epsilon_{ijkl}$$

Using the rules previously described, the expected mean squares are derived in Table 8-4.

We notice by examining the expected mean squares in Table 8-4, that if A, B, and C are all random factors, then no exact test exists for the main effects. That is, if we wish to test the hypothesis $\sigma_\tau^2 = 0$, we cannot form a ratio of two expected mean squares such that the only term in the numerator that is not in the denominator is $bcn\sigma_\tau^2$. The same phenomenon occurs for the main effects of B and C. Notice that proper tests do exist for the two-factor and three-factor interactions. However, it is likely that tests on the main effects are of central importance to the experimenter. Therefore, how should the main effects be tested? This problem is considered in the next section.

■

8-3   APPROXIMATE *F* TESTS

In factorial experiments with three or more factors involving a random or mixed model and certain other, more complex designs, there are frequently no exact

Table 8-2   Expected Mean Square Derivation, Two-Factor Random Effects Model

Factor	R a i	R b j	R n k	Expected Mean Square
τ_i	1	b	n	$\sigma^2 + n\sigma_{\tau\beta}^2 + bn\sigma_\tau^2$
β_j	a	1	n	$\sigma^2 + n\sigma_{\tau\beta}^2 + an\sigma_\beta^2$
$(\tau\beta)_{ij}$	1	1	n	$\sigma^2 + n\sigma_{\tau\beta}^2$
$\epsilon_{(ij)k}$	1	1	1	σ^2

Table 8-3 Expected Mean Square Derivation, Two-Factor Mixed Model

Factor	F a i	R b j	R n k	Expected Mean Square
τ_i	0	b	n	$\sigma^2 + n\sigma_{\tau\beta}^2 + \dfrac{bn\Sigma\tau_i^2}{a-1}$
β_j	a	1	n	$\sigma^2 + an\sigma_\beta^2$
$(\tau\beta)_{ij}$	0	1	n	$\sigma^2 + n\sigma_{\tau\beta}^2$
$\epsilon_{(ij)k}$	1	1	1	σ^2

test statistics for certain effects in the models. One possible solution to this dilemma is to assume that certain interactions are negligible. To illustrate, if we could reasonably assume that all the two-factor interactions in Example 8-1, are negligible, then we could put $\sigma_{\tau\beta}^2 = \sigma_{\tau\gamma}^2 = \sigma_{\beta\gamma}^2 = 0$, and tests for main effects could be conducted.

Although this seems to be an attractive possibility, we must point out that there must be something in the nature of the process—or some strong prior knowledge—in order for us to assume that one or more of the interactions are negligible. In general, this assumption is not easily made, nor should it be taken lightly. We should not eliminate certain interactions from the model without conclusive evidence that it is appropriate to do so. A procedure advocated by some experimenters is to test the interactions first, then set at zero those interactions found to be insignificant, and then assume that these interactions are zero when testing other effects in the same experiment. Although sometimes done in practice, this procedure can be dangerous because any decision regarding an interaction is subject to both type I and type II errors.

A variation of this idea is to pool the mean squares in the analysis of variance

Table 8-4 Expected Mean Square Derivation for Three-Factor Random Effects Model

Factor	R a i	R b j	R c k	R n l	Expected Mean Squares
τ_i	1	b	c	n	$\sigma^2 + cn\sigma_{\tau\beta}^2 + bn\sigma_{\tau\gamma}^2 + n\sigma_{\tau\beta\gamma}^2 + bcn\sigma_\tau^2$
β_j	a	1	c	n	$\sigma^2 + cn\sigma_{\tau\beta}^2 + an\sigma_{\beta\gamma}^2 + n\sigma_{\tau\beta\gamma}^2 + acn\sigma_\beta^2$
γ_k	a	b	1	n	$\sigma^2 + bn\sigma_{\tau\gamma}^2 + an\sigma_{\beta\gamma}^2 + n\sigma_{\tau\beta\gamma}^2 + abn\sigma_\gamma^2$
$(\tau\beta)_{ij}$	1	1	c	n	$\sigma^2 + n\sigma_{\tau\beta\gamma}^2 + cn\sigma_{\tau\beta}^2$
$(\tau\gamma)_{ik}$	1	b	1	n	$\sigma^2 + n\sigma_{\tau\beta\gamma}^2 + bn\sigma_{\tau\gamma}^2$
$(\beta\gamma)_{jk}$	a	1	1	n	$\sigma^2 + n\sigma_{\tau\beta\gamma}^2 + an\sigma_{\beta\gamma}^2$
$(\tau\beta\gamma)_{ijk}$	1	1	1	n	$\sigma^2 + n\sigma_{\tau\beta\gamma}^2$
$\epsilon_{(ijk)l}$	1	1	1	1	σ^2

to obtain an estimate of error with more degrees of freedom. For instance, suppose that in Example 8-1 the test statistic $F_0 = MS_{ABC}/MS_E$ was not significant. Thus, $H_0: \sigma^2_{\tau\theta\gamma} = 0$ is not rejected, and *both* MS_{ABC} and MS_E estimate the error variance σ^2. The experimenter might consider pooling or combining MS_{ABC} and MS_E according to

$$MS_{E'} = \frac{abc(n-1)MS_E + (a-1)(b-1)(c-1)MS_{ABC}}{abc(n-1) + (a-1)(b-1)(c-1)}$$

so that $E(MS_{E'}) = \sigma^2$. Note that $MS_{E'}$ has $abc(n-1) + (a-1)(b-1)(c-1)$ degrees of freedom, compared to $abc(n-1)$ degrees of freedom for the original MS_E. The danger of pooling is that one may make a type II error and combine the mean square for a factor that really *is* significant with error, thus obtaining a new residual mean square ($MS_{E'}$) that is too large. This will make other significant effects more difficult to detect. On the other hand, if the original error mean square has a very small number of degrees of freedom (e.g., less than 6), then the experimenter may have much to gain by pooling because it could potentially increase the precision of further tests considerably. A reasonably practical procedure is as follows. If the original error mean square has more than 6 degrees of freedom, do not pool. If the original error mean square has fewer than 6 degrees of freedom, pool only if the F statistic for the mean square to be pooled is not significant at a large value of α, such as $\alpha = .25$.

If we cannot assume that certain interactions are negligible and we still need to make inferences about those effects for which exact tests do not exist, a procedure attributed to Satterthwaite (1946) can be employed. Satterthwaite's method uses linear combinations of mean squares, for example,

$$MS' = MS_r + \cdots + MS_s \tag{8-1}$$

and

$$MS'' = MS_u + \cdots + MS_v \tag{8-2}$$

where the mean squares in Equations 8-1 and 8-2 are chosen so that $E(MS') - E(MS'')$ is equal to the effect (the model parameter or variance component) considered in the null hypothesis. Then the test statistic would be

$$F = \frac{MS'}{MS''} \tag{8-3}$$

which is distributed approximately as $F_{p,q}$, where

$$p = \frac{(MS_r + \cdots + MS_s)^2}{MS_r^2/f_r + \cdots + MS_s^2/f_s} \tag{8-4}$$

and

$$q = \frac{(MS_u + \cdots + MS_v)^2}{MS_u^2/f_u + \cdots + MS_v^2/f_v} \tag{8-5}$$

In p and q, f_i is the number of degrees of freedom associated with the mean square MS_i. There is no assurance that p and q will be integers, so it will be necessary to interpolate in the tables of the F distribution. For example, in the three-factor random effects model (Table 8-4), it is relatively easy to see that an appropriate test statistic for $H_0: \sigma_\tau^2 = 0$ would be $F = MS'/MS''$, with

$$MS' = MS_A + MS_{ABC}$$

and

$$MS'' = MS_{AB} + MS_{AC}$$

The degrees of freedom for F would be computed from Equations 8-4 and 8-5.

The theory underlying this test is that both the numerator and the denominator of the test statistic (Equation 8-3) are distributed approximately as multiples of chi-square random variables, and because no mean square appears in both the numerator or denominator of Equation 8-3, the numerator and denominator are independent. Thus F in Equation 8-3 is distributed approximately as $F_{p,q}$. Satterthwaite remarks that caution should be used in applying the procedure when some of the mean squares in MS' and MS'' are involved negatively. Gaylor and Hopper (1969) report that if $MS' = MS_1 - MS_2$, then Satterthwaite's approximation holds reasonably well if

$$\frac{MS_1}{MS_2} > F_{.025, f_2, f_1} \times F_{.50, f_2, f_2}$$

if $f_1 \le 100$ and $f_2 \ge f_1/2$.

Example 8-2

The pressure drop experienced across an expansion valve in a turbine is being studied. The design engineer considers the important variables that influence pressure drop to be gas temperature on the inlet side (A), turbine speed (B), and inlet gas pressure (C). These three factors are arranged in a factorial design, with gas temperature fixed, and turbine speed and pressure random. The coded data for two replicates are shown in Table 8-5. The linear model for this design is

$$y_{ijkl} = \mu + \tau_i + \beta_j + \gamma_k + (\tau\beta)_{ij} + (\tau\gamma)_{ik} + (\beta\gamma)_{jk} + (\tau\beta\gamma)_{ijk} + \epsilon_{ijkl}$$

where τ_i is the effect of the gas temperature (A), β_j is the effect of the turbine speed (B), and γ_k is the effect of the gas pressure (C).

The analysis of variance is shown in Table 8-6. A column entitled "Expected Mean Squares" has been added to this table, and the entries in this column are derived by the methods discussed in Section 8-2. From the expected mean square column, we observe that exact tests exist for all effects except the main effect A. Results for these tests are shown in Table 8-6. To test $H_0: \tau_i = 0$, we could use the statistic

$$F = \frac{MS'}{MS''}$$

where

$$MS' = MS_A + MS_{ABC}$$

Table 8-5   Coded Pressure Drop Data for the Turbine Experiment

Pressure (C)	Gas Temperature (A)											
	60°F				75°F				90°F			
	Speed (B)				Speed (B)				Speed (B)			
	150	200	225	300	150	200	225	300	150	200	225	300
50 psi	-2	0	-1	4	14	6	1	-7	-8	-2	-1	-2
	-3	-9	-8	4	14	0	2	6	-8	20	-2	1
75 psi	-6	-5	-8	-3	22	8	6	-5	-8	1	-9	-8
	4	-1	-2	-7	24	6	2	2	3	-7	-8	3
85 psi	-1	-4	0	-2	20	2	3	-5	-2	-1	-4	1
	-2	-8	-7	4	16	0	0	-1	-1	-2	-7	3

and

$$MS'' = MS_{AB} + MS_{AC}$$

since

$$E(MS') - E(MS'') = \frac{bcn \sum \tau_i^2}{a - 1}$$

To determine the test statistic for $H_0 : \tau_i = 0$, we compute

$$MS' = MS_A + MS_{ABC}$$
$$= 308.39 + 19.26 = 327.65$$
$$MS'' = MS_{AB} + MS_{AC}$$
$$= 134.91 + 44.77 = 179.68$$

Table 8-6   Analysis of Variance for the Pressure Drop Data

Source of Variation	Sum of Squares	Degrees of Freedom	Expected Mean Squares	Mean Square	F_0
Temperature, A	616.78	2	$\sigma^2 + bn\sigma_{\tau\gamma}^2 + cn\sigma_{\tau\beta}^2 + n\sigma_{\tau\beta\gamma}^2 + \dfrac{bcn\sum\tau_i^2}{a-1}$	308.39	1.82
Speed, B	175.56	3	$\sigma^2 + an\sigma_{\beta\gamma}^2 + acn\sigma_\beta^2$	58.52	1.45
Pressure, C	5.03	2	$\sigma^2 + an\sigma_{\beta\gamma}^2 + abn\sigma_\gamma^2$	2.52	0.06
AB	809.44	6	$\sigma^2 + n\sigma_{\tau\beta\gamma}^2 + cn\sigma_{\tau\beta}^2$	134.91	7.00[a]
AC	179.06	4	$\sigma^2 + n\sigma_{\tau\beta\gamma}^2 + bn\sigma_{\tau\gamma}^2$	44.77	2.32[b]
BC	242.19	6	$\sigma^2 + an\sigma_{\beta\gamma}^2$	40.37	1.16
ABC	231.07	12	$\sigma^2 + n\sigma_{\tau\beta\gamma}^2$	19.26	0.56
Error	1248.00	36	σ^2	34.67	
Total	3507.11	71			

[a] Significant at 1 percent.

[b] Significant at 25 percent.

Figure 8-1.   Interactions in pressure drop experiment.

and

$$F = \frac{MS'}{MS''} = \frac{327.65}{179.68} = 1.82$$

The degrees of freedom for this statistic are

$$p = \frac{(MS_A + MS_{ABC})^2}{MS_A^2/2 + MS_{ABC}^2/12}$$

$$= \frac{(327.65)^2}{(308.39)^2/2 + (19.26)^2/12} = 2.26$$

and

$$q = \frac{(MS_{AB} + MS_{AC})^2}{MS_{AB}^2/6 + MS_{AC}^2/4}$$

$$= \frac{(179.68)^2}{(134.91)^2/6 + (44.77)^2/4} = 6.34$$

Comparing $F = 1.82$ to $F_{.05,2.26,6.34}$, we cannot reject H_0.

The AB interaction is large, and there is some indication of an AC interaction. The graphical analysis of the AB and AC interactions, shown in Figure 8-1, indicates that the effect of A may be large at low levels of B and $C = 75$. Thus, it seems possible that the main effects of A and B are masked by the large AB interaction.

∎

8-4 PROBLEMS

8-1 Use the procedure discussed in Section 8-2 to determine the expected mean squares for the three-factor factorial, fixed effects model.

8-2 By application of the expectation operator, develop the expected mean squares for the two-factor factorial, mixed model. Check your results with the expected mean squares given in Table 8-3 to see that they agree.

8-3 Consider the three-factor factorial design in Example 8-1. Propose appropriate test statistics for all main effects and interactions. Repeat for the case where A and B are fixed and C is random.

8-4 Consider the data in Example 8-2. Analyze the data for the case where A, B, and C are random.

8-5 Derive the expected mean squares shown in Table 8-6.

8-6 Consider a four-factor factorial experiment where factor A is at a levels, factor B is at b levels, factor C is at c levels, factor D is at d levels, and there are n replicates. Write down the sums of squares, the degrees of freedom, and the expected mean squares for the following cases.

(a) A, B, C, and D are fixed factors.

(b) A, B, C, and D are random factors.

(c) A is fixed and B, C, and D are random.

(d) A and B are fixed and C and D are random.

(e) A, B, C, are fixed and D is random.

Do exact tests exist for all effects? If not, propose test statistics for those effects that cannot be directly tested.

8-7 Consider the data in Problem 7-20. Suppose that hardwood concentration and cooking time are fixed whereas pressure is random. Analyze the data under these conditions and draw appropriate conclusions.

8-8   In Problem 7-21, assume that operators, cycle times, and temperatures were selected at random. Analyze the data under these conditions and draw conclusions.
Estimate the variance components.

8-9   Consider the three-factor factorial model

$$y_{ijk} = \mu + \tau_i + \beta_j + \gamma_k + (\tau\beta)_{ij} + (\beta\gamma)_{jk} + \epsilon_{ijk} \begin{cases} i = 1,2, \ldots, a \\ j = 1,2, \ldots, b \\ k = 1,2, \ldots, c \end{cases}$$

Assuming that all the factors are random, develop the analysis of variance table,
including the expected mean squares. Propose appropriate test statistics for all
effects.

8-10   The three-factor factorial model for a single replicate is

$$y_{ijk} = \mu + \tau_i + \beta_j + \gamma_k + (\tau\beta)_{ij} + (\beta\gamma)_{jk} + (\tau\gamma)_{ik} + (\tau\beta\gamma)_{ijk} + \epsilon_{ijk}$$

If all the factors are random, can any effects be tested? If the three-factor interaction
and the $(\tau\beta)_{ij}$ interaction do not exist, can all the remaining effects be tested?

Chapter 9
The 2^k Factorial Design

9-1 INTRODUCTION

Factorial designs are widely used in experiments involving several factors where it is necessary to study the joint effect of the factors on a response. Chapter 7 presented general methods for the analysis of factorial designs. However, there are several special cases of the general factorial design that are important because they are widely used in research work and also because they form the basis of other designs of considerable practical value.

The most important of these special cases is that of k factors, each at only two levels. These levels may be quantitative, such as two values of temperature, pressure, or time; or they may be qualitative, such as two machines, two operators, the "high" and "low" levels of a factor, or perhaps the presence and absence of a factor. A complete replicate of such a design requires $2 \times 2 \times \cdots \times 2 = 2^k$ observations and is called a 2^k *factorial design*.

This chapter presents special methods for the analysis of this useful series of designs. Throughout this chapter we assume that (1) the factors are fixed, (2) the designs are completely randomized, and (3) the usual normality assumptions are satisfied.

The 2^k design is particularly useful in the early stages of experimental work, when there are likely to be many factors to be investigated. It provides the smallest number of runs with which k factors can be studied in a complete factorial design. Because there are only two levels for each factor, we must assume that the response is approximately linear over the range of the factor levels chosen.

9-2 THE 2^2 DESIGN

The first design in the 2^k series is one with only two factors, say A and B, each run at two levels. This design is called a 2^2 *factorial design*. The levels of the factors may be arbitrarily called "low" and "high." As an example, consider an investigation into the effect of the concentration of the reactant and the amount of the catalyst on the reaction time of a chemical process. Let the reactant concentration be factor A, and let the two levels of interest be 15 percent and 25 percent. The catalyst is factor B, with the high level denoting the use of two bags of the catalyst and the low level denoting the use of only one bag. The experiment is replicated three times, and the data are as follows:

Treatment Combination	Replicate			Total
	I	II	III	
A low, B low	28	25	27	80
A high, B low	36	32	32	100
A low, B high	18	19	23	60
A high, B high	31	30	29	90

The treatment combinations in this design are shown graphically in Figure 9-1. By convention, we denote the effect of a factor by a capital Latin letter. Thus "A" refers to the effect of a factor A, "B" refers to the effect of factor B, and "AB" refers to the AB interaction. In the 2^2 design the low and high levels of A and B are denoted by "$-$" and "$+$", respectively, on the A and B axes. Thus, $-$ on the A axis represents the low level of concentration (15%) whereas $+$ represents the high level (25%), and $-$ on the B axis represents the low level of catalyst whereas $+$ denotes the high level.

The four treatment combinations in the design are usually represented by lowercase letters, as shown in Figure 9-1. We see from the figure that the high level of any factor in the treatment combination is denoted by the corresponding lowercase letter and that the low level of a factor in the treatment combination is denoted by the absence of the corresponding letter. Thus, a represents the

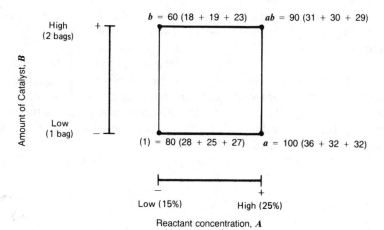

Figure 9-1. Treatment combinations in the 2^2 design.

treatment combination of A at the high level and B at the low level, b represents A at the low level and B at the high level, and ab represents both factors at the high level. By convention, (1) is used to denote both factors at the low level. This notation is used throughout the 2^k series.

Define the average effect of a factor as the change in response produced by a change in the level of that factor averaged over the levels of the other factor. Also, the notations (1), a, b, and ab now represent the *total* of all n replicates taken at the treatment combination, as illustrated in Figure 9-1. Now the effect of A at the low level of B is $[a - (1)]/n$ and the effect of A at the high level of B is $[ab - b]/n$. Averaging these two quantities yields

$$A = \frac{1}{2n}\{[ab - b] + [a - (1)]\}$$

$$= \frac{1}{2n}[ab + a - b - (1)]$$

(9-1)

The average B effect is found from the effect of B at the low level of A (i.e., $[b - (1)]/n$) and at the high level of A (i.e., $[ab - a]/n$) as

$$B = \frac{1}{2n}\{[ab - a] + [b - (1)]\}$$

$$= \frac{1}{2n}[ab + b - a - (1)]$$

(9-2)

Define the interaction effect AB as the average difference between the effect of A at the high level of B and the effect of A at the low level of B. Thus,

$$AB = \frac{1}{2n}\{[ab - b] - [a - (1)]\}$$

$$= \frac{1}{2n}[ab + (1) - a - b]$$

(9-3)

Alternatively, we may define AB as the average difference between the effect of B at the high level of A and the effect of B at the low level of A. This will also lead to Equation 9-3.

The formulas for the effects of A, B, and AB may be derived by another method. The effect of A can be found as the difference in the average response of the two treatment combinations on the right-hand side of the square in Figure 9-1 (call this average $\bar{y}_{A+}$, since it is the average response at the treatment combinations where A is at the high level) and the two treatment combinations on the left-hand side (or $\bar{y}_{A-}$). That is,

$$A = \bar{y}_{A+} - \bar{y}_{A-}$$

$$= \frac{ab + a}{2n} - \frac{b + (1)}{2n}$$

$$= \frac{1}{2n}[ab + a - b - (1)]$$

This is exactly the same result as in Equation 9-1. The effect of B, Equation 9-2, is found as the difference between the average of the two treatment combinations on the top of the square $(\overline{y}_{B+})$ and the average of the two treatment combinations on the bottom $(\overline{y}_{B-})$, or

$$B = \overline{y}_{B+} - \overline{y}_{B-}$$

$$= \frac{ab + b}{2n} - \frac{a + (1)}{2n}$$

$$= \frac{1}{2n}[ab + b - a - (1)]$$

Finally, the interaction effect AB is the average of the right-to-left diagonal treatment combinations in the square [ab and (1)] minus the average of the · left-to-right diagonal treatment combinations (a and b), or

$$AB = \frac{ab + (1)}{2n} - \frac{a + b}{2n}$$

$$= \frac{1}{2n}[ab + (1) - a - b]$$

which is identical to Equation 9-3.

Using the example data in Figure 9-1, we may estimate the average effects as

$$A = \frac{1}{2(3)}(90 + 100 - 60 - 80) = 8.33$$

$$B = \frac{1}{2(3)}(90 + 60 - 100 - 80) = -5.00$$

$$AB = \frac{1}{2(3)}(90 + 80 - 100 - 60) = 1.67$$

The effect of A (reactant concentration) is positive; this suggests that increasing A from the low level (15%) to the high level (25%) will increase the yield. The effect of B (catalyst) is negative; this suggests that increasing the amount of catalyst added to the process will decrease the yield. The interaction effect appears to be small relative to the two main effects.

In many experiments involving 2^k designs, we will examine the *magnitude* and *direction* of the factor effects to determine which variables are likely to be important. The analysis of variance can generally be used to confirm this interpretation. In the 2^k design, there are some special, time-saving methods for performing the calculations in the analysis of variance.

Consider the sums of squares for A, B, and AB. Note from Equation 9-1 that a contrast is used in estimating A, namely

$$\text{Contrast}_A = ab + a - b - (1) \tag{9-4}$$

We usually call this contrast the *total* effect of A. From Equations 9-2 and 9-3, we see that contrasts are also used to estimate B and AB. Furthermore, these three contrasts are orthogonal. The sum of squares for any contrast can be computed from Equation 3-19, which states that the contrast sum of squares is equal to the contrast squared divided by the number of observations in each total in the contrast times the sum of the squares of the contrast coefficients. Consequently, we have

$$SS_A = \frac{[ab + a - b - (1)]^2}{n \cdot 4} \tag{9-5}$$

$$SS_B = \frac{[ab + b - a - (1)]^2}{n \cdot 4} \tag{9-6}$$

and

$$SS_{AB} = \frac{[ab + (1) - a - b]^2}{n \cdot 4} \tag{9-7}$$

as the sums of squares for A, B, and AB.

Using the data in Figure 9-1, we may find the sums of squares from Equations 9-5, 9-6, and 9-7 as

$$SS_A = \frac{(50)^2}{4(3)} = 208.33$$

$$SS_B = \frac{(-30)^2}{4(3)} = 75.00 \tag{9-8}$$

and

$$SS_{AB} = \frac{(10)^2}{4(3)} = 8.33$$

The total sum of squares is found in the usual way, that is,

$$SS_T = \sum_{i=1}^{2} \sum_{j=1}^{2} \sum_{k=1}^{n} y_{ijk}^2 - \frac{y_{...}^2}{4n} \tag{9-9}$$

In general, SS_T has $4n - 1$ degrees of freedom. The error sum of squares, with $4(n - 1)$ degrees of freedom, is usually computed by subtraction as

$$SS_E = SS_T - SS_A - SS_B - SS_{AB} \tag{9-10}$$

For the data in Figure 9-1, we obtain

$$SS_T = \sum_{i=1}^{2} \sum_{j=1}^{2} \sum_{k=1}^{3} y_{ijk}^2 - \frac{y_{...}^2}{4(3)}$$

$$= 9398.00 - 9075.00 = 323.00$$

and

$$SS_E = SS_T - SS_A - SS_B - SS_{AB}$$

$$= 323.00 - 208.33 - 75.00 - 8.33$$

$$= 31.34$$

Table 9-1   Analysis of Variance for Data in Figure 9-1

Source of Variation	Sum of Squares	Degrees of Freedom	Mean Square	F_0
A	208.33	1	208.33	53.15[a]
B	75.00	1	75.00	19.13[a]
AB	8.33	1	8.33	2.13
Error	31.34	8	3.92	
Total	323.00	11		

[a] Significant at 1 percent.

using SS_A, SS_B, and SS_{AB} from Equations 9-8. The complete analysis of variance is summarized in Table 9-1. Both main effects are statistically significant at 1 percent. This confirms our initial interpretation of the data based on the magnitudes of the factor effects.

It is often convenient to write down the treatment combinations in the order (1), a, b, ab. This is referred to as *standard order*. Using this standard order, we see that the contrast coefficients used in estimating the effects are

Effects	(1)	a	b	ab
A:	−1	+1	−1	+1
B:	−1	−1	+1	+1
AB:	+1	−1	−1	+1

Note that the contrast coefficients for estimating the interaction effect are just the product of the corresponding coefficients for the two main effects. The contrast coefficient is always either +1 or −1, and a table of plus and minus signs such as in Table 9-2 can be used to determine the proper sign for each treatment combination. The column headings in Table 9-2 are the main effects (A and B), the AB interaction, and I, which represents the total or average of the entire experiment. Notice that the column corresponding to I has only plus signs. The

Table 9-2   Algebraic Signs for Calculating Effects in the 2^2 Design

Treatment Combination	Factorial Effect			
	I	A	B	AB
(1)	+	−	−	+
a	+	+	−	−
b	+	−	+	−
ab	+	+	+	+

row designators are the treatment combinations. To find the contrast for estimating any effect, simply multiply the signs in the appropriate column of the table by the corresponding treatment combination and add. For example, to estimate A, the contrast is $-(1) + a - b + ab$, which agrees with Equation 9-1.

Residual Analysis ▪ It is easy to compute the residuals from a 2^k design via a regression model. For the chemical process experiment, the regression model is

$$y = \beta_0 + \beta_1 x_1 + \beta_2 x_2 + \epsilon$$

where x_1 is a coded variable that represents the reactant concentration and x_2 is a coded variable that represents the amount of catalyst and the β's are regression coefficients. The relationship between the *natural* variables, the reactant concentration and the amount of catalyst, and the coded variables is

$$x_1 = \frac{Conc - (Conc_{low} + Conc_{high})/2}{(Conc_{high} - Conc_{low})/2}$$

and

$$x_2 = \frac{Catalyst - (Catalyst_{low} + Catalyst_{high})/2}{(Catalyst_{high} - Catalyst_{low})/2}$$

When the natural variables have only two levels, this coding will produce the familiar ± 1 notation for the levels of the coded variables. To illustrate this for our example, note that

$$x_1 = \frac{Conc - (15 + 25)/2}{(25 - 15)/2}$$
$$= \frac{Conc - 20}{5}$$

Thus, if the concentration is at the high level (*Conc* = 25%), then $x_1 = +1$, whereas if the concentration is at the low level (*Conc* = 15%), then $x_1 = -1$. Furthermore,

$$x_2 = \frac{Catalyst - (1 + 2)/2}{(2 - 1)/2}$$
$$= \frac{Catalyst - 1.5}{0.5}$$

Thus, if the catalyst is at the high level (*Catalyst* = 2 bags), then $x_2 = +1$, whereas if the catalyst is at the low level (*Catalyst* = 1 bag), then $x_2 = -1$.
   The fitted regression model is

$$\hat{y} = 27.5 + \left(\frac{8.33}{2}\right)x_1 + \left(\frac{-5.00}{2}\right)x_2$$

where the intercept is the grand average of all 12 observations, and the regression coefficients $\hat{\beta}_1$ and $\hat{\beta}_2$ are one-half the corresponding factor effect estimates. The reason that the regression coefficient is one-half the effect estimate is that a regression coefficient measures the effect of a unit change in x on the mean of y, and the effect estimate is based on a two-unit change (from -1 to $+1$).

This model can be used to generate the predicted values of y at the four points in the design. For example, when the reactant concentration is at the low level ($x_1 = -1$) and the catalyst is at the low level ($x_2 = -1$), the predicted yield is

$$\hat{y} = 27.5 + \left(\frac{8.33}{2}\right)(-1) + \left(\frac{-5.00}{2}\right)(-1)$$

$$= 25.835$$

There are three observations at this treatment combination, and the residuals are

$$e_1 = 28 - 25.835 = 2.165$$
$$e_2 = 25 - 25.835 = -0.835$$
$$e_3 = 27 - 25.835 = 1.165$$

The remaining predicted values and residuals are calculated similarly. For the high level of the reactant concentration and the low level of the catalyst:

$$\hat{y} = 27.5 + \left(\frac{8.33}{2}\right)(+1) + \left(\frac{-5.00}{2}\right)(-1)$$

$$= 34.165$$

and

$$e_4 = 36 - 34.165 = 1.835$$
$$e_5 = 32 - 34.165 = -2.165$$
$$e_6 = 32 - 34.165 = -2.165$$

For the low level of the reactant concentration and the high level of the catalyst:

$$\hat{y} = 27.5 + \left(\frac{8.33}{2}\right)(-1) + \left(\frac{-5.00}{2}\right)(+1)$$

$$= 20.835$$

and

$$e_7 = 18 - 20.835 = -2.835$$
$$e_8 = 19 - 20.835 = -1.835$$
$$e_9 = 23 - 20.835 = 2.165$$

Finally, for the high level of both factors:

$$\hat{y} = 27.5 + \left(\frac{8.33}{2}\right)(+1) + \left(\frac{-5.00}{2}\right)(+1)$$

$$= 29.165$$

and

$$e_{10} = 31 - 29.165 = 1.835$$
$$e_{11} = 30 - 29.165 = 0.835$$
$$e_{12} = 29 - 29.165 = -0.165$$

Figure 9-2 presents a normal probability plot of these residuals and plots of the residuals versus the two factors, reactant concentration and amount of catalyst. These plots appear satisfactory, so we have no reason to suspect problems with the validity of our conclusions.

9-3   THE 2^3 DESIGN

Suppose that three factors, A, B, and C, each at two levels, are of interest. The design is called a 2^3 *factorial design* and the eight treatment combinations can

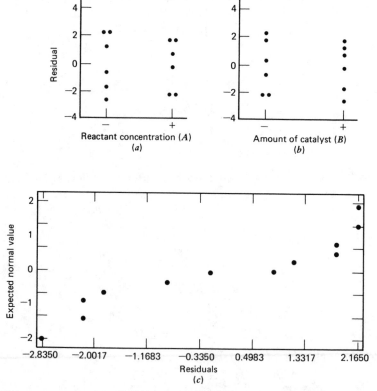

Figure 9-2.   Residual plots for the chemical process experiment. (a) Reactant concentration (A). (b) Amount of catalyst (B). (c) Residuals.

now be displayed graphically as a cube, as shown in Figure 9-3. Extending the notation discussed in Section 9-2, we write the treatment combinations in standard order as (1), a, b, ab, c, ac, bc, and abc. Remember that these symbols also represent the *total* of all n observations taken at that particular treatment combination.

There are actually three different notations that are widely used for the runs in the 2^k design. The first is the "+ and −" notation, often called the "geometric notation." The second is the use of lowercase letters to identify the treatment combinations. The final notation uses 1 and 0 to denote high and low factor levels, respectively, instead of + and −. These different notations are illustrated below for the 2^3 design:

Run	A	B	C	Treatment Combinations	A	B	C
1	−	−	−	(1)	0	0	0
2	+	−	−	a	1	0	0
3	−	+	−	b	0	1	0
4	+	+	−	ab	1	1	0
5	−	−	+	c	0	0	1
6	+	−	+	ac	1	0	1
7	−	+	+	bc	0	1	1
8	+	+	+	abc	1	1	1

There are seven degrees of freedom between the eight treatment combinations in the 2^3 design. Three degrees of freedom are associated with the main effects of A, B, and C. Four degrees of freedom are associated with interactions; one each with AB, AC, and BC and one with ABC.

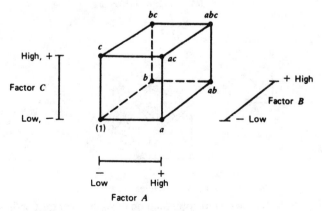

Figure 9-3. The 2^3 factorial design.

Consider estimating the main effects. First, consider estimating the main effect A. The effect of A when B and C are at the low level is $[a - (1)]/n$. Similarly, the effect of A when B is at the high level and C is at the low level is $[ab - b]/n$. The effect of A when C is at the high level and B is at the low level is $[ac - c]/n$. Finally, the effect of A when both B and C are at the high level is $[abc - bc]/n$. Thus, the average effect of A is just the average of these four, or

$$A = \frac{1}{4n}[a - (1) + ab - b + ac - c + abc - bc] \qquad (9\text{-}11)$$

This equation can also be developed as a contrast between the four treatment combinations in the right face of the cube in Figure 9-4a (where A is at the high

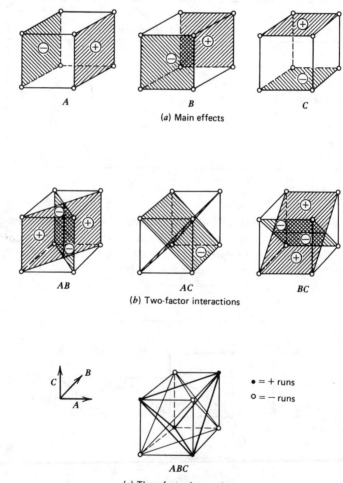

(a) Main effects

(b) Two-factor interactions

$\bullet = +$ runs

$\circ = -$ runs

(c) Three-factor interaction

Figure 9-4. Geometric presentation of contrasts corresponding to the main effects and interaction in the 2^3 design. (a) Main effects. (b) Two-factor interactions. (c) Three-factor interaction.

level) and the four in the left face (where A is at the low level). That is, the A effect is just the average of the four runs where A is at the high level $(\bar{y}_{A+})$ minus the average of the four runs where A is at the low level $(\bar{y}_{A-})$, or

$$A = \bar{y}_{A+} - \bar{y}_{A-}$$
$$= \frac{a + ab + ac + abc}{4n} - \frac{(1) + b + c + bc}{4n}$$

This equation can be rearranged as

$$A = \frac{1}{4n}[a + ab + ac + abc - (1) - b - c - bc]$$

which is identical to Equation 9-11.

In a similar manner, the effect of B is the difference in averages between the four treatment combinations in the front face of the cube and the four in the back. This yields

$$B = \bar{y}_{B+} - \bar{y}_{B-} \qquad (9-12)$$
$$= \frac{1}{4n}[b + ab + bc + abc - (1) - a - c - ac]$$

The effect of C is the difference in averages between the four treatment combinations in the top face of the cube and the four in the bottom, that is,

$$C = \bar{y}_{C+} - \bar{y}_{C-} \qquad (9-13)$$
$$= \frac{1}{4n}[c + ac + bc + abc - (1) - a - b - ab]$$

The two-factor interaction effects may be computed easily. A measure of the AB interaction is the difference between the average A effects at the two levels of B. By convention, one-half of this difference is called the AB interaction. Symbolically,

$\underline{B}$	Average A Effect
High (+)	$\dfrac{[(abc - bc) + (ab - b)]}{2n}$
Low (−)	$\dfrac{\{(ac - c) + [a - (1)]\}}{2n}$
Difference	$\dfrac{[abc - bc + ab - b - ac + c - a + (1)]}{2n}$

Since the AB interaction is one-half of this difference,

$$AB = \frac{[abc - bc + ab - b - ac + c - a + (1)]}{4n} \qquad (9-14)$$

We could write Equation 9-14 as follows:

$$AB = \frac{abc + ab + c + (1)}{4n} - \frac{bc + b + ac + a}{4n}$$

In this form, the AB interaction is easily seen to be the difference in averages between runs on two diagonal planes in the cube in Figure 9-4b. Using similar logic and referring to Figure 9-4b, the AC and BC interactions are

$$AC = \frac{1}{4n}[(1) - a + b - ab - c + ac - bc + abc] \qquad (9\text{-}15)$$

and

$$BC = \frac{1}{4n}[(1) + a - b - ab - c - ac + bc + abc] \qquad (9\text{-}16)$$

The ABC interaction is defined as the average difference between the AB interaction for the two different levels of C. Thus,

$$
\begin{aligned}
ABC &= \frac{1}{4n}\{[abc - bc] - [ac - c] - [ab - b] + [a - (1)]\} \\
&= \frac{1}{4n}[abc - bc - ac + c - ab + b + a - (1)]
\end{aligned}
\qquad (9\text{-}17)
$$

As before, we can think of the ABC interaction as the difference in two averages. If the runs in the two averages are isolated, they define the vertices of the two tetrahedra that comprise the cube in Figure 9-4c.

In Equations 9-11 through 9-17, the quantities in brackets are contrasts in the treatment combinations. A table of plus and minus signs can be developed from the contrasts and is shown in Table 9-3. Signs for the main effects are determined by associating a plus with the high level and a minus with the low level. Once the signs for the main effects have been established, the signs for the remaining columns can be obtained by multiplying the appropriate preceding

Table 9-3 Algebraic Signs for Calculating Effects in the 2^3 Design

Treatment Combination	Factorial Effect							
	I	*A*	*B*	*AB*	*C*	*AC*	*BC*	*ABC*
(1)	+	−	−	+	−	+	+	−
a	+	+	−	−	−	−	+	+
b	+	−	+	−	−	+	−	+
ab	+	+	+	+	−	−	−	−
c	+	−	−	+	+	−	−	+
ac	+	+	−	−	+	+	−	−
bc	+	−	+	−	+	−	+	−
abc	+	+	+	+	+	+	+	+

columns, row by row. For example, the signs in the AB column are the product of the A and B column signs in each row. The contrast for any effect can be obtained easily from this table.

Table 9-3 has several interesting properties: (1) Except for column I, every column has an equal number of plus and minus signs. (2) The sum of the products of the signs in any two columns is zero. (3) Column I multiplied times any column leaves that column unchanged. That is, I is an *identity element*. (4) The product of any two columns yields a column in the table. For example, $A \times B = AB$, and

$$AB \times B = AB^2 = A$$

We see that the exponents in the products are formed by using *modulus* 2 arithmetic. (That is, the exponent can only be zero or one; if it is greater than one, it is reduced by multiples of two until it is either zero or one.) All of these properties are implied by the orthogonality of the contrasts used to estimate the effects.

Sums of squares for the effects are easily computed, since each effect has a corresponding single-degree-of-freedom contrast. In the 2^3 design with n replicates, the sum of squares for any effect is

$$SS = \frac{(Contrast)^2}{8n} \tag{9-18}$$

Example 9-1

Recall Example 7-4, which presented a study on the effect of percent carbonation, operating pressure, and line speed on the fill height of a carbonated beverage. Suppose that only two levels of carbonation are used so that the experiment is a 2^3 factorial design with two replicates. The data, deviations from the target fill height, are shown in Table 9-4, and the design is shown in Figure 9-5.

Table 9-4 Data for the Fill Height Problem, Example 9-1

	Operating Pressure (B)			
	25 psi Line Speed (C)		30 psi Line Speed (C)	
Percent Carbonation (A)	200	250	200	250
10	-3 -1	-1 0	-1 0	1 1
	$-4 = (1)$	$-1 = c$	$-1 = b$	$2 = bc$
12	0 1	2 1	2 3	6 5
	$1 = a$	$3 = ac$	$5 = ab$	$11 = abc$

Using the totals under the treatment combinations shown in Table 9-4, we may estimate the factor effects as follows:

$$A = \frac{1}{4n}[a - (1) + ab - b + ac - c + abc - bc]$$

$$= \frac{1}{8}[1 - (-4) + 5 - (-1) + 3 - (-1) + 11 - 2]$$

$$= \frac{1}{8}[24] = 3.00$$

$$B = \frac{1}{4n}[b + ab + bc + abc - (1) - a - c - ac]$$

$$= \frac{1}{8}[-1 + 5 + 2 + 11 - (-4) - 1 - (-1) - 3]$$

$$= \frac{1}{8}[18] = 2.25$$

$$C = \frac{1}{4n}[c + ac + bc + abc - (1) - a - b - ab]$$

$$= \frac{1}{8}[-1 + 3 + 2 + 11 - (-4) - 1 - (-1) - 5]$$

$$= \frac{1}{8}[14] = 1.75$$

$$AB = \frac{1}{4n}[ab - a - b + (1) + abc - bc - ac + c]$$

$$= \frac{1}{8}[5 - 1 - (-1) + (-4) + 11 - 2 - 3 + (-1)]$$

$$= \frac{1}{8}[6] = 0.75$$

$$AC = \frac{1}{4n}[(1) - a + b - ab - c + ac - bc + abc]$$

$$= \frac{1}{8}[-4 - 1 + (-1) - 5 - (-1) + 3 - 2 + 11]$$

$$= \frac{1}{8}[2] = 0.25$$

$$BC = \frac{1}{4n}[(1) + a - b - ab - c - ac + bc + abc]$$

$$= \frac{1}{8}[-4 + 1 - (-1) - 5 - (-1) - 3 + 2 + 11]$$

$$= \frac{1}{8}[4] = 0.50$$

and

$$ABC = \frac{1}{4n}[abc - bc - ac + c - ab + b + a - (1)]$$

$$= \frac{1}{8}[11 - 2 - 3 + (-1) - 5 + (-1) + 1 - (-4)]$$

$$= \frac{1}{8}[4] = 0.50$$

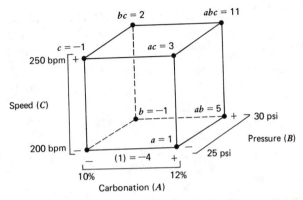

Figure 9-5. The 2^3 design for the fill height deviation experiment for Example 9-1.

The largest effects are for carbonation ($A = 3.00$), pressure ($B = 2.25$), speed ($C = 1.75$) and the carbonation–pressure interaction ($AB = 0.75$), although the interaction effect does not appear to have as large an impact on fill height deviation as the main effects.

The analysis of variance may be used to confirm the magnitude of these effects. From Equation 9-18, the sums of squares are

$$SS_A = \frac{(24)^2}{16} = 36.00$$

$$SS_B = \frac{(18)^2}{16} = 20.25$$

$$SS_C = \frac{(14)^2}{16} = 12.25$$

$$SS_{AB} = \frac{(6)^2}{16} = 2.25$$

$$SS_{AC} = \frac{(2)^2}{16} = 0.25$$

$$SS_{BC} = \frac{(4)^2}{16} = 1.00$$

and

$$SS_{ABC} = \frac{(4)^2}{16} = 1.00$$

The total sum of squares is $SS_T = 78.00$, and by subtraction, $SS_E = 5.00$. The analysis of variance is summarized in Table 9-5, and it confirms the significance of the main effects. The AB interaction is significant at about the 10 percent level; thus there is some mild interaction between carbonation and pressure.

The reader may wish to refer to Example 7-4 for the practical interpretation of this experiment. The process developers decided to run the process at low pressure and high line speed and to reduce the variability in carbonation by controlling the temperature more precisely. This resulted in a substantial reduction in the deviation of fill height from the target value.

■

Table 9-5   Analysis of Variance for the Fill Height Data

Source of Variation	Sum of Squares	Degrees of Freedom	Mean Square	F_0
Percent Carbonation (A)	36.00	1	36.00	57.60[a]
Pressure (B)	20.25	1	20.25	32.40[a]
Line speed (C)	12.25	1	12.25	19.60[a]
AB	2.25	1	2.25	3.60
AC	0.25	1	0.25	0.40
BC	1.00	1	1.00	1.60
ABC	1.00	1	1.00	1.60
Error	5.00	8	0.625	
Total	78.00	15		

[a] Significant at 1 percent.

Residuals and Diagnostic Checking ▪ The residuals from this experiment may be generated by the regression model approach used previously. The regression model for predicting fill height deviation is

$$\hat{y} = \hat{\beta}_0 + \hat{\beta}_1 x_1 + \hat{\beta}_2 x_2 + \hat{\beta}_3 x_3 + \hat{\beta}_{12} x_1 x_2$$
$$= 1.00 + \left(\frac{3.00}{2}\right) x_1 + \left(\frac{2.25}{2}\right) x_2 + \left(\frac{1.75}{2}\right) x_3 + \left(\frac{0.75}{2}\right) x_1 x_2,$$

where the coded variables x_1, x_2, and x_3 represent A, B, and C, respectively. The $x_1 x_2$ term is the AB interaction. Residuals can be obtained as the difference between observed and predicted fill height deviations. We leave the analysis of these residuals as an exercise for the reader.

Other Methods for Judging the Significance of Effects ▪ The analysis of variance is a formal way to determine which factor effects are nonzero. There are two other methods that are useful. In the first method, we calculate the standard error of the effects and compare the magnitude of the effects to their standard errors. The second method, which we will illustrate in Section 9-5, uses normal probability plots to assess the importance of the effects.

The standard error of an effect is easy to find. If we assume that there are n replicates at each of the 2^k runs in the design, and if $y_{i1}, y_{i2}, \ldots, y_{in}$ are the observations at the ith run, then

$$S_i^2 = \frac{1}{n-1} \sum_{j=1}^{n} (y_{ij} - \bar{y}_i)^2 \quad i = 1, 2, \ldots, 2^k$$

is an estimate of the variance at the ith run. The 2^k variance estimates can be combined to give an overall variance estimate

$$S^2 = \frac{1}{2^k(n-1)} \sum_{i=1}^{2^k} \sum_{j=1}^{n} (y_{ij} - \bar{y}_i)^2$$

This is also the variance estimate given by the error mean square in the analysis of variance. The variance of each effect estimate is

$$V(\text{Effect}) = V\left(\frac{\text{Contrast}}{n2^{k-1}}\right)$$

$$= \frac{1}{(n2^{k-1})^2} V(\text{Contrast})$$

Each contrast is a linear combination of 2^k treatment totals, and each total consists of n observations. Therefore,

$$V(\text{Contrast}) = n2^k\sigma^2$$

and the variance of an effect is

$$V(\text{Effect}) = \frac{1}{(n2^{k-1})^2}n2^k\sigma^2$$

$$= \frac{1}{n2^{k-2}}\sigma^2 \tag{9-19}$$

The estimated standard error would be found by replacing σ^2 by its estimate S^2 and taking the square root of Equation 9-19.

To illustrate this method, consider the fill height deviation experiment. The mean square error is $MS_E = 0.625$. Therefore, the standard error of each effect is (using $S^2 = MS_E$)

$$SE(\text{Effect}) = \sqrt{\frac{1}{n2^{k-2}}S^2}$$

$$= \sqrt{\frac{1}{2(2^{3-2})}0.625}$$

$$= 0.40$$

Two standard error limits on the effect estimates are then

$$A: 3.00 \pm 0.80$$
$$B: 2.25 \pm 0.80$$
$$C: 1.75 \pm 0.80$$
$$AB: 0.75 \pm 0.80$$
$$AC: 0.25 \pm 0.80$$
$$BC: 0.50 \pm 0.80$$
$$ABC: 0.50 \pm 0.80$$

These intervals are approximate 95 percent confidence intervals. This analysis indicates that A, B, and C are important factors, as they are the only factor effect estimates for which the intervals do not include zero.

Dispersion Effects ▪ The process engineer working on the filling process was also interested in *dispersion effects;* that is, do any of the factors affect variability in fill height deviation from run to run? One way to answer the question is to look at the *range* of fill height deviations for each of the eight runs in the 2^3 design. These ranges are plotted on the cube in Figure 9-6. Notice that the ranges are about the same for all eight runs in the design. Consequently, there is no strong evidence indicating that some of the process variables directly affect the variability in fill height deviation in the process.

9-4 THE GENERAL 2^k DESIGN

The methods of analysis that we have presented thus far may be generalized to the case of a 2^k factorial design, that is, a design with k factors each at two levels. The statistical model for a 2^k design would include k main effects, $\binom{k}{2}$ two-factor interactions, $\binom{k}{3}$ three-factor interactions, . . . , and one k-factor interaction. That is, for a 2^k design the complete model would contain $2^k - 1$ effects. The notation introduced earlier for treatment combinations is also used here. For example, in a 2^5 design *abd* denotes the treatment combination with factors A, B, and D at the high level and factors C and E at the low level. The treatment combinations may be written in standard order by introducing the factors one at a time, with each new factor being successively combined with those that precede it. For example, the standard order for a 2^4 design is (1), a, b, ab, c, ac, bc, abc, d, ad, bd, abd, cd, acd, bcd, and abcd.

To estimate an effect or to compute the sum of squares for an effect, we must first determine the contrast associated with that effect. This can always be done by using a table of plus and minus signs, such as Table 9-2 or Table 9-3. However, for large values of k this is awkward, and we can use an alternate method. In general, we determine the contrast for effect $AB \cdot \cdot \cdot K$ by expanding the right-

Figure 9-6. Ranges of fill height deviation for Example 9-1.

hand side of

$$\text{Contrast}_{AB...K} = (a \pm 1)(b \pm 1) \cdots (k \pm 1) \qquad (9\text{-}20)$$

In expanding Equation 9-20, ordinary algebra is used with "1" being replaced by (1) in the final expression. The sign in each set of parentheses is negative if the factor is included in the effect and positive if the factor is not included.

To illustrate the use of Equation 9-20, consider a 2^3 factorial. The contrast for AB would be

$$\text{Contrast}_{AB} = (a - 1)(b - 1)(c + 1)$$
$$= abc + ab + c + (1) - ac - bc - a - b$$

As a further example, in a 2^5 design, the contrast for $ABCD$ would be

$$\text{Contrast}_{ABCD} = (a - 1)(b - 1)(c - 1)(d - 1)(e + 1)$$
$$= abcde + cde + bde + ade + bce$$
$$+ ace + abe + e + abcd + cd + bd$$
$$+ ad + bc + ac + ab + (1) - a - b - c$$
$$- abc - d - abd - acd - bcd - ae$$
$$- be - ce - abce - de - abde - acde - bcde$$

Once the contrasts for the effects have been computed, we may estimate the effects and compute the sums of squares according to

$$AB \cdots K = \frac{2}{n2^k}(\text{Contrast}_{AB...K}) \qquad (9\text{-}21)$$

and

$$SS_{AB...K} = \frac{1}{n2^k}(\text{Contrast}_{AB...K})^2 \qquad (9\text{-}22)$$

respectively, where n denotes the number of replicates. The analysis of variance for the 2^k design is summarized in Table 9-6. We present another method for estimating the effects in the 2^k design in Section 9-6.

9-5 A SINGLE REPLICATE OF THE 2^k DESIGN

For even a moderate number of factors, the total number of treatment combinations in a 2^k factorial design is large. For example, a 2^5 design has 32 treatment combinations, a 2^6 design has 64 treatment combinations, and so on. Since resources are usually limited, the number of replicates that the experimenter can employ may be restricted. Frequently, available resources only allow a single replicate of the design to be run, unless the experimenter is willing to omit some of the original factors.

Table 9-6   Analysis of Variance for a 2^k Design

Source of Variation	Sum of Squares	Degrees of Freedom
k main effects		
A	SS_A	1
B	SS_B	1
.	.	.
.	.	.
.	.	.
K	SS_K	1
$\binom{k}{2}$ two-factor interactions		
AB	SS_{AB}	1
AC	SS_{AC}	1
.	.	.
.	.	.
JK	SS_{JK}	1
$\binom{k}{3}$ three-factor interactions		
ABC	SS_{ABC}	1
ABD	SS_{ABD}	1
.	.	.
.	.	.
IJK	SS_{IJK}	1
.	.	.
.	.	.
$\binom{k}{k} = 1$ k-factor interaction		
$ABC\ldots K$	$SS_{ABC\ldots K}$	1
Error	SS_E	$2^k(n-1)$
Total	SS_T	$n2^k - 1$

A single replicate of a 2^k design is sometimes called an *unreplicated factorial*. With only one replicate, there is no estimate of error. One approach to the analysis of an unreplicated factorial is to assume that certain high-order interactions are negligible and combine their mean squares to estimate the error. This is an appeal to the *sparsity of effects principle;* that is, most systems are dominated by some of the main effects and low-order interactions, and most high-order interactions are negligible.

When analyzing data from unreplicated factorial designs, occasionally real high-order interactions occur. The use of an error mean square obtained by pooling high-order interactions is inappropriate in these cases. A method of analysis attributed to Daniel (1959) provides a simple way to overcome this problem. Daniel suggests plotting the estimates of the effects on normal probability paper.

The effects that are negligible are normally distributed, with mean zero and variance σ^2 and will tend to fall along a straight line on this plot, whereas significant effects will have nonzero means and will not lie along the straight line. We will illustrate this method in the next example.

Example 9-2

A Single Replicate of the 2^4 Design

A chemical product is produced in a pressure vessel. A factorial experiment is carried out in the pilot plant to study the factors thought to influence the filtration rate of this product. The four factors are temperature (A), pressure (B), concentration of formaldehyde (C), and stirring rate (D). Each factor is present at two levels, and the data obtained from a single replicate of the 2^4 experiment are shown in Table 9-7 and Figure 9-7. The 16 runs are made in random order. The process engineer is interested in maximizing the filtration rate. Current process conditions give filtration rates of around 75 gal/h. The process also currently uses the concentration of formaldehyde, factor C, at the high level. The engineer would like to reduce the formaldehyde concentration as much as possible but has been unable to do so because it always results in lower filtration rates.

We will begin the analysis of this data by plotting the effect estimates on normal probability paper. The table of plus and minus signs for the contrast constants for the 2^4 design are shown in Table 9-8. From these contrasts, we may estimate the 15

Table 9-7   Pilot Plant Filtration Rate Experiment

Run Number	Factor				Treatment Combination	Filtration Rate (gal/h)
	A	B	C	D		
1	−	−	−	−	(1)	45
2	+	−	−	−	a	71
3	−	+	−	−	b	48
4	+	+	−	−	ab	65
5	−	−	+	−	c	68
6	+	−	+	−	ac	60
7	−	+	+	−	bc	80
8	+	+	+	−	abc	65
9	−	−	−	+	d	43
10	+	−	−	+	ad	100
11	−	+	−	+	bd	45
12	+	+	−	+	abd	104
13	−	−	+	+	cd	75
14	+	−	+	+	acd	86
15	−	+	+	+	bcd	70
16	+	+	+	+	abcd	96

Figure 9-7. Data from the pilot plant filtration rate experiment for Example 9-2.

factorial effects that follow:

Order (j)	Effect	Estimate	($j - .5$)/15
15	A	21.63	.9667
14	AD	16.63	.9000
13	D	14.63	.8333
12	C	9.88	.7667
11	ABD	4.13	.7000
10	B	3.13	.6333
9	BC	2.38	.5667
8	ABC	1.88	.5000
7	ABCD	1.38	.4333
6	AB	0.13	.3667
5	CD	−0.38	.3000
4	BD	−1.13	.2333
3	ACD	−1.63	.1667
2	BCD	−2.63	.1000
1	AC	−18.13	.0333

The normal probability plot of these effects is shown in Figure 9-8. All of the effects that lie along the line are negligible, whereas the large effects are far from the line. The important effects that emerge from this analysis are the main effects of A, C, and D and the AC and AD interactions.

The main effects of A, C, and D are plotted in Figure 9-9a. All three effects are positive, and if we considered only these main effects, we would run all three factors at the high level to maximize the filtration rate. However, it is always necessary to examine any interactions that are important. Remember that main effects do not have much meaning when they are involved in significant interactions.

The AC and AD interactions are plotted in Figure 9-9b. These interactions are

Table 9-8   Contrast Constants for the 2^4 Design

	A	B	AB	C	AC	BC	ABC	D	AD	BD	ABD	CD	ACD	BCD	ABCD
(1)	−	−	+	−	+	+	−	−	+	+	−	+	−	−	+
a	+	−	−	−	−	+	+	−	−	+	+	+	+	−	−
b	−	+	−	−	+	−	+	−	+	−	+	+	−	+	−
ab	+	+	+	−	−	−	−	−	−	−	−	+	+	+	+
c	−	−	+	+	−	−	+	−	+	+	−	−	+	+	−
ac	+	−	−	+	+	−	−	−	−	+	+	−	−	+	+
bc	−	+	−	+	−	+	−	−	+	−	+	−	+	−	+
abc	+	+	+	+	+	+	+	−	−	−	−	−	−	−	−
d	−	−	+	−	+	+	−	+	−	−	+	−	+	+	−
ad	+	−	−	−	−	+	+	+	+	−	−	−	−	+	+
bd	−	+	−	−	+	−	+	+	−	+	−	−	+	−	+
abd	+	+	+	−	−	−	−	+	+	+	+	−	−	−	−
cd	−	−	+	+	−	−	+	+	−	−	+	+	−	−	+
acd	+	−	−	+	+	−	−	+	+	−	−	+	+	−	−
bcd	−	+	−	+	−	+	−	+	−	+	−	+	−	+	−
abcd	+	+	+	+	+	+	+	+	+	+	+	+	+	+	+

Figure 9-8.   Ordered effects for the 2^4 factorial in Example 9-2.

(a) Main effect plots

(b) Interaction plots

Figure 9-9.   Main effect and interaction plots for Example 9-2. (a) Main effect plots. (b) Interaction plots.

the key to solving the problem. Note from the AC interaction that the temperature effect is very small when the concentration is at the high level and very large when the concentration is at the low level, with the best results obtained with low concentration and high temperature. The AD interaction indicates that stirring rate D has little effect at low temperature but a large positive effect at high temperature. Therefore, the best filtration rates would appear to be obtained when A and D are at the high level and C is at the low level. This would allow the reduction of the formaldehyde concentration to a lower level, another objective of the experimenter.

∎

Design Projection ▪ Another interpretation of the data in Figure 9-8 is possible. Since B (pressure) is not significant and all interactions involving B are negligible, we may discard B from the experiment so that the design becomes a 2^3 factorial in A, C, and D with two replicates. This is easily seen from examining only columns A, C, and D in Table 9-7 and noting that those columns form two replicates of a 2^3 design. The analysis of variance for the data using this simplifying assumption is summarized in Table 9-9. The conclusions that we would draw from this analysis are essentially unchanged from those of Example 9-2. Note that by projecting the single replicate of the 2^4 into a replicated 2^3, we now have both an estimate of the ACD interaction and an estimate of error based on replication.

The concept of projecting an unreplicated factorial into a replicated factorial in fewer factors is very useful. In general, if we have a single replicate of a 2^k design and if $h(h < k)$ factors are negligible and can be dropped, then the original data correspond to a full two-level factorial in the remaining $k - h$ factors with 2^h replicates.

Diagnostic Checking ▪ The usual diagnostic checks should be applied to the residuals of a 2^k design. Our analysis indicates that the only significant effects are $A = 21.63$, $C = 9.88$, $D = 14.63$, $AC = -18.13$, and $AD = 16.63$. If this is true, then the estimated filtration rates at the vertices of the design are given by

$$\hat{y} = 70.06 + \left(\frac{21.63}{2}\right)x_1 + \left(\frac{9.88}{2}\right)x_3 + \left(\frac{14.63}{2}\right)x_4 - \left(\frac{18.13}{2}\right)x_1x_3$$
$$+ \left(\frac{16.63}{2}\right)x_1x_4$$

where 70.06 is the average response and the coded variables x_1, x_3, x_4 take on the values $+1$ or -1. The predicted filtration rate at run (1) is

$$\hat{y} = 70.06 + \left(\frac{21.63}{2}\right)(-1) + \left(\frac{9.88}{2}\right)(-1) + \left(\frac{14.63}{2}\right)(-1)$$
$$- \left(\frac{18.13}{2}\right)(-1)(-1) + \left(\frac{16.63}{2}\right)(-1)(-1)$$
$$= 46.22$$

Table 9-9 Analysis of Variance for the Pilot Plant Filtration Rate Experiment in A, C, and D

Source of Variation	Sum of Squares	Degrees of Freedom	Mean Square	F_0
A	1870.56	1	1870.56	83.36[a]
C	390.06	1	390.06	17.38[a]
D	855.56	1	855.56	38.13[a]
AC	1314.06	1	1314.06	58.56[a]
AD	1105.56	1	1105.56	49.27[a]
CD	5.06	1	5.06	<1
ACD	10.56	1	10.56	<1
Error	179.52	8	22.44	
Total	5730.94	15		

[a] Significant at the 1 percent level.

Since the observed value is 45, the residual is $e = y - \hat{y} = 45 - 46.22 = -1.22$. The values of y, $\hat{y}$, and $e = y - \hat{y}$ for all 16 observations follow:

	y	$\hat{y}$	$e = y - \hat{y}$
(1)	45	46.22	−1.22
a	71	69.39	1.61
b	48	46.22	1.78
ab	65	69.39	−4.39
c	68	74.23	−6.23
ac	60	61.14	−1.14
bc	80	74.23	5.77
abc	65	61.14	3.86
d	43	44.22	−1.22
ad	100	100.65	−0.65
bd	45	44.22	0.78
abd	104	100.65	3.35
cd	75	72.23	2.77
acd	86	92.40	−6.40
bcd	70	72.23	−2.23
abcd	96	92.40	3.60

These residuals are plotted on normal probability paper in Figure 9-10. The points on this plot lie reasonably close to a straight line, lending support to our conclusion that A, C, D, AC, and AD are the only significant effects and that the underlying assumptions of the analysis are satisfied.

Example 9-3

Data Transformation in a Factorial Design

Daniel (1976) describes a 2^4 factorial used to study the advance rate of a drill as a function of four factors: drill load (A), flow rate (B), rotational speed (C), and the type of drilling mud used (D). The data from the experiment are shown in Figure 9-11.

The normal probability plot of the effect estimates from this experiment is shown in Figure 9-12. Based on this plot, factors B, C, and D along with the BC and CD interactions require interpretation. Figure 9-13 is the normal probability plot of the residuals and Figure 9-14 is the plot of the residuals versus the predicted advance rate from the model containing the identified factors. There are clearly problems with normality and equality of variance. A data transformation is often used to deal with such problems. As the response variable is a rate, the log transformation seems a reasonable candidate. (The methods described in Chapter 4 could be used to select an appropriate transformation, if desired.)

Figure 9-15 presents a normal probability plot of the effect estimates following the transformation $y^* = \ln y$. Notice that a much simpler interpretation now seems

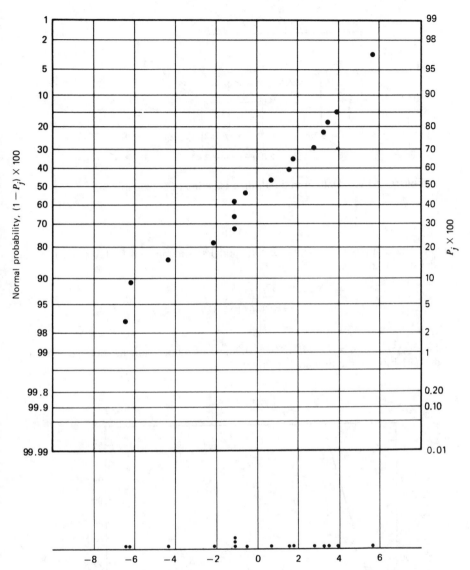

Figure 9-10.   Normal probability plot of residuals for Example 9-2.

possible, as only factors *B*, *C*, and *D* are active. That is, expressing the data in the correct metric has simplified its structure to the point that the two interactions are no longer required in the explanatory model.

Figures 9-16 and 9-17 present, respectively a normal probability plot of the residuals and a plot of the residuals versus the predicted advance rate for the model in the log scale containing *B*, *C*, and *D*. These plots are now satisfactory. We conclude that the model for $y^* = \ln y$ requires only factors *B*, *C*, and *D* for adequate interpre-

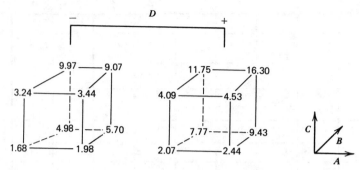

Figure 9-11.   Data from the drilling experiment of Example 9-3.

tation. The analysis of variance for this model is summarized in Table 9-10. The model sum of squares is

$$SS_{model} = SS_B + SS_C + SS_D$$
$$= 1.346 + 5.331 + 0.427$$
$$= 7.104$$

and $R^2 = SS_{Model}/SS_T = 7.104/7.277 = 0.98$, so the model explains about 98 percent of the variability in the drill advance rate.

■

Figure 9-12.   Normal probability plot of effects for Example 9-3.

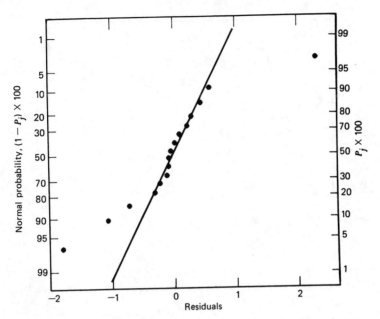

Figure 9-13. Normal probability plot of residuals for Example 9-3.

Figure 9-14. Plot of residuals versus predicted for Example 9-3.

Figure 9-15.   Normal probability plot of effects for Example 9-3 following log transformation.

Figure 9-16.   Normal probability plot of residuals for Example 9-3 following log transform.

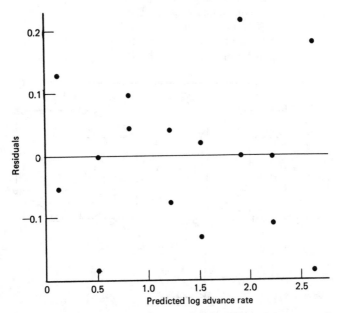

Figure 9-17. Plot of residuals versus predicted for Example 9-3 following log transform.

Example 9-4

Location and Dispersion Effects in an Unreplicated Factorial

A 2^4 design was run in a manufacturing process producing interior side-wall and window panels for commercial aircraft. The panels are formed in a press, and under present conditions the average number of defects per panel in a press load is much too high. (The current process average is 5.5 defects per panel.) Four factors are investigated using a single replicate of a 2^4 design, with each replicate corresponding to a single press load. The factors are temperature (A), clamp time (B), resin flow (C), and press closing time (D). The data for this experiment are shown in Figure 9-18.

A normal probability plot of the factor effects is shown in Figure 9-19. Clearly the two largest effects are $A = 5.75$ and $C = -4.25$. No other factor effects appear

Table 9-10 Analysis of Variance for Example 9-3 Following the Log Transformation

Source of Variation	Sums of Squares	Degrees of Freedom	Mean Square	F_0
B (Flow)	1.346	1	1.346	96.14
C (Speed)	5.331	1	5.331	380.79
D (Mud)	0.427	1	0.427	30.50
Error	0.173	12	0.014	
Total	7.277	15		

Factors		Low (−)	High (+)
A =	Temperature (°F)	295	325
B =	Clamp time (min)	7	9
C =	Resin flow	10	20
D =	Closing time (s)	15	30

Figure 9-18. Data for the panel process experiment of Example 9-4.

Figure 9-19. Normal probability plot of the factor effects for the panel process experiment of Example 9-4.

Figure 9-20. Plot of residuals versus clamp time for Example 9-4.

to be large, and A and C explain about 77 percent of the total variability, so we conclude that lower temperature (A) and higher resin flow (C) would reduce the incidence of panel defects.

Careful residual analysis is an important aspect of any experiment. A normal probability plot of the residuals showed no anomalies, but when the experimenter plotted the residuals versus each of the factors A through D, the plot of residuals versus B (clamp time) presented the pattern in Figure 9-20. This factor, which is unimportant insofar as the average number of defects per panel is concerned, is very important in its affect on process variability, with the lower clamp time resulting in less variability in the average number of defects per panel in a press load.

The dispersion effect of clamp time is also very evident from the *cube plot* in Figure 9-21, which plots the average number of defects per panel and the range of the number of defects at each point in the cube defined by factors A, B, and C. The average range when B is at the high level (the back face of the cube in Figure 9-21) is $\overline{R}_{B+} = 4.75$, and when B is at the low level it is $\overline{R}_{B-} = 1.25$.

As a result of this experiment, the engineer decided to run the process at low temperature and high resin flow to reduce the average number of defects, at low clamp time to reduce the variability in the number of defects per panel, and at low press closing time (which had no effect on either location or dispersion). The new set of operating conditions resulted in a new process average of less than one defect per panel.

■

Figure 9-21. Cube plot of temperature, clamp time, and resin flow for Example 9-4.

The residuals from a 2^k design provide much information about the problem under study. Since residuals can be thought of as observed values of the noise or error, they often give insight into process variability. We can systematically examine the residuals from an unreplicated 2^k to provide information about process variability.

Consider the residual plot in Figure 9-20. The standard deviation of the eight residuals where B is at the low level is $S(B^-) = 0.83$, and the standard deviation of the eight residuals where B is at the high level is $S(B^+) = 2.72$. The statistic

$$F_B^* = \ln \frac{S^2(B^+)}{S^2(B^-)} \tag{9-23}$$

has an approximate standard normal distribution if the two variances $\sigma^2(B^+)$ and $\sigma^2(B^-)$ are equal. The value of F_B^* is

$$F_B^* = \ln \frac{S^2(B^+)}{S^2(B^-)}$$

$$= \ln \frac{(2.72)^2}{(0.83)^2}$$

$$= 2.37$$

which exceeds the upper one-percent point of the standard normal distribution ($Z_{.01} = 2.33$ from Table I in the Appendix).

Table 9-11 presents the complete set of contrasts for the 2^4 design along with the residuals for each run from the panel process experiment in Example 9-4. Each column in this table contains an equal number of plus and minus signs, and we can calculate the standard deviation of the residuals for each group of signs in each column, say $S(i^+)$ and $S(i^-)$, $i = 1, 2, \ldots, 15$. Then

$$F_i^* = \ln \frac{S^2(i^+)}{S^2(i^-)} \qquad i = 1, 2, \ldots, 15 \tag{9-24}$$

is a statistic that can be used to assess the magnitude of the *dispersion effects* in the experiment. If the variance of the residuals for the runs where factor i is positive equals the variance of the residuals for the runs where factor i is negative, then F_i^* has an approximate standard normal distribution. The values of F_i^* are shown below each column in Table 9-11.

Figure 9-22 plots the dispersion effects F_i^* and compares them to the .025 and .01 points of the standard normal distribution. Clearly, B is an important factor with respect to process dispersion. An alternative to Figure 9-22 would be to plot the F_i^* on normal probability paper, as shown in Figure 9-23. For more discussion of this procedure, see Box and Meyer (1986).

9-6   THE ADDITION OF CENTER POINTS TO THE 2^k DESIGN

A potential concern in the use of two-level factorial designs is the assumption of linearity in the factor effects. Of course, perfect linearity is unnecessary, and the

Table 9-11   Calculation of Dispersion Effects for Example 9-4

Run	A	B	AB	C	AC	BC	ABC	D	AD	BD	ABD	CD	ACD	BCD	ABCD	Residual
1	−	−	+	−	+	+	−	−	+	+	−	+	−	−	+	−0.94
2	+	−	−	−	−	+	+	−	−	+	+	+	+	−	−	−0.69
3	−	+	−	−	+	−	+	−	+	−	+	+	−	+	−	−2.44
4	+	+	+	−	−	−	−	−	−	−	−	+	+	+	+	−2.69
5	−	−	+	+	−	−	+	−	+	+	−	−	+	+	−	−1.19
6	+	−	−	+	+	−	−	−	−	+	+	−	−	+	+	0.56
7	−	+	−	+	−	+	−	−	+	−	+	−	+	−	+	−0.19
8	+	+	+	+	+	+	+	−	−	−	−	−	−	−	−	2.06
9	−	−	+	−	+	+	−	+	−	−	+	−	+	+	−	0.06
10	+	−	−	−	−	+	+	+	+	−	−	−	−	+	+	0.81
11	−	+	−	−	+	−	+	+	−	+	−	−	+	−	+	2.06
12	+	+	+	−	−	−	−	+	+	+	+	−	−	−	−	3.81
13	−	−	+	+	−	−	+	+	−	−	+	+	−	−	+	−0.69
14	+	−	−	+	+	−	−	+	+	−	−	+	+	−	−	−1.44
15	−	+	−	+	−	+	−	+	−	+	−	+	−	+	−	3.31
16	+	+	+	+	+	+	+	+	+	+	+	+	+	+	+	−2.44
$S(i^+)$	2.25	2.72	2.21	1.91	1.81	1.80	1.80	2.24	2.05	2.28	1.97	1.93	1.52	2.09	1.61	
$S(i^-)$	1.85	0.83	1.86	2.20	2.24	2.26	2.24	1.55	1.93	1.61	2.11	1.58	2.16	1.89	2.33	
F_i^*	0.39	2.37	0.34	−0.28	−0.43	−0.46	−0.44	0.74	0.12	0.70	−0.14	0.40	−0.70	0.28	−0.74	

Figure 9-22. Dispersion effects for Example 9-4.

2^k system will work quite well even when the linearity assumption holds only very approximately. There is, however, a method of replicating certain points in a 2^k factorial that will provide protection against curvature as well as allow an independent estimate of error to be obtained. The method consists of adding *center points* to the 2^k design. These consist of n replicates run at the points $x_i = 0$ ($i = 1, 2, \ldots, k$). One important reason for adding the replicate runs at the design center is that center points do not impact the usual effects estimates in a 2^k design. We assume that the k factors are quantitative.

To illustrate the approach, consider a 2^2 design with one observation at each of the factorial points $(-, -)$, $(+, -)$, $(-, +)$ and $(+, +)$ and n_C observations at the center points $(0, 0)$. Figure 9-24 illustrates the situation. Let $\bar{y}_F$ be the average of the four runs at the four factorial points and let $\bar{y}_C$ be the average of the n_C runs at the center point. If the difference $\bar{y}_F - \bar{y}_C$ is small, then the center points lie on or near the plane passing through the factorial points, and there is no curvature. On the other hand, if $\bar{y}_F - \bar{y}_C$ is large, then curvature is present. A single-degree-of-freedom sum of squares for curvature is given by

$$SS_{\text{Curvature}} = \frac{n_F n_C (\bar{y}_F - \bar{y}_C)^2}{n_F + n_C} \tag{9-25}$$

Figure 9-23. Normal probability plot of the dispersion effects F_i^* for Example 9-4.

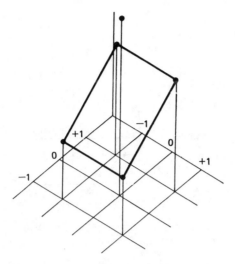

Figure 9-24. A 2^2 design with center points.

where, in general, n_F is the number of factorial design points. This quantity may be compared to the error mean square to test for curvature. More specifically, when points are added to the center of the 2^k design, then the model we may entertain is

$$y = \beta_0 + \sum_{j=1}^{k} \beta_j x_j + \sum\sum_{i<j} \beta_{ij} x_i x_j$$
$$+ \sum_{j=1}^{k} \beta_{jj} x_j^2 + \epsilon$$

where the β_{jj} are pure quadratic effects. The test for curvature actually tests the hypotheses

$$H_0 : \sum_{j=1}^{k} \beta_{jj} = 0$$

$$H_1 : \sum_{j=1}^{k} \beta_{jj} \neq 0$$

Furthermore, if the factorial points in the design are unreplicated, one may use the n_C center points to construct an estimate of error with $n_C - 1$ degrees of freedom.

Example 9-5

A chemical engineer is studying the yield of a process. There are two variables of interest, reaction time and reaction temperature. Because she is uncertain about the assumption of linearity over the region of exploration, the engineer decides to conduct

Figure 9-25. The 2^2 design with five center points for Example 9-5.

a 2^2 design (with a single replicate of each factorial run) argumented with five center points. The design and the yield data are shown in Figure 9-25.

Table 9-12 summarizes the analysis of variance for this experiment. The mean square error is calculated from the center points as follows:

$$MS_E = \frac{SS_E}{n_C - 1}$$

$$= \frac{\sum\limits_{Center\ points} (y_i - \bar{y})^2}{n_C - 1}$$

$$= \frac{\sum\limits_{i=1}^{5} (y_i - 40.46)^2}{4}$$

$$= \frac{0.1720}{4}$$

$$= 0.0430$$

Table 9-12 Analysis of Variance for Example 9-5

Source of Variation	Sum of Squares	Degrees of Freedom	Mean Square	F_0
A (Time)	2.4025	1	2.4025	55.87
B (Temperature)	0.4225	1	0.4225	9.83
AB	0.0025	1	0.0025	0.06
Curvature	0.0027	1	0.0027	0.06
Error	0.1720	4	0.0430	
Total	3.0022	8		

The average of the points in the factorial portion of the design is $\bar{y}_F = 40.425$, and the average of the points at the center is $\bar{y}_C = 40.46$. The difference $\bar{y}_F - \bar{y}_C = 40.425 - 40.46 = -0.035$ appears to be small. The curvature sum of squares in the analysis of variance table is computed from Equation 9-25 as follows:

$$SS_{Curvature} = \frac{n_F n_C (\bar{y}_F - \bar{y}_C)^2}{n_F + n_C}$$

$$= \frac{(4)(5)(-0.035)^2}{4 + 5}$$

$$= 0.0027$$

The analysis of variance indicates that both factors exhibit significant main effects, that there is no interaction, and that there is no evidence of curvature in the response over the region of exploration. That is, the null hypothesis $H_0 : \Sigma_{j=1}^k \beta_{jj} = 0$ cannot be rejected.

■

9-7 YATES' ALGORITHM FOR THE 2^k DESIGN

There is a very simple technique devised by Yates (1937) for estimating the effects and determining the sums of squares in a 2^k factorial design. The procedure is best learned through the study of a numerical example.

Consider the data for the 2^3 design in Example 9-1. These data have been entered in Table 9-13. The treatment combinations are always written down in standard order, and the column labeled "Response" contains the corresponding observation (or total of all observations) at that treatment combination. The first half of column (1) is obtained by adding the responses in adjacent pairs. The second half of column (1) is obtained by changing the sign of the first entry in each of the pairs in the Response column and adding the adjacent pairs. For example, in column (1) we obtain for the fifth entry $5 = -(-4) + 1$, for the sixth entry $6 = -(-1) + 5$, and so on.

Column (2) is obtained from column (1) just as column (1) is obtained from the Response column. Column (3) is obtained from column (2) similarly. In gen-

Table 9-13   Yates' Algorithm for the Data in Example 9-1

Treatment Combination	Response	(1)	(2)	(3)	Effect	Estimate of Effect $(3) \div n2^{k-1}$	Sum of Squares $(3)^2 \div n2^k$
(1)	−4	−3	1	16	I	—	—
a	1	4	15	24	A	3.00	36.00
b	−1	2	11	18	B	2.25	20.25
ab	5	13	13	6	AB	0.75	2.25
c	−1	5	7	14	C	1.75	12.25
ac	3	6	11	2	AC	0.25	0.25
bc	2	4	1	4	BC	0.50	1.00
abc	11	9	5	4	ABC	0.50	1.00

eral, for a 2^k design we would construct k columns of this type. Column (3) [in general, column (k)] is the contrast for the effect designated at the beginning of the row. To obtain the estimate of the effect, we divide the entries in column (3) by $n2^{k-1}$ (in our example, $n2^{k-1} = 8$). Finally, the sums of squares for the effects are obtained by squaring the entries in column (3) and dividing by $n2^k$ (in our example, $n2^k = 16$).

The estimates of the effects and sums of squares obtained by Yates' algorithm for the data in Example 9-1 are in agreement with the results found there by the usual methods. Note that the entry in column (3) [in general, column (k)] for the row corresponding to (1) is always equal to the grand total of the observations.

In spite of its apparent simplicity, it is notoriously easy to make numerical errors in Yates' algorithm, and we should be extremely careful in executing the procedure. As a partial check on the computations, we may use the fact that the sum of the squares of the elements in the jth column is 2^j times the sum of the squares of the elements in the response column. Note, however, that this check is subject to errors in sign in column j. See Davies (1956), Good (1955, 1958), Kempthorne (1952), and Rayner (1967), for other error-checking techniques.

9-8 PROBLEMS

9-1   An engineer is interested in the effects of cutting speed (A), tool geometry (B), and cutting angle (C) on the life (in hours) of a machine tool. Two levels of each factor are chosen, and three replicates of a 2^3 factorial design are run. The results follow:

A	B	C	Treatment Combination	Replicate I	II	III
−	−	−	(1)	22	31	25
+	−	−	a	32	43	29
−	+	−	b	35	34	50
+	+	−	ab	55	47	46
−	−	+	c	44	45	38
+	−	+	ac	40	37	36
−	+	+	bc	60	50	54
+	+	+	abc	39	41	47

(a) Estimate the factor effects. Which effects appear to be large?

(b) Use the analysis of variance to confirm your conclusions for part (a).

(c) Analyze the residuals. Are there any obvious problems?

(d) What levels of A, B, and C would you recommend?

9-2 A router is used to cut locating notches on a printed circuit board. The vibration level at the surface of the board as it is cut is considered to be a major source of dimensional variation in the notches. Two factors are thought to influence vibration: bit size (A) and cutting speed (B). Two bit sizes ($\frac{1}{16}$ and $\frac{1}{8}$ inch) and two speeds (40 and 90 rpm) are selected, and four boards are cut at each set of conditions shown below. The response variable is vibration measured as the resultant vector of three accelerometers (x, y, and z) on each test circuit board.

			Replicate			
A	B	Treatment Combination	I	II	III	IV
−	−	(1)	18.2	18.9	12.9	14.4
+	−	a	27.2	24.0	22.4	22.5
−	+	b	15.9	14.5	15.1	14.2
+	+	ab	41.0	43.9	36.3	39.9

(a) Analyze the data from this experiment.
(b) Plot the residuals on normal probability paper and versus the predicted vibration level. Interpret these plots.
(c) Draw the AB interaction plot. Interpret this plot. What levels of bit size and speed would you recommend for routine operation?

9-3 Find the two standard error limits for the factor effects in Problem 9-1. Does the results of this analysis agree with the conclusions from the analysis of variance?

9-4 Plot the factor effects from Problem 9-1 on a graph relative to an appropriately scaled t distribution. Does this graphical display adequately identify the important factors? Compare the conclusions from this plot with the results from the analysis of variance.

9-5 An experiment was performed to improve the yield of a chemical process. Four factors were selected, and two replicates of a completely randomized experiment were run. The results are shown in the following table:

Treatment Combination	Replicate		Treatment Combination	Replicate	
	I	II		I	II
(1)	90	93	d	98	95
a	74	78	ad	72	76
b	81	85	bd	87	83
ab	83	80	abd	85	86
c	77	78	cd	99	90
ac	81	80	acd	79	75
bc	88	82	bcd	87	84
abc	73	70	abcd	80	80

(a) Estimate the factor effects.

(b) Prepare an analysis of variance table, and determine which factors are important in explaining yield.

(c) Plot the residuals versus the predicted yield and on normal probability paper. Does the residual analysis appear satisfactory?

(d) Two three-factor interactions, ABC and ABD, apparently have large effects. Draw a cube plot in the factors A, B, and C with the average yields shown at each corner. Repeat using the factors A, B, and D. Do these two plots aid in data interpretation? Where would you recommend that the process be run with respect to the four variables?

9-6 A bacteriologist is interested in the effects of two different culture media and two different times on the growth of a particular virus. She performs six replicates of a 2^2 design, making the runs in random order. Analyze the bacterial growth data that follow and draw appropriate conclusions. Analyze the residuals and comment on the model's adequacy.

		Culture Medium			
Time		1		2	
12 hr	21	22	25	26	
	23	28	24	25	
	20	26	29	27	
18 hr	37	39	31	34	
	38	38	29	33	
	35	36	30	35	

9-7 An industrial engineer employed by a beverage bottler is interested in the effects of two different types of 32-ounce bottles on the time to deliver 12-bottle cases of the product. The two bottle types are glass and plastic. Two workers are used to perform a task consisting of moving 40 cases of the product 50 feet on a standard type of hand truck and stacking the cases in a display. Four replicates of a 2^2 factorial design are performed, and the times observed are listed in the following table. Analyze the data and draw appropriate conclusions. Analyse the residuals and comment on the model's adequacy.

		Worker		
Bottle Type	1		2	
Glass	5.12	4.89	6.65	6.24
	4.98	5.00	5.49	5.55
Plastic	4.95	4.43	5.28	4.91
	4.27	4.25	4.75	4.71

9-8 In Problem 9-7, the engineer was also interested in potential fatigue differences between the two types of bottles. As a measure of the amount of effort required, he measured the elevation of the heart rate (pulse) induced by the task. The results follow. Analyze the data and draw conclusions. Analyze the residuals and comment on the model's adequacy.

Bottle Type	Worker 1		Worker 2	
Glass	39	45	20	13
	58	35	16	11
Plastic	44	35	13	10
	42	21	16	15

9-9 Calculate the two standard error limits on the factor effects in Problem 9-8. Does the results of this analysis agree with the analysis of variance performed in Problem 9-8?

9-10 An article in the *AT&T Technical Journal* (March/April 1986, Vol. 65, pp. 39–50) describes the application of two-level factorial designs to integrated circuit manufacturing. A basic processing step is to grow an epitaxial layer on polished silicon wafers. The wafers mounted on a susceptor are positioned inside a bell jar, and chemical vapors are introduced. The susceptor is rotated and heat is applied until the epitaxial layer is thick enough. An experiment was run using two factors: arsenic flow rate (A) and deposition time (B). Four replicates were run, and the epitaxial layer thickness was measured (in μm). The data are shown below.

A	B	Replicate I	II	III	IV	Factor Levels	Low ($-$)	High ($+$)
$-$	$-$	14.037	16.165	13.972	13.907	A	55%	59%
$+$	$-$	13.880	13.860	14.032	13.914			
$-$	$+$	14.821	14.757	14.843	14.878	B	Short	Long
$+$	$+$	14.888	14.921	14.415	14.932			

(a) Estimate the factor effects.

(b) Conduct the analysis of variance. Which factors are important?

(c) Analyze the residuals. Are there any residuals that should cause concern?

(d) Discuss how you might deal with the potential outlier found in part (c).

9-11   A nickel–titanium alloy is used to make components for jet turbine aircraft engines. Cracking is a potentially serious problem in the final part, as it can lead to non-recoverable failure. A test is run at the parts producer to determine the effect of four factors on cracks. The four factors are pouring temperature (A), titanium content (B), heat treatment method (C), and amount of grain refiner used (D). Two replicates of a 2^4 design are run, and the length of crack (in mm) induced in a sample coupon subjected to a standard test is measured. The data are shown below:

A	B	C	D	Treatment Combination	Replicate I	Replicate II
−	−	−	−	(1)	1.71	1.91
+	−	−	−	a	1.42	1.48
−	+	−	−	b	1.35	1.53
+	+	−	−	ab	1.67	1.55
−	−	+	−	c	1.23	1.38
+	−	+	−	ac	1.25	1.26
−	+	+	−	bc	1.46	1.42
+	+	+	−	abc	1.29	1.27
−	−	−	−	d	2.04	2.19
+	−	−	+	ad	1.86	1.85
−	+	−	+	bd	1.79	1.95
+	+	−	+	abd	1.42	1.59
−	−	+	+	cd	1.81	1.92
+	−	+	+	acd	1.34	1.29
−	+	+	+	bcd	1.46	1.53
+	+	+	+	abcd	1.38	1.35

(a) Estimate the factor effects. Which factor effects appear to be large?

(b) Conduct an analysis of variance. Do any of the factors affect cracking?

(c) Analyze the residuals from this experiments.

(d) Is there an indication that any of the factors affect the variability in cracking?

(e) What recommendations would you make regarding process operations?

9-12   An article in *Solid State Technology* ("Orthogonal Design for Process Optimization and its Application in Plasma Etching", May 1987, pp. 127–132) describes the application of factorial designs in developing a nitride etch process on a single-wafer plasma etcher. The process uses C_2F_6 as the reactant gas. Four factors are of interest: anode–cathode gap (A), pressure in the reactor chamber (B), C_2F_6 gas flow (C), and power applied to the cathode (D). The response variable of interest is the etch rate for silicon nitride. A single replicate of a 2^4 design is run, and the data are shown below:

Run Number	Actual Run Order	A	B	C	D	Etch Rate (Å/min)	Factor Levels Low (−)		High (+)
1	13	−	−	−	−	550	A (cm)	0.80	1.20
2	8	+	−	−	−	669	B (mTorr)	4.50	550
3	12	−	+	−	−	604	C (SCCM)	125	200
4	9	+	+	−	−	650	D (W)	275	325
5	4	−	−	+	−	633			
6	15	+	−	+	−	642			
7	16	−	+	+	−	601			
8	3	+	+	+	−	635			
9	1	−	−	−	+	1037			
10	14	+	−	−	+	749			
11	5	−	+	−	+	1052			
12	10	+	+	−	+	868			
13	11	−	−	+	+	1075			
14	2	+	−	+	+	860			
15	7	−	+	+	+	1063			
16	6	+	+	+	+	729			

(a) Estimate the factor effects. Construct a normal probability plot of the factor effects. Which effects appear large?

(b) Conduct an analysis of variance to confirm your findings for part (a).

(c) Analyze the residuals from this experiment. Comment on the model's adequacy.

(d) If not all the factors are important, project the 2^4 design into a 2^k design with $k < 4$ and conduct the analysis of variance.

(e) Draw graphs to interpret any significant interactions.

(f) Plot the residuals versus the actual run order. What problems might be revealed by this plot?

9-13 Consider the single replicate of the 2^4 design in Example 9-2. Suppose we had arbitrarily decided to analyze the data assuming that all three- and four-factor interactions were negligible. Conduct this analysis and compare your results with those obtained in the example. Do you think that it is a good idea to arbitrarily assume interactions to be negligible even if they are relatively high-order ones?

9-14 An experiment was run in a semiconductor fabrication plant in an effort to increase yield. Five factors, each at two levels, were studied. The factors (and levels) were A = aperture setting (small, large), B = exposure time (20% below nominal, 20% above nominal), C = development time (30 s, 45 s), D = mask dimension (small, large), and E = etch time (14.5 min, 15.5 min). The unreplicated 2^5 design shown below was run.

(1) = 7	d = 8	e = 8	de = 6
a = 9	ad = 10	ae = 12	ade = 10
b = 34	bd = 32	be = 35	bde = 30
ab = 55	abd = 50	abe = 52	abde = 53
c = 16	cd = 18	ce = 15	cde = 15
ac = 20	acd = 21	ace = 22	acde = 20
bc = 40	bcd = 44	bce = 45	bcde = 41
abc = 60	abcd = 61	abce = 65	abcde = 63

(a) Construct a normal probability plot of the effect estimates. Which effects appear to be large?

(b) Conduct an analysis of variance to confirm your findings for part (a).

(c) Plot the residuals on normal probability paper. Is the plot satisfactory?

(d) Plot the residuals versus the predicted yields and versus each of the five factors. Comment on the plots.

(e) Interpret any significant interactions.

(f) What are your recommendations regarding process operating conditions?

(g) Project the 2^5 design in this problem into a 2^k design in the important factors. Sketch the design and show the average and range of yields at each run. Does this sketch aid in data interpretation?

9-15 In a process development study on yield, four factors were studied, each at two levels: time (A), concentration (B), pressure (C), and temperature (D). A single replicate of a 2^4 design was run, and the resulting data are shown in the following table:

Run Number	Actual Run Order	A	B	C	D	Yield (lbs)	Factor Levels Low (−)		High (+)
1	5	−	−	−	−	12	A (h)	2.5	3
2	9	+	−	−	−	18	B (%)	14	18
3	8	−	+	−	−	13	C (psi)	60	80
4	13	+	+	−	−	16	D (°C)	225	250
5	3	−	−	+	−	17			
6	7	+	−	+	−	15			
7	14	−	+	+	−	20			
8	1	+	+	+	−	15			
9	6	−	−	−	+	10			
10	11	+	−	−	+	25			
11	2	−	+	−	+	13			
12	15	+	+	−	+	24			
13	4	−	−	+	+	19			
14	16	+	−	+	+	21			
15	10	−	+	+	+	17			
16	12	+	+	+	+	23			

(a) Plot the effect estimates on normal probability paper. Which factors appear to have large effects?

(b) Conduct an analysis of variance using the normal probability plot in part (a) for guidance in forming an error term. What are your conclusions?

(c) Analyze the residuals from this experiment. Does your analysis indicate any potential problems?

(d) Can this design be collapsed into a 2^3 design with two replicates? If so, sketch the design with the average and range of yield shown at each point in the cube. Interpret the results.

9-16 *The Scrumptious Brownie Experiment.* The author is an engineer by training and a firm believer in learning by doing. I have taught experimental design for many years to a wide variety of audiences and have always assigned the planning, conduct, and analysis of an actual experiment to the class participants. The participants seem to enjoy this practical experience and always learn a great deal from it. This problem uses the results of an experiment performed by Gretchen Krueger at Arizona State University.

There are many different ways to bake brownies. The purpose of this experiment was to determine how the pan material, the brand of brownie mix, and the stirring method affect the scrumptiousness of brownies. The factor levels were

Factor	Low (−)	High (+)
A = pan material	Glass	Aluminum
B = stirring method	Spoon	Mixer
C = brand of mix	Expensive	Cheap

The response variable was scrumptiousness, a subjective measure derived from a questionnaire given to the subjects who sampled each batch of brownies. (The questionnaire dealt with such issues as taste, appearance, consistency, aroma, and so forth). An eight-person test panel sampled each batch and filled out the questionnaire. The complete set of data are shown below.

Brownie Batch	A	B	C	Test Panel Results							
				1	2	3	4	5	6	7	8
1	−	−	−	11	9	10	10	11	10	8	9
2	+	−	−	15	10	16	14	12	9	6	15
3	−	+	−	9	12	11	11	11	11	11	12
4	+	+	−	16	17	15	12	13	13	11	11
5	−	−	+	10	11	15	8	6	8	9	14
6	+	−	+	12	13	14	13	9	13	14	9
7	−	+	+	10	12	13	10	7	7	17	13
8	+	+	+	15	12	15	13	12	12	9	14

(a) Analyze the data from this experiment as if there were eight replicates of a 2^3 design. Comment on the results.

(b) Is the analysis in part (a) the correct approach? There are only eight batches. Do we really have eight replicates of a 2^3 factorial?

(c) Analyze the average and standard deviation of the scrumptiousness ratings. Comment on the results. Is this analysis more appropriate than the one in part (a)? Why or why not?

9-17   A factor that appears at 4 levels can be considered as two pseudo-factors, each at two levels. In general, a factor with 2^k levels can be thought of as a k pseudo-factors at two levels each. Yates' method can then be used on these designs, and the sums of squares associated with the pseudo-factors can be combined to yield the sums of squares for the main effects and interactions for the original factors. Use the data in Problem 7-8 to demonstrate this technique.

9-18   Analyze the data in Example 9-2 using Yates' algorithm.

9-19   Consider the single replicate of the 2^4 design in Example 9-2. Suppose that we ran five points at the center (0, 0, 0, 0) and observed the following responses: 73, 75, 71, 69, and 76. Test for curvature in this experiment. Interpret the results.

9-20   *A Missing Value in a 2^k Factorial.* It is not unusual to find that one of the observations in a 2^k design is missing due to faulty measuring equipment, a spoiled test, or some other reason. If the design is replicated n times ($n > 1$) some of the techniques discussed in Chapter 7 can be employed. However, for an unreplicated factorial ($n = 1$) some other method must be used. One logical approach is to estimate the missing value with a number that makes the highest-order interaction contrast zero. Apply this technique to the experiment in Example 9-2 assuming that run ab is missing. Compare the results with the results of Example 9-2.

Chapter 10
Confounding in The 2^k Factorial

10-1 INTRODUCTION

There are many problems for which it is impossible to perform a complete replicate of a factorial design in one block, where the block might be one day, one homogeneous batch of raw material, one laboratory, or the like. *Confounding* is a design technique for arranging a complete factorial experiment in blocks, where the block size is smaller than the number of treatment combinations in one replicate. The technique causes information about certain treatment effects (usually high-order interactions) to be *indistinguishable from,* or *confounded with,* blocks. In this chapter we concentrate on confounding systems for the 2^k factorial design. Note that even though the designs presented are incomplete block designs because each block does not contain all the treatments or treatment combinations, the special structure of the 2^k factorial system allows a simplified method of analysis. As in Chapter 9, we continue to assume that all factors are fixed effects.

We consider the construction and analysis of the 2^k factorial design in 2^p incomplete blocks, where $p < k$. Consequently, these designs can be run in two blocks, four blocks, eight blocks, and so on.

10-2 THE 2^k FACTORIAL DESIGN IN TWO BLOCKS

Suppose that we wish to run a single replicate of the 2^2 design. Each of the 2^2 = 4 treatment combinations requires a quantity of raw material, for example, and each batch of raw material is only large enough for two treatment combinations to be tested. Thus, two batches of raw material are required. If batches of raw material are considered as blocks, then we must assign two of the four treatment combinations to each block.

Figure 10-1 shows one possible design for this problem. The geometric view, Figure 10-1(a), indicates that treatment combinations on opposing diagonals are assigned to different blocks. Notice from Figure 10-1(b) that block 1 contains the treatment combinations (1) and ab and that block 2 contains a and b. Of course, the *order* in which the treatment combinations are run within a block is randomly determined. We would also randomly decide which block to run first. Suppose we estimate the main effects of A and B just as if no blocking had occurred. From Equations 9-1 and 9-2, we obtain

$$A = \tfrac{1}{2}[ab + a - b - (1)]$$
$$B = \tfrac{1}{2}[ab + b - a - (1)]$$

Note that both A and B are unaffected by blocking since in each estimate there is one plus and one minus treatment combination from each block. That is, any difference between block 1 and block 2 will cancel out.

Now consider the AB interaction

$$AB = \tfrac{1}{2}[ab + (1) - a - b]$$

Since the two treatment combinations with the plus sign [ab and (1)] are in block 1 and the two with the minus sign [a and b] are in block 2, the block effect and the AB interaction are identical. That is, AB is *confounded* with blocks.

The reason for this is apparent from the table of plus and minus signs for the 2^2 design. This was originally given as Table 9-2, but for convenience it is reproduced as Table 10-1. From this table, we see that all treatment combinations that have a plus on AB are assigned to block 1, whereas all treatment combinations that have a minus sign on AB are assigned to block 2. This approach can be used to confound any effect (A, B, or AB) with blocks. For example, if (1)

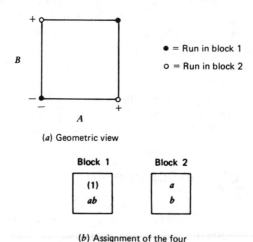

(a) Geometric view

Block 1   Block 2

(1) a
ab b

(b) Assignment of the four
runs to two blocks

Figure 10-1. A 2^2 design in two blocks. (a) Geometric view. (b) Assignment of the four runs to two blocks.

and *b* had been assigned to block 1 and *a* and *ab* to block 2, then the main effect *A* would have been confounded with blocks. The usual practice is to confound the highest-order interaction with blocks.

This scheme can be used to confound any 2^k design in two blocks. As a second example, consider a 2^3 design run in two blocks. Suppose we wish to confound the three-factor interaction *ABC* with blocks. From the table of plus and minus signs shown in Table 10-2, we assign the treatment combinations that are minus on *ABC* to block 1 and those that are plus on *ABC* to block 2. The resulting design is shown in Figure 10-2. Once again, we emphasize that the treatment combinations *within* a block are run in random order.

Other Methods for Constructing the Blocks ▪ There is another method for constructing these designs. The method uses the linear combination

$$L = \alpha_1 x_1 + \alpha_2 x_2 + \cdots + \alpha_k x_k \qquad (10\text{-}1)$$

where x_i is the level of the *i*th factor appearing in a particular treatment combination and α_i is the exponent appearing on the *i*th factor in the effect to be confounded. For the 2^k system, we have $\alpha_i = 0$ or 1 and $x_i = 0$ (low level) or $x_i = 1$ (high level). Equation 10-1 is called a *defining contrast*. Treatment combinations that produce the same value of *L* (mod 2) will be placed in the same block. Since the only possible values of *L* (mod 2) are 0 and 1, this will assign the 2^k treatment combinations to exactly two blocks.

To illustrate the approach, consider a 2^3 design with *ABC* confounded with blocks. Here x_1 corresponds to *A*, x_2 to *B*, x_3 to *C*, and $\alpha_1 = \alpha_2 = \alpha_3 = 1$. Thus, the defining contrast corresponding to *ABC* is

$$L = x_1 + x_2 + x_3$$

The treatment combination (1) is written 000 in the (0, 1) notation; therefore,

$$L = 1(0) + 1(0) + 1(0) = 0 = 0 \text{ (mod 2)}$$

Similarly, the treatment combination *a* is 100, yielding

$$L = 1(1) + 1(0) + 1(0) = 1 = 1 \text{ (mod 2)}$$

Thus, (1) and *a* would be run in different blocks. For the remaining treatment

Table 10-1   Table of Plus and Minus Signs for the 2^2 Design

Treatment Combination	Factorial Effect			
	I	*A*	*B*	*AB*
(1)	+	−	−	+
a	+	+	−	−
b	+	−	+	−
ab	+	+	+	+

Table 10-2 Table of Plus and Minus Signs for the 2^3 Design

Treatment Combination	Factorial Effect							
	I	*A*	*B*	*AB*	*C*	*AC*	*BC*	*ABC*
(1)	+	−	−	+	−	+	+	−
a	+	+	−	−	−	−	+	+
b	+	−	+	−	−	+	−	+
ab	+	+	+	+	−	−	−	−
c	+	−	−	+	+	−	−	+
ac	+	+	−	−	+	+	−	−
bc	+	−	+	−	+	−	+	−
abc	+	+	+	+	+	+	+	+

combinations, we have

$$b: \quad L = 1(0) + 1(1) + 1(0) = 1 = 1 \text{ (mod 2)}$$
$$ab: \quad L = 1(1) + 1(1) + 1(0) = 2 = 0 \text{ (mod 2)}$$
$$c: \quad L = 1(0) + 1(0) + 1(1) = 1 = 1 \text{ (mod 2)}$$
$$ac: \quad L = 1(1) + 1(0) + 1(1) = 2 = 0 \text{ (mod 2)}$$
$$bc: \quad L = 1(0) + 1(1) + 1(1) = 2 = 0 \text{ (mod 2)}$$
$$abc: \quad L = 1(1) + 1(1) + 1(1) = 3 = 1 \text{ (mod 2)}$$

Thus (1), *ab*, *ac*, and *bc* are run in block 1 and *a*, *b*, *c*, and *abc* are run in block

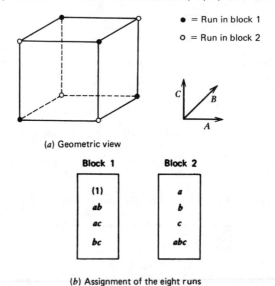

● = Run in block 1
○ = Run in block 2

(*a*) Geometric view

Block 1	Block 2
(1)	***a***
ab	***b***
ac	***c***
bc	***abc***

(*b*) Assignment of the eight runs to two blocks

Figure 10-2. The 2^3 design in two blocks with *ABC* confounded. (*a*) Geometric view. (*b*) Assignment of the eight runs to two blocks.

2. This is the same design shown in Figure 10-2 that was generated from the table of plus and minus signs.

Another method may be used to construct these designs. The block containing the treatment combination (1) is called the *principal block*. The treatment combinations in this block have a useful group-theoretic property; namely, they form a group with respect to multiplication modulus 2. This implies that any element [except (1)] in the principal block may be generated by multiplying two other elements in the principal block modulus 2. For example, consider the principal block of the 2^3 design with ABC confounded, as shown in Figure 10-2. Note that

$$ab \cdot ac = a^2bc = bc$$
$$ab \cdot bc = ab^2c = ac$$
$$ac \cdot bc = abc^2 = ab$$

Treatment combinations in the other block (or blocks) may be generated by multiplying one element in the new block by each element in the principal block modulus 2. For the 2^3 with ABC confounded, since the principal block is (1), ab, ac, and bc, we know that b is in the other block. Thus, the elements of this second block are

$$b \cdot (1) \qquad\quad = b$$
$$b \cdot ab = ab^2 = a$$
$$b \cdot ac \qquad\quad = abc$$
$$b \cdot bc = b^2c = c$$

This agrees with the results obtained previously.

Estimation of Error ▪ When the number of variables is small, say $k = 2$ or 3, it is usually necessary to replicate the experiment to obtain an estimate of error. For example, suppose that a 2^3 factorial must be run in two blocks with ABC confounded, and the experimenter decides to replicate the design four times. The resulting design might appear as in Figure 10-3. Note that ABC is confounded in each replicate.

The analysis of variance for this design is shown in Table 10-3. There are 32

Replicate I		Replicate II		Replicate III		Replicate IV	
Block 1	Block 2	Block 1	Block 2	Block 1	Block 2	Block 1	Block 2
(1)	abc	(1)	abc	(1)	abc	(1)	abc
ac	a	ac	a	ac	a	ac	a
ab	b	ab	b	ab	b	ab	b
bc	c	bc	c	bc	c	bc	c

Figure 10-3. Four replicates of the 2^3 design with ABC confounded.

observations and 31 total degrees of freedom. Furthermore, since there are eight blocks, 7 degrees of freedom must be associated with these blocks. One break-down of those 7 degrees of freedom is shown in Table 10-3. The error sum of squares actually consists of the two-factor interactions between replicates and each of the effects (A, B, C, AB, AC, BC). It is usually safe to consider the interactions to be zero and to treat the resulting mean square as an estimate of error. Main effects and two-factor interactions are tested against the mean square error. Cochran and Cox (1957) state that the block or ABC mean square could be compared to the error for the ABC mean square, which is really replicates × blocks. This test usually is very insensitive.

If resources are sufficient to allow the replication of confounded designs, it is generally better to use a slightly different method of designing the blocks in each replicate. This approach consists of confounding a different effect in each replicate so that some information on all effects is obtained. Such a procedure is called *partial confounding* and is discussed in Section 10-5. If k is moderately large, say $k \geq 4$, we frequently can afford only a single replicate. The experimenter usually assumes higher-order interactions to be negligible and combines their sums of squares as error. The normal probability plot of factor effects can be very helpful in this regard.

Example 10-1

Consider the situation described in Example 9-2. Recall that four factors—temperature (A), pressure (B), concentration of formaldehyde (C), and stirring rate (D)—are studied in a pilot plant to determine their effect on product filtration rate. Suppose now that the $2^4 = 16$ treatment combinations cannot all be run on the same day. The experimenter can run 8 treatment combinations in one day, so a 2^4 design confounded in two blocks seems appropriate. It is logical to confound the highest-order interaction

Table 10-3   Analysis of Variance for Four Replicates of a 2^3 Design with ABC Confounded

Source of Variation	Degrees of Freedom
Replicates	3
Blocks (ABC)	1
Error for ABC (replicates × blocks)	3
A	1
B	1
C	1
AB	1
AC	1
BC	1
Error (or replicates × effects)	18
Total	31

ABCD with blocks. The defining contrast is

$$L = x_1 + x_2 + x_3 + x_4$$

and it is easy to verify that the design is as shown in Figure 10-4.

The data are analyzed by Yates' algorithm in Table 10-4. Since the block totals are 566 and 555, the sum of squares for blocks is

$$SS_{\text{Blocks}} = \frac{(566)^2 + (555)^2}{8} - \frac{(1121)^2}{16} = 7.5625$$

which is identical to the sum of squares for *ABCD* in Table 10-4.

The sums of squares for the main effects and two-factor interactions are taken directly from the Yates' algorithm. A normal probability plot of effects indicates that three-factor interactions are negligible, so the error sum of squares is

$$SS_E = SS_{ABC} + SS_{ABD} + SS_{ACD} + SS_{BCD}$$
$$= 14.0625 + 68.0625 + 10.5625 + 27.5625 = 120.2500$$

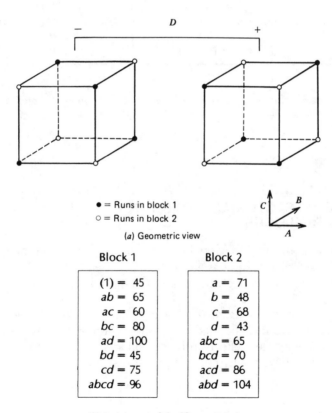

• = Runs in block 1
○ = Runs in block 2

(a) Geometric view

Block 1	Block 2
(1) = 45	a = 71
ab = 65	b = 48
ac = 60	c = 68
bc = 80	d = 43
ad = 100	abc = 65
bd = 45	bcd = 70
cd = 75	acd = 86
abcd = 96	abd = 104

(b) Assignment of the 16 runs to two blocks

Figure 10-4. The 2^4 design in two blocks for Example 10-1. (a) Geometric view. (b) Assignment of the 16 runs to two blocks.

Table 10-4 Yates' Algorithm for Example 10-1

Treatment Combination	Response	(1)	(2)	(3)	(4)	Effects	Sum of Squares
(1)	45	116	229	502	1121	—	—
a	71	113	273	619	173	A	1870.5625
b	48	128	292	20	25	B	39.0625
ab	65	145	327	153	1	AB	0.0625
c	68	143	43	14	79	C	390.0625
ac	60	149	−23	11	−145	AC	1314.0625
bc	80	161	116	−16	19	BC	22.5625
abc	65	166	37	17	15	ABC	14.0625
d	43	26	−3	44	117	D	855.5625
ad	100	17	17	35	133	AD	1105.5625
bd	45	−8	6	−66	−3	BD	0.5625
abd	104	−15	5	−79	33	ABD	68.0625
cd	75	57	−9	20	−9	CD	5.0625
acd	86	59	−7	−1	−13	ACD	10.5625
bcd	70	11	2	2	−21	BCD	27.5625
abcd	96	26	15	13	11	ABCD	7.5625

with 4 degrees of freedom. The complete analysis of variance is summarized in Table 10-5. For further details of the interpretation of this experiment, refer to Example 9-2.

∎

10-3 THE 2^k FACTORIAL DESIGN IN FOUR BLOCKS

It is possible to construct 2^k factorial designs confounded in four blocks of 2^{k-2} observations each. These designs are particularly useful in situations where the number of factors is moderately large, say $k \geq 4$, and block sizes are relatively small.

As an example, consider the 2^5 design. If each block will hold only eight runs, then four blocks must be used. The construction of this design is relatively straightforward. Select *two* effects to be confounded with blocks, say ADE and BCE. These effects have the two defining contrasts

$$L_1 = x_1 + x_4 + x_5$$
$$L_2 = x_2 + x_3 + x_5$$

associated with them. Now every treatment combination will yield a particular pair of values of L_1 (mod 2) and L_2 (mod 2); that is, either $(L_1, L_2) = (0, 0)$, $(0, 1)$, $(1, 0)$, or $(1, 1)$. Treatment combinations yielding the same values of (L_1, L_2)

Table 10-5   Analysis of Variance for Example 10-1

Source of Variation	Sum of Squares	Degrees of Freedom	Mean Square	F_0
Blocks ($ABCD$)	7.5625	1	—	—
A	1870.5625	1	1870.5625	62.22[b]
B	39.0625	1	39.0625	1.30
C	390.0625	1	390.0625	12.98[a]
D	855.5625	1	855.5625	28.46[b]
AB	0.0625	1	0.0625	< 1
AC	1314.0625	1	1314.0625	43.71[b]
AD	1105.5625	1	1105.5625	36.78[b]
BC	22.5625	1	22.5625	< 1
BD	0.5625	1	0.5625	< 1
CD	5.0625	1	5.0625	< 1
Error (or $ABC + ABD + ACD + BCD$)	120.2500	4	30.0625	
Total		15		

[a] Significant at 5 percent.
[b] Significant at 1 percent.

are assigned to the same block. In our example we find

$$L_1 = 0, L_2 = 0 \quad \text{for} \quad (1), ad, bc, abcd, abe, ace, cde, bde$$

$$L_1 = 1, L_2 = 0 \quad \text{for} \quad a, d, abc, bcd, be, abde, ce, acde$$

$$L_1 = 0, L_2 = 1 \quad \text{for} \quad b, abd, c, acd, ae, de, abce, bcde$$

$$L_1 = 1, L_2 = 1 \quad \text{for} \quad e, ade, bce, abcde, ab, bd, ac, cd$$

These treatment combinations would be assigned to different blocks. The complete design is as shown in Figure 10-5.

With a little reflection we realize that another effect in addition to ADE and BCE must be confounded with blocks. Since there are four blocks with three degrees of freedom between them, and since ADE and BCE have only one degree of freedom each, clearly an additional effect with one degree of freedom must

Block 1 $L_1 = 0$ $L_2 = 0$		Block 2 $L_1 = 1$ $L_2 = 0$		Block 3 $L_1 = 0$ $L_2 = 1$		Block 4 $L_1 = 1$ $L_2 = 1$	
(1)	abe	a	be	b	abce	e	abcde
ad	ace	d	abde	abd	ae	ade	bd
bc	cde	abc	ce	c	bcde	bce	ac
abcd	bde	bcd	acde	acd	de	ab	cd

Figure 10-5.   The 2^5 design in four blocks with ADE, BCE, and $ABCD$ confounded.

be confounded. This effect is the *generalized interaction* of *ADE* and *BCE*, which is defined as the product of *ADE* and *BCE* modulus 2. Thus, in our example the generalized interaction $(ADE)(BCE) = ABCDE^2 = ABCD$ is also confounded with blocks. It is easy to verify this by referring to a table of plus and minus signs for the 2^5 design, such as in Davies (1956). Inspection of such a table reveals that the treatment combinations are assigned to the blocks as follows:

Treatment Combinations in	Sign on *ADE*	Sign on *BCE*	Sign on *ABCD*
Block 1	−	−	+
Block 2	+	−	−
Block 3	−	+	−
Block 4	+	+	+

Notice that the product of signs of any two effects for a particular block (e.g., *ADE* and *BCE*) yields the sign of the other effect for that block (in this case *ABCD*). Thus, *ADE*, *BCE*, and *ABCD* are all confounded with blocks.

The group-theoretic properties of the principal block mentioned in Section 10-2 still hold. For example, we see that the product of two treatment combinations in the principal block yields another element of the principal block. That is,

$$ad \cdot bc = abcd \qquad \text{and} \qquad abe \cdot bde = ab^2de^2 = ad$$

and so forth. To construct another block, select a treatment combination that is not in the principal block (e.g., *b*) and multiply *b* by all the treatment combinations in the principal block. This yields

$$b \cdot (1) = b \qquad b \cdot ad = abd \qquad b \cdot bc = c \qquad b \cdot abcd = ab^2cd = acd$$

and so forth, which will produce the eight treatment combinations in block 3. In practice, the principal block can be obtained from the defining contrasts and the group-theoretic property, and the remaining blocks can be determined from these treatment combinations by the method shown above.

The general procedure for constructing a 2^k design confounded in four blocks is to choose two effects to generate the blocks, automatically confounding a third effect that is the generalized interaction of the first two. Then, the design is constructed by using the two defining contrasts (L_1, L_2) and the group-theoretic properties of the principal block. In selecting effects to be confounded with blocks, care must be exercised to obtain a design that does not confound effects that may be of interest. For example, in a 2^5 design we might choose to confound *ABCDE* and *ABD*, which automatically confounds *CE*, an effect that is probably of interest. A better choice is to confound *ADE* and *BCE*, which automatically confounds *ABCD*. It is preferable to sacrifice information on the three-factor interactions *ADE* and *BCE* instead of the two-factor interaction *CE*.

10-4 THE 2^k FACTORIAL DESIGN IN 2^p BLOCKS

The methods described above may be extended to the construction of a 2^k factorial design confounded in 2^p blocks ($p < k$), where each block contains exactly 2^{k-p} runs. We select p independent effects to be confounded, where by "independent" we mean that no effect chosen is the generalized interaction of the others. The blocks may be generated by use of the p defining contrasts L_1, $L_2, \ldots, L_p$ associated with these effects. In addition, exactly $2^p - p - 1$ other effects will be confounded with blocks, these being the generalized interactions of those p independent effects initially chosen. Care should be exercised in selecting effects to be confounded so that information on effects that may be of potential interest is not sacrificed.

The statistical analysis of these designs is straightforward. Sums of squares for all the effects are computed as if no blocking had occurred. Then, the block sum of squares is found by adding the sums of squares for all the effects confounded with blocks.

Obviously, the choice of the p effects used to generate the block is critical since the confounding structure of the design directly depends on them. Table 10-6 presents a list of useful designs. To illustrate the use of this table, suppose we wish to construct a 2^6 design confounded in $2^3 = 8$ blocks of $2^6 = 8$ runs each. Table 10-6 indicates that we would choose $ABEF$, $ABCD$, and ACE as the $p = 3$ independent effects to generate the blocks. The remaining $2^p - p - 1 = 2^3 - 3 - 1 = 4$ effects that are confounded are the generalized interactions of these three; that is,

$$(ABEF)(ABCD) = A^2B^2CDEF = CDEF$$
$$(ABEF)(ACE) = A^2BCE^2F = BCF$$
$$(ABCD)(ACE) = A^2BC^2ED = BDE$$
$$(ABEF)(ABCD)(ACE) = A^3B^2C^2DE^2F = ADF$$

The reader is asked to generate the eight blocks for this design in Problem 10-8.

10-5 PARTIAL CONFOUNDING

We remarked in Section 10-2 that, unless experimenters have a prior estimate of error or are willing to assume certain interactions to be negligible, they must replicate the design to obtain an estimate of error. Figure 10-3 shows a 2^3 factorial in two blocks with ABC confounded, replicated four times. From the analysis of variance for this design, shown in Table 10-3, we note that information on the ABC interaction cannot be retrieved since ABC is confounded with blocks *in each replicate*. This design is said to be *completely confounded*.

Consider the alternative shown in Figure 10-6. Once again, there are four replicates of the 2^3 design, but a *different* interaction has been confounded in

Table 10-6   Suggested Blocking Arrangements for the 2^k Factorial Design

Number of Factors, k	Number of Blocks, 2^p	Block Size, 2^{k-p}	Effects Chosen to Generate the Blocks	Interactions Confounded with Blocks
3	2	4	ABC	ABC
	4	2	AB, AC	AB, AC, BC
4	2	8	ABCD	ABCD
	4	4	ABC, ACD	ABC, ACD, BD
	8	2	AB, BC, CD	AB, BC, CD, AC, BD, AD, ABCD
5	2	16	ABCDE	ABCDE
	4	8	ABC, CDE	ABC, CDE, ABDE
	8	4	ABE, BCE, CDE	ABE, BCE, CDE, AC, ABCD, BD, ADE
	16	2	AB, AC, CD, DE	All 2-factor and 4-factor interactions (15 effects)
6	2	32	ABCDEF	ABCDEF
	4	16	ABCF, CDEF	ABCF, CDEF, ABDE
	8	8	ABEF, ABCD, ACE	ABEF, ABCD, ACE, BCF, BDE, CDEF, ADF
	16	4	ABF, ACF, BDF, DEF	ABE, ACF, BDF, DEF, BC, ABCD, ABDE, AD, ACDE, CE, BDF, BCDEF, ABCEF, AEF, BE
	32	2	AB, BC, CD, DE, EF	All 2-factor, 4-factor, and 6-factor interactions (31 effects)
7	2	64	ABCDEFG	ABCDEFG
	4	32	ABCFG, CDEFG	ABCFG, CDEFG, ABDE
	8	16	ABC, DEF, AFG	ABC, DEF, AFG, ABCDEF, DCFG, ADEG, BCDEG
	16	8	ABCD, EFG, CDE, ADG	ABCD, EFG, CDE, ADG, ABCDEFG, ABE, BCG, CDFG, ADEF, ACEG, ABFG, BCEF, BDEG, ACF, BDF
	32	4	ABG, BCG, CDG, DEG, EFG	ABG, BCG, CDG, DEG, EFG, AC, BD, CE, DF, AE, BE, ABCD, ABDE, ABEF, BCDE, BCEF, CDEF, ABCDEFG, ADG, ACDEG, ACEFG, ABDFG, ABCEG, BEG, BDEFG, CFG, ADEF, ACDF, ABCF, AFG
	64	2	AB, BC, CD, DE, EF, FG	All 2-factor, 4-factor, and 6-factor interactions (63 effects)

Figure 10-6. Partial confounding in the 2^3 design.

each replicate. That is, ABC is confounded in replicate I, AB is confounded in replicate II, BC is confounded in replicate III, and AC is confounded in replicate IV. As a result, information on ABC can be obtained from the data in replicates II, III, and IV; information on AB can be obtained from replicates I, III, and IV; information on AC can be obtained from replicates I, II, and III; and information on BC can be obtained from replicates I, II, and IV. We say that three-quarters information can be obtained on the interactions because they are unconfounded in only three replicates. Yates (1937) calls the ratio 3/4 the *relative information for the confounded effects*. This design is said to be *partially confounded*.

The analysis of variance for this design is shown in Table 10-7. In calculating the interaction sums of squares, only data from the replicates in which an interaction is unconfounded are used. The error sum of squares consists of replicates × main effect sums of squares plus replicates × interaction sums of squares for each replicate in which that interaction is unconfounded (e.g., replicates × ABC for replicates II, III, and IV). Furthermore, there are seven degrees of freedom among the eight blocks. This is usually partitioned into three degrees of freedom for replicates and four degrees of freedom for blocks within replicates. The composition of the sum of squares for blocks is shown in Table 10-7 and follows directly from the choice of the effect confounded in each replicate.

Table 10-7   Analysis of Variance for a Partially Confounded 2^3 Design

Source of Variation	Degrees of Freedom
Replicates	3
Blocks within replicates [or ABC (rep. I) + AB (rep. II) + BC (rep. III) + AC (rep. IV)]	4
A	1
B	1
C	1
AB (from replicates I, III, and IV)	1
AC (from replicates I, II, and III)	1
BC (from replicates I, II, and IV)	1
ABC (from replicates II, III, and IV)	1
Error	17
Total	31

Example 10-2

A 2^3 with Partial Confounding

Consider Example 9-1, in which a study was performed to determine the effect of percent carbonation (A), operating pressure (B), and line speed (C) on the fill height of a carbonated beverage. Suppose that each batch of syrup is only large enough to test four treatment combinations. Thus, each replicate of the 2^3 design must be run in two blocks. Two replicates are run, with ABC confounded in replicate I and AB confounded in replicate II. The data are as follows:

Replicate I ABC Confounded		Replicate II AB Confounded	
$(1) = -3$	$a = 0$	$(1) = -1$	$a = 1$
$ab = 2$	$b = -1$	$c = 0$	$b = 0$
$ac = 2$	$c = -1$	$ab = 3$	$ac = 1$
$bc = 1$	$abc = 6$	$abc = 5$	$bc = 1$

The sums of squares for A, B, C, AC, and BC may be calculated in the usual manner. Yates' algorithm may also be employed, as shown in Table 10-8. Note that in this table, the calculations of the sums of squares for AB and ABC are not completed. We must find SS_{ABC} using only the data in replicate II and SS_{AB} using only the data in replicate I as follows:

$$SS_{ABC} = \frac{[a + b + c + abc - ab - ac - bc - (1)]^2}{n2^k}$$

$$= \frac{[1 + 0 + 0 + 5 - 3 - 1 - 1 - (-1)]^2}{(1)(8)} = 0.50$$

$$S_{AB} = \frac{[(1) + abc - ac + c - a - b + ab - bc]^2}{n2^k}$$

$$= \frac{[-3 + 6 - 2 + (-1) - 0 - (-1) + 2 - 1]^2}{(1)(8)} = 0.50$$

Table 10-8   Yates' Algorithm for the Data in Example 10-2

Treatment Combination	Response	(1)	(2)	(3)	Effect	Sum of Squares
(1)	-4	-3	1	16	I	—
a	1	4	15	24	A	36.00
b	-1	2	11	18	B	20.25
ab	5	13	13	6	AB	Not direct
c	-1	5	7	14	C	12.25
ac	3	6	11	2	AC	0.25
bc	2	4	1	4	BC	1.00
abc	11	9	5	4	ABC	Not direct

Table 10-9 Analysis of Variance for Example 10-2

Source of Variation	Sum of Squares	Degrees of Freedom	Mean Square	F_0
Replicates	1.00	1	1.00	—
Blocks within replicates	2.50	2	1.25	—
A	36.00	1	36.00	48.00[a]
B	20.25	1	20.25	27.00[a]
C	12.25	1	12.25	16.33[a]
AB (rep. I only)	0.50	1	0.50	0.67
AC	0.25	1	0.25	0.33
BC	1.00	1	1.00	1.33
ABC (rep. II only)	0.50	1	0.50	0.67
Error	3.75	5	0.75	—
Total	78.00	15	—	—

[a] Significant at 1 percent.

The sum of squares for the replicates is, in general,

$$SS_{Rep} = \sum_{h=1}^{n} \frac{R_h^2}{2^k} - \frac{y_{...}^2}{N}$$

$$= \frac{(6)^2 + (10)^2}{8} - \frac{(16)^2}{16} = 1.00$$

where R_h is the total of the observations in the hth replicate. The block sum of squares is the sum of SS_{ABC} from replicate I and SS_{AB} from replicate II, or $SS_{Blocks} = 2.50$.

The analysis of variance is summarized in Table 10-9. All three main effects are significant at 1 percent. ∎

10-6 PROBLEMS

10-1 Consider the data from the first replicate of Problem 9-1. Suppose that these observations could not all be run using the same bar stock. Set up a design to run these observations in two blocks of four observations each with ABC confounded. Analyze the data.

10-2 Consider the data from the first replicate of Problem 9-5. Construct a design with two blocks of eight observations each with $ABCD$ confounded. Analyze the data.

10-3 Repeat Problem 10-2 assuming that four blocks are required. Confound ABD and ABC (and consequently CD) with blocks.

10-4 Using the data from the 2^5 design in Problem 9-14, construct and analyze a design in two blocks with $ABCDE$ confounded with blocks.

10-5 Repeat Problem 10-4 assuming that four blocks are necessary. Suggest a reasonable confounding scheme.

10-6 Consider the data from the 2^5 design in Problem 9-14. Suppose that it was necessary to run this design in four blocks with $ACDE$ and BCD (and consequently ABE) confounded. Analyze the data from this design.

10-7 Design an experiment for confounding a 2^6 factorial in four blocks. Suggest an appropriate confounding scheme, different from the one shown in Table 10-6.

10-8 Consider the 2^6 design in eight blocks of eight runs each with $ABCD$, ACE, and $ABEF$ as the independent effects chosen to be confounded with blocks. Generate the design. Find the other effects confounded with blocks.

10-9 Consider the 2^2 design in two blocks with AB confounded. Prove algebraically that $SS_{AB} = SS_{Blocks}$.

10-10 Consider the data in Example 10-1. Suppose that all the observations in block 2 are increased by 20. Analyze the data that would result. Estimate the block effect. Can you explain its magnitude? Do blocks now appear to be an important factor? Are any other effect estimates impacted by the change you made to the data?

10-11 Suppose that in Problem 9-1 we had confounded ABC in replicate I, AB in replicate II, and BC in replicate III. Construct the analysis of variance table.

10-12 Repeat Problem 9-1 assuming that ABC was confounded with blocks in each replicate.

10-13 Suppose that in Problem 9-5 $ABCD$ was confounded in replicate I and ABC was confounded in replicate II. Perform the statistical analysis of this design.

10-14 Construct a 2^3 design with ABC confounded in the first two replicates and BC confounded in the third. Outline the analysis of variance and comment on the information obtained.

Chapter 11
Two-Level Fractional Factorial Designs

11-1 INTRODUCTION

As the number of factors in a 2^k factorial design increases, the number of runs required for a complete replicate of the design rapidly outgrows the resources of most experimenters. For example, a complete replicate of the 2^6 design requires 64 runs. In this design only 6 of the 63 degrees of freedom correspond to main effects, and only 15 degrees of freedom correspond to two-factor interactions. The remaining 42 degrees of freedom are associated with three-factor and higher interactions.

If the experimenter can reasonably assume that certain high-order interactions are negligible, then information on the main effects and low-order interactions may be obtained by running only a fraction of the complete factorial experiment. These *fractional factorial designs* are among the most widely used types of designs for product and process design and for process trouble shooting.

A major use of fractional factorials is in *screening experiments*. These are experiments in which many factors are considered with the purpose of identifying those factors (if any) that have large effects. Screening experiments are usually performed in the early stages of a project when it is likely that many of the factors initially considered have little or no effect on the response. The factors that are identified as important are then investigated more thoroughly in subsequent experiments.

The successful use of fractional factorial designs is based on three key ideas:

1. **The sparsity of effects principle.** When there are several variables, the system or process is likely to be driven primarily by some of the main effects and low-order interactions.

2. **The projective property.** Fractional factorial designs can be projected into stronger (larger) designs in the subset of significant factors.

3. **Sequential experimentation.** It is possible to combine the runs of two (or

more) fractional factorials to assemble sequentially a larger design to estimate the factor effects and interactions of interest.

We will focus on these principles in this chapter and illustrate them with several examples.

11-2   THE ONE-HALF FRACTION OF THE 2^k DESIGN

Consider a situation in which three factors, each at two levels, are of interest, but the experimenters cannot afford to run all $2^3 = 8$ treatment combinations. They can, however, afford four runs. This suggests a one-half fraction of a 2^3 design. Because the design contains $2^{3-1} = 4$ treatment combinations, a one-half fraction of the 2^3 design is often called a 2^{3-1} design.

The table of plus and minus signs for the 2^3 design is shown in Table 11-1. Suppose we select the four treatment combinations a, b, c, and abc as our one-half fraction. These runs are shown in the top half of Table 11-1 and in Figure 11-1a. We use both the lowercase letter notation (a, b, c, . . .) and the plus and minus notation for the runs. The equivalence between the two notations is as follows:

Notation 1	Notation 2
a	$+ - -$
b	$- + -$
c	$- - +$
abc	$+ + +$

Notice that the 2^{3-1} design is formed by selecting only those treatment combinations that have a plus in the ABC column. Thus, ABC is called the *generator*

Table 11-1   Plus and Minus Signs for the 2^3 Factorial Design

Treatment Combination	Factorial Effect							
	I	A	B	C	AB	AC	BC	ABC
a	$+$	$+$	$-$	$-$	$-$	$-$	$+$	$+$
b	$+$	$-$	$+$	$-$	$-$	$+$	$-$	$+$
c	$+$	$-$	$-$	$+$	$+$	$-$	$-$	$+$
abc	$+$	$+$	$+$	$+$	$+$	$+$	$+$	$+$
ab	$+$	$+$	$+$	$-$	$+$	$-$	$-$	$-$
ac	$+$	$+$	$-$	$+$	$-$	$+$	$-$	$-$
bc	$+$	$-$	$+$	$+$	$-$	$-$	$+$	$-$
(1)	$+$	$-$	$-$	$-$	$+$	$+$	$+$	$-$

(a) The principal fraction, $I = +ABC$

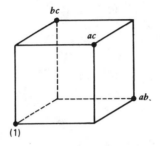

(b) The alternate fraction, $I = -ABC$

Figure 11-1. The two one-half fractions of the 2^3 design. (a) The principal fraction, $I = +ABC$. (b) The alternate fraction, $I = -ABC$.

of this particular fraction. Sometimes we will refer to a generator such as ABC as a *word*. Furthermore, the identity column I is also always plus, so we call

$$I = ABC$$

the *defining relation* for our design. In general, the defining relation for a fractional factorial will always be the set of all columns that are equal to the identity column I.

The treatment combinations in the 2^{3-1} design yield three degrees of freedom that we may use to estimate the main effects. Referring to Table 11-1, we note that the linear combinations of the observations used to estimate the main effects of A, B, and C are

$$\ell_A = \tfrac{1}{2}(a - b - c + abc)$$
$$\ell_B = \tfrac{1}{2}(-a + b - c + abc)$$
$$\ell_C = \tfrac{1}{2}(-a - b + c + abc)$$

It is also easy to verify that the linear combinations of the observations used to estimate the two-factor interactions are

$$\ell_{BC} = \tfrac{1}{2}(a - b - c + abc)$$
$$\ell_{AC} = \tfrac{1}{2}(-a + b - c + abc)$$
$$\ell_{AB} = \tfrac{1}{2}(-a - b + c + abc)$$

Thus, $\ell_A = \ell_{BC}$, $\ell_B = \ell_{AC}$, and $\ell_C = \ell_{AB}$; consequently, it is impossible to differentiate between A and BC, B and AC, and C and AB. In fact, when we estimate A, B, and C we are *really* estimating $A + BC$, $B + AC$, and $C + AB$. Two or more effects that have this property are called *aliases*. In our example A and BC are aliases, B and AC are aliases, and C and AB are aliases. We indicate this by the notation $\ell_A \to A + BC$, $\ell_B \to B + AC$, and $\ell_C \to C + AB$.

The alias structure for this design may be easily determined by using the defining relation $I = ABC$. Multiplying any column by the defining relation yields the aliases for that effect. In our example, this yields as the alias of A

$$A \cdot I = A \cdot ABC = A^2BC$$

or, since the square of any column is just the identity I,

$$A = BC$$

Similarly, we find the aliases of B and C as

$$B \cdot I = B \cdot ABC$$
$$B = AB^2C = AC$$

and

$$C \cdot I = C \cdot ABC$$
$$C = ABC^2 = AB$$

This one-half fraction, with $I = +ABC$, is usually called the *principal fraction*.

Now suppose that we had chosen the *other* one-half fraction, that is, the treatment combinations in Table 11-1 associated with minus in the ABC column. This *alternate* or *complimentary* one-half fraction is shown in Figure 11-1b. Notice that it consists of the runs

Notation 1	Notation 2
(1)	$-\ -\ -$
ab	$+\ +\ -$
ac	$+\ -\ +$
bc	$-\ +\ +$

The defining relation for this design is

$$I = -ABC$$

The linear combination of the observations, say ℓ'_A, ℓ'_B, and ℓ'_C, from the alternate fraction give us

$$\ell'_A \to A - BC$$
$$\ell'_B \to B - AC$$
$$\ell'_C \to C - AB$$

Thus, when we estimate A, B, and C with this particular fraction, we are really estimating $A - BC$, $B - AC$, and $C - AB$.

In practice, it does not matter which fraction is actually used. Both fractions belong to the same *family;* that is, the two one-half fractions form a complete 2^3 design. This is easily seen by reference to the a and b parts of Figure 11-1.

Suppose that after running one of the one-half fractions of the 2^3 design, the other one was also run. Thus, all 8 runs associated with the full 2^3 are now available. We may now obtain de-aliased estimates of all the effects by analyzing the 8 runs as a full 2^3 design in two blocks of 4 runs each. This could also be done by adding and subtracting the linear combination of effects from the two individual fractions. For example, consider $\ell_A \rightarrow A + BC$ and $\ell'_A \rightarrow A - BC$. This implies that

$$\tfrac{1}{2}(\ell_A + \ell'_A) = \tfrac{1}{2}(A + BC + A - BC) \rightarrow A$$

and that

$$\tfrac{1}{2}(\ell_A - \ell'_A) = \tfrac{1}{2}(A + BC - A + BC) \rightarrow BC$$

Thus, for all three pairs of linear combinations, we would obtain the following:

i	From $\tfrac{1}{2}(\ell_i + \ell'_i)$	From $\tfrac{1}{2}(\ell_i - \ell'_i)$
A	A	BC
B	B	AC
C	C	AB

Design Resolution ▪ The preceding 2^{3-1} design is called a *resolution III design.* In such a design, main effects are aliased with two-factor interactions. A design is of resolution R if no p-factor effect is aliased with another effect containing less than $R - p$ factors. We usually employ a Roman numeral subscript to denote design resolution; thus, the one-half fraction of the 2^3 design with the defining relation $I = ABC$ (or $I = -ABC$) is a 2_{III}^{3-1} design.

Designs of resolution III, IV, and V are particularly important. The definitions of these designs and an example of each follow.

1. **Resolution III designs.** These are designs in which no main effects are aliased with any other main effect, but main effects are aliased with two-factor interactions and two-factor interactions may be aliased with each other. The 2^{3-1} design in Table 11-1 is of resolution III (2_{III}^{3-1}).

2. **Resolution IV designs.** These are designs in which no main effect is aliased with any other main effect *or* with any two-factor interaction, but two-factor interactions are aliased with other. A 2^{4-1} design with $I = ABCD$ is of resolution IV (2_{IV}^{4-1}).

3. **Resolution V designs.** These are designs in which no main effect or two-factor interaction is aliased with any other main effect or two-factor interaction, but two-factor interactions are aliased with three-factor interactions. A 2^{5-1} design with $I = ABCDE$ is of resolution V (2_V^{5-1}).

In general, the resolution of a two-level fractional factorial design is equal to the smallest number of letters in any word in the defining relation. Consequently, we could call the preceding design types three-letter, four-letter, and five-letter designs, respectively. We usually like to employ fractional designs that have the highest possible resolution consistent with the degree of fractionation required. The higher the resolution, the less restrictive the assumptions that are required regarding which interactions are negligible in order to obtain a unique interpretation of the data.

Constructing One-Half Fractions ▪ A one-half fraction of the 2^k design of the highest resolution may be constructed by writing down a *basic design* consisting of the runs for a *full* 2^{k-1} factorial and then adding the kth factor by identifying its plus and minus levels with the plus and minus signs of the highest-order interaction $ABC \ldots (K - 1)$. Therefore, the 2_{III}^{3-1} fractional factorial is obtained by writing down the full 2^2 factorial as the basic design and then equating factor C to the AB interaction. The alternate fraction would be obtained by equating factor C to the $-AB$ interaction. This approach is illustrated in Table 11-2. Notice that the basic design always has the right number of runs (rows), but it is missing one column. The generator $I = ABC \ldots K$ is then solved for the missing column (K) so that $K = ABC \ldots (K - 1)$ defines the product of plus and minus signs to use in each row to produce the levels for the kth factor.

Note that *any* interaction effect could be used to generate the column for the kth factor. However, using any effect other than $ABC \ldots (K - 1)$ will not produce a design of the highest possible resolution.

Another way to view the construction of a one-half fraction is to partition the runs into two blocks with the highest-order interaction $ABC \ldots K$ confounded. Each block is a 2^{k-1} fractional factorial design of the highest resolution.

Table 11-2 The Two One-Half Fractions of the 2^3 Design

Run	Full 2^2 Factorial (Basic Design)		2_{III}^{3-1}, $I = ABC$			2_{III}^{3-1}, $I = -ABC$		
	A	B	A	B	$C = AB$	A	B	$C = -AB$
1	−	−	−	−	+	−	−	−
2	+	−	+	−	−	+	−	+
3	−	+	−	+	−	−	+	+
4	+	+	+	+	+	+	+	−

Projection of Fractions into Factorials ▪ Any fractional factorial design of resolution R contains complete factorial designs (possibly replicated factorials) in any subset of $R - 1$ factors. This is an important and useful concept. For example, if an experimenter has several factors of potential interest but believes that only $R - 1$ of them have important effects, then a fractional factorial design of resolution R is the appropriate choice of design. If the experimenter is correct, then the fractional factorial design of resolution R will project into a full factorial in the $R - 1$ significant factors. This process is illustrated in Figure 11-2 for the 2_{III}^{3-1} design, which projects into a 2^2 design in every subset of two factors.

Since the maximum possible resolution of a one-half fraction of the 2^k design is $R = k$, every 2^{k-1} design will project into a full factorial in any $(k - 1)$ of the original k factors. Furthermore, a 2^{k-1} design may be projected into two replicates of a full factorial in any subset of $k - 2$ factors, four replicates of a full factorial in any subset of $k - 3$ factors, and so on.

Sequences of Fractional Factorials ▪ Using fractional factorial designs often leads to great economy and efficiency in experimentation, particularly if the runs can be made sequentially. For example, suppose that we were investigating $k = 4$ factors ($2^4 = 16$ runs). It is almost always preferable to run a 2_{IV}^{4-1} fractional design (8 runs), analyze the results, and then decide on the best set of runs to perform next. If it is necessary to resolve ambiguities, we can always run the alternate fraction and complete the 2^4 design. When this method is used to complete the design, both one-half fractions represent blocks of the complete design with the highest-order interaction confounded with blocks (here $ABCD$ would be confounded). Thus, sequential experimentation has the result of losing information only on the highest-order interaction. Alternatively, in many cases we learn enough from the one-half fraction to proceed to the next stage of experimentation, which might involve adding or removing factors, changing responses, or varying some of the factors over new ranges.

Figure 11-2. Projection of a 2_{III}^{3-1} design into three 2^2 designs.

Table 11-3   The 2_{IV}^{4-1} Design with the Defining Relation $I = ABCD$

Run	Basic Design				Treatment Combination	Filtration Rate
	A	B	C	D = ABC		
1	−	−	−	−	(1)	45
2	+	−	−	+	ad	100
3	−	+	−	+	bd	45
4	+	+	−	−	ab	65
5	−	−	+	+	cd	75
6	+	−	+	−	ac	60
7	−	+	+	−	bc	80
8	+	+	+	+	abcd	96

Example 11-1

Consider the filtration rate experiment in Example 9-2. The original design, shown in Table 9-7, is a single replicate of the 2^4 design. In this example, we found that the main effects A, C, and D and the interactions AC and AD were different from zero. We will now return to this experiment and simulate what would have happened if a half-fraction of the 2^4 design had been run instead of the full factorial.

We will use the 2^{4-1} design with $I = ABCD$, as this choice of generator will result in a design of the highest possible resolution (IV). To construct the design, we first write down the basic design, which is a 2^3 design, as shown in the first three columns of Table 11-3. This basic design has the necessary number of runs (8), but only three columns (factors). To find the fourth factor levels, solve $I = ABCD$ for D, or $D = ABC$. Thus, the level of D in each run is the product of the plus and minus signs in columns A, B, and C. The process is illustrated in Table 11-3. Since the generator $ABCD$ is positive, this 2_{IV}^{4-1} design is the principal fraction. The design is shown graphically in Figure 11-3.

Using the defining relation, we note that each main effect is aliased with a three-factor interaction; that is, $A = A^2BCD = BCD$, $B = AB^2CD = ACD$, $C = ABC^2D = ABD$, and $D = ABCD^2 = ABC$. Furthermore, every two-factor interaction is aliased with another two-factor interaction. These alias relationships are $AB = CD$, $AC = BD$, and $BC = AD$. The four main effects plus the three two-factor interaction alias pairs account for the seven degrees of freedom for the design.

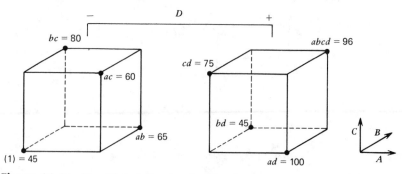

Figure 11-3.   The 2_{IV}^{4-1} design for the filtration rate experiment of Example 11-1.

Table 11-4   Estimates of Effects and Aliases from Example 11-1[a]

Estimate	Alias Structure
$\ell_A = \quad 19.00$	$\ell_A \rightarrow$ **A** $+ BCD$
$\ell_B = \quad 1.50$	$\ell_B \rightarrow B + ACD$
$\ell_C = \quad 14.00$	$\ell_C \rightarrow$ **C** $+ ABD$
$\ell_D = \quad 16.50$	$\ell_D \rightarrow$ **D** $+ ABC$
$\ell_{AB} = \quad -1.00$	$\ell_{AB} \rightarrow AB + CD$
$\ell_{AC} = \quad -18.50$	$\ell_{AC} \rightarrow$ **AC** $+ BD$
$\ell_{BC} = \quad 19.00$	$\ell_{BC} \rightarrow BC +$ **AD**

[a] Significant effects are shown in boldface type.

At this point, we would normally randomize the eight runs and perform the experiment. Since we have already run the full 2^4 design, we will simply select the eight observed filtration rates from Example 9-2 that correspond to the runs in the 2_{IV}^{4-1} design. These observations are shown in the last column of Table 11-3 and are also shown in Figure 11-3.

The estimates of the effects obtained from this 2_{IV}^{4-1} design are shown in Table 11-4. To illustrate the calculations, the linear combination of observations associated with the A effect is

$$\ell_A = \tfrac{1}{4}(-45 + 100 - 45 + 65 - 75 + 60 - 80 + 96) = 19.00 \rightarrow A + BCD$$

whereas for the AB effect, we would obtain

$$\ell_{AB} = \tfrac{1}{4}(45 - 100 - 45 + 65 + 75 - 60 - 80 + 96) = -1.00 \rightarrow AB + CD$$

From inspection of the information in Table 11-4, it is not unreasonable to conclude that the main effects A, C, and D are large and that the AC and AD interactions are also significant. This agrees with the conclusions from the analysis of the complete 2^4 design in Example 9-2.

Since factor B is not significant, we may drop it from consideration. Consequently, we may project this 2_{IV}^{4-1} design into a single replicate of the 2^3 design in factors A, C, and D, as shown in Figure 11-4. Visual examination of this cube plot makes us more comfortable with the conclusions reached above. Notice that if the temperature (A) is at the low level, the concentration (C) has a large positive effect, whereas if the temperature is at the high level, the concentration has a very small effect. This is likely due to an AC interaction. Furthermore, if the temperature is at the low level, the effect of the stirring rate (D) is negligible, whereas if the temperature is at the high level, the stirring rate has a large positive effect. This is likely due to the AD interaction tentatively identified previously.

Now suppose that the experimenter decided to run the alternate fraction, given by $I = -ABCD$. It is straightforward to show that the design and the responses are as follows:

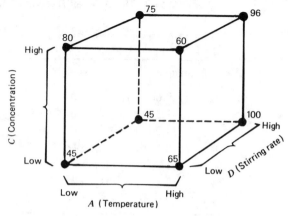

Figure 11-4. Projection of the 2_{IV}^{4-1} design into a 2^3 design in A, C, and D for Example 11-1.

Run	A	B	C	D = −ABC	Treatment Combination	Filtration Rate
1	−	−	−	+	d	43
2	+	−	−	−	a	71
3	−	+	−	−	b	48
4	+	+	−	+	abd	104
5	−	−	+	−	c	68
6	+	−	+	+	acd	86
7	−	+	+	+	bcd	70
8	+	+	+	−	abc	65

The linear combinations of observations obtained from this alternate fraction are

$$\ell'_A = 24.25 \rightarrow A - BCD$$
$$\ell'_B = 4.75 \rightarrow B - ACD$$
$$\ell'_C = 5.75 \rightarrow C - ABD$$
$$\ell'_D = 12.75 \rightarrow D - ABC$$
$$\ell'_{AB} = 1.25 \rightarrow AB - CD$$
$$\ell'_{AC} = -17.75 \rightarrow AC - BD$$
$$\ell'_{BC} = -14.25 \rightarrow BC - AD$$

These estimates may be combined with those obtained from the original one-half fraction to yield the following estimates of the effects:

i	From $\frac{1}{2}(\ell_i + \ell'_i)$	From $\frac{1}{2}(\ell_i - \ell'_i)$
A	21.63 → A	−2.63 → BCD
B	3.13 → B	−1.63 → ACD
C	9.75 → C	4.13 → ABD
D	14.63 → D	1.88 → ABC
AB	0.13 → AB	−1.13 → CD
AC	−18.13 → AC	−0.38 → BD
BC	2.38 → BC	16.63 → AD

These estimates agree exactly with those from the original analysis of the data as a single replicate of a 2^4 factorial design, as reported in Example 9-2.

∎

Example 11-2

A 2^{5-1} Design Used for Process Improvement

Five factors in a manufacturing process for an integrated circuit were investigated in a 2^{5-1} design with the objective of improving the process yield. The five factors were A = aperture setting (small, large), B = exposure time (20 percent below nominal, 20 percent above nominal), C = development time (30 s, 45 s), D = mask dimension (small, large), and E = etch time (14.5 min, 15.5 min). The construction of the 2^{5-1} design is shown in Table 11-5. Notice that the design was constructed by writing down the basic design having 16 runs (a 2^4 design in A, B, C, and D), selecting $ABCDE$ as the generator, and then setting the levels of the fifth factor $E = ABCD$. Figure 11-5 gives a pictorial representation of the design.

The defining relation for the design is $I = ABCDE$. Consequently, every main effect is aliased with a four-factor interaction (for example, $\ell_A \rightarrow A + BCDE$), and every two-factor interaction is aliased with a three-factor interaction (for example, $\ell_{AB} \rightarrow AB + CDE$). Thus, the design is of resolution V. We would expect this 2^{5-1} design to provide excellent information concerning the main effects and two-factor interactions.

Figure 11-6 presents a normal probability plot of the effect estimates from this experiment. The main effects of A, B, and C and the AB interaction are large. Remember that, because of aliasing, these effects are really $A + BCDE$, $B + ACDE$, $C + ABDE$, and $AB + CDE$. However, since it seems plausible that three-factor and higher interactions are negligible, we feel safe in concluding that only A, B, C, and AB are important effects.

Table 11-5 A 2^{5-1} Design for Example 11-2

Run	A	B	C	D	$E = ABCD$	Treatment Combination	Yield
1	−	−	−	−	+	e	8
2	+	−	−	−	−	a	9
3	−	+	−	−	−	b	34
4	+	+	−	−	+	abe	52
5	−	−	+	−	−	c	16
6	+	−	+	−	+	ace	22
7	−	+	+	−	+	bce	45
8	+	+	+	−	−	abc	60
9	−	−	−	+	−	d	6
10	+	−	−	+	+	ade	10
11	−	+	−	+	+	bde	30
12	+	+	−	+	−	abd	50
13	−	−	+	+	+	cde	15
14	+	−	+	+	−	acd	21
15	−	+	+	+	−	bcd	44
16	+	+	+	+	+	$abcde$	63

Table 11-6 summarizes the analysis of variance for this experiment. The model sum of squares is $SS_{Model} = SS_A + SS_B + SS_C + SS_{AB} = 5747.25$, and this accounts for over 99 percent of the total variability in yield. Figure 11-7 presents a normal probability plot of the residuals, and Figure 11-8 is a plot of the residuals versus the predicted values. Both plots are satisfactory.

The three factors A, B, and C have large positive effects. The AB or aperture–exposure time interaction is plotted in Figure 11-9. This plot confirms that the yields are higher when both A and B are at the high level.

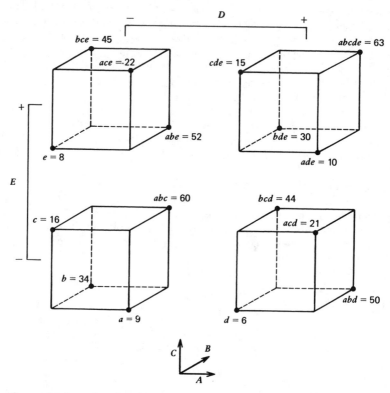

Figure 11-5. The 2_V^{5-1} design for Example 11-2.

Table 11-6 Analysis of Variance for Example 11-2

Source of Variation	Sum of Squares	Degrees of Freedom	Mean Square	F_0
A (Aperture)	495.0625	1	495.0625	193.20
B (Exposure time)	4590.0625	1	4590.0625	1791.24
C (Development time)	473.0625	1	473.0625	184.61
AB	189.0625	1	189.0625	73.78
Error	28.1875	11	2.5625	
Total	5775.4375	15		

The 2^{5-1} design will collapse into two replicates of a 2^3 design in any three of the original five factors. (Looking at Figure 11-5 will help you visualize this.) Figure 11-10 is a cube plot in the factors A, B, and C with the average yields superimposed on the eight corners. It is clear from inspection of the cube plot that highest yields

Figure 11-6.   Normal probability plot of effects for Example 11-2.

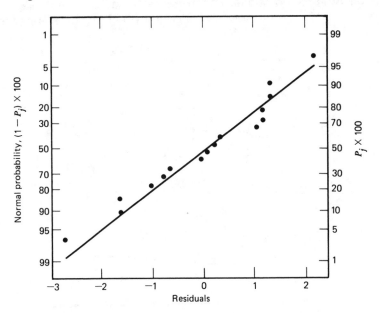

Figure 11-7.   Normal probability plot of the residuals for Example 11-2.

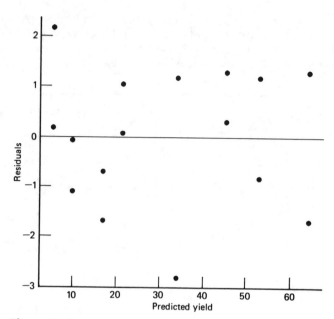

Figure 11-8. Plot of residuals versus predicted yield for Example 11-2.

are achieved with A, B, and C all at the high level. Factors D and E have little effect on average process yield and may be set to values that optimize other objectives (such as cost).

■

Yates' Algorithm for Fractional Factorials ▪ We may use Yates' algorithm for the analysis of a 2^{k-1} fractional factorial design by initially considering the data

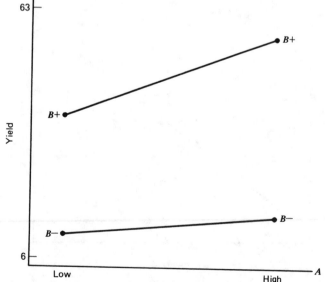

Figure 11-9. Aperture–exposure time interaction for Example 11-2.

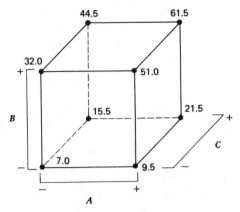

Figure 11-10. Projection of the 2_V^{5-1} design in Example 11-2 into two replicates of a 2^3 design in the factors A, B, and C.

as having been obtained from a full factorial in $k - 1$ variables. The treatment combinations for this full factorial are listed in standard order, and then an additional letter (or letters) is added in parentheses to these treatment combinations to produce the actual treatment combinations run. Yates' algorithm then proceeds as usual. The actual effects estimated are identified by multiplying the effects associated with the treatment combinations in the full 2^{k-1} design by the defining relation of the 2^{k-1} fractional factorial.

The procedure is demonstrated in Table 11-7 using the data from Example 11-1. The data are arranged as a full 2^3 design in the factors A, B, and C. Then the letter d is added in parentheses to yield the actual treatment combinations. The effect estimated by, say, the second row in this table is $A + BCD$ since A and BCD are aliases.

11-3 THE ONE-QUARTER FRACTION OF THE 2^k DESIGN

For a moderately large number of factors, smaller fractions of the 2^k design are frequently useful. Consider a one-quarter fraction of the 2^k design. This design contains 2^{k-2} runs and is usually called a 2^{k-2} fractional factorial.

Table 11-7 Yates' Algorithm for the 2_{IV}^{4-1} Fractional Factorial in Example 11-1

Treatment Combination	Response	(1)	(2)	(3)	Effect	Estimate of Effect $2 \times (3)/N$
(1)	45	145	255	566	—	—
a(d)	100	110	311	76	$A + BCD$	19.00
b(d)	45	135	75	6	$B + ACD$	1.50
ab	65	176	1	−4	$AB + CD$	−1.00
c(d)	75	55	−35	56	$C + ABD$	14.00
ac	60	20	41	−74	$AC + BD$	−18.50
bc	80	−15	−35	76	$BC + AD$	19.00
abc(d)	96	16	31	66	$ABC + D$	16.50

The 2^{k-2} design may be constructed by first writing down a basic design consisting of the runs associated with a full factorial in $k-2$ factors and then associating the two additional columns with appropriately chosen interactions involving the first $k-2$ factors. Thus, a one-quarter fraction of the 2^k design has two generators. If P and Q represent the generators chosen, then $I = P$ and $I = Q$ are called the *generating relations* for the design. The signs of P and Q (either $+$ or $-$) determine which one of the one-quarter fractions is produced. All four fractions associated with the choice of generators $\pm P$ and $\pm Q$ are members of the same *family*. The fraction for which *both* P and Q are positive is the principal fraction.

The *complete* defining relation for the design consists of all the columns that are equal to the identity column I. These will consist of P, Q, and their *generalized interaction PQ*; that is, the defining relation is $I = P = Q = PQ$. We call the elements P, Q, and PQ in the defining relation *words*. The aliases of any effect are produced by the multiplication of the column for that effect by each word in the defining relation. Clearly, each effect has *three* aliases. The experimenter should be careful in choosing the generators so that potentially important effects are not aliased with each other.

As an example, consider the 2^{6-2} design. Suppose we choose $I = ABCE$ and $I = BCDF$ as the design generators. Now the generalized interaction of the generators $ABCE$ and $BCDF$ is $ADEF$; therefore, the complete defining relation for this design is

$$I = ABCE = BCDF = ADEF$$

Consequently, this design is of resolution IV. To find the aliases of any effect (e.g., A), multiply that effect by each word in the defining relation. For A, this produces

$$A = BCE = ABCDF = DEF$$

It is easy to verify that every main effect is aliased by three-factor and five-factor interactions, whereas two-factor interactions are aliased with each other and with higher-order interactions. Thus, when we estimate A, for example, we are really estimating $A + BCE + DEF + ABCDF$. The complete alias structure of this design is shown in Table 11-8. If three-factor and higher interactions are negli-

Table 11-8   Alias Structure for the 2_{IV}^{6-2} Design with $I = ABCE = BCDF = ADEF$

$A = BCE = DEF = ABCDF$	$AB = CE = ACDF = BDEF$
$B = ACE = CDF = ABDEF$	$AC = BE = ABDF = CDEF$
$C = ABE = BDF = ACDEF$	$AD = EF = BCDE = ABCF$
$D = BCF = AEF = ABCDE$	$AE = BC = DF = ABCDEF$
$E = ABC = ADF = BCDEF$	$AF = DE = BCEF = ABCD$
$F = BCD = ADE = ABCEF$	$BD = CF = ACDE = ABEF$
	$BF = CD = ACEF = ABDE$
$ABD = CDE = ACF = BEF$	
$ACD = BDE = ABF = CEF$	

gible, this design gives clear estimates of the main effects.

To construct the design, first write down the *basic design,* which consists of the 16 runs for a full $2^{6-2} = 2^4$ design in A, B, C, and D. Then the two factors E and F are added by associating their plus and minus levels with the plus and minus signs of the interactions ABC and BCD, respectively. This procedure is shown in Table 11-9.

Another way to construct this design is to derive the four blocks of the 2^6 design with $ABCE$ and $BCDF$ confounded and then choose the block with treatment combinations that are positive on $ABCE$ and $BCDF$. This would be a 2^{6-2} fractional factorial with generating relations $I = ABCE$ and $I = BCDF$, and since both generators $ABCE$ and $BCDF$ are positive, this is the principal fraction.

There are, of course, three alternate fractions of this particular 2_{IV}^{6-2} design. They are the fractions with generating relationships $I = ABCE$ and $I = -BCDF$; $I = -ABCE$ and $I = BCDF$; and $I = -ABCE$ and $I = -BCDF$. These fractions may be easily constructed by the method shown in Table 11-9. For example, if we wish to find the fraction for which $I = ABCE$ and $I = -BCDF$, then in the last column of Table 11-9 we set $F = -BCD$, and the column of levels for factor F becomes

$$+ + - - - - + + - - + + + + - -$$

The complete defining relation for this alternative fraction is $I = ABCE = -BCDF = -ADEF$. Certain signs in the alias structure in Table 11-8 are now changed; for instance, the aliases of A are $A = BCE = -DEF = -ABCDF$.

Table 11-9 Construction of the 2_{IV}^{6-2} Design with the Generators $I = ABCE$ and $I = BCDF$

Run	A	B	C	D	$E = ABC$	$F = BCD$
1	−	−	−	−	−	−
2	+	−	−	−	+	−
3	−	+	−	−	+	+
4	+	+	−	−	−	+
5	−	−	+	−	+	+
6	+	−	+	−	−	+
7	−	+	+	−	−	−
8	+	+	+	−	+	−
9	−	−	−	+	−	+
10	+	−	−	+	+	+
11	−	+	−	+	+	−
12	+	+	−	+	−	−
13	−	−	+	+	+	−
14	+	−	+	+	−	−
15	−	+	+	+	−	+
16	+	+	+	+	+	+

Thus, the linear combination of the observations ℓ_A actually estimates $A + BCE - DEF - ABCDF$.

Finally, note that the 2_{IV}^{6-2} fractional factorial collapses to a single replicate of a 2^4 design in any subset of four factors that is not a word in the defining relation. It also collapses to a replicated one-half fraction of a 2^4 in any subset of four factors that is a word in the defining relation. Thus, the design in Table 11-9 becomes two replicates of a 2^{4-1} in the factors $ABCE$, $BCDF$, and $ADEF$, since these are the words in the defining relation. There are 12 other combinations of the six factors, such as $ABCD$, $ABCF$, and so on, for which the design projects to a single replicate of the 2^4. This design also collapses to two replicates of a 2^3 in *any* subset of three of the six factors or four replicates of a 2^2 in any subset of two factors.

In general, any 2^{k-2} fractional factorial design can be collapsed into either a full factorial or a fractional factorial in some subset of $r \leq k - 2$ of the original factors. Those subsets of variables that form full factorials are not words in the complete defining relation.

Example 11-3

Parts manufactured in an injection molding process are showing excessive shrinkage. This is causing problems in assembly operations upstream from the injection molding area. A quality improvement team has decided to use a designed experiment to study the injection molding process so that shrinkage can be reduced. The team decides to investigate six factors—mold temperature (A), screw speed (B), holding time (C), cycle time (D), gate size (E), and holding pressure (F)—each at two levels, with the

Table 11-10   A 2_{IV}^{6-2} Design for the Injection Molding Experiment in Example 11-3

Run	A	B	C	D	E = ABC	F = BCD	Observed Shrinkage (× 10)
1	−	−	−	−	−	−	6
2	+	−	−	−	+	−	10
3	−	+	−	−	+	+	32
4	+	+	−	−	−	+	60
5	−	−	+	−	+	+	4
6	+	−	+	−	−	+	15
7	−	+	+	−	−	−	26
8	+	+	+	−	+	−	60
9	−	−	−	+	−	+	8
10	+	−	−	+	+	+	12
11	−	+	−	+	+	−	34
12	+	+	−	+	−	−	60
13	−	−	+	+	+	−	16
14	+	−	+	+	−	−	5
15	−	+	+	+	−	+	37
16	+	+	+	+	+	+	52

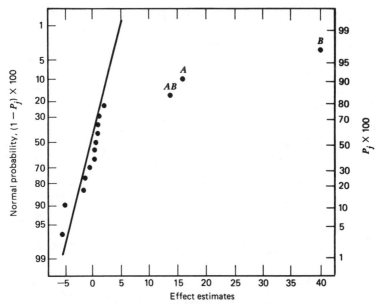

Figure 11-11. Normal probability plot of effects for Example 11-3.

objective of learning how each factor affects shrinkage and also, something about how the factors interact.

The team decides to use the 16-run two-level fractional factorial design in Table 11-9. The design is shown again in Table 11-10, along with the observed shrinkage ($\times$ 10) for the test part produced at each of the 16 runs in the design.

A normal probability plot of the effect estimates from this experiment is shown in Figure 11-11. The only large effects are A (mold temperature), B (screw speed), and the AB interaction. In light of the alias relationships in Table 11-8, it seems reasonable to adopt these conclusions tentatively. The plot of the AB interaction in Figure 11-12 shows that the process is very insensitive to temperature if the screw speed is at the low level but very sensitive to temperature if the screw speed is at the high level. With the screw speed at the low level, the process should produce an average shrinkage of around 10 percent regardless of the temperature level chosen.

Based on this initial analysis, the team decides to set both the mold temperature and the screw speed at the low level. This set of conditions will reduce the *mean* shrinkage of parts to around 10 percent. However, the variability in shrinkage from part to part is still a potential problem. In effect, the mean shrinkage can be adequately reduced by the above modifications; however, the part-to-part variability in shrinkage over a production run could still cause problems in assembly. One way to address this issue is to see if any of the process factors affect the *variability* in parts shrinkage.

Figure 11-13 presents the normal probability plot of the residuals. This plot appears satisfactory. The plots of residuals versus each factor were then constructed. One of these plots, that for residuals versus factor C (holding time) is shown in Figure 11-14. The plot reveals that there is much less scatter in the residuals at the low holding time than at the high holding ltime. These residuals were obtained in the

Figure 11-12.   Plot of *AB* (mold temperature–screw speed) interaction for Example 11-3.

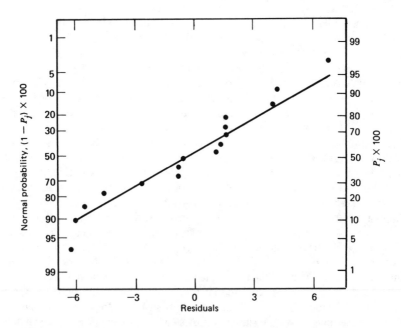

Figure 11-13.   Normal probability plot of residuals for Example 11-3.

usual way from a model for predicted shrinkage

$$\hat{y} = \hat{\beta}_0 + \hat{\beta}_1 x_1 + \hat{\beta}_2 x_2 + \hat{\beta}_{12} x_1 x_2$$
$$= 27.3125 + 6.9375 x_1 + 17.8125 x_2 + 5.9375 x_1 x_2$$

where x_1, x_2, and $x_1 x_2$ are coded variables that correspond to the factors A and B and the AB interaction. The residuals are then

$$e = y - \hat{y}$$

The regression model used to produce the residuals essentially removes the *location* effects of A, B, and AB from the data; the residuals therefore contain information about unexplained variability. Figure 11-14 indicates that there is a *pattern* in the variability and that the variability in the shrinkage of parts may be smaller when the holding time is at the low level.

This is further amplified by the analysis of residuals shown in Table 11-11. In this table, the residuals are arranged at the low ($-$) and high ($+$) levels of each factor, and the standard deviation of the residuals at the low and high levels of each factor have been calculated. Note that the standard deviation of the residuals with C at the low level $[S(C^-) = 1.63]$ is considerably smaller than the standard deviation of the residuals with C at the high level $[S(C^+) = 5.70]$.

The bottom line of Table 11-12 presents the statistic

$$F_i^* = \ln \frac{S^2(i^+)}{S^2(i^-)}$$

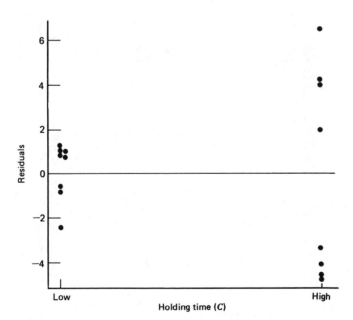

Figure 11-14.   Residuals versus holding time (c) for Example 11-3.

Table 11-11   Calculation of Dispersion Effects for Example 11-3

Run	A	B	AB = CE	C	AC = BE	AE = BC	E	D	AD = EF	BD = CE	ABD	BF = CD	ACD	F	AF = DE	Residual
1	−	−	+	−	+	+	−	−	+	+	−	+	−	−	+	−2.50
2	+	−	−	−	−	+	+	−	−	+	+	+	+	−	−	−0.50
3	−	+	−	−	+	−	+	−	+	−	+	+	−	+	−	−0.25
4	+	+	+	−	−	−	−	−	−	−	−	+	+	+	+	2.00
5	−	−	+	+	−	−	−	−	+	+	+	−	+	+	−	−4.50
6	+	−	−	+	+	+	−	−	−	+	+	−	+	−	+	4.50
7	−	+	−	+	−	+	+	−	+	−	−	−	−	−	+	−6.25
8	+	+	+	+	+	+	+	−	−	−	+	−	+	−	+	2.00
9	−	−	+	−	+	+	+	+	−	−	+	−	+	+	−	−0.50
10	+	−	−	−	−	−	−	+	+	+	−	−	−	+	+	1.50
11	−	+	−	−	+	−	+	+	−	+	−	−	+	−	+	1.75
12	+	+	+	−	−	−	−	+	+	−	+	−	−	−	−	2.00
13	−	−	+	+	+	−	+	+	+	−	−	+	+	−	−	7.50
14	+	−	−	+	+	+	−	+	−	−	+	+	+	+	+	−5.50
15	−	+	−	+	−	+	−	+	−	+	−	+	−	+	−	4.75
16	+	+	+	+	+	+	+	+	+	+	+	+	+	+	+	−6.00
$S(i^+)$	3.80	4.01	4.33	5.70	3.68	3.85	4.17	4.64	3.39	4.01	4.72	4.71	3.50	3.88	4.87	
$S(i^-)$	4.60	4.41	4.10	1.63	4.53	4.33	4.25	3.59	2.75	4.41	3.51	3.65	3.12	4.52	3.40	
F_i^*	−0.38	−0.19	0.11	2.50	−0.42	−0.23	−0.04	0.51	0.42	−0.19	0.59	0.51	0.23	−0.31	0.72	

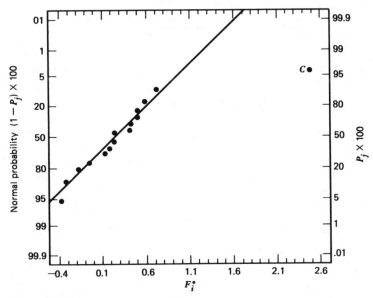

Figure 11-15.   Normal probability plot of the dispersion effects F_i^* for Example 11-3.

Recall that if the variances of the residuals at the high $(+)$ and low $(-)$ levels of factor i are equal, then this ratio is approximately normally distributed with mean zero and unit standard deviation, so it can be used to judge the difference in the response variability at the two levels of factor i. Since the ratio F_C^* exceeds the upper $2\frac{1}{2}$ percent point $Z_{.025} = 1.96$, we would conclude that the apparent *dispersion* or variability effect observed in Figure 11-14 is real. Thus, setting the holding time at its low level would contribute to reducing the variability in shrinkage from part to part during a production run. Figure 11-15 presents a normal probability plot of the F_i^* values in Table 11-12; this also indicates that factor C has a large dispersion effect.

Figure 11-16 shows the data from this experiment projected onto a cube in the factors A, B, and C. The average observed shrinkage and the range of observed shrinkage are shown at each corner of the cube. From inspection of this figure, we

Figure 11-16.   Average shrinkage and range of shrinkage in factors A, B, and C for Example 11-3.

Table 11-12 Selected 2^{k-p} Fractional Factorial Designs

Number of Factors k	Fraction	Number of Runs	Design Generators
3	2_{III}^{3-1}	4	$C = \pm AB$
4	2_{IV}^{4-1}	8	$D = \pm ABC$
5	2_{V}^{5-1}	16	$E = \pm ABCD$
	2_{III}^{5-2}	8	$D = \pm AB$
			$E = \pm AC$
6	2_{VI}^{6-1}	32	$F = \pm ABCDE$
	2_{IV}^{6-2}	16	$E = \pm ABC$
			$F = \pm BCD$
	2_{III}^{6-3}	8	$D = \pm AB$
			$E = \pm AC$
			$F = \pm BC$
7	2_{VII}^{7-1}	64	$G = \pm ABCDEF$
	2_{IV}^{7-2}	32	$F = \pm ABCD$
			$G = \pm ABDE$
	2_{IV}^{7-3}	16	$E = \pm ABC$
			$F = \pm BCD$
			$G = \pm ACD$
	2_{III}^{7-4}	8	$D = \pm AB$
			$E = \pm AC$
			$F = \pm BC$
			$G = \pm ABC$
8	2_{V}^{8-2}	64	$G = \pm ABCD$
			$H = \pm ABEF$
	2_{IV}^{8-3}	32	$F = \pm ABC$
			$G = \pm ABD$
			$H = \pm BCDE$
	2_{IV}^{8-4}	16	$E = \pm BCD$
			$F = \pm ACD$
			$G = \pm ABC$
			$H = \pm ABD$
9	2_{VI}^{9-2}	128	$H = \pm ACDFG$
			$J = \pm BCEFG$
	2_{IV}^{9-3}	64	$G = \pm ABCD$
			$H = \pm ACEF$
			$J = \pm CDEF$
	2_{IV}^{9-4}	32	$F = \pm BCDE$
			$G = \pm ACDE$
			$H = \pm ABDE$
			$J = \pm ABCE$
	2_{III}^{9-5}	16	$E = \pm ABC$
			$F = \pm BCD$
			$G = \pm ACD$
			$H = \pm ABD$
			$J = \pm ABCD$

Table 11-12   Continued

Number of Factors k	Fraction	Number of Runs	Design Generators
10	2_V^{10-3}	128	$H = \pm ABCG$
			$J = \pm ACDE$
			$K = \pm ACDF$
	2_{IV}^{10-4}	64	$G = \pm BCDF$
			$H = \pm ACDF$
			$J = \pm ABDE$
			$K = \pm ABCE$
	2_{IV}^{10-5}	32	$F = \pm ABCD$
			$G = \pm ABCE$
			$H = \pm ABDE$
			$J = \pm ACDE$
			$K = \pm BCDE$
	2_{III}^{10-6}	16	$E = \pm ABC$
			$F = \pm BCD$
			$G = \pm ACD$
			$H = \pm ABD$
			$J = \pm ABCD$
			$K = \pm AB$
11	2_{IV}^{11-5}	64	$G = \pm CDE$
			$H = \pm ABCD$
			$J = \pm ABF$
			$K = \pm BDEF$
			$L = \pm ADEF$
	2_{IV}^{11-6}	32	$F = \pm ABC$
			$G = \pm BCD$
			$H = \pm CDE$
			$J = \pm ACD$
			$K = \pm ADE$
			$L = \pm BDE$
	2_{III}^{11-7}	16	$E = \pm ABC$
			$F = \pm BCD$
			$G = \pm ACD$
			$H = \pm ABD$
			$J = \pm ABCD$
			$K = \pm AB$
			$L = \pm AC$

see that running the process with the screw speed (B) at the low level is the key to reducing average parts shrinkage. If B is low, virtually any combination of temperature (A) and holding time (C) will result in low values of average parts shrinkage. However, from examining the ranges of the shrinkage values at each corner of the cube, it is immediately clear that setting the holding time (C) at the low level is the only reasonable choice if we wish to keep the part-to-part variability in shrinkage low during a production run.   ∎

11-4   THE GENERAL 2^{k-p} FRACTIONAL FACTORIAL DESIGN

A 2^k fractional factorial design containing 2^{k-p} runs is called a $1/2^p$ fraction of the 2^k design or, more simply, a 2^{k-p} fractional factorial design. These designs require the selection of p independent generators. The defining relation for the design consists of the p generators initially chosen and their $2^p - p - 1$ generalized interactions. In this section we discuss the construction and analysis of these designs.

The alias structure may be found by multiplying each effect column by the defining relation. Care should be exercised in choosing the generators so that effects of potential interest are not aliased with each other. Each effect has $2^p - 1$ aliases. For moderately large values of k, we usually assume higher-order interactions (say, third- or fourth-order and higher) to be negligible, and this greatly simplifies the alias structure.

It is important to select the p generators for a 2^{k-p} fractional factorial design in such a way that we obtain the best possible alias relationships. A reasonable criterion is to select the generators such that the resulting 2^{k-p} design has the highest possible design resolution. Table 11-12 presents a selection of 2^{k-p} fractional factorial designs for $k \leq 11$ factors and up to $n \leq 128$ runs. The suggested generators in this table will result in a design of the highest possible resolution.

The alias relationships for all of the designs in Table 11-12 for which $n \leq 64$ are given in Appendix Table XII(a–v). The alias relationships presented in this table focus on main effects and two-factor and three-factor interactions. The complete defining relation is given for each design. This appendix table makes it very easy to select a design of sufficient resolution to ensure that any interactions of potential interest can be estimated.

Example 11-4

To illustrate the use of Table 11-12, suppose that we have seven factors and that we are interested in estimating the seven main effects and getting some insight regarding the two-factor interactions. We are willing to assume that three-factor and higher interactions are negligible. This information suggests that a resolution IV design would be appropriate.

Table 11-12 shows that there are two resolution IV fractions available: the 2_{IV}^{7-2} with 32 runs and the 2_{IV}^{7-3} with 16 runs. Appendix Table XII contains the complete alias relationships for these two designs. The aliases for the 2_{IV}^{7-3} 16-run design are in Appendix Table XII(i). Notice that all seven main effects are aliased with three-factor interactions. The two-factor interactions are all aliased in groups of three. Therefore, this design will satisfy our objectives; that is, it will allow the estimation of the main effects, and it will give some insight regarding two-factor interactions. It is not necessary to run the 2_{IV}^{7-2} design, which would require 32 runs. Appendix Table XII(j) shows that this design would allow the estimation of all seven main effects and that 15 of the 21 two-factor interactions could also be uniquely estimated. (Recall that three-factor and higher interactions are negligible.) This is more information about interactions than is necessary. The complete design is shown in Table 11-13. Notice that it was constructed by starting with the 16-run 2^4 design in A, B, C, and D as the basic design and then adding the three columns $E = ABC$, $F = BCD$, and $G = ACD$. The generators are $I = ABCE$, $I = BCDF$, and $I = ACDG$ (Table 11-12). The complete defining relation is $I = ABCE = BCDF = ADEF = ACDG = BDEG = CEFG = ABFG$.

Analysis of 2^{k-p} Fractional Factorials ▪ The 2^{k-p} fractional factorial design may be analyzed by Yates' algorithm. This is accomplished by considering the data as from a full 2^r factorial, where $r = k - p$ is a subset of the original k factors. The additional factors are incorporated when identifying the actual effects estimated, as discussed in Section 11-2. Usually, one may select the subset of r factors to be those columns in the basic design.

The design may also be analyzed by resorting to first principles; the ith effect is estimated by

$$\ell_i = \frac{2(Contrast_i)}{N} = \frac{Contrast_i}{(N/2)}$$

where the $Contrast_i$ is found using the plus and minus signs in column i and $N = 2^{k-p}$ is the total number of observations. The 2^{k-p} design allows only $2^{k-p} - 1$ effects (and their aliases) to be estimated.

Projection of the 2^{k-p} Fractional Factorial ▪ The 2^{k-p} design collapses into either a full factorial or a fractional factorial in any subset of $r \leq k - p$ of the

Table 11-13   A 2_{IV}^{7-3} Fractional Factorial Design

	Basic Design						
Run	A	B	C	D	$E = ABC$	$F = BCD$	$G = ACD$
1	−	−	−	−	−	−	−
2	+	−	−	−	+	−	+
3	−	+	−	−	+	+	−
4	+	+	−	−	−	+	+
5	−	−	+	−	+	+	+
6	+	−	+	−	−	+	−
7	−	+	+	−	−	−	+
8	+	+	+	−	+	−	−
9	−	−	−	+	−	+	−
10	+	−	−	+	+	+	−
11	−	+	−	+	+	−	+
12	+	+	−	+	−	−	−
13	−	−	+	+	+	−	−
14	+	−	+	+	−	−	+
15	−	+	+	+	−	+	−
16	+	+	+	+	+	+	+

original factors. Those subsets of factors providing fractional factorials are subsets appearing as words in the complete defining relation. This is particularly useful in screening experiments when we suspect at the outset of the experiment that most of the original factors will have small effects. The original 2^{k-p} fractional factorial can then be projected into a full factorial, say, in the most interesting factors. Conclusions drawn from designs of this type should be considered tentative and subject to further analysis. It is usually possible to find alternative explanations of the data involving higher-order interactions.

As an example, consider the 2_{IV}^{7-3} design from Example 11-4. This is a 16-run design involving seven factors. It will project into a full factorial in any four of the original seven factors that is not a word in the defining relation. There are 35 subsets of four factors, seven of which appear in the complete defining relation (see Table 11-13). Thus, there are 28 subsets of four factors that would form 2^4 designs. One combination that is obvious upon inspecting Table 11-13 is A, B, C, and D.

To illustrate the usefulness of this projection properly, suppose that we are conducting an experiment to improve the efficiency of a ball mill and the seven factors are

1. Motor speed.
2. Gain.
3. Feed mode.
4. Feed sizing.
5. Material type.
6. Screen angle.
7. Screen vibration level.

We are fairly certain that motor speed, feed mode, feed sizing, and material type will affect efficiency and that these factors may interact. The role of the other three factors is less well-known, but it is likely that they are negligible. A reasonable strategy would be to assign motor speed, feed mode, feed sizing, and material type to columns A, B, C, and D, respectively, in Table 11-13. Gain, screen angle, and screen vibration level would be assigned to columns E, F, and G, respectively. If we are correct and the "minor variables" E, F, and G are negligible, we will be left with a full 2^4 design in the key process variables.

Blocking Fractional Factorials ▪ Occasionally, a fractional factorial design requires so many runs that all of them cannot be made under homogeneous conditions. In these situations, fractional factorials may be confounded in blocks. Appendix Table XII contains recommended blocking arrangements for many of the fractional factorial designs in Table 11-12. The minimum block size for these designs is 8 runs.

To illustrate the general procedure, consider the 2_{IV}^{6-2} fractional factorial design with the defining relation $I = ABCE = BCDF = ADEF$ shown in Table

11-9. This fractional design contains 16 treatment combinations. Suppose we wish to run the design in two blocks of 8 treatment combinations each. In selecting an interaction to confound with blocks, we note from examining the alias structure in Appendix Table XII(i) that there are two alias sets involving only three-factor interactions. The table suggests selecting ABD (and its aliases) to be confounded with blocks. This would give the two blocks shown in Figure 11-17. Notice that the principal block contains those treatment combinations that have an even number of letters in common with ABD. These are also the treatment combinations for which $L = x_1 + x_2 + x_4 = 0$ (mod 2).

Example 11-5

A five-axis CNC machine is used to machine an impeller used in a jet turbine engine. The blade profiles are an important quality characteristic. Specifically, the deviation of the blade profile from the profile specified on the engineering drawing is of interest. An experiment is run to determine which machine parameters affect profile deviation. The eight factors selected for the design are as follows:

Factor	Low Level ($-$)	High Level ($+$)
A = x-Axis shift (0.001 in)	0	15
B = y-Axis shift (0.001 in)	0	15
C = z-Axis shift (0.001 in)	0	15
D = Tool vendor	1	2
E = a-Axis shift (0.001 deg)	0	30
F = Spindle speed (%)	90	110
G = Fixture height (0.001 in)	0	15
H = Feed rate (%)	90	110

One test blade on each part is selected for inspection. The profile deviation is measured using a coordinate measuring machine, and the standard deviation of the difference between the actual profile and the specified profile is used as a response variable.

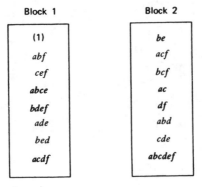

Block 1	Block 2
(1)	be
abf	acf
cef	bcf
abce	ac
bdef	df
ade	abd
bed	cde
acdf	abcdef

Figure 11-17.   The 2_{IV}^{6-2} design in two blocks with ABD confounded.

The machine has four spindles. Because there may be differences in the spindles, the process engineers feel that the spindles should be treated as blocks.

The engineers feel confident that three-factor and higher interactions are not too important, but they are reluctant to ignore the two-factor interactions. From Table 11-12, two designs initially appear appropriate: the 2_{IV}^{8-4} design with 16 runs, and the 2_{IV}^{8-3} design with 32 runs. Appendix Table XII(l) indicates that if the 16-run design is used there will be fairly extensive aliasing of two-factor interactions. Furthermore, this design cannot be run in four blocks without confounding four two-factor interactions with blocks. Therefore, the experimenters decide to use the 2_{IV}^{8-3} design in

Table 11-14 The 2^{8-3} Design in Four Blocks for Example 11-5

Run	A	B	C	D	E	F = ABC	G = ABD	H = BCDE	Block	Actual Run Order	Standard Deviation ($\times 10^3$ in
1	−	−	−	−	−	−	−	+	3	18	2.76
2	+	−	−	−	−	+	+	+	2	16	6.18
3	−	+	−	−	−	+	−	−	4	29	2.43
4	+	+	−	−	−	−	−	−	1	4	4.01
5	−	−	+	−	−	+	−	−	1	6	2.48
6	+	−	+	−	−	−	−	−	4	26	5.91
7	−	+	+	−	−	−	+	+	2	14	2.39
8	+	+	+	−	−	+	−	+	3	22	3.35
9	−	−	−	+	−	−	+	−	1	8	4.40
10	+	−	−	+	−	+	−	−	4	32	4.10
11	−	+	−	+	−	+	−	+	2	15	3.22
12	+	+	−	+	−	−	+	+	3	19	3.78
13	−	−	+	+	−	+	+	+	3	24	5.32
14	+	−	+	+	−	−	−	+	2	11	3.87
15	−	+	+	+	−	−	−	−	4	27	3.03
16	+	+	+	+	−	+	+	−	1	3	2.95
17	−	−	−	−	+	−	−	−	2	10	2.64
18	+	−	−	−	+	+	+	−	3	21	5.50
19	−	+	−	−	+	+	+	+	1	7	2.24
20	+	+	−	−	+	−	−	+	4	28	4.28
21	−	−	+	−	+	+	−	+	4	30	2.57
22	+	−	+	−	+	−	+	+	1	2	5.37
23	−	+	+	−	+	−	+	−	3	17	2.11
24	+	+	+	−	+	+	−	−	2	13	4.18
25	−	−	−	+	+	−	+	+	4	25	3.96
26	+	−	−	+	+	+	−	+	1	1	3.27
27	−	+	−	+	+	+	−	−	3	23	3.41
28	+	+	−	+	+	−	+	−	2	12	4.30
29	−	−	+	+	+	+	+	−	2	9	4.44
30	+	−	+	+	+	−	−	−	3	20	3.65
31	−	+	+	+	+	−	−	+	1	5	4.41
32	+	+	+	+	+	+	+	+	4	31	3.40

Figure 11-18. Normal probability plot of the effect estimates for Example 11-5.

four blocks. This confounds one three-factor interaction alias string and one two-factor interaction (EH) and its three-factor interaction aliases with blocks. The EH interaction is the interaction between the a-axis shift and the feed rate, and the engineers consider an interaction between these two variables to be fairly unlikely. (One could always assign other factors that are unlikely to interact to E and H in the event that the original assignment is unsatisfactory.)

Table 11-14 contains the design and the resulting responses in standard deviation $\times 10^3$ in. Since the response variable is a standard deviation, it is often best to perform the analysis following a log transformation. Figure 11-18 is a normal probability plot of the effect estimates, using ln (standard deviation $\times 10^3$) as the response variable. The only large effects are A = x-axis shift, B = y-axis shift, and the AD (x-axis shift–tool vendor) interaction. Table 11-15 is the analysis of variance for this model. Notice that the block effect appears small, suggesting that the machine spindles are not different.

Table 11-15 Analysis of Variance for Example 11-5

Source of Variation	Sum of Squares	Degrees of Freedom	Mean Square	F_0
A	0.6740	1	0.6740	33.53
B	0.3217	1	0.3217	16.00
AD	1.1197	1	1.1197	55.71
Blocks	0.0201	3	0.0067	
Error	0.5034	25	0.0201	
Total	2.6389	31		

Figure 11-19. Normal probability plot of the residuals for Example 11-5.

Figure 11-19 is a normal probability plot of the residuals from this experiment. This plot is suggestive of slightly heavier than normal tails, so possibly other transformations should be considered. The *AD* interaction plot is in Figure 11-20. Notice that tool vendor (*D*) and the magnitude of the *x*-axis shift (*A*) have a profound impact on the variability of the blade profile from design specifications. Running *A* at the

Figure 11-20. Plot of *AD* interaction for Example 11-5.

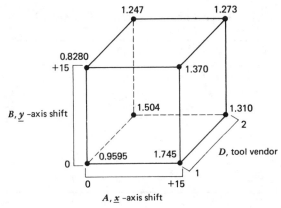

Figure 11-21. The 2_{IV}^{8-3} design in Example 11-5 projected into four replicates of a 2^3 design in factors A, B, and D.

low level (zero offset) and buying tools from vendor 1 gives the best results. Figure 11-21 shows the projection of this 2_{IV}^{8-3} design into four replicates of a 2^3 design in factors A, B, and D. The best combination of operating conditions is A at the low level (0 offset), B at the high level (0.015 in offset), and D at the low level (tool vendor 1).

11-5 RESOLUTION III DESIGNS

As indicated earlier, the sequential use of fractional factorial designs is very useful, often leading to great economy and efficiency in experimentation. We now illustrate these ideas using the class of resolution III designs.

It is possible to construct resolution III designs for investigating up to $k = N - 1$ factors in only N runs, where N is a multiple of 4. These designs are frequently useful in industrial experimentation. Designs in which N is a power of 2 can be constructed by the methods presented earlier in this chapter, and these are presented first. Of particular importance are designs requiring 4 runs for up to 3 factors, 8 runs for up to 7 factors and 16 runs for up to 15 factors. If $k = N - 1$, the fractional factorial design is said to be *saturated*.

A design for analyzing up to three factors in four runs is the 2_{III}^{3-1} design, presented in Section 11-2. Another very useful saturated fractional factorial is a design for studying seven factors in eight runs; that is, the 2_{III}^{7-4} design. This design is a one-sixteenth fraction of the 2^7. It may be constructed by first writing down as the basic design the plus and minus levels for a full 2^3 design in A, B, and C and then associating the levels of four additional factors with the interactions of the original three as follows: $D = AB$, $E = AC$, $F = BC$, and $G = ABC$. Thus, the generators for this design are $I = ABD$, $I = ACE$, $I = BCF$, and $I = ABCG$. The design is shown in Table 11-16.

The complete defining relation for this design is obtained by multiplying the four generators ABD, ACE, BCF, and $ABCG$ together two at a time, three at a

Table 11-16   The 2_{III}^{7-4} Design with the Generators $I = ABD$, $I = ACE$, $I = BCF$, and $I = ABCG$

Run	A	B	C	D = AB	E = AC	F = BC	G = ABC	
1	−	−	−	+	+	+	−	def
2	+	−	−	−	−	+	+	afg
3	−	+	−	−	+	−	+	beg
4	+	+	−	+	−	−	−	abd
5	−	−	+	+	−	−	+	cdg
6	+	−	+	−	+	−	−	ace
7	−	+	+	−	−	+	−	bcf
8	+	+	+	+	+	+	+	abcdefg

time, and four at a time, yielding

$$I = ABD = ACE = BCF = ABCG = BCDE = ACDF = CDG$$
$$= ABEF = BEG = AFG = DEF = ADEG = CEFG = BDFG = ABCDEFG$$

To find the aliases of any effect, simply multiply the effect by each word in the defining relation. For example, the aliases of B are

$$B = AD = ABCE = CF = ACG = CDE = ABCDF = BCDG = AEF = EG$$
$$= ABFG = BDEF = ABDEG = BCEFG = DFG = ACDEFG$$

This design is a one-sixteenth fraction, and since the signs chosen for the generators are positive, this is the principal fraction. It is also of resolution III since the smallest number of letters in any word of the defining contrast is three. Any one of the 16 different 2_{III}^{7-4} designs in this family could be constructed by using the generators with one of the 16 possible arrangements of signs in $I = \pm ABD$, $I = \pm ACE$, $I = \pm BCF$, $I = \pm ABCG$.

The seven degrees of freedom in this design may be used to estimate the seven main effects. Each of these effects has 15 aliases; however, if we assume that three-factor and higher interactions are negligible, then considerable simplification in the alias structure results. Making this assumption, each of the linear combinations associated with the seven main effects in this design actually estimates the main effect and three two-factor interactions:

$$\ell_A \rightarrow A + BD + CE + FG$$
$$\ell_B \rightarrow B + AD + CF + EG$$
$$\ell_C \rightarrow C + AE + BF + DG$$
$$\ell_D \rightarrow D + AB + CG + EF \qquad (11\text{-}1)$$
$$\ell_E \rightarrow E + AC + BG + DF$$
$$\ell_F \rightarrow F + BC + AG + DE$$
$$\ell_G \rightarrow G + CD + BE + AF$$

These aliases are found in Appendix Table XII(h), ignoring three-factor and higher interactions.

The saturated 2_{III}^{7-4} design in Table 11-16 can be used to obtain resolution III designs for studying fewer than seven factors in eight runs. For example, to generate a design for six factors in eight runs, simply drop any one column in Table 11-16, for example, column G. This produces the design shown in Table 11-17.

It is easy to verify that this design is also of resolution III; in fact, it is a 2_{III}^{6-3} or a one-eighth fraction of the 2^6 design. The defining relation for the 2_{III}^{6-3} design is equal to the defining relation for the original 2_{III}^{7-4} design with any words containing the letter G deleted. Thus, the defining relation for our new design is

$$I = ABD = ACE = BCF = BCDE = ACDF = ABEF = DEF$$

In general, when d factors are dropped to produce a new design, the new defining relation is obtained as those words in the original defining relation that do not contain any dropped letters. When constructing designs by this method, care should be exercised to obtain the best arrangement possible. If we drop columns B, D, F, and G from Table 11-16, we obtain a design for three factors in eight runs, yet the treatment combinations correspond to two replicates of a 2^2 design. The experimenter would probably prefer to run a full 2^3 design in A, C, and E.

It is also possible to obtain a resolution III design for studying up to 15 factors in 16 runs. This saturated 2_{III}^{15-11} design can be generated by first writing down the 16 treatment combinations associated with a 2^4 design in A, B, C, and D and then equating 11 new factors with the 2-, 3-, and 4-factor interactions of the original 4. In this design, each of the 15 main effects is aliased with seven 2-factor interactions. A similar procedure can be used for the 2_{III}^{31-26} design, which allows up to 31 factors to be studied in 32 runs.

Sequential Assembly of Fractions to Separate Effects ▪ By combining fractional factorial designs in which certain signs are switched, we can systematically isolate effects of potential interest. The alias structure for any fraction with the signs for one or more factors reversed is obtained by making changes of sign on the appropriate factors in the alias structure of the original fraction.

Table 11-17   A 2_{III}^{6-3} Design with the Generators $I = ABD$, $I = ACE$, and $I = BCF$

Run	A	B	C	$D = AB$	$E = AC$	$F = BC$	
1	−	−	−	+	+	+	def
2	+	−	−	−	−	+	af
3	−	+	−	−	+	−	be
4	+	+	−	+	−	−	abd
5	−	−	+	+	−	−	cd
6	+	−	+	−	+	−	ace
7	−	+	+	−	−	+	bcf
8	+	+	+	+	+	+	abcdef

Consider the 2_{III}^{7-4} design in Table 11-16. Suppose that along with this principal fraction a second fractional design with the signs reversed in the column for factor D is also run. That is, the column for D in the second fraction is

$$- + + - - + + -$$

The effects that may be estimated from the first fraction are shown in Equation 11-1, and from the second fraction we obtain

$$\ell'_A \rightarrow A - BD + CE + FG$$
$$\ell'_B \rightarrow B - AD + CF + EG$$
$$\ell'_C \rightarrow C + AE + BF - DG$$
$$\ell'_D \rightarrow D - AB - CG - EF \qquad (11\text{-}2)$$
$$\text{i.e., } \ell'_{-D} \rightarrow -D + AB + CG + EF$$
$$\ell'_E \rightarrow E + AC + BG - DF$$
$$\ell'_F \rightarrow F + BC + AG - DE$$
$$\ell'_G \rightarrow G - CD + BE + AF$$

assuming that three-factor and higher interactions are insignificant. Now from the two linear combinations of effects $\frac{1}{2}(\ell_i + \ell'_i)$ and $\frac{1}{2}(\ell_i - \ell'_i)$ we obtain

i	From $\frac{1}{2}(\ell_i + \ell'_i)$	From $\frac{1}{2}(\ell_i - \ell'_i)$
A	$A + CE + FG$	BD
B	$B + CF + EG$	AD
C	$C + AE + BF$	DG
D	D	$AB + CG + EF$
E	$E + AC + BG$	DF
F	$F + BC + AG$	DE
G	$G + BE + AF$	CD

Thus, we have isolated the main effect of D and all of its two-factor interactions. In general, if we add to a fractional factorial design of resolution III or higher a further fraction with the signs of a *single factor* reversed, then the combined design will provide estimates of the main effect of that factor and its two-factor interactions.

Now suppose we add to a resolution III fractional a second fraction in which the signs for *all the factors* are reversed. This procedure, which is called *folding over* the design, breaks the alias links between main effects and two-factor interactions. That is, we may use the *combined* design to estimate all the main effects clear of any two-factor interactions. The following example illustrates the technique.

Example 11-6

A human performance analyst is conducting an experiment to study eye focus time and has built an apparatus in which several factors can be controlled during the test. The factors he initially regards as important are acuity or sharpness of vision (A), distance from target to eye (B), target shape (C), illumination level (D), target size (E), target density (F), and subject (G). Two levels of each factor are considered. He suspects that only a few of these seven factors are of major importance, and that high-order interactions between the factors can be neglected. On the basis of this assumption, the analyst decides to run a screening experiment to identify the most important factors and then to concentrate further study on those. To screen these seven factors, he runs the treatment combinations from the 2_{III}^{7-4} design in Table 11-16 in random order, obtaining the focus times in milliseconds, as shown in Table 11-18.

Seven main effects and their aliases may be estimated from these data. From Equation 11-1, we see that the effects and their aliases are

$$\ell_A = 20.63 \rightarrow A + BD + CE + FG$$

$$\ell_B = 38.38 \rightarrow B + AD + CF + EG$$

$$\ell_C = -0.28 \rightarrow C + AE + BF + DG$$

$$\ell_D = 28.88 \rightarrow D + AB + CG + EF$$

$$\ell_E = -0.28 \rightarrow E + AC + BG + DF$$

$$\ell_F = -0.63 \rightarrow F + BC + AG + DE$$

$$\ell_G = -2.43 \rightarrow G + CD + BE + AF$$

For example,

$$\ell_A = \tfrac{1}{4}(-85.5 + 75.1 - 93.2 + 145.4 - 83.7 + 77.6 - 95.0 + 141.8) = 20.63$$

The three largest effects are ℓ_A, ℓ_B, and ℓ_D. The simplest interpretation of the data is that the main effects of A, B, and D are all significant. However, this interpretation is not unique, since one could also logically conclude that A, B, and the AB interaction, or perhaps B, D, and BD interaction, or perhaps A, D, and the AD interaction are the true effects.

Notice that ABD is a word in the defining relation for this design. Therefore, this

Table 11-18   A 2_{III}^{7-4} Design for the Eye Focus Time Experiment

Run	A	B	C	D = AB	E = AC	F = BC	G = ABC		Time
1	−	−	−	+	+	+	−	def	85.5
2	+	−	−	−	−	+	+	afg	75.1
3	−	+	−	−	+	−	+	beg	93.2
4	+	+	−	+	−	−	−	abd	145.4
5	−	−	+	+	−	−	+	cdg	83.7
6	+	−	+	−	+	−	−	ace	77.6
7	−	+	+	−	−	+	−	bcf	95.0
8	+	+	+	+	+	+	+	abcdefg	141.8

Figure 11-22. The 2_{III}^{7-4} design projected into two replicates of a 2_{III}^{3-1} design in A, B, and D.

2_{III}^{7-4} design does not project into a full 2^3 factorial in ABD; instead, it projects into two replicates of a 2^{3-1} design, as shown in Figure 11-22. Since the 2^{3-1} design is a resolution III design, A will be aliased with BD, B will be aliased with AD, and D will be aliased with AB, so the interactions cannot be separated from the main effects. The analyst here may have been unlucky. If he had assigned illumination level to C instead of D, the design would have projected into a full 2^3 design.

To separate the main effects and the two-factor interactions, a second fraction is run with all the signs reversed. This fold-over design is shown in Table 11-19 along with the observed responses. Notice that when we fold over a resolution III design in this manner, we (in effect) change the signs on the generators that have an odd number of letters. The effects estimated by this fraction are

$$\ell'_A = -17.68 \rightarrow A - BD - CE - FG$$
$$\ell'_B = 37.73 \rightarrow B - AD - CF - EG$$
$$\ell'_C = -3.33 \rightarrow C - AE - BF - DG$$
$$\ell'_D = 29.88 \rightarrow D - AB - CG - EF$$
$$\ell'_E = 0.53 \rightarrow E - AC - BG - DF$$
$$\ell'_F = 1.63 \rightarrow F - BC - AG - DE$$
$$\ell'_G = 2.68 \rightarrow G - CD - BE - AF$$

Table 11-19   A Second 2_{III}^{7-4} Design for the Eye Focus Time Experiment

Run	A	B	C	$D = -AB$	$E = -AC$	$F = -BC$	$G = ABC$		Time
1	+	+	+	−	−	−	+	$abcg$	91.3
2	−	+	+	+	+	−	−	$bcde$	136.7
3	+	−	+	+	−	+	−	$acdf$	82.4
4	−	−	+	−	+	+	+	$cefg$	73.4
5	+	+	−	−	+	+	−	$abef$	94.1
6	−	+	−	+	−	+	+	$bdfg$	143.8
7	+	−	−	+	+	−	+	$adeg$	87.3
8	−	−	−	−	−	−	−	(1)	71.9

By combining this second fraction with the original one, we obtain the following estimates of the effects:

i	From $\frac{1}{2}(\ell_i + \ell_i')$	From $\frac{1}{2}(\ell_i - \ell_i')$
A	A = 1.48	**BD** + CE + FG = 19.15
B	**B** = 38.05	AD + CF + EG = 0.33
C	C = −1.80	AE + BF + DG = 1.53
D	**D** = 29.38	AB + CG + EF = −0.50
E	E = 0.13	AC + BG + DF = −0.40
F	F = 0.50	BC + AG + DE = −1.53
G	G = 0.13	CD + BE + AF = −2.55

The two largest effects are B and D. Furthermore, the third largest effect is $BD + CE + FG$, so it seems reasonable to attribute this to the BD interaction. The analyst used the two factors distance (B) and illumination level (D) in subsequent experiments with the other factors A, C, E, and F at standard settings and verified the results obtained here. He decided to use subjects as blocks in these new experiments rather than ignore a potential subject effect, since several different subjects had to be used to complete the experiment.

■

The Defining Relation for a Fold-Over Design ▪ Combining fractional factorial designs via fold over as demonstrated in Example 11-6 is a very useful technique. It is often of interest to know the defining relation for the combined design. It can be easily determined. Each separate fraction will have $L + U$ words used as generators: L words of like sign and U words of unlike sign. The combined design will have $L + U - 1$ words used as generators. These will be the L words of like sign and the $U - 1$ words consisting of independent even products of the words of unlike sign. (*Even products* are words taken two at a time, four at a time, and so forth.)

To illustrate this procedure, consider the design in Example 11-6. For the first fraction, the generators are

$$I = ABD, I = ACE, I = BCF, \text{ and } I = ABCG$$

and for the second fraction, they are

$$I = -ABD, I = -ACE, I = -BCF, \text{ and } I = ABCG$$

Notice that in the second fraction we have switched the signs on the generators with an odd number of letters. Also, notice that $L + U = 1 + 3 = 4$. The combined design will have $I = ABCG$ (the like sign word) as a generator, and two words that are independent even products of the words of unlike sign. For example, take $I = ABD$ and $I = ACE$; then $I = (ABD)(ACE) = BCDE$ is a generator of the combined design. Also, take $I = ABD$ and $I = BCF$; then $I = (ABD)(BCF) = ACDF$ is a generator of the combined design. The complete defining relation for the combined design is

$$I = ABCG = BCDE = ACDF = ADEG = BDFG = ABEF = CEFG$$

Plackett-Burman Designs ▪ These designs, attributed to Plackett and Burman (1946), are two-level fractional factorial designs for studying $K = N - 1$ variables in N runs, where N is a multiple of 4. If N is a power of 2, these designs are identical to those presented earlier in this section. However, for $N = 12, 20,$ 24, 28, and 36, the Plackett-Burman designs are sometimes of interest.

The upper half of Table 11-20 presents rows of plus and minus signs that are used to construct the Plackett-Burman designs for $N = 12, 20, 24,$ and 36, whereas the lower half of the table presents blocks of plus and minus signs for constructing the design for $N = 28$. The designs for $N = 12, 20, 24,$ and 36 are obtained by writing the appropriate row in Table 11-20 as a column (or row). A second column (or row) is then generated from this first one by moving the elements of the column (or row) down (or to the right) one position and placing the last element in the first position. A third column (or row) is produced from the second similarly, and the process continued until column (or row) k is generated. A row of minus signs is then added, completing the design. For $N = 28$, the three blocks X, Y, and Z are written down in the order

$$
\begin{array}{ccc}
X & Y & Z \\
Z & X & Y \\
Y & Z & X
\end{array}
$$

and a row of minus signs is added to these 27 rows. The design for $N = 12$ runs and $k = 11$ factors is shown in Table 11-21.

The Plackett-Burman designs for $N = 12, 20, 24, 28,$ and 36 have very

Table 11-20   Plus and Minus Signs for the Plackett-Burman Designs

$k = 11\ N = 12\ + + - + + + - - - + -$
$k = 19\ N = 20\ + + - - + + + + - + - - - - + + -$
$k = 23\ N = 24\ + + + + + - + - + + - - + + - - + - + - - - -$
$k = 35\ N = 36\ - + - + + + - - - + + + + + - + + + - - + - - - - + - + - + + - - + -$

$k = 27, N = 28$		
+ - + + + + - - -	- + - - - + - - +	+ + - + - + + - +
+ + - + + + - - -	- - + + - - + - -	- + + + + - + + -
- + + + + + - - -	+ - - - + - - + -	+ - + - + + - + +
- - - + - + + + +	- - + - + - - - +	+ - + + + - + - +
- - - + + - + + +	+ - - - - + + - -	+ + - - + + + + -
- - - - + + + + +	- + - + - - - + -	- + + + - + - + +
+ + + - - - + - +	- - + - - + - + -	+ - + + - + + + -
+ + + - - - + + -	+ - - + - - - - +	+ + - + + - - + +
+ + + - - - - + +	- + - - + - + - -	- + + - + + + - +

Table 11-21   Plackett-Burman Design for $N = 12, k = 11$

Run	A	B	C	D	E	F	G	H	I	J	K
1	+	−	+	−	−	−	+	+	+	−	+
2	+	+	−	+	−	−	−	+	+	+	−
3	−	+	+	−	+	−	−	−	+	+	+
4	+	−	+	+	−	+	−	−	−	+	+
5	+	+	−	+	+	−	+	−	−	−	+
6	+	+	+	−	+	+	−	+	−	−	−
7	−	+	+	+	−	+	+	−	+	−	−
8	−	−	+	+	+	−	+	+	−	+	−
9	−	−	−	+	+	+	−	+	+	−	+
10	+	−	−	−	+	+	+	−	+	+	−
11	−	+	−	−	−	+	+	+	−	+	+
12	−	−	−	−	−	−	−	−	−	−	−

messy alias structures. For example, in the 12-run design every main effect is partially aliased with every two-factor interaction not involving itself. For example, the AB interaction is aliased with the nine main effects $C, D, \ldots, K$. Furthermore, each main effect is partially aliased with 45 two-factor interactions. In the larger designs, the situation is even more complex. We advise the experimenter to use these designs *very* carefully.

The projective properties of the Plackett-Burman designs are not terribly attractive. For example, consider the 12-run design in Table 11-21. This design will project into three replicates of a full 2^2 design in any 2 of the original 11 factors. However, in 3 factors, the projected design is a full 2^3 factorial plus a 2_{III}^{3-1} fractional factorial (see Figure 11-23a). The four-dimensional projections are shown in Figure 11-23b. Notice that these 3- and 4-factor projections are not balanced designs. This would complicate their interpretation.

11-6 RESOLUTION IV AND V DESIGNS

A 2^{k-p} fractional factorial design is of resolution IV if the main effects are clear of two-factor interactions and some two-factor interactions are aliased with each other. Thus, if three-factor and higher interactions are suppressed, the main effects may be estimated directly in a 2_{IV}^{k-p} design. An example is the 2_{IV}^{6-2} design in Table 11-9. Furthermore, the two combined fractions of the 2_{III}^{7-4} design in Example 11-6 yields a 2_{IV}^{7-3} design.

Any 2_{IV}^{k-p} design must contain at least $2k$ runs. Resolution IV designs that contain exactly $2k$ runs are called *minimal* designs. Resolution IV designs may be obtained from resolution III designs by the process of *fold over*. Recall that to fold over a 2_{III}^{k-p} design, simply add to the original fraction a second fraction with

(a) Projection into three factors

(b) Projection into four factors

Figure 11-23. Projection of the 12-run Plackett-Burman design into three- and four-factor designs. (a) Projection into three factors. (b) Projection into four factors.

all the signs reversed. Then the plus signs in the identity column I in the first fraction could be switched in the second fraction, and a $(k + 1)$st factor associated with this column. The result is a 2_{IV}^{k+1-p} fractional factorial design. The process is demonstrated in Table 11-22 for the 2_{III}^{3-1} design. It is easy to verify that the resulting design is a 2_{IV}^{4-1} design with defining relation $I = ABCD$.

It is also possible to fold over resolution IV designs to separate two-factor interactions that are aliased with each other. To fold over a resolution IV design, run a second fraction in which the sign is reversed on every design generator that has an even number of letters. To illustrate, consider the 2_{IV}^{6-2} design used for the injection molding experiment in Example 11-3. The generators for the design in Table 11-10 are $I = ABCE$ and $I = BCDF$. The second fraction would use the generators $I = -ABCE$ and $I = -BCDF$, and the single generator for the combined design would be $I = ADEF$. Thus, the combined design is still a resolution IV fractional factorial design. However, the alias relationships will be much simpler than in the original 2_{IV}^{6-2} fractional. In fact, the only two-factor interactions that will be aliased are $AD = EF$, $AE = DF$, and $AF = DE$. All the other two-factor interactions can be estimated from the combined design.

As another example, consider the 2_{IV}^{8-3} design with 32 runs. Table 11-12

Table 11-22 A 2_{IV}^{4-1} Design Obtained by Fold Over

D			
I	A	B	C

Original 2_{III}^{3-1} $I = ABC$

+	−	−	+
+	+	−	−
+	−	+	−
+	+	+	+

Second 2_{III}^{3-1} with signs switched

−	+	+	−
−	−	+	+
−	+	−	+
−	−	−	−

indicates that the best set of generators for this design is $I = ABCF$, $I = ABDG$, and $I = BCDEH$. Appendix Table XII(m) shows the aliases for this design. Notice that there are six pairs of two-factor interactions and one group of three two-factor interactions that are aliases. If we fold over this design, the second fraction would have the generators $I = -ABCF$, $I = -ABDG$, and $I = BCDEH$. The combined design has the generators $I = CDFG$ and $I = BCDEH$, and the complete defining relation is

$$I = CDFG = BCDEH = BEFGH$$

The combined design is of resolution IV, but the only two-factor interactions that are still aliased are $CD = FG$, $CF = DG$, and $CG = DF$. This is a considerable simplification of the aliasing in the original fraction.

Notice that when you start with a resolution III design, the fold-over procedure guarantees that the combined design will be of resolution IV, thereby ensuring that all the main effects can be separated from their two-factor interaction aliases. When folding over a resolution IV design, we will not necessarily separate *all* the two-factor interactions. In fact, if the *original* fraction has an alias structure with more than two two-factor interactions in any alias chain, folding over will not completely separate all the two-factor interactions. Both of the foregoing examples, the 2_{IV}^{6-2} and the 2_{IV}^{8-3}, have at least one such two-factor interaction alias chain.

Resolution V designs are fractional factorials in which the main effects and the two-factor interactions do not have other main effects and two-factor interactions as their aliases. These are very powerful designs, allowing the unique estimation of all the main effects and two-factor interactions provided that all the three-factor and higher interactions are negligible. The smallest word in the defining relation of such a design must have five letters. The 2^{5-1} design with the

Table 11-23 Useful Factorial and Fractional Factorial Designs from the 2^{k-p} System. The Numbers in the Cells are the Number of Factors in the Experiment

Design Type	Number of Runs			
	4	8	16	32
Full Factorial	2	3	4	5
Half-Fraction	3	4	5	6
Resolution IV Fraction	—	4	6–8	7–16
Resolution III Fraction	3	5–7	9–15	17–31

generating relation $I = ABCDE$ is of resolution V. Another example is the 2_V^{8-2} design with the generating relations $I = ABCDG$ and $I = ABEFH$. Further examples of these designs are given by Box and Hunter (1961b).

11-7 SUMMARY

This chapter has introduced the 2^{k-p} fractional factorial design. We have emphasized the use of these designs in screening experiments to identify quickly and efficiently the subset of factors that are active and to provide some information on interaction. The projective property of these designs makes it possible in many cases to examine the active factors in more detail. Sequential assembly of these designs via fold over is a very effective way to gain additional information about interactions that an initial experiment may identify as possibly important.

In practice, 2^{k-p} fractional factorial designs with $N = 4$, 8, 16, and 32 runs are highly useful. Table 11-23 summarizes these designs, identifying how many factors can be used with each design to obtain various types of screening experiments. For example, the 16-run design is a full factorial for 4 factors, a one-half fraction for 5 factors, a resolution IV fraction for 6-8 factors, and a resolution III fraction for 9 to 15 factors. All of these designs may be constructed using the methods discussed in this chapter, and many of their alias structures are shown in Appendix Table XII.

11-8 PROBLEMS

11-1 Suppose that in the chemical process development experiment described in Problem 9-5, it was only possible to run a one-half fraction of the 2^4 design. Construct the design and perform the statistical analysis, using the data from replicate I.

11-2 Suppose that in Problem 9-11, only a one-half fraction of the 2^4 design could be run. Construct the design and perform the analysis, using the data from replicate I.

11-3 Consider the plasma etch experiment described in Problem 9-12. Suppose that only a one-half fraction of the design could be run. Set up the design and analyze the data.

11-4 Example 11-2 describes a process improvement study in the manufacturing process of an integrated circuit. Suppose that only eight runs could be made in this process.

Set up an appropriate 2^{5-2} design and find the alias structure. Use the data from Example 11-2 as the observations in this design and estimate the factor effects. What conclusions can you draw?

11-5 **Continuation of Problem 11-4.** Suppose you have made the eight runs in the 2^{5-2} design in Problem 11-4. What additional runs would be required to identify the factor effects that are of interest? What are the alias relationships in the combined design?

11-6 R. D. Snee ("Experimenting with a Large Number of Variables," in *Experiments in Industry: Design, Analysis and Interpretation of Results*, by R. D. Snee, L. B. Hare, and J. B. Trout, Editors, ASQC, 1985) describes an experiment in which a 2^{5-1} design with $I = ABCDE$ was used to investigate that effects of five factors on the color of a chemical product. The factors are A = solvent/reactant, B = catalyst/reactant, C = temperature, D = reactant purity, and D = reactant pH. The results obtained were as follows:

$$
\begin{array}{ll}
e = -0.63 & d = 6.79 \\
a = 2.51 & ade = 5.47 \\
b = -2.68 & bde = 3.45 \\
abe = 1.66 & abd = 5.68 \\
c = 2.06 & cde = 5.22 \\
ace = 1.22 & acd = 4.38 \\
bce = -2.09 & bcd = 4.30 \\
abc = 1.93 & abcde = 4.05
\end{array}
$$

(a) Prepare a normal probability plot of the effects. Which effects seem active?

(b) Calculate the residuals. Construct a normal probability plot of the residuals and plot the residuals versus the fitted values. Comment on the plots.

(c) If any factors are negligible, collapse the 2^{5-1} design into a full factorial in the active factors. Comment on the resulting design, and interpret the results.

11-7 An article in the *Journal of Quality Technology*, (Vol. 17, 1985, pp. 198–206) describes the use of a replicated fractional factorial to investigate the effect of five factors on the free height of leaf springs used in an automotive application. The factors are A = furnace temperature, B = heating time, C = transfer time, D = hold down time, and E = quench oil temperature. The data are shown below:

A	B	C	D	E	Free Height		
−	−	−	−	−	7.78	7.78	7.81
+	−	−	+	−	8.15	8.18	7.88
−	+	−	+	−	7.50	7.56	7.50
+	+	−	−	−	7.59	7.56	7.75
−	−	+	+	−	7.54	8.00	7.88
+	−	+	−	−	7.69	8.09	8.06
−	+	+	−	−	7.56	7.52	7.44
+	+	+	+	−	7.56	7.81	7.69
−	−	−	−	+	7.50	7.25	7.12
+	−	−	+	+	7.88	7.88	7.44
−	+	−	+	+	7.50	7.56	7.50
+	+	−	−	+	7.63	7.75	7.56
−	−	+	+	+	7.32	7.44	7.44
+	−	+	−	+	7.56	7.69	7.62
−	+	+	−	+	7.18	7.18	7.25
+	+	+	+	+	7.81	7.50	7.59

(a) Write out the alias structure for this design. What is the resolution of this design?

(b) Analyze the data. What factors influence the mean free height?

(c) Calculate the range and standard deviation of the free height for each run. Is there any indication that any of these factors affects variability in the free height?

(d) Analyze the residuals from this experiment, and comment on your findings.

(e) Is this the best possible design for five factors in 16 runs? Specifically, can you find a fractional design for five factors in 16 runs with a higher resolution than this one?

11-8   An article in *Industrial and Engineering Chemistry* ("More on Planning Experiments to Increase Research Efficiency," 1970, pp. 60–65) uses a 2^{5-2} design to investigate the effect of A = condensation temperature, B = amount of materials, C = solvent volume, D = condensation time, and E = amount of material 2 on yield. The results obtained are as follows:

$$e = 23.2 \qquad ad = 16.9 \qquad cd = 23.8 \qquad bde = 16.8$$
$$ab = 15.5 \qquad bc = 16.2 \qquad ace = 23.4 \qquad abcde = 18.1$$

(a) Verify that the design generators used were $I = ACE$ and $I = BDE$.

(b) Write down the complete defining relation and the aliases for this design.

(c) Estimate the main effects.

(d) Prepare an analysis of variance table. Verify that the AB and AD interactions are available to use as error.

(e) Plot the residuals versus the fitted values. Also construct a normal probability plot of the residuals. Comment on the results.

11-9   Consider the leaf spring experiment in Problem 11-7. Suppose that factor E (quench oil temperature) is very difficult to control during manufacturing. Where would you set factors A, B, C, and D to reduce variability in the free height as much as possible regardless of the quench oil temperature used?

11-10 Construct a 2^{7-2} design by choosing two four-factor interactions as the independent generators. Write down the complete alias structure for this design. Outline the analysis of variance table. What is the resolution of this design?

11-11 Consider the 2^5 design in Problem 9-14. Suppose that only a one-half fraction could be run. Furthermore, two days were required to take the 16 observations, and it was necessary to confound the 2^{5-1} design in two blocks. Construct the design and analyze the data.

11-12 Analyze the data in Problem 9-15 as if it came from a 2_{IV}^{4-1} design with $I = ABCD$. Project the design into a full factorial in the subset of the original four factors that appear to be significant.

11-13 Repeat Problem 11-12 using $I = -ABCD$. Does use of the alternate fraction change your interpretation of the data?

11-14 Project the 2_{IV}^{4-1} design in Example 11-1 into two replicates of a 2^2 design in the factors A and B. Analyze the data and draw conclusions.

11-15 Construct a 2_{III}^{6-3} design. Determine the effects that may be estimated if a second fraction of this design is run with all signs reversed.

11-16 Consider the 2_{III}^{6-3} design in Problem 11-15. Determine the effects that may be estimated if a second fraction of this design is run with the signs for factor A reversed.

11-17 Fold over the 2_{II}^{7-4} design in Table 11-19. Verify that the resulting design is a 2_{IV}^{8-4} design. Is this a minimal design?

11-18 Fold over a 2_{III}^{5-2} design. Verify that the resulting design is a 2_{IV}^{6-2} design. Compare this design to the 2_{IV}^{6-2} design in Table 11-9.

11-19 An industrial engineer is conducting an experiment using a Monte Carlo simulation model of an inventory system. The independent variables in her model are the order quantity (A), the reorder point (B), the setup cost (C), the backorder cost (D), and the carrying cost rate (E). The response variable is average annual cost. To conserve computer time, she decides to investigate these factors using a 2_{III}^{5-2} design with $I = ABD$ and $I = BCE$. The results she obtains are $de = 95$, $ae = 134$, $b = 158$, $abd = 190$, $cd = 92$, $ac = 187$, $bce = 155$, and $abcde = 185$.

(a) Verify that the treatment combinations given are correct. Estimate the effects, assuming three-factor and higher interactions are negligible.

(b) Suppose that a second fraction is added to the first, for example, $ade = 136$, $e = 93$, $ab = 187$, $bd = 153$, $acd = 139$, $c = 99$, $abce = 191$, and $bcde = 150$. How was this second fraction obtained? Add this data to the original fraction, and estimate the effects.

(c) Suppose that the fraction $abc = 189$, $ce = 96$, $bcd = 154$, $acde = 135$, $abe = 193$, $bde = 152$, $ad = 137$, and $(1) = 98$ was run. How was this fraction obtained? Add this data to the original fraction and estimate the effects.

11-20 Construct a 2^{5-1} design. Show how the design may be run in two blocks of eight observations each. Are any main effects or two-factor interactions confounded with blocks?

11-21 Construct a 2^{7-2} design. Show how the design may be run in four blocks of eight observations each. Are any main effects or two-factor interactions confounded with blocks?

11-22 *Irregular Fractions of the 2^k* [John (1971)]. Consider a 2^4 design. We must estimate the four main effects and the six two-factor interactions, but the full 2^4 factorial cannot be run. The largest possible block size contains 12 runs. These 12 runs can be obtained from the four one-quarter replicates defined by $I = \pm AB = \pm ACD = \pm BCD$ by omitting the principal fraction. Show how the remaining three 2^{4-2} fractions can be combined to estimate the required effects, assuming three-factor and higher interactions are negligible. This design could be thought of as a three-quarter fraction.

11-23 Carbon anodes used in a smelting process are baked in a ring furnace. An experiment is run in the furnace to determine which factors influence the weight of packing material that is stuck to the anodes after baking. Six variables are of interest, each at two levels: A = pitch/fines ratio (0.45, 0.55); B = packing material type (1, 2); C = packing material temperature (ambient, 325°C); D = flue location (inside, outside); E = pit temperature (ambient, 195°C); and F = delay time before packing (zero, 24 hours). A 2^{6-3} design is run, and three replicates are obtained at each of the design points. The weight of packing material stuck to the anodes is measured in grams. The data in run order are as follows: abd = (984, 826, 936); $abcdef$ = (1275, 976, 1457); be = (1217, 1201, 890); af = (1474, 1164, 1541); def = (1320, 1156, 913); cd = (765, 705, 821); ace = (1338, 1254, 1294); and bcf = (1325, 1299, 1253). We wish to minimize the amount of stuck packing material

(a) Verify that the eight runs correspond to a 2^{6-3}_{III} design. What is the alias structure?

(b) Use the average weight as a response. What factors appear to be influential?

(c) Use the range of the weights as a response. What factors appear to be influential?

(d) What recommendations would you make to the process engineers?

11-24 A 16-run experiment was performed in a semiconductor manufacturing plant to study the effects of six factors on the curvature or camber of the substrate devices produced. The six variables and their levels are shown below:

Run	Lamination Temperature (°C)	Lamination Time (s)	Lamination Pressure (Ton)	Firing Temperature (°C)	Firing Cycle Time (h)	Firing Dew Point (°C)
1	55	10	5	1580	17.5	20
2	75	10	5	1580	29	26
3	55	25	5	1580	29	20
4	75	25	5	1580	17.5	26
5	55	10	10	1580	29	26
6	75	10	10	1580	17.5	20
7	55	25	10	1580	17.5	26
8	75	25	10	1580	29	20
9	55	10	5	1620	17.5	26
10	75	10	5	1620	29	20
11	55	25	5	1620	29	26
12	75	25	5	1620	17.5	20
13	55	10	10	1620	29	20
14	75	10	10	1620	17.5	26
15	55	25	10	1620	17.5	20
16	75	25	10	1620	29	26

Each run was replicated four times, and a camber measurement was taken on the substrate. The data are shown below:

	Camber for Replicate (in/in)				Total	Mean	Standard
Run	1	2	3	4	$(10^{-4}$ in/in)	$(10^{-4}$ in/in)	Deviation
1	0.0167	0.0128	0.0149	0.185	629	157.25	24.418
2	0.0062	0.0066	0.0044	0.0020	192	48.00	20.976
3	0.0041	0.0043	0.0042	0.0050	176	44.00	4.0825
4	0.0073	0.0081	0.0039	0.0030	223	55.75	25.025
5	0.0047	0.0047	0.0040	0.0089	223	55.75	22.410
6	0.0219	0.0258	0.0147	0.0296	920	230.00	63.639
7	0.0121	0.0090	0.0092	0.0086	389	97.25	16.029
8	0.0255	0.0250	0.0226	0.0169	900	161.13	73.749
9	0.0032	0.0023	0.0077	0.0069	201	50.25	26.725
10	0.0078	0.0158	0.0060	0.0045	341	85.25	50.341
11	0.0043	0.0027	0.0028	0.0028	126	31.50	7.681
12	0.0186	0.0137	0.0158	0.0159	640	160.00	20.083
13	0.0110	0.0086	0.0101	0.0158	455	113.75	31.12
14	0.0065	0.0109	0.0126	0.0071	371	92.75	29.51
15	0.0155	0.0158	0.0145	0.0145	603	150.75	6.75
16	0.0093	0.0124	0.0110	0.0133	460	115.00	17.45

(a) What type of design did the experimenters use?

(b) What are the alias relationships in this design?

(c) Do any of the process variables affect average camber?

(d) Do any of the process variables affect the variability in camber measurements?

(e) If it is important to reduce camber as much as possible, what recommendations would you make?

11-25 A spin coater is used to apply photoresist to a bare silicon wafer. This operation usually occurs early in the semiconductor manufacturing process, and the average coating thickness and the variability in the coating thickness has an important impact on downstream manufacturing steps. Six variables are used in the experiment. The variables and their high and low levels are as follows:

Factor	Low Level	High Level
Final spin speed	7300 rpm	6650 rpm
Acceleration rate	5	20
Volume of resist applied	3 cc	5 cc
Time of spin	14 s	6 s
Resist batch variation	Batch 1	Batch 2
Exhaust pressure	Cover off	Cover on

The experimenter decides to use a 2^{6-1} design, and to make three readings on resist thickness on each test wafer. The data are shown in Table 11-24.

(a) Verify that this is a 2^{6-1} design. Discuss the alias relationships in this design.

(b) What factors appear to affect average resist thickness?

(c) Since the volume of resist applied has little affect on average thickness, does this have any important practical implications for the process engineers?

(d) Project this design into a smaller design involving only the significant factors. Graphically display the results. Does this aid in interpretation?

(e) Use the range of resist thickness as a response variable. Is there any indication that any of these factors affect the variability in resist thickness?

(f) Where would you recommend that the process engineers run the process?

Table 11-24   Data for Problem 11-25

Run	A Volume	B Batch	C Time	D Speed	E Acc.	F Cover	Resist thickness Left	Center	Right	Avg.
1	5	Batch 2	14 sec	7350	5	Off	4531	4531	4515	4525.7
2	5	Batch 1	6 sec	7350	5	Off	4446	4464	4428	4446
3	3	Batch 1	6 sec	6650	5	Off	4452	4490	4452	4464.7
4	3	Batch 2	14 sec	7350	20	Off	4316	4328	4308	4317.3
5	3	Batch 1	14 sec	7350	5	Off	4307	4295	4289	4297
6	5	Batch 1	6 sec	6650	20	Off	4470	4492	4495	4485.7
7	3	Batch 1	6 sec	7350	5	On	4496	4502	4482	4493.3
8	5	Batch 2	14 sec	6650	20	Off	4542	4547	4538	4542.3
9	5	Batch 1	14 sec	6650	5	Off	4621	4643	4613	4625.7
10	3	Batch 1	14 sec	6650	5	On	4653	4670	4645	4656
11	3	Batch 2	14 sec	6650	20	On	4480	4486	4470	4478.7
12	3	Batch 1	6 sec	7350	20	Off	4221	4233	4217	4223.7
13	5	Batch 1	6 sec	6650	5	On	4620	4641	4619	4626.7
14	3	Batch 1	6 sec	6650	20	On	4455	4480	4466	4467
15	5	Batch 2	14 sec	7350	20	On	4255	4288	4243	4262
16	5	Batch 2	6 sec	7350	5	On	4490	4534	4523	4515.7
17	3	Batch 2	14 sec	7350	5	On	4514	4551	4540	4535
18	3	Batch 1	14 sec	6650	20	Off	4494	4503	4496	4497.7
19	5	Batch 2	6 sec	7350	20	Off	4293	4306	4302	4300.3
20	3	Batch 2	6 sec	7350	5	Off	4534	4545	4512	4530.3
21	5	Batch 1	14 sec	6650	20	On	4460	4457	4436	4451
22	3	Batch 2	6 sec	6650	5	On	4650	4688	4656	4664.7
23	5	Batch 1	14 sec	7350	20	Off	4231	4244	4230	4235
24	3	Batch 2	6 sec	7350	20	On	4225	4228	4208	4220.3
25	5	Batch 1	14 sec	7350	5	On	4381	4391	4376	4382.7
26	3	Batch 2	6 sec	6650	20	Off	4533	4521	4511	4521.7
27	3	Batch 1	14 sec	7350	20	On	4194	4230	4172	4198.7
28	5	Batch 2	6 sec	6650	5	Off	4666	4695	4672	4677.7
29	5	Batch 1	6 sec	7350	20	On	4180	4213	4197	4196.7
30	5	Batch 2	6 sec	6650	20	On	4465	4496	4463	4474.7
31	5	Batch 2	14 sec	6650	5	On	4653	4685	4665	4667.7
32	3	Batch 2	14 sec	6650	5	Off	4683	4712	4677	4690.7

11-26 Harry and Judy Peterson-Nedry (two friends of the author) own a vineyard in Oregon. They grow several varieties of grapes and manufacture wine. Harry and Judy have used factorial designs for process and product development in the winemaking segment of their business. This problem describes the experiment conducted for their 1985 Pinot Noir. Eight variables, shown below, were originally studied in this experiment:

Variable	Low Level ($-$)	High Level ($+$)
A = Pinot Noir clone	Pommard	Wadenswil
B = Oak type	Allier	Troncais
C = Age of barrel	Old	New
D = Yeast/skin contact	Champagne	Montrachet
E = Stems	None	All
F = Barrel toast	Light	Medium
G = Whole cluster	None	10%
H = Fermentation temperature	Low (75°F max)	High (92°F max)

Harry and Judy decided to use a 2^{8-4}_{IV} design with 16 runs. The wine was taste-tested by a panel of experts on 8 March 1986. Each expert ranked the 16 samples of wine tasted, with rank 1 being the best. The design and the taste-test panel results are shown in the Table 11-25.

(a) What are the alias relationships in the design selected by Harry and Judy?

(b) Use the average ranks ($\bar{y}$) as a response variable. Analyze the data and draw conclusions. You will find it helpful to examine a normal probability plot of the effect estimates.

(c) Use the standard deviation of the ranks (or some appropriate transformation such as log s) as a response variable. What conclusions can you draw about the effects of the eight variables on variability in wine quality?

(d) After looking at the results, Harry and Judy decide that one of the panel members (DCM) knows more about beer than he does about wine, so they decide to delete his ranking. What affect would this have on the results and conclusions from parts (b) and (c)?

(e) Suppose that just before the start of the experiment, Harry and Judy discovered that the eight new barrels they ordered from France for use in the experiment would not arrive in time, and all 16 runs would have to be made with old barrels. If Harry and Judy just drop column C from their design, what does this do to the alias relationships? Do they need to start over and construct a new design?

(f) Harry and Judy know from experience that some treatment combinations are unlikely to produce good results. For example, the run with all eight variables at the high level generally results in a poorly rated wine. This was confirmed in the 8 March 1986 taste test. They want to set up a new design for their 1986 Pinot Noir using these same eight variables, but they do not want to make the run with all eight factors at the high level. What design would you suggest?

Table 11-25   Design and Results for Wine Tasting Experiment

Run	Variable								Panel rankings					Summary	
	A	B	C	D	E	F	G	H	HPN	JPN	CAL	DCM	RGB	$\bar{y}$	s
1	−	−	−	−	−	−	−	−	12	6	13	10	7	9.6	3.05
2	+	−	−	−	−	+	+	+	10	7	14	14	9	10.8	3.11
3	−	+	−	−	+	−	+	+	14	13	10	11	15	12.6	2.07
4	+	+	−	−	+	+	−	−	9	9	7	9	12	9.2	1.79
5	−	−	+	−	+	+	+	−	8	8	11	8	10	9.0	1.41
6	+	−	+	−	+	−	−	+	16	12	15	16	16	15.0	1.73
7	−	+	+	−	−	+	−	+	6	5	6	5	3	5.0	1.22
8	+	+	+	−	−	−	+	−	15	16	16	15	14	15.2	0.84
9	−	−	−	+	−	−	+	+	1	2	3	3	2	2.2	0.84
10	+	−	−	+	−	+	−	−	7	11	4	7	6	7.0	2.55
11	−	+	−	+	+	−	−	−	13	3	8	12	8	8.8	3.96
12	+	+	−	+	+	+	+	+	3	1	5	1	4	2.8	1.79
13	−	−	+	+	−	+	−	+	2	10	2	4	5	9.6	3.29
14	+	−	+	+	−	−	+	−	4	4	1	2	1	2.4	1.52
15	−	+	+	+	+	−	−	−	5	15	9	6	11	9.2	4.02
16	+	+	+	+	+	+	+	+	11	14	12	13	13	12.6	1.14

Chapter 12
Some Other Topics Regarding Factorial and Fractional Factorial Designs

The two-level series of factorial and fractional factorial designs discussed in Chapters 9, 10, and 11 are widely used in industrial research and development. There are some extensions and variations of these designs that are occasionally useful, such as the designs for cases where all the factors are present at three levels. These 3^k designs will be discussed in this chapter. We will also consider cases where some factors have two levels and other factors have either three or four levels.

The last few years have seen a growing awareness of and interest in some ideas of Professor Genechi Taguchi regarding experimental design and their application in quality improvement. This chapter will present some of these ideas and will provide a brief evaluation of Taguchi's contributions to quality engineering.

12-1 THE 3^k FACTORIAL DESIGN

12-1.1 Notation and Motivation for the 3^k Design

We now discuss the 3^k factorial design; that is, a factorial arrangement with k factors each at three levels. Factors and interactions will be denoted by capital letters. Without loss of generality, we may refer to the three levels of the factors as low, intermediate, and high. These levels will be designated by the digits 0 (low), 1 (intermediate), and 2 (high). Each treatment combination in the 3^k design will be denoted by k digits, where the first digit indicates the level of factor A, the second digit indicates the level of factor B, . . . , and the kth digit indicates the level of factor K. For example, in a 3^2 design, 00 denotes the treatment combination corresponding to A and B both at the low level, and 01 denotes the treatment combination corresponding to A at the low level and B at the intermediate level.

This system of notation could have been used for the 2^k designs presented previously, with 0 and 1 used in place of the minus and plus signs, respectively. In the 2^k design, we prefer the $\pm$ notation because it facilitates the geometric view of the design and because it is directly applicable to regression modeling, blocking, and the construction of fractional factorials.

The 3^k design is often considered by an experimenter who is concerned about curvature in the response function. The addition of a third level allows the relationship between the response and each factor to be modeled with a quadratic relationship. However, two points need to be considered:

1. The 3^k design is not the best way to model a quadratic relationship; the *response surface* designs discussed in Chapter 16 are superior alternatives.
2. The 2^k design augmented with center points, as discussed in Chapter 9, is an excellent way to obtain an indication of curvature. It allows one to keep the size and complexity of the design low and simultaneously obtain some protection against curvature.

12-1.2 The 3^2 Design

The simplest design in the 3^k system is the 3^2 design, which has two factors, each at three levels. The treatment combinations for this design are shown in Figure 12-1. Since there are $3^2 = 9$ treatment combinations, there are eight

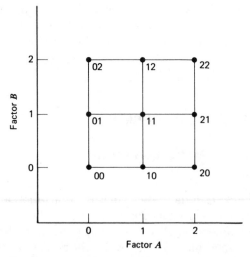

Figure 12-1. Treatment combinations in a 3^2 design.

degrees of freedom between these treatment combinations. The main effects of A and B each have two degrees of freedom, and the AB interaction has four degrees of freedom. If there are n replicates, there will be $n3^2 - 1$ total degrees of freedom and $3^2(n - 1)$ degrees of freedom for error.

The sums of squares for A, B, and AB may be computed by the usual methods for factorial designs discussed in Chapter 7. Alternatively, in Section 12-1.5 we give an extension of Yates' algorithm to the 3^k series. The sum of squares for any main effect may be partitioned into a linear and a quadratic component, each with a single degree of freedom, using the orthogonal contrast constants as demonstrated in Chapter 7. Of course, this is only meaningful if the factor is quantitative and if the three levels are equally spaced.

As an illustration, consider the tool life problem in Example 7-6. The objective of this experiment was to investigate the effect of cutting speed and tool angle on tool life in a numerically controlled machine. Three cutting speeds and three tool angles were selected for study. This is a 3^2 design with $n = 2$ replicates. The analysis of variance is repeated for convenience in Table 12-1. For details of the sum of squares calculations and the partitioning of the main effects into linear and quadratic components, refer to Chapter 7, Section 7-6.

The two-factor interaction AB may be partitioned in two ways. The first method consists of subdividing AB into the four single-degree-of-freedom components corresponding to $AB_{L\times L}$, $AB_{L\times Q}$, $AB_{Q\times L}$, and $AB_{Q\times Q}$. This is done using orthogonal contrast constants, as demonstrated in Example 7-6. For the tool life data, this yields $SS_{AB_{L\times L}} = 8.00$ $SS_{AB_{L\times Q}} = 42.67$, $SS_{AB_{Q\times L}} = 2.67$, and $SS_{AB_{Q\times Q}} = 8.00$. Since this is an orthogonal partitioning of AB, note that $SS_{AB} = SS_{AB_{L\times L}} + SS_{AB_{L\times Q}} + SS_{AB_{Q\times L}} + SS_{AB_{Q\times Q}}$.

The second method is based on orthogonal Latin squares. Consider the totals of the treatment combinations for the data in Example 7-6. These totals are shown in Figure 12-2 as the circled numbers in the squares. The two factors A and B correspond to the rows and columns, respectively, of a 3 × 3 Latin square. In Figure 12-2, two particular 3 × 3 Latin squares are shown superimposed on the cell totals.

Table 12-1   Analysis of Variance for Example 7-6

Source of Variation	Sum of Squares	Degrees of Freedom	Mean Square	F_0
Tool Angle (A)	24.33	2	12.17	8.45[a]
(A_L)	(8.33)	1	(8.33)	5.78[a]
(A_Q)	(16.00)	1	(16.00)	11.11[a]
Cutting Speed (B)	25.33	2	12.67	8.80[a]
(B_L)	(21.33)	1	(21.33)	14.81[a]
(B_Q)	(4.00)	1	(4.00)	2.77
AB interaction	61.34	4	6.82	4.75[a]
Error	13.00	9	1.44	
Total	124.00			

[a]Significant at 5 percent.

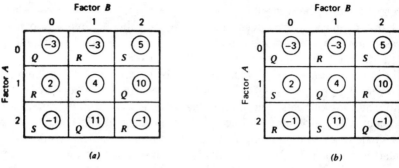

Figure 12-2. Treatment combination totals from Example 7-6 with two orthogonal Latin squares superimposed.

These two Latin squares are *orthogonal;* that is, if one square is superimposed on the other, each letter in the first square will appear exactly once with each letter in the second square. The totals for the letters in the (a) square are $Q = 18$, $R = -2$, and $S = 8$, and the sum of squares between these totals is $[18^2 + (-2)^2 + 8^2]/(3)(2) - [24^2/(9)(2)] = 33.34$, with two degrees of freedom. Similarly, the letter totals in the (b) square are $Q = 0$, $R = 6$, and $S = 18$, and the sum of squares between these totals is $[0^2 + 6^2 + 18^2]/(3)(2) - [24^2/(9)(2)] = 28.00$, with two degrees of freedom. Note that the sum of these two components is

$$33.34 + 28.00 = 61.34 = SS_{AB}$$

with $2 + 2 = 4$ degrees of freedom.

In general, the sum of squares computed from square (a) is called the AB component of interaction, and the sum of squares computed from square (b) is called the AB^2 component of interaction. The components AB and AB^2 each have two degrees of freedom. This terminology is used because if we denote the levels (0, 1, 2) for A and B by x_1 and x_2, respectively, then we find that the letters occupy cells according to the following pattern:

Square (a)	Square (b)
$Q: x_1 + x_2 = 0$ (mod 3)	$Q: x_1 + 2x_2 = 0$ (mod 3)
$R: x_1 + x_2 = 1$ (mod 3)	$S: x_1 + 2x_2 = 1$ (mod 3)
$S: x_1 + x_2 = 2$ (mod 3)	$R: x_1 + 2x_2 = 2$ (mod 3)

For example, in square (b), note that the middle cell corresponds to $x_1 = 1$ and $x_2 = 1$; thus, $x_1 + 2x_2 = 1 + (2)(1) = 3 = 0$ (mod 3), and Q would occupy the middle cell. When considering expressions of the form $A^p B^q$, we establish the convention that the only exponent allowed on the first letter is one. If the first letter is not one, the entire expression is squared and the exponents are reduced modulus 3. For example, $A^2 B$ is the same as AB^2 since

$$A^2 B = (A^2 B)^2 = A^4 B^2 = AB^2$$

The AB and AB^2 components of the AB interaction have no actual meaning and are usually not displayed in the analysis of variance table. However, this rather arbitrary partitioning of the AB interaction into two orthogonal two-degree-of-freedom components is very useful in constructing more complex designs. Also, there is no connection between the AB and AB^2 components of interaction and the sums of squares for $AB_{L \times L}$, $AB_{L \times Q}$, $AB_{Q \times L}$, and $AB_{Q \times Q}$.

The AB and AB^2 components of interaction may be computed another way. Consider the treatment combination totals in either square in Figure 12-2. If we add the data by diagonals downward from left to right, we obtain the totals $-3 + 4 - 1 = 0$, $-3 + 10 - 1 = 6$, and $5 + 11 + 2 = 18$. The sum of squares between these totals is 28.00 (AB^2). Similarly, the diagonal totals downward from right to left are $5 + 4 - 1 = 8$, $-3 + 2 - 1 = -2$, and $-3 + 11 + 10 = 18$. The sum of squares between these totals is 33.34 (AB). Yates called these components of interaction the I and J components of interaction, respectively. We use both notations interchangeably; that is,

$$I(AB) = AB^2$$
$$J(AB) = AB$$

12-1.3 The 3^3 Design

Now suppose there are three factors (A, B, and C) under study, and each factor is at three levels arranged in a factorial experiment. This is a 3^3 factorial design, and the experimental layout and treatment combination notation are shown in Figure 12-3. The 27 treatment combinations have 26 degrees of freedom. Each main effect has 2 degrees of freedom, each two-factor interaction has 4 degrees of freedom, and the three-factor interaction has 8 degrees of freedom. If there are n replicates, there are $n3^3 - 1$ total degrees of freedom and $3^3(n - 1)$ degrees of freedom for error.

The sums of squares may be calculated using the standard methods for factorial designs. In addition, if the factors are quantitative and equally spaced, the main effects may be partitioned into linear and quadratic components, each with a single degree of freedom. The two-factor interactions may be decomposed into linear × linear, linear × quadratic, quadratic × linear, and quadratic × quadratic effects. Finally, the three-factor interaction ABC can be partitioned into eight single-degree-of-freedom components corresponding to linear × linear × linear, linear × linear × quadratic, and so on. Such a breakdown for the three-factor interaction is generally not very useful.

It is also possible to partition the two-factor interactions into their I and J components. These would be designated AB, AB^2, AC, AC^2, BC, and BC^2, and each component would have 2 degrees of freedom. As in the 3^2 design, these components have no physical significance.

The three-factor interaction ABC may be partitioned into four orthogonal two-degrees-of-freedom components, which are usually called the W, X, Y, and Z

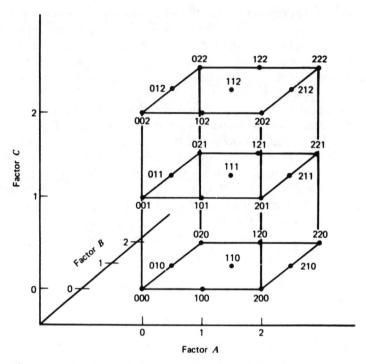

Figure 12-3.   Treatment combinations in a 3^3 design.

components of the interaction. They are also referred to as the AB^2C^2, AB^2C, ABC^2, and ABC components of the ABC interaction, respectively. The two notations are used interchangably; that is,

$$W(ABC) = AB^2C^2$$
$$X(ABC) = AB^2C$$
$$Y(ABC) = ABC^2$$
$$Z(ABC) = ABC$$

Note that no first letter can have an exponent other than 1. Like the I and J components, the W, X, Y, and Z components have no practical interpretation. They are, however, useful in constructing more complex designs.

Example 12-1

A machine is used to fill 5 gallon metal containers with soft drink syrup. The variable of interest is the amount of syrup loss due to frothing. Three factors are thought to influence frothing: the nozzle configuration (A), the operator (B), and the operating pressure (C). Three nozzles, three operators, and three pressures are chosen and two replicates of a 3^3 factorial experiment are run. The coded data are shown in Table 12-2.

Table 12-2   Syrup Loss Data for Example 12-1 (units are cubic centimeters − 70)

	Nozzle (A)								
	0			1			2		
	Operator (B)								
Pressure (C)	0	1	2	0	1	2	0	1	2
0	−35	−45	−40	17	−65	20	−39	−55	15
	−25	−60	15	24	−58	4	−35	−67	−30
1	100	−10	80	55	−55	110	90	−28	110
	75	30	54	120	−44	44	113	−26	135
2	4	−40	31	−23	−64	−20	−30	−61	54
	5	−30	36	−5	−62	−31	−55	−52	4

The analysis of variance for the syrup loss data is shown in Table 12-3. The sums of squares have been computed by the usual methods. We see that the operator and operating pressure are significant at the 1 percent level. All three two-factor interactions are also significant. The two-factor interactions are analyzed graphically in Figure 12-4. Operator 1 is superior to operators 0 and 2, and nozzle type 1 and the high pressure (2) seem most effective in reducing syrup loss.

■

We will now demonstrate numerically the partitioning of the ABC interaction into its four orthogonal two-degree-of-freedom components using the data in Example 12-1. The general procedure has been described by Cochran and Cox (1957), Davies (1956), and Hicks (1973). First, select any two of the three factors, say AB, and compute the I and J totals of the AB interaction at each level of the

Table 12-3   Analysis of Variance for Syrup Loss Data

Source of Variation	Sum of Squares	Degrees of Freedom	Mean Square	F_0
A, nozzle	993.77	2	496.89	1.17
B, operator	61,190.33	2	30,595.17	71.74[b]
C, pressure	69,105.33	2	34,552.67	81.01[b]
AB	6,300.90	4	1,575.22	3.69[a]
AC	7,513.90	4	1,878.47	4.40[b]
BC	12,854.34	4	3,213.58	7.53[b]
ABC	4,628.76	8	578.60	1.36
Error	11,515.50	27	426.50	
Total	174,102.83	53		

[a]Significant at 5 percent.
[b]Significant at 1 percent.

third factor C. These calculations follow:

C	B	A 0	1	2	Totals I	J
	0	−60	41	−74	−198	−222
0	1	−105	−123	−122	−106	−79
	2	−25	24	−15	−155	−158
	0	185	175	203	331	238
1	1	20	−99	−54	255	440
	2	134	154	245	377	285
	0	9	−28	−85	−59	−144
2	1	−70	−126	−113	−74	−40
	2	67	−51	58	−206	−155

The $I(AB)$ and $J(AB)$ totals are now arranged in a two-way table with factor C, and the I and J diagonal totals of this new display are computed:

C	$I(AB)$			Totals I	J	C	$J(AB)$			Totals I	J
0	−198	−106	−155	−149	41	0	−222	−79	−158	63	138
1	331	255	377	212	19	1	238	440	285	62	4
2	−59	−74	−206	102	105	2	−144	−40	−155	40	23

The I and J diagonal totals computed above are actually the totals representing the quantities $I[I(AB) \times C] = AB^2C^2$, $J[I(AB) \times C] = AB^2C$, $I[J(AB) \times C] = ABC^2$, and $J[J(AB) \times C] = ABC$, or the W, X, Y, and Z components of ABC. The sums of squares are found in the usual way; that is,

$$I[I(AB) \times C] = AB^2C^2 = W(ABC)$$
$$= \frac{(-149)^2 + (212)^2 + (102)^2}{18} - \frac{(165)^2}{54} = 3804.11$$

$$J[I(AB) \times C] = AB^2C = X(ABC)$$
$$= \frac{(41)^2 + (19)^2 + (105)^2}{18} - \frac{(165)^2}{54} = 221.77$$

$$I[J(AB) \times C] = ABC^2 = Y(ABC)$$
$$= \frac{(63)^2 + (62)^2 + (40)^2}{18} - \frac{(165)^2}{54} = 18.77$$

$$J[J(AB) \times C] = ABC = Z(ABC)$$
$$= \frac{(138)^2 + (4)^2 + (23)^2}{18} - \frac{(165)^2}{54} = 584.11$$

Figure 12-4.   Two-factor interactions for Example 12-1.

Although this is an orthogonal partitioning of SS_{ABC}, we point out again that it is not customarily displayed in the analysis of variance table. In subsequent sections, we discuss the occasional need for the computation of one or more of these components.

12-1.4   The General 3^k Design

The concepts utilized in the 3^2 and 3^3 designs can be readily extended to the case of k factors, each at three levels, that is, to a 3^k factorial design. The usual digital notation is employed for the treatment combinations, so 0120 represents a treatment combination in a 3^4 design with A and D at the low levels, B at the intermediate level, and C at the high level. Threre are 3^k treatment combinations, with $3^k - 1$ degrees of freedom between them. These treatment combinations allow sums of squares to be determined for k main effects, each with two degrees of freedom; $\binom{k}{2}$ two-factor interactions, each with four degrees of freedom; . . . ; and one k-factor interaction with 2^k degrees of freedom. In general, an h-factor interaction has 2^h degrees of freedom. If there are n replicates, there are $n3^k - 1$ total degrees of freedom and $3^k(n - 1)$ degrees of freedom for error. The analysis of variance is outlined in Table 12-4.

Sums of squares for effects and interactions are computed by the usual methods for factorial designs. As a general rule, three-factor and higher interactions are not broken down any further. However, any h-factor interaction has 2^{h-1} orthogonal two-degrees-of-freedom components. For example, the four-factor interaction $ABCD$ has $2^{4-1} = 8$ orthogonal two-degrees-of-freedom components, denoted by $ABCD^2$, ABC^2D, AB^2CD, $ABCD$, ABC^2D^2, AB^2C^2D, AB^2CD^2, and $AB^2C^2D^2$. In writing these components, note that the only exponent allowed on the first letter is 1. If the exponent on the first letter is not 1, then the entire expression must be squared and the exponents reduced modulus 3. To demon-

Table 12-4   Analysis of Variance for a 3^k Design

Source of Variation	Sum of Squares	Degrees of Freedom
k main effects		
A	SS_A	2
B	SS_B	2
.	.	.
.	.	.
.	.	.
K	SS_K	2
$\binom{k}{2}$ two-factor interactions		
AB	SS_{AB}	4
AC	SS_{AC}	4
.	.	.
.	.	.
.	.	.
JK	SS_{JK}	4
$\binom{k}{3}$ three-factor interactions		
ABC	SS_{ABC}	8
ABD	SS_{ABD}	8
.	.	.
.	.	.
.	.	.
IJK	SS_{IJK}	8
.	.	.
.	.	.
$\binom{k}{k} = 1$ k-factor interaction		
$ABC . . . K$	$SS_{ABC...K}$	2^k
Error	SS_E	$3^k(n - 1)$
Total	SS_T	$n3^k - 1$

strate, consider

$$A^2BCD = (A^2BCD)^2 = A^4B^2C^2D^2 = AB^2C^2D^2$$

These interaction components have no physical interpretation, but they are useful in constructing more complex designs.

The size of the design increases rapidly with k. For example, a 3^3 design has 27 treatment combinations per replication, a 3^4 design has 81, a 3^5 design has 243, and so on. Therefore, frequently only a single replicate of the 3^k design is considered, and higher-order interactions are combined to provide an estimate of error. As an illustration, if three-factor and higher interactions are negligible, then a single replicate of the 3^3 design provides 8 degrees of freedom for error, and a single replicate of the 3^4 design provides 48 degrees of freedom for error. These are still large designs for $k \geq 3$ factors, and, consequently, not too useful.

12-1.5 Yates' Algorithm for the 3^k Design

Yates' algorithm can be modified for use in the 3^k factorial design. We illustrate the procedure using the data in Example 7-1. The data for this example are given in Table 7-6. This is a 3^2 design used to investigate the effect of material type (A) and temperature (B) on the life of a battery. There are $n = 4$ replicates.

The procedure is displayed in Table 12-5. The treatment combinations are written down in standard order; that is, the factors are introduced one at a time, each level being combined successively with every set of factor levels above it in the table. (The standard order for a 3^3 design would be 000, 100, 200, 010, 110, 210, 020, 120, 220, 001, ...) The Response column contains the total of all observations taken under the corresponding treatment combination. The entries in column (1) are computed as follows: The first third of the column consists of the sums of each of the three sets of three values in the Response column. The second third of the column is the third minus the first observation

Table 12-5 Yates' Algorithm for the 3^2 Design in Example 7-1

Treatment Combination	Response	(1)	(2)	Effect	Divisor	Sum of Squares
00	539	1738	3799	—	—	—
10	623	1291	503	A_L	$2^1 \times 3^1 \times 4$	10,542.04
20	576	770	−101	A_Q	$2^1 \times 3^2 \times 4$	141.68
01	229	37	−968	B_L	$2^1 \times 3^1 \times 4$	39,042.66
11	479	354	75	$AB_{L \times L}$	$2^2 \times 3^0 \times 4$	351.56
21	583	112	307	$AB_{Q \times L}$	$2^2 \times 3^1 \times 4$	1,963.52
02	230	−131	−74	B_Q	$2^1 \times 3^2 \times 4$	76.06
12	198	−146	−559	$AB_{L \times Q}$	$2^2 \times 3^1 \times 4$	6,510.02
22	342	176	337	$AB_{Q \times Q}$	$2^2 \times 3^2 \times 4$	788.67

in the same set of three. This operation computes the linear component of the effect. The last third of the column is obtained by taking the sum of the first and third value minus twice the second in each set of three observations. This computes the quadratic component. For example, in column (1), the second, fifth, and eighth entries are $229 + 479 + 583 = 1291$, $-229 + 583 = 354$, and $229 - (2)(479) + 583 = -146$, respectively. Column (2) is obtained similarly from column (1). In general, k columns must be constructed.

The Effect column is determined by converting the treatment combinations at the left of the row into corresponding effects. That is, 10 represents the linear effect of A, A_L, and 11 represents the $AB_{L \times L}$ component of the AB interaction. The entries in the divisor column are found from

$$2^r 3^t n$$

where r is the number of factors in the effect considered, t is the number of factors in the experiment minus the number of linear terms in this effect, and n is the number of replicates. For example, B_L has the divisor $2^1 \times 3^1 \times 4 = 24$.

The sums of squares are obtained by squaring the element in column (2) and dividing by the corresponding entry in the divisor column. The sum of squares column now contains all of the required quantities to construct an analysis of variance table if both factors A and B are quantitative. However, in this example, factor A (material type) is qualitative; thus, the linear and quadratic partitioning of A is not appropriate. Individual observations are required to compute the total sum of squares, and the error sum of squares is obtained by subtraction.

The analysis of variance is summarized in Table 12-6. A comparison of it with Table 7-22 shows that substantially the same results were obtained by conventional analysis of variance methods. Examples 7-1 and 7-5 provide an interpretation of this experiment.

Table 12-6   Analysis of Variance for the 3^2 Design in Example 7-1

Source of Variation	Sum of Squares	Degrees of Freedom	Mean Square	F_0
$A = ``A_L" + ``A_Q"$	10,683.72	2	5,341.86	7.91[b]
B, temperature	39,118.72	2	19,558.36	28.97[b]
(B_L)	(39,042.67)	1	39,042.67	57.82[b]
(B_Q)	(76.05)	1	76.05	0.12
AB	9,613.78	4	2,403.44	3.56[a]
$(A \times B_L = ``AB_{L \times L}" + ``AB_{Q \times L}")$	(2,315.08)	2	1,157.54	1.71
$(A \times B_Q = ``AB_{L \times Q}" + ``AB_{Q \times Q}")$	(7,298.70)	2	3,649.75	5.41[a]
Error	18,230.75	27	675.21	
Total	77,646.97	35		

[a]Significant at 5 percent.
[b]Significant at 1 percent.

12-2 CONFOUNDING IN THE 3^k FACTORIAL DESIGN

Even when a single replicate of the 3^k design is considered, the design requires so many runs that it is unlikely that all 3^k runs can be made under uniform conditions. Thus, confounding in blocks is often necessary. The 3^k design may be confounded in 3^p incomplete blocks, where $p < k$. Thus, these designs may be confounded in three blocks, nine blocks, and so on.

12-2.1 The 3^k Factorial Design in Three Blocks

Suppose that we wish to confound the 3^k design in three incomplete blocks. These three blocks have two degrees of freedom among them; thus, there must be two degrees of freedom confounded with blocks. Recall that in the 3^k factorial series each main effect has two degrees of freedom. Furthermore, every two-factor interaction has four degrees of freedom and can be decomposed into two components of interaction (e.g., AB and AB^2), each with two degrees of freedom; every three-factor interaction has eight degrees of freedom and can be decomposed into four components of interaction (e.g., ABC, ABC^2, AB^2C, and AB^2C^2), each with two degrees of freedom; and so on. Therefore, it is convenient to confound a component of interaction with blocks.

The general procedure is to construct a defining contrast

$$L = \alpha_1 x_1 + \alpha_2 x_2 + \cdots + \alpha_k x_k$$

where α_i represents the exponent on the ith factor in the effect to be confounded and x_i is the level of the ith factor in a particular treatment combination. For the 3^k series, we have $\alpha_i = 0$, 1, or 2 with the first nonzero α_i being unity, and $x_i = 0$ (low level), 1 (intermediate level), or 2 (high level). The treatment combinations in the 3^k design are assigned to blocks based on the value of L (mod 3). Since L (mod 3) can take on only the values 0, 1, or 2, three blocks are uniquely defined. The treatment combinations satisfying $L = 0$ (mod 3) constitute the *principal block*. This block will always contain the treatment combination 00 . . . 0.

For example, suppose we wish to construct a 3^2 factorial design in three blocks. Either component of the AB interaction, AB or AB^2, may be confounded with blocks. Arbitrarily choosing AB^2, we obtain the defining contrast

$$L = x_1 + 2x_2$$

The value of L (mod 3) of each treatment combination may be found as follows.

00: $L = 1(0) + 2(0) = 0 = 0$ (mod 3)   11: $L = 1(1) + 2(1) = 3 = 0$ (mod 3)
01: $L = 1(0) + 2(1) = 2 = 2$ (mod 3)   21: $L = 1(2) + 2(1) = 4 = 1$ (mod 3)
02: $L = 1(0) + 2(2) = 4 = 1$ (mod 3)   12: $L = 1(1) + 2(2) = 5 = 2$ (mod 3)
10: $L = 1(1) + 2(0) = 1 = 1$ (mod 3)   22: $L = 1(2) + 2(2) = 6 = 0$ (mod 3)

20: $L = 1(2) + 2(0) = 2 = 2$ (mod 3)

The blocks are shown in Figure 12-5.

The elements in the principal block form a group with respect to addition modulus 3. Referring to Figure 12-5, we see that $11 + 11 = 22$, and $11 + 22 = 00$. Treatment combinations in the other two blocks may be generated by adding modulus 3, any element in the new block, to the elements of the principal block. Thus, for block 2 we use 10 and obtain

$$10 + 00 = 10 \qquad 10 + 11 = 21 \qquad \text{and} \qquad 10 + 22 = 02$$

To generate block 3, using 01 we find

$$01 + 00 = 01 \qquad 01 + 11 = 12 \qquad \text{and} \qquad 01 + 22 = 20$$

Example 12-2

We illustrate the statistical analysis of the 3^2 design confounded in three blocks by using the following data, which come from the single replicate of the 3^2 design shown in Figure 12-5. Using conventional methods for the analysis of factorials, we find that $SS_A = 131.56$ and $SS_B = 0.22$. We also find that

$$SS_{\text{Blocks}} = \frac{(0)^2 + (7)^2 + (0)^2}{3} - \frac{(7)^2}{9} = 10.89$$

	Block 1	Block 2	Block 3
	$00 = \;\;\;4$ $11 = -4$ $22 = \;\;\;0$	$10 = -2$ $21 = \;\;\;1$ $02 = \;\;\;8$	$01 = \;\;\;5$ $12 = -5$ $20 = \;\;\;0$
Block Totals =	0	7	0

However, SS_{Blocks} is exactly equal to the AB^2 component of interaction. To see this, write the observation as follows:

		\multicolumn{3}{c}{Factor B}		
		0	1	2
	0	4	5	8
Factor A	1	-2	-4	-5
	2	0	1	0

Block 1	Block 2	Block 3
00 11 22	10 21 02	01 12 20

Figure 12-5. The 3^2 design in three blocks with AB^2 confounded.

Recall from Section 12-1.2 that the I or AB^2 component of the AB interaction may be found by computing the sum of squares between the left-to-right diagonal totals in the above layout. This yields

$$SS_{AB2} = \frac{(0)^2 + (0)^2 + (7)^2}{3} - \frac{(7)^2}{9} = 10.89$$

which is identical to SS_{Blocks}.

The analysis of variance is shown in Table 12-7. Because there is only one replicate, no formal tests can be performed. Note that it is probably unsafe to use the AB component of interaction as an estimate of error.

■

We now look at a slightly more complicated design—a 3^3 factorial confounded in three blocks of nine runs each. A component of the three-factor interaction (e.g., AB^2C^2) is to be confounded with blocks. The defining contrast is

$$L = x_1 + 2x_2 + 2x_3$$

It is easy to verify that the treatment combinations 000, 012, and 101 belong in the principal block. The remaining elements in the principal block are generated as follows:

(1) 000	(4) 101 + 101 = 202	(7) 101 + 021 = 122
(2) 012	(5) 012 + 012 = 021	(8) 012 + 202 = 211
(3) 101	(6) 101 + 012 = 110	(9) 021 + 202 = 220

To find the elements in another block, note that the treatment combination 200 is not in the principal block. Thus, the elements of block 2 are

(1) 200 + 000 = 200	(4) 200 + 202 = 102	(7) 200 + 122 = 022
(2) 200 + 012 = 212	(5) 200 + 021 = 221	(8) 200 + 211 = 111
(3) 200 + 101 = 001	(6) 200 + 110 = 010	(9) 200 + 220 = 120

Notice that these elements all satisfy $L = 2 \pmod 3$. The final block is found by observing that 100 does not belong in block 1 or 2. Using 100 as above yields

(1) 100 + 000 = 100	(4) 100 + 202 = 002	(7) 100 + 122 = 222
(2) 100 + 012 = 112	(5) 100 + 021 = 121	(8) 100 + 211 = 011
(3) 100 + 101 = 201	(6) 100 + 110 = 210	(9) 100 + 220 = 020

The blocks are shown in Figure 12-6.

The analysis of variance for this design is shown in Table 12-8. Using this confounding scheme, information on all the main effects and two-factor interactions is available. The remaining components of the three-factor interaction

Table 12-7   Analysis of Variance for Data in Example 12-2

Source of Variation	Sum of Squares	Degrees of Freedom
Blocks (AB^2)	10.89	2
A	131.56	2
B	0.22	2
AB	2.89	2
Total	145.56	8

(ABC, AB^2C, and ABC^2) are combined as an estimate of error. The sum of squares for those three components could be obtained by subtraction. In general, for the 3^k design in three blocks, we would always select a component of the highest-order interaction to confound with blocks. The remaining unconfounded components of this interaction could be obtained by computing the k-factor interaction in the usual way and subtracting from this quantity the sum of squares for blocks.

12-2.2 The 3^k Factorial Design in Nine Blocks

In some experimental situations it may be necessary to confound the 3^k design in nine blocks. Thus, eight degrees of freedom will be confounded with blocks. To construct these designs, we choose *two* components of interaction and, as a result, two more will be confounded automatically, to yielding the required eight degrees of freedom. These two are the generalized interactions of the two effects originally chosen. In the 3^k system, the *generalized interactions* of two effects (e.g., P and Q) are defined as PQ and PQ^2 (or P^2Q).

Block 1	Block 2	Block 3
000	200	100
012	212	112
101	001	201
202	102	002
021	221	121
110	010	210
122	022	222
211	111	011
220	120	020

Figure 12-6. The 3^3 design in three blocks with AB^2C^2 confounded.

Table 12-8 Analysis of Variance for a 3^3 Design with AB^2C^2 Confounded

Source of Variation	Degrees of Freedom
Blocks (AB^2C^2)	2
A	2
B	2
C	2
AB	4
AC	4
BC	4
Error ($ABC + AB^2C + ABC^2$)	6
Total	26

The two components of interaction initially chosen yield *two* defining contrasts

$$L_1 = \alpha_1 x_1 + \alpha_2 x_2 + \cdots + \alpha_k x_k = u \ (\text{mod } 3) \qquad u = 0,1,2$$
$$L_2 = \beta_1 x_1 + \beta_2 x_2 + \cdots + \beta_k x_k = h \ (\text{mod } 3) \qquad h = 0,1,2 \qquad (12\text{-}2)$$

where $\{\alpha_i\}$ and $\{\beta_i\}$ are the exponents in the first and second generalized interactions, respectively, with the convention that the first nonzero α_i and β_i are unity. The defining contrasts in Equation 12-2 imply nine simultaneous equations specified by the pair of values for L_1 and L_2. Treatment combinations having the same pair of values for (L_1, L_2) are assigned to the same block.

The principal block consists of treatment combinations satisfying $L_1 = L_2 = 0 \ (\text{mod } 3)$. The elements of this block form a group with respect to addition modulus 3; thus, the scheme given in Section 12-2.1 can be used to generate the blocks.

As an example, consider the 3^4 factorial design confounded in nine blocks of nine runs each. Suppose we choose to confound ABC and AB^2D^2. Their generalized interactions

$$(ABC)(AB^2D^2) = A^2B^3CD^2 = (A^2B^3CD^2)^2 = AC^2D$$

$$(ABC)(AB^2D^2)^2 = A^3B^5CD^4 = B^2CD = (B^2CD)^2 = BC^2D^2$$

are also confounded with blocks. The defining contrasts for ABC and AB^2D^2 are

$$L_1 = x_1 + x_2 + x_3$$
$$L_2 = x_1 + 2x_2 + 2x_4 \qquad (12\text{-}3)$$

The nine blocks may be constructed by using the defining contrasts (Equation 12-3) and the group-theoretic property of the principal block. The design is shown in Figure 12-7.

For the 3^k design in nine blocks, there will be four components of interaction confounded. The remaining unconfounded components of these interactions can

Block 1	Block 2	Block 3	Block 4	Block 5	Block 6	Block 7	Block 8	Block 9
0000	0001	2000	0200	0020	0010	1000	0100	0002
0122	0120	2122	0022	0112	0102	1122	0222	0121
0211	0212	2211	0111	0201	0221	1211	0011	0210
1021	1022	0021	0221	1011	1001	2021	1121	1020
1110	1111	0110	1010	1100	1120	2110	1210	1112
1201	1200	0202	1102	1222	1212	2202	1002	1201
2012	2010	1012	2212	2002	2022	0012	2112	2011
2101	2012	1101	2001	2121	2111	0101	2201	2100
2220	2221	1220	2120	2210	2200	0220	2020	2222

| $(L_1, L_2) =$ (0,0) | (0,1) | (2,2) | (2,0) | (2,1) | (1,2) | (1,1) | (1,0) | (0,2) |

Figure 12-7. The 3^4 design in nine blocks with ABC, AB^2D^2, AC^2D, and BC^2D^2 confounded.

be determined by subtracting the sum of squares for the confounded component from the sum of squares for the entire interaction. The method described in Section 12-1.3 may be useful in computing the components of interaction.

12-2.3 The 3^k Factorial Design in 3^p Blocks

The 3^k factorial design may be confounded in 3^p blocks of 3^{k-p} observations each, where $p < k$. The procedure is to select p independent effects to be confounded with blocks. As a result, exactly $(3^p - 2p - 1)/2$ other effects are automatically confounded. These effects are the generalized interactions of those effects originally chosen.

As an illustration, consider a 3^7 design to be confounded in 27 blocks. Since $p = 3$, we would select three independent components of interaction and automatically confound $[3^3 - 2(3) - 1]/2 = 10$ others. Suppose we choose ABC^2DG, BCE^2F^2G, and $BDEFG$. Three defining contrasts can be constructed from these effects and the 27 blocks can be generated by the methods previously described. The other 10 effects confounded with blocks are

$$(ABC^2DG)(BCE^2F^2G) = AB^2DE^2F^2G^2$$

$$(ABC^2DG)(BCE^2F^2G)^2 = AB^3C^4DE^4F^4G^3 = ACDEF$$

$$(ABC^2DG)(BDEFG) = AB^2C^2D^2EFG^2$$

$$(ABC^2DG)(BDEFG)^2 = AB^3C^2D^3E^2F^2\,G^3 = AC^2E^2F^2$$

$$(BCE^2F^2G)(BDEFG) = B^2CDE^3F^3G^2 = BC^2D^2G$$

$$(BCE^2F^2G)(BDEFG)^2 = B^3CD^2E^4F^4G^3 = CD^2EF$$

$$(ABC^2DG)(BCE^2F^2G)(BDEFG) = AB^3C^3D^2E^3F^3G^3 = AD^2$$

$$(ABC^2DG)^2(BCE^2F^2G)(BDEFG) = A^2B^4C^5D^3G^4 = AB^2CG^2$$

$$(ABC^2DG)(BCE^2F^2G)^2(BDEFG) = ABCD^2E^2F^2G$$

$$(ABC^2DG)(BCE^2F^2G)(BDEFG)^2 = ABC^3D^3E^4F^4G^4 = ABEFG$$

This is a very large design requiring $3^7 = 2187$ observations arranged in 27 blocks of 81 observations each. In practice, we would usually resort to a fractional factorial design, which would require fewer runs.

12-3 FRACTIONAL REPLICATION OF THE 3^k FACTORIAL DESIGN

The concept of fractional replication can be extended to the 3^k factorial designs. Because a complete replicate of the 3^k design can require a rather large number of runs even for moderate values of k, fractional replication of these designs is of interest. As we shall see, however, some of these designs have unattractive alias structures.

12-3.1 The One-Third Fraction of the 3^k Design

The largest fraction of the 3^k design is a one-third fraction containing 3^{k-1} runs. Consequently, we refer to this as a 3^{k-1} fractional factorial design. To construct a 3^{k-1} fractional factorial design select a two-degrees-of-freedom component of interaction (generally, the highest-order interaction) and partition the *full* 3^k design into three blocks. Each of the three resulting blocks is a 3^{k-1} fractional design, and any one of the blocks may be selected for use. If $AB^{\alpha 2}C^{\alpha 3} \cdots K^{\alpha k}$ is the component of interaction used to define the blocks, then $I = AB^{\alpha 2}C^{\alpha 3} \cdots K^{\alpha k}$ is called the *defining relation* of the fractional factorial design. Each main effect or component of interaction estimated from the 3^{k-1} design has two aliases, which may be found by multiplying the effect by *both* I and I^2 modulus 3.

As an example, consider a one-third fraction of the 3^3 design. We may select any component of the ABC interaction to construct the design, that is, ABC, AB^2C, ABC^2, or AB^2C^2. Thus, there are actually 12 *different* one-third fractions of the 3^3 design defined by

$$x_1 + \alpha_2 x_2 + \alpha_3 x_3 = u(\text{mod } 3)$$

where $\alpha_i = 1$ or 2 and $u = 0$, 1, or 2. Suppose we select the component AB^2C^2. Each fraction of the resulting 3^{3-1} design will contain exactly $3^2 = 9$ treatment combinations that must satisfy

$$x_1 + 2x_2 + 2x_3 = u(\text{mod } 3)$$

where $u = 0$, 1, or 2. It is easy to verify that the three one-third fractions are as shown in Figure 12-8.

If any one of the 3^{3-1} designs in Figure 12-8 is run, the resulting alias structure is

$$A = A(AB^2C^2) = A^2B^2C^2 = ABC$$

$$A = A(AB^2C^2)^2 = A^3B^4C^4 = BC$$

$$B = B(AB^2C^2) = AB^3C^2 = AC^2$$

$$B = B(AB^2C^2)^2 = A^2B^5C^4 = ABC^2$$

$$C = C(AB^2C^2) = AB^2C^3 = AB^2$$

$$C = C(AB^2C^2)^2 = A^2B^4C^5 = AB^2C$$

$$AB = AB(AB^2C^2) = A^2B^3C^2 = AC$$

$$AB = AB(AB^2C^2)^2 = A^3B^5C^4 = BC^2$$

Consequently, the four effects that are actually estimated from the eight degrees of freedom in the design are $A + BC + ABC$, $B + AC^2 + ABC^2$, $C + AB^2 + AB^2C$, and $AB + AC + BC^2$. This design would be of practical value only if all the interactions were small relative to the main effects. Since the main effects are aliased with two-factor interactions, this is a resolution III design. Notice how complex the alias relationships are in this design. Each main effect is aliased with a *component* of interaction. If, for example, the two-factor interaction BC is large, this will potentially distort the estimate of the main effect of A and make the $AB + AC + BC^2$ effect very difficult to interpret. It is very difficult to see how this design could be useful unless we assume that all interactions are negligible.

Before leaving the 3_{III}^{3-1} design, note that if we let A denote the row and B

Design 1	Design 2	Design 3
000	200	100
012	212	112
101	001	201
202	102	002
021	221	121
110	010	210
122	022	222
211	111	011
220	120	020

Figure 12-8. The three one-third fractions of the 3^3 design with defining relation $I = AB^2C^2$.

denote the column, then the design can be written as

000	012	021
101	110	122
202	211	220

which is a 3 × 3 Latin square. The assumption of negligible interactions required for unique interpretations of the 3_{III}^{3-1} design is paralleled in the Latin square design. However, the two designs arise from different motives, one as a consequence of fractional replication and the other from randomization restrictions. From Table 5-13 we observe that there are only twelve 3 × 3 Latin squares and that each one corresponds to one of the twelve different 3^{3-1} fractional factorial designs.

The treatment combinations in a 3^{k-1} design with the defining relation $I = AB^{\alpha_2}C^{\alpha_3} \cdots K^{\alpha_k}$ can be constructed using a method similar to that employed in the 2^{k-p} series. First, write down the 3^{k-1} runs for a *full* three-level factorial design in $k - 1$ factors, with the usual 0,1,2 notation. This is the *basic design* in the terminology of chapter 11. Then introduce the kth factor by equating its levels x_k to the appropriate component of the highest-order interaction, say $AB^{\alpha_2}C^{\alpha_3} \cdots (K - 1)^{\alpha_{k-1}}$, through the relationship

$$x_k = \beta_1 x_1 + \beta_2 x_2 + \cdots + \beta_{k-1} x_{k-1} \qquad (12\text{-}4)$$

where $\beta_i = (3 - \alpha_k)\alpha_i \pmod 3$ for $1 \le i \le k - 1$. This yields a design of the highest possible resolution.

As an illustration, we use this method to generate the 3_{IV}^{4-1} design with the defining relation $I = AB^2CD$ shown in Table 12-9. It is easy to verify that the first three digits of each treatment combination in this table are the 27 runs of a full 3^3 design. This is the basic design. For AB^2CD, we have $\alpha_1 = \alpha_3 = \alpha_4 = 1$ and $\alpha_2 = 2$. This implies that $\beta_1 = (3 - 1)\alpha_1 \pmod 3 = (3 - 1)(1) = 2$, $\beta_2 = (3 - 1)\alpha_2 \pmod 3 = (3 - 1)(2) = 4 = 1 \pmod 3$, and $\beta_3 = (3 - 1)\alpha_3 \pmod 3 = (3 - 1)(1) = 2$. Thus, Equation 12-4 becomes

$$x_4 = 2x_1 + x_2 + 2x_3 \qquad (12\text{-}5)$$

Table 12-9 A 3_{IV}^{4-1} Design
with $I = AB^2CD$

0000	0012	2221
0101	0110	0021
1100	0211	0122
1002	1011	0220
0202	1112	1020
1201	1210	1121
2001	2010	1222
2102	2111	2022
2200	2212	2121

The levels of the fourth factor satisfy Equation 12-5. For example, we have $2(0) + 1(0) + 2(0) = 0$, $2(0) + 1(1) + 2(0) = 1$, $2(1) + 1(1) + 2(0) = 3 = 0$, and so on.

The resulting 3_{IV}^{4-1} design has 26 degrees of freedom that may be used to compute the sums of squares for the 13 main effects and components of interactions (and their aliases). The aliases of any effect are found in the usual manner; for example, the aliases of A are $A(AB^2CD) = ABC^2D^2$ and $A(AB^2CD)^2 = BC^2D^2$. One may verify that the four main effects are clear of any two-factor interaction components, but that some two-factor interaction components are aliased with each other. Once again, we notice the complexity of the alias structure. If any two-factor interactions are large, it will likely be very difficult to isolate them with this design.

The statistical analysis of a 3^{k-1} design is accomplished by the usual analysis of variance procedures for factorial experiments. The sums of squares for the components of interaction may be computed as in Section 12-1. Remember when interpreting results that the components of interactions have no practical interpretation.

12-3.2 Other 3^{k-p} Fractional Factorial Designs

For moderate to large values of k, even further fractionation of the 3^k design is potentially desirable. In general, we may construct a $(\frac{1}{3})^p$ fraction of the 3^k design for $p < k$, where the fraction contains 3^{k-p} runs. Such a design is called a 3^{k-p} fractional factorial design. Thus, a 3^{k-2} design is a one-ninth fraction, a 3^{k-3} design is a one–twenty-seventh fraction, and so on.

The procedure for constructing a 3^{k-p} fractional factorial design is to select p components of interaction and use these effects to partition the 3^k treatment combinations into 3^p blocks. Each block is then a 3^{k-p} fractional factorial design. The defining relation I of any fraction consists of the p effects initially chosen and their $(3^p - 2p - 1)/2$ generalized interactions. The alias of any main effect or component of interaction is produced by multiplication modulus 3 of the effect by I and I^2.

We may also generate the runs defining a 3^{k-p} fractional factorial design by first writing down the treatment combinations of a *full* 3^{k-p} factorial design and then introducing the additional p factors by equating them to components of interaction, as we did in Section 12-3.1.

We now illustrate the procedure by constructing a 3^{4-2} design, that is, a one-ninth fraction of the 3^4 design. Let AB^2C and BCD be the two components of interaction chosen to construct the design. Their generalized interactions are $(AB^2C)(BCD) = AC^2D$ and $(AB^2C)(BCD)^2 = ABD^2$. Thus, the defining relation for this design is $I = AB^2C = BCD = AC^2D = ABD^2$, and the design is of resolution III. The nine treatment combinations in the design are found by writing down a 3^2 design in the factors A and B, and then adding two new factors by

Table 12-10 A 3_{III}^{4-2} Design with $I = AB^2C$ and $I = BCD$

0000	0111	0222
1021	1102	1210
2012	2120	2201

setting

$$x_3 = 2x_1 + x_2$$
$$x_4 = 2x_2 + 2x_3$$

This is equivalent to using AB^2C and BCD to partition the full 3^4 design into nine blocks and then selecting one of these blocks as the desired fraction. The complete design is shown in Table 12-10.

This design has eight degrees of freedom that may be used to estimate four main effects and their aliases. The aliases of any effect may be found by multiplying the effect modulus 3 by AB^2C, BCD, AC^2D, ABD^2, and their squares. The complete alias structure for the design is given in Table 12-11.

From the alias structure, we see that this design is useful only in the absence of interaction. Furthermore, if A denotes the rows and B denotes the columns, then from examining Table 12-10 we see that the 3_{III}^{4-2} design is also a Graeco–Latin square.

The publication by Connor and Zelen (1959) contains an extensive selection of designs for $4 \le k \le 10$. This pamphlet was prepared for the National Bureau of Standards and is the most complete table of fractional 3^{k-p} plans available.

In this section, we have noted several times the complexity of the alias relationships in 3^{k-p} fractional factorial designs. In general, if k is moderately large, say $k \ge 4$ or 5, the size of the 3^k design will drive most experimenters to consider fairly small fractions. Unfortunately, these designs have alias relationships that involve the *partial* aliasing of two-degrees-of-freedom components of interaction. This, in turn, results in a design that will be difficult if not impossible to interpret if interactions are not negligible. Furthermore, there are no simple augmentation schemes (such as fold over) that can be used to combine two or more fractions to isolate significant interactions. The 3^k design is often suggested

Table 12-11 Alias Structure for the 3_{III}^{4-2} Design in Table 12-10

Effect								
			I				I^2	
A	ABC^2	$ABCD$	ACD^2	AB^2D	BC^2	$AB^2C^2D^2$	CD^2	BD^2
B	AC	BC^2D^2	ABC^2D	AB^2D^2	ABC	CD	AB^2C^2D	AD^2
C	AB^2C^2	BC^2D	AD	$ABCD^2$	AB^2	BD	ACD	ABC^2D^2
D	AB^2CD	BCD^2	AC^2D^2	AB	AB^2CD^2	BC	AC^2	ABD

as appropriate when curvature is present. However, these are more efficient alterantives (see Chapter 16). For these reasons, we feel that the 3^{k-p} fractional factorial designs are solutions looking for a problem—they are not generally good designs.

12-4 FACTORIALS WITH MIXED LEVELS

We have emphasized factorial and fractional factorial designs in which all the factors have the same number of levels. The two-level system discussed in Chapters 9, 10, and 11 are particularly useful. The three-level system presented earlier in this chapter is much less useful because the designs are relatively large even for a modest number of factors, and most of the small fractions have complex alias relationships that would require very restrictive assumptions regarding interactions to be useful.

It is our belief that the two-level factorial and fractional factorial designs should be the cornerstone of industrial experimentation for product and process development, troubleshooting, and improvement. There are, however, some situations in which it is necessary to include a factor (or a few factors) that have more than two levels. In this section we show how some three-level and four-level factors can be accommodated in a 2^k design.

12-4.1 Factors at Two and Three Levels

Designs in which some factors have two levels and other factors have three levels can be derived from the table of plus and minus signs for the usual 2^k design. The general procedure is best illustrated with an example. Suppose we have two variables, with A at two levels and X at three levels. Consider a table of plus and minus signs for the usual eight-run 2^3 design. The signs in columns B and C have the pattern shown on the left side of Table 12-12. Let the levels of X be represented by x_1, x_2, and x_3. The right side of Table 12-12 shows how the sign patterns for B and C are combined to form the levels of the three-level factor.

Table 12-12 Use of Two-Level Factors to Form a Three-Level Factor

Two-Level Factors		Three-Level Factor
B	C	X
$-$	$-$	x_1
$+$	$-$	x_2
$-$	$+$	x_2
$+$	$+$	x_3

ble 12-13   One Two-Level and One Three-Level Factor in a 2^3 Design

	A	X_L	X_L	$A \times X_L$	$A \times X_L$	X_Q	$A \times X_Q$	Actual Treatment Combinations	
un	A	B	C	AB	AC	BC	ABC	A	X
1	−	−	−	+	+	+	−	Low	Low
2	+	−	−	−	−	+	+	High	Low
3	−	+	−	−	+	−	+	Low	Med
4	+	+	−	+	−	−	−	High	Med
5	−	−	+	+	−	−	+	Low	Med
6	+	−	+	−	+	−	−	High	Med
7	−	+	+	−	−	+	−	Low	High
8	+	+	+	+	+	+	+	High	High

Now factor X has two degrees of freedom and can generally be thought of as consisting of a linear and a quadratic component, each component having one degree of freedom. Table 12-13 shows the 2^3 design with the columns labeled to show the actual effects that they estimate, with X_L and X_Q denoting the linear and quadratic effects of X, respectively. Note that the linear effect of X is the sum of the two effect estimates computed from the columns usually associated with B and C, and that the effect of A can only be computed from the runs where X is at either the low or high levels; namely runs 1, 2, 7, and 8. Similarly, the $A \times X_L$ effect is the sum of the two effects that would be computed from the columns usually labeled AB and AC. Furthermore, note that runs 3 and 5 are replicates. Therefore, a one-degree-of-freedom estimate of error can be made using these two runs. Similarly, runs 4 and 6 are replicates, and this would lead to a second one-degree-of-freedom estimate of error. The average variance at these two pairs of runs could be used as a mean square for error with two degrees of freedom. The complete analysis of variance is summarized in Table 12-14.

If we are willing to assume that the two-factor and higher interactions are negligible, we can convert the design in Table 12-13 into a resolution III fraction with up to four two-level factors and a single three-level factor. This would be

Table 12-14   Analysis of Variance for the Design in Table 12-13

Source of Variation	Sum of Squares	Degrees of Freedom	Mean Square
A	SS_A	1	MS_A
X ($X_L + X_Q$)	SS_X	2	MS_B
AX ($A \times X_L + A \times X_Q$)	SS_{AX}	2	MS_{AX}
Error (from runs 3 and 5 and runs 4 and 6)	SS_E	2	MS_E
Total	SS_T	7	

accomplished by associating the two-level factors with columns A, AB, AC, and ABC. Column BC cannot be used for a two-level factor because it contains the quadratic effect of the three-level factor X.

This same procedure can be applied to the 16-run, 32-run, and 64-run 2^k designs. For 16 runs, it is possible to construct resolution V fractional factorials with two two-level factors and either two or three factors at three levels. A 16-run resolution V fraction can also be obtained with three two-level factors and one three-level factor. If we include four two-level factors and a single three-level factor in 16 runs, the design will be of resolution III. The 32- and 64-run designs allow similar arrangements. For additional discussion of some of these designs, see Addleman (1962).

12-4.2   Factors at Two and Four Levels

It is very easy to accommodate a four-level factor in a 2^k design. The procedure for doing this involves using two two-level factors to represent the four-level factor. For example, suppose that A is a four-level factor with levels a_1, a_2, a_3, and a_4. Consider two columns of the usual table of plus and minus signs, say columns P and Q. The pattern of signs in these two columns is as shown on the left side of Table 12-15. The right side of this table shows how these four sign patterns would correspond to the four levels of factor A. The effects represented by columns P and Q and the PQ interaction are mutually orthogonal and correspond to the three-degrees-of-freedom A effect.

To illustrate this idea more completely, suppose that we have one four-level factor and two two-level factors and that we need to estimate all the main effects and interactions involving these factors. This can be done with a 16-run design. Table 12-16 shows the usual table of plus and minus signs for the 16-run 2^4 design, with columns A and B used to form the four-level factor, say X, with levels x_1, x_2, x_3, and x_4. Sums of squares would be calculated for each column A, B, . . . , $ABCD$ just as in the usual 2^k system. Then the sums of squares for

Table 12-15   Four-Level Factor A Expressed as Two Two-Level Factors

	Two-Level Factors		Four-Level Factor
Run	P	Q	A
1	$-$	$-$	a_1
2	$+$	$-$	a_2
3	$-$	$+$	a_3
4	$+$	$+$	a_4

Table 12-16 A Single Four-Level Factor and Two Two-Level Factors in 16 Runs

Run	(A	B)	= X	C	D	AB	AC	BC	ABC	AD	BD	ABD	CD	ACD	BCD	ABCD
1	−	−	x_1	−	−	+	+	+	−	+	+	−	+	−	−	+
2	+	−	x_2	−	−	−	−	+	+	−	+	+	+	+	−	−
3	−	+	x_3	−	−	−	+	−	+	+	−	+	+	−	+	−
4	+	+	x_4	−	−	+	−	−	−	−	−	−	+	+	+	+
5	−	−	x_1	+	−	+	−	−	+	+	+	−	−	+	+	−
6	+	−	x_2	+	−	−	+	−	−	−	+	+	−	−	+	+
7	−	+	x_3	+	−	−	−	+	−	+	−	+	−	+	−	+
8	+	+	x_4	+	−	+	+	+	+	−	−	−	−	−	−	−
9	−	−	x_1	−	+	+	+	+	−	−	−	+	−	+	+	−
10	+	−	x_2	−	+	−	−	+	+	+	−	−	−	−	+	+
11	−	+	x_3	−	+	−	+	−	+	−	+	−	−	+	−	+
12	+	+	x_4	−	+	+	−	−	−	+	+	+	−	−	−	−
13	−	−	x_1	+	+	+	−	−	+	−	−	+	+	−	−	+
14	+	−	x_2	+	+	−	+	−	−	+	−	−	+	+	−	−
15	−	+	x_3	+	+	−	−	+	−	−	+	−	+	−	+	−
16	+	+	x_4	+	+	+	+	+	+	+	+	+	+	+	+	+

all factors X, C, D and their interactions are formed as follows:

$$SS_X = SS_A + SS_B + SS_{AB} \qquad \text{(3 degrees of freedom)}$$
$$SS_C = SS_C \qquad \text{(1 degree of freedom)}$$
$$SS_D = SS_D \qquad \text{(1 degree of freedom)}$$
$$SS_{CD} = SS_{CD} \qquad \text{(1 degree of freedom)}$$
$$SS_{XC} = SS_{AC} + SS_{BC} + SS_{ABC} \qquad \text{(3 degrees of freedom)}$$
$$SS_{XD} = SS_{AD} + SS_{BD} + SS_{ABD} \qquad \text{(3 degrees of freedom)}$$
$$SS_{XCD} = SS_{ACD} + SS_{BCD} + SS_{ABCD} \qquad \text{(3 degrees of freedom)}$$

This could be called a 4×2^2 design. If we are willing to ignore two-factor interactions, up to nine additional two-level factors can be associated with the two-factor interaction (except AB), three-factor interaction, and four-factor interaction columns.

12-5 TAGUCHI'S CONTRIBUTIONS TO EXPERIMENTAL DESIGN AND QUALITY ENGINEERING

Throughout this book, we have emphasized the importance of using designed experiments for product and process improvement. One important goal of quality improvement is to design quality into every product and the processes that build them. Statistically designed experiments are a major element of this activity.

Unfortunately, the principles of experimental design (and statistical methods, in general) have not been as widely used in the West as in Japan. Japanese engineers have had much greater exposure to these concepts, and consequently, experimental design methods have become more of an engineering tool in Japan than they have in the United States. In the early 1980s, Professor Genechi Taguchi introduced his approach to using experimental design for

1. Designing products or processes so that they are robust to environmental conditions.

2. Designing/developing products so that they are robust to component variation.

3. Minimizing variation around a target value.

[See Taguchi and Wu (1980) and Taguchi (1986).] By *robust*, we mean that the product or process performs consistently on target and is relatively insensitive to factors that are difficult to control. Taguchi refers to the three goals of his approach as *parameter design*.

The philosophy that Taguchi recommends is sound and should be included in the quality improvement process of any organization. However, he has advocated some novel methods of statistical data analysis and some approaches to the design of experiments that are unnecessarily complicated, inefficient, and sometimes ineffective. In this section, we will briefly overview Taguchi's philosophy regarding quality engineering and experimental design. We will present some examples of his approach to parameter design, and we will use these examples to highlight the problems with his technical methods. As we will see, it is possible to combine his sound engineering concepts with more efficient and effective experimental design and statistical data analysis methods.

12-5.1   The Taguchi Philosophy

Professor Taguchi advocates a philosophy of quality engineering that is broadly applicable. He considers three stages in a product's (or process's) development: system design, parameter design, and tolerance design. In system design, the engineer uses scientific and engineering principles to determine the basic configuration. For example, if we wish to measure an unknown resistance, we may use our knowledge of electrical circuits to determine that the basic system should be configured as a Wheatstone bridge. If we are designing a process to assemble printed circuit boards, we will determine the need for specific types of axial insertion machines, surface-mount placement machines, flow solder machines, and so forth.

In the parameter design stage, the specific values for the system parameters are determined. This would involve choosing the nominal resistor and power supply values for the Wheatstone bridge, the number and type of component placement machines for the printed circuit board assembly process, and so forth. Usually, the objective is to specify these nominal parameter values such that the variability transmitted from uncontrollable (or noise) variables is minimized.

Tolerance design is used to determine the best tolerances for the parameters. For example, in the Wheatstone bridge, tolerance design methods would reveal which components in the design were most sensitive and where the tolerances should be set. If a component does not have much effect on the performance of the circuit, it can be specified with a wide tolerance.

Taguchi recommends that statistical experimental design methods be employed to assist in quality improvement, particularly during parameter design and tolerance design. In this section, we will focus on parameter design. Experimental design methods can be used to find a best product or process design, where by "best" we mean a product or process that is robust or insensitive to uncontrollable factors that will influence the product or process once it is in routine operation.

The notion of *robust design* is not new. Engineers have always tried to design products so that they will work well under uncontrollable conditions. For example, commercial transport aircraft fly about as well in a thunderstorm as they do in clear air. Taguchi deserves recognition for realizing that experimental design can be used as a formal part of the *engineering design process* to help accomplish this objective.

A key component of Taguchi's philosophy is the *reduction of variability*.

Generally, each product or process performance characteristic will have a target or *nominal* value. The objective is to reduce the variability around this target value. Taguchi models the departures that may occur from this target value with a *loss function*. The loss refers to the cost that is incurred by *society* when the consumer uses a product whose quality characteristics differ from the nominal. The concept of societal loss is a departure from traditional thinking. Taguchi imposes a quadratic loss function of the form

$$L(y) = k(y - T)^2 \qquad (12\text{-}6)$$

shown in Figure 12-9. Clearly this type of function will penalize even small departures of y from the target T. Again, this is a departure from traditional Western thinking, which usually attaches penalties only to cases where y is outside of the upper and lower specifications (say $y > USL$ or $y < LSL$ in Figure 12-9). However, the Taguchi philosophy regarding reduction of variability and the emphasis on minimizing costs is entirely consistent with the continuous improvement philosophy of Deming and Juran.

In summary, Taguchi's philosophy involves three central ideas:

1. Products and processes should be designed so that they are robust to external sources of variability.

2. Experimental design methods are an engineering tool to help accomplish this objective.

3. Operation on-target is more important than conformance to specifications.

These are sound concepts, and their value should be readily apparent. Furthermore, experimental design methods can play a major role in translating these ideas into practice.

We now turn to a discussion of the specific methods that Professor Taguchi recommends for applying his concepts in practice. As we will see, his approach to experimental design and data analysis can be improved.

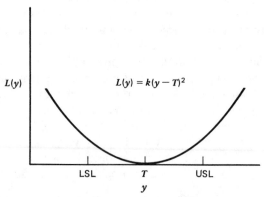

Figure 12-9. Taguchi's quadratic loss function.

12-5.2   The Taguchi Approach to Parameter Design

We will use an example to illustrate the approach to parameter design suggested by Professor Taguchi. The example was published in *Quality Progress* in December 1987 (see "The Taguchi Approach to Parameter Design," by D. M. Byrne and S. Taguchi, *Quality Progress*, December 1987, pp. 19–26).

The Problem ▪ The experiment involves finding a method to assemble an elastometric connector to a nylon tube that would deliver the required pull-off performance to be suitable for use in an automotive engine application. The specific objective of the experiment is to maximize the pull-off force. Four controllable and three uncontrollable noise factors were identified. These factors are defined in Table 12-17. We want to find the levels of the controllable factors that are the least influenced by the noise factors and that provide the maximum pull-off force. Notice that although the noise factors are not controllable during routine operations, they can be controlled for the purposes of a test. Each controllable factor is tested at three levels, and each noise factor is tested at two levels.

The Experimental Design ▪ In the Taguchi parameter design methodology, one experimental design is selected for the controllable factors and another experimental design is selected for the noise or uncontrollable factors. These designs are shown in Table 12-18. Panel (a) of Table 12-18 contains an L_9 *orthogonal array*—a table of integers whose column elements (1, 2, and 3) represent the low, medium, and high levels of the column factors. Each row of the orthogonal array represents a run, that is, a specific set of factor levels to be tested. The L_9 orthogonal array will accommodate four factors at three levels each in nine runs. Panel (b) of Table 12-18 contains the L_8 orthogonal array—a design for up to seven factors at two levels each in eight runs. The L_8 array in this example contains only three factors—E, F, and G—so the remaining columns can be used to estimate interactions. The purpose of the noise factor array (L_8) is to create

Table 12-17   Factors and Levels for the Parameter Design Example

Controllable Factors		Levels	
A = Interference	Low	Medium	High
B = Connector wall thickness	Thin	Medium	Thick
C = Insertion depth	Shallow	Medium	Deep
D = Percent adhesive in connector pre-dip	Low	Medium	High

Uncontrollable Factors	Levels	
E = Conditioning time	24 h	120 h
F = Conditioning temperature	72°F	150°F
G = Conditioning relative humidity	25%	75%

Table 12-18   Designs for the Controllable and Uncontrollable Factors

(a) L_9 Orthogonal Array for the Controllable Factors				(b) L_8 Orthogonal Array for the Uncontrollable Factors							
	Variable					Variable					
Run	A	B	C	D	Run	E	F	E × F	G	E × G	F × G	e
1	1	1	1	1	1	1	1	1	1	1	1	1
2	1	2	2	2	2	1	1	1	2	2	2	2
3	1	3	3	3	3	1	2	2	1	1	2	2
4	2	1	2	3	4	1	2	2	2	2	1	1
5	2	2	3	1	5	2	1	2	1	2	1	2
6	2	3	1	2	6	2	1	2	2	1	2	1
7	3	1	3	2	7	2	2	1	1	2	2	1
8	3	2	1	3	8	2	2	1	2	1	1	2
9	3	3	2	1								

variation so that we can identify the controllable factor levels that are the least sensitive to the noise factors.

The two designs are combined as shown in Table 12-19. In this complete *parameter design layout,* the L_9 array containing the controllable factors is called the *inner array,* and the L_8 array containing the noise factors is called the *outer array.* Literally, each of the 9 runs from the inner array is tested across the 8 runs from the outer array, for a total sample size of 72 runs. The observed pull-off force is reported in Table 12-19.

Data Analysis and Conclusions ▪ The data from this experiment may now be analyzed. Taguchi recommends analyzing the mean response for each run in the inner array (see Table 12-19), and he also suggests analyzing variation using an appropriately chosen signal-to-noise ratio (SN). These signal-to-noise ratios are derived from the quadratic loss function, and three of them are considered to be "standard" and widely applicable. They are

1. Nominal the best:

$$SN_T = 10 \log \left(\frac{\bar{y}^2}{S^2} \right) \tag{12-7}$$

2. Larger the better:

$$SN_L = -10 \log \left(\frac{1}{n} \sum_{i=1}^{n} \frac{1}{y_i^2} \right) \tag{12-8}$$

3. Smaller the better:

$$SN_S = -10 \log \left(\frac{1}{n} \sum_{i=1}^{n} y_i^2 \right) \tag{12-9}$$

Table 12-19   Parameter Design with Both Inner and Outer Arrays

Run	Inner Array (L$_9$)				Outer Array (L$_8$)								Responses	
					E	1	1	1	1	2	2	2	2	
					F	1	1	2	2	1	1	2	2	
					G	1	2	1	2	1	2	1	2	
	A	B	C	D										$\bar{y}$	SN_L
1	1	1	1	1	15.6	9.5	16.9	19.9	19.6	19.6	20.0	19.1	17.525	24.025
2	1	2	2	2	15.0	16.2	19.4	19.2	19.7	19.8	24.2	21.9	19.475	25.522
3	1	3	3	3	16.3	16.7	19.1	15.6	22.6	18.2	23.3	20.4	19.025	25.335
4	2	1	2	3	18.3	17.4	18.9	18.6	21.0	18.9	23.2	24.7	20.125	25.904
5	2	2	3	1	19.7	18.6	19.4	25.1	25.6	21.4	27.5	25.3	22.825	26.908
6	2	3	1	2	16.2	16.3	20.0	19.8	14.7	19.6	22.5	24.7	19.225	25.326
7	3	1	3	2	16.4	19.1	18.4	23.6	16.8	18.6	24.3	21.6	19.850	25.711
8	3	2	1	3	14.2	15.6	15.1	16.8	17.8	19.6	23.2	24.2	18.338	24.852
9	3	3	2	1	16.1	19.9	19.3	17.3	23.1	22.7	22.6	28.6	21.200	26.152

Notice that these SN ratios are expressed in a decibel scale. We would use SN_T if the objective is to reduce variability around a specific target, SN_L if the system is optimized when the response is as large as possible, and SN_S if the system is optimized when the response is as small as possible. Factor levels that maximize the appropriate SN ratio are optimal.

In this problem, we would use SN_L because the objective is to maximize the pull-off force. The last two columns of the Table 12-19 contain $\bar{y}$ and SN_L values for each of the nine inner-array runs. Taguchi-oriented practitioners often use the analysis of variance to determine the factors that influence $\bar{y}$ and the factors that influence the signal-to-noise ratio. They also employ graphs of the "marginal means" of each factor, such as the ones shown in Figures 12-10 and 12-11. The usual approach is to examine the graphs and "pick the winner." In this case, factors A and C have larger effects than do B and D. In terms of maximizing SN_L, we would select A_{Medium}, C_{Deep}, B_{Medium}, and D_{Low}. In terms of maximizing the average pull-off force $\bar{y}$, we would choose A_{Medium}, C_{Medium}, B_{Medium}, and D_{Low}. Notice that there is almost no difference between C_{Medium} and C_{Deep}. The implication is that this choice of levels will maximize the mean pull-off force and reduce variability in the pull-off force.

Taguchi advocates claim that the use of the SN ratio generally eliminates the need for examining specific interactions between the controllable and noise factors, although sometimes looking at these interactions improves process understanding. The authors of this study found that the AG and DE interactions were large. Analysis of these interactions, shown in Figure 12-12, suggests that A_{Medium} is best. (It gives the highest pull-off force and a slope close to zero, indicating that if we choose A_{Medium}, the effect of relative humidity is minimized. The analysis also suggests that D_{Low} gives the highest pull-off force regardless of the conditioning time.

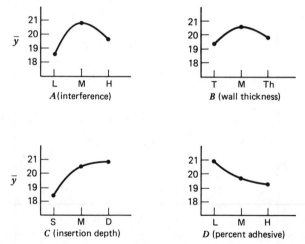

Figure 12-10.   The effects of controllable factors on mean response.

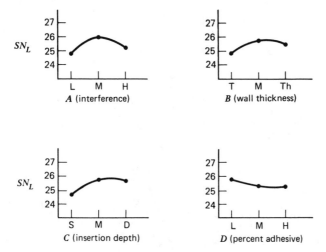

Figure 12-11.   The effects of controllable factors on SN_L.

When cost and other factors were taken into account, the experimenters in this example finally decided to use A_{Medium}, B_{Thin}, C_{Medium}, and D_{Low}. (B_{Thin} was much less expensive than B_{Medium}, and C_{Medium} was felt to give slightly less variability that C_{Deep}.) Since this combination was not a run in the original nine inner array trials, five additional tests were made at this set of conditions as a *confirmation experiment*. For this confirmation experiment, the levels used on the noise variables were E_{Low}, F_{Low}, and G_{Low}. The authors report that good results were obtained from the confirmation test.

Critique of the Experimental Design ▪ The advocates of Taguchi's approach to parameter design utilize the orthogonal array designs, two of which (the L_8 and the L_9) were presented in the foregoing example. Figure 12-13 presents several additional orthogonal arrays: the L_4, L_{12}, L_{16}, L_{18}, and L_{27}. These designs were not developed by Taguchi; for example, the L_8 is a 2_{III}^{7-4} fractional factorial, the L_9 is a 3_{III}^{4-2} fractional factorial, the L_{12} is a Plackett-Burman design, the L_{16} is a 2_{III}^{15-11} factorial factorial, and so on. Box, Bisgaard, and Fung (1988) trace the origin of these designs. Some of these designs have very complex alias structures. In particular, the L_{12} and all of the designs that use three-level factors will involve partial aliasing of two-factor interactions with main effects. If any two-factor interactions are large, this may lead to a situation in which the experimenter does not get the correct answer.

Taguchi argues that we do not need to consider two-factor interactions explicitly. He claims that it is possible to eliminate these interactions either by correctly specifying the response and design factors or by using a sliding setting approach to choose factor levels. As an example of the latter approach, consider the two factors pressure and temperature. Varying these factors independently will probably produce an interaction. However, if temperature levels are chosen contingent on the pressure levels, then the interaction effect can be minimized.

Figure 12-12. AG and DE interactions.

In practice, these two approaches are usually difficult to implement unless we have an unusually high level of process knowledge. The lack of provision for adequately dealing with potential interactions between the controllable process factors is a major weakness of the Taguchi approach to parameter design.

Instead of designing the experiment to investigate potential interactions, Taguchi prefers to use three-level factors to estimate curvature. For example, in the inner and outer array design used by Byrne and Taguchi, all four controllable factors were run at three levels. Let x_1, x_2, x_3, and x_4 represent the controllable factors and let z_1, z_2, and z_3 represent the three noise factors. Recall that the noise factors were run at two levels in a complete factorial design. The design they used allows us to fit the following model:

$$
\begin{aligned}
y = \beta_0 &+ \sum_{j=1}^{4} \beta_j x_j + \sum_{j=1}^{4} \beta_{jj} x_j^2 \\
&+ \sum_{i=1}^{3} \gamma_i z_i + \sum_{i<j}^{3} \sum_{j=2}^{3} \gamma_{ij} z_i z_j \\
&+ \sum_{i=1}^{3} \sum_{j=1}^{4} \delta_{ij} z_i x_j + \epsilon
\end{aligned}
$$

Notice that we can fit the linear and quadratic effects of the controllable factors but not their two-factor interactions (which are aliased with the main effects). We

$L_4 (2^3)$

No. \ Col.	1	2	3
1	1	1	1
2	1	2	2
3	2	1	2
4	2	2	1

$L_{16} (2^{15})$

No. \ Col.	1	2	3	4	5	6	7	8	9	10	11	12	13	14	15
1	1	1	1	1	1	1	1	1	1	1	1	1	1	1	1
2	1	1	1	1	1	1	1	2	2	2	2	2	2	2	2
3	1	1	1	2	2	2	2	1	1	1	1	2	2	2	2
4	1	1	1	2	2	2	2	2	2	2	2	1	1	1	1
5	1	2	2	1	1	2	2	1	1	2	2	1	1	2	2
6	1	2	2	1	1	2	2	2	2	1	1	2	2	1	1
7	1	2	2	2	2	1	1	1	1	2	2	2	2	1	1
8	1	2	2	2	2	1	1	2	2	1	1	1	1	2	2
9	2	1	2	1	2	1	2	1	2	1	2	1	2	1	2
10	2	1	2	1	2	1	2	2	1	2	1	2	1	2	1
11	2	1	2	2	1	2	1	1	2	1	2	2	1	2	1
12	2	1	2	2	1	2	1	2	1	2	1	1	2	1	2
13	2	2	1	1	2	2	1	1	2	2	1	1	2	2	1
14	2	2	1	1	2	2	1	2	1	1	2	2	1	1	2
15	2	2	1	2	1	1	2	1	2	2	1	2	1	1	2
16	2	2	1	2	1	1	2	2	1	1	2	1	2	2	1

$L_{12} (2^{11})$

No. \ Col.	1	2	3	4	5	6	7	8	9	10	11
1	1	1	1	1	1	1	1	1	1	1	1
2	1	1	1	1	1	2	2	2	2	2	2
3	1	1	2	2	2	1	1	1	2	2	2
4	1	2	1	2	2	1	2	2	1	1	2
5	1	2	2	1	2	2	1	2	1	2	1
6	1	2	2	2	1	2	2	1	2	1	1
7	2	1	2	2	1	1	2	2	1	2	1
8	2	1	2	1	2	2	2	1	1	1	2
9	2	1	1	2	2	2	1	2	2	1	1
10	2	2	2	1	1	1	1	2	2	1	2
11	2	2	1	2	1	2	1	1	1	2	2
12	2	2	1	1	2	1	2	1	2	2	1

$L_{27} (3^{13})$

No. \ Col.	1	2	3	4	5	6	7	8	9	10	11	12	13
1	1	1	1	1	1	1	1	1	1	1	1	1	1
2	1	1	1	1	2	2	2	2	2	2	2	2	2
3	1	1	1	1	3	3	3	3	3	3	3	3	3
4	1	2	2	2	1	1	1	2	2	2	3	3	3
5	1	2	2	2	2	2	2	3	3	3	1	1	1
6	1	2	2	2	3	3	3	1	1	1	2	2	2
7	1	3	3	3	1	1	1	3	3	3	2	2	2
8	1	3	3	3	2	2	2	1	1	1	3	3	3
9	1	3	3	3	3	3	3	2	2	2	1	1	1
10	2	1	2	3	1	2	3	1	2	3	1	2	3
11	2	1	2	3	2	3	1	2	3	1	2	3	1
12	2	1	2	3	3	1	2	3	1	2	3	1	2
13	2	2	3	1	1	2	3	2	3	1	3	1	2
14	2	2	3	1	2	3	1	3	1	2	1	2	3
15	2	2	3	1	3	1	2	1	2	3	2	3	1
16	2	3	1	2	1	2	3	3	1	2	2	3	1
17	2	3	1	2	2	3	1	1	2	3	3	1	2
18	2	3	1	2	3	1	2	2	3	1	1	2	3
19	3	1	3	2	1	3	2	1	3	2	1	3	2
20	3	1	3	2	2	1	3	2	1	3	2	1	3
21	3	1	3	2	3	2	1	3	2	1	3	2	1
22	3	2	1	3	1	3	2	2	1	3	3	2	1
23	3	2	1	3	2	1	3	3	2	1	1	3	2
24	3	2	1	3	3	2	1	1	3	2	2	1	3
25	3	3	2	1	1	3	2	3	2	1	2	1	3
26	3	3	2	1	2	1	3	1	3	2	3	2	1
27	3	3	2	1	3	2	1	2	1	3	1	3	2

$L_{18} (2^1 \times 3^7)$

No. \ Col.	1	2	3	4	5	6	7	8
1	1	1	1	1	1	1	1	1
2	1	1	2	2	2	2	2	2
3	1	1	3	3	3	3	3	3
4	1	2	1	1	2	2	3	3
5	1	2	2	2	3	3	1	1
6	1	2	3	3	1	1	2	2
7	1	3	1	2	1	3	2	3
8	1	3	2	3	2	1	3	1
9	1	3	3	1	3	2	1	2
10	2	1	1	3	3	2	2	1
11	2	1	2	1	1	3	3	2
12	2	1	3	2	2	1	1	3
13	2	2	1	2	3	1	3	2
14	2	2	2	3	1	2	1	3
15	2	2	3	1	2	3	2	1
16	2	3	1	3	2	3	1	2
17	2	3	2	1	3	1	2	3
18	2	3	3	2	1	2	2	1

Figure 12-13. Some orthogonal array designs.

can also fit the linear effects of the noise factors and all the two-factor interactions involving the noise factors. Finally, we can fit the two-factor interactions involving the controllable factors and the noise factors. It may be unwise to ignore potential interactions in the controllable factors. A much safer strategy is to identify potential effects and interactions that may be important and then consider curvature only in the important variables if there is evidence that the curvature is important. This will usually lead to fewer experiments, simpler interpretation of the data, and better overall process understanding.

Another criticism of the Taguchi approach to parameter design is that the inner and outer array structure usually leads to a very large experiment. For example, in the foregoing application, the authors used 72 tests to investigate only seven factors, and they still could not estimate any of the two-factor interactions among the four controllable factors. There are several alternative experimental designs that would be superior to the inner and outer method used in this example. Suppose that we run all seven factors at two levels. Consider the 2_{IV}^{7-2} fractional factorial design. The alias relationships for this design are shown in the top half of Table 12-20. Notice that this design requires only 32 runs (as

Table 12-20   An Alternative Parameter Design

A one-quarter fraction of 7 factors in 32 runs. Resolution IV.
$$I = ABCDF = ABDEG = CEFG.$$

Aliases:

A	$AF = BCD$	$CG = EF$
B	$AG = BDE$	$DE = ABG$
$C = EFG$	$BC = ADF$	$DF = ABC$
D	$BD = ACF = AEG$	$DG = ABE$
$E = CFG$	$BE = ADG$	$ACE = AFG$
$F = CEG$	$BF = ACD$	$ACG = AEF$
$G = CEF$	$BG = ADE$	$BCE = BFG$
$AB = CDF = DEG$	$CD = ABF$	$BCG = BEF$
$AC = BDF$	$CE = FG$	$CDE = DFG$
$AD = BCF = BEG$	$CF = ABD = EG$	$CDG = DEF$
$AF = BDG$		

Factor Assignment Schemes:

1.   Controllable factors are assigned to the letters C, E, F, and G. Noise factors are assigned to the letters A, B, and D. All interactions between controllable factors and noise factors can be estimated, and all controllable factor main effects can be estimated clear of two-factor interactions.

2.   Controllable factors are assigned to the letters A, B, C, and D. Noise factors are assigned to the letters E, F, and G. All controllable factor main effects and two-factor interactions can be estimated; only the CE, CF, and CG interactions are aliased with interactions of the noise factors.

compared to 72). In the bottom half of Table 12-20, two different possible schemes for assigning process controllable variables and noise variables to the letters A through G are given. The first assignment scheme allows all the interactions between controllable factors and noise factors to be estimated, and it allows main effect estimates to be made that are clear of two-factor interactions. The second assignment scheme allows all the controllable factor main effects and their two-factor interactions to be estimated, it allows all noise factor main effects to be estimated clear of two-factor interactions; and it aliases only three interactions between controllable factors and noise factors with a two-factor interaction between two noise factors. Both of these arrangements present much cleaner alias relationships than are obtained from the inner and outer array parameter design, which also required over twice as many runs.

In general, the inner and outer array approach is often unnecessary. A better strategy is to use a single design (Taguchi advocates would call it an inner array only) that incorporates both controllable factors and noise factors. Of course, the design chosen must have a high enough resolution to allow all of the interactions of interest to be estimated. (Usually, the design must be of resolution IV or higher.) This approach will amost always lead to a dramatic reduction in the size of the experiment, and at the same time, it will produce information that is more likely to improve process understanding.

A final aspect of Taguchi's parameter design is the use of linear graphs to assign factors to the columns of the orthogonal array. A set of linear graphs for the L_8 design is shown in Figure 12-14. In these graphs, each number is a column

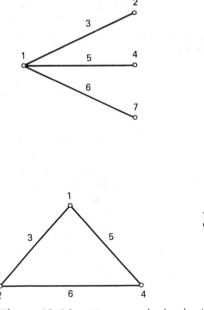

Figure 12-14.   Linear graphs for the L_8 design.

in the design. A line segment on the graph corresponds to an interaction between the nodes it connects. To assign variables to columns in an orthogonal array, assign the variables to nodes first; then when the nodes are used up, assign the variables to the line segments. When you assign variables to the nodes, strike out any line segments that correspond to interactions that might be important. The linear graphs in Figure 12-14 imply that column 3 in the L_8 design contains the interaction between columns 1 and 2, column 5 contains the interaction between columns 1 and 4, and so forth. If we had four factors, we should assign them to columns 1, 2, 4, and 7. This would ensure that each main effect is clear of two-factor interactions. What is *not* clear is the two-factor interaction aliasing. If the main effects are in columns 1, 2, 4, and 7, then column 3 contains the 1–2 *and* the 4–7 interaction, column 5 contains the 1–4 *and* the 2–7 interaction, and column 6 contains the 1–7 *and* the 2–4 interaction. This is clearly the case because four variables in eight runs is a resolution IV plan with all pairs of two-factor interactions aliased. In order to understand fully the two-factor interaction aliasing, Taguchi would refer the experiment designer to a supplementary interaction table.

Taguchi (1986) gives a collection of linear graphs for each of his recommended orthogonal array designs. These linear graphs seem to have been developed heuristically. Unfortunately, their use can lead to inefficient designs. For examples, see his car engine experiment [Taguchi and Wu (1980)] and his cutting tool experiment [Taguchi (1986)]. Both of these are 16-run designs that he sets up as resolution III designs in which main effects are aliased with two-factor interactions. Conventional methods for setting up these designs would have resulted in resolution IV plans in which the main effects are clear of the two-factor interactions. For the experimenter who simply wants to generate a good design, the linear graph approach may not produce the best result. A better approach is to use a simple table that presents the design and its full alias structure such as one of those in Appendix Table XII. These tables are easy to construct, or they can be generated by several widely available and inexpensive computer programs.

Critique of the Data Analysis Methods ▪ Several of Taguchi's data analysis methods are questionable. For example, he recommends some variations of the analysis of variance that are known to produce spurious results, and he also proposes some unique methods for the analysis of attribute and life testing data. For a discussion and critique of these methods, refer to Box, Bisgaard, and Fung (1988) and the references contained therein. In this section we focus on three aspects of data analysis Taguchi-style—the use of "marginal means" plots to optimize factor settings, the use of signal-to-noise ratios, and some of his uses of the analysis of variance

Consider the use of "marginal means" plots and the associated "pick the winner" optimization that was demonstrated previously in the pull-off force problem. To keep the situation simple, suppose that we have two factors A and B, each at three levels, as shown in Table 12-21. The "marginal means" plots are shown in Figure 12-15. From looking at these graphs, we would select A_3 and

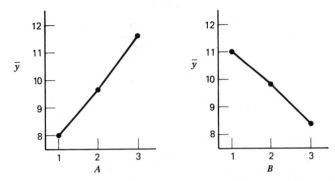

Figure 12-15. "Marginal means" plots for the data in Table 12-21.

B_1 as the optimum combination, assuming that we wish to maximize y. However, this is the wrong answer. Direct inspection of Table 12-21 or the AB interaction plot in Figure 12-16 shows that the combination of A_3 and B_2 produces the maximum value of y. In general, playing "pick the winner" with marginal averages can never be guaranteed to produce the optimum. The Taguchi advocates recommend that a confirmation experiment be run, although this offers no guarantees either. We might be confirming a response that differs dramatically from the optimum. The best way to find a set of optimum conditions is with response surface methods, as discussed in Chapter 16.

Taguchi's signal-to-noise ratios SN_T, SN_L, and SN_S (see Equations 12-7, 12-8, and 12-9) are his recommended performance measures in a wide variety of situations. By maximizing the appropriate SN ratio, he claims that variability is minimized.

Consider first the SN_T ratio,

$$SN_T = 10 \log \left(\frac{\bar{y}^2}{S^2} \right)$$

This ratio would be used if we wish to minimize variability around a fixed target value. It has been suggested by Taguchi that it is preferable to work with SN_T instead of the standard deviation because in many cases the process mean and standard deviation are related. (As μ gets larger, σ gets larger, for example.) In such cases, he argues that we cannot directly minimize the standard deviation

Table 12-21   Data for the "Marginal Means" Plots in Figure 12-15

		A			
		1	2	3	Averages for B
	1	10	10	13	11.00
B	2	8	10	14	9.67
	3	6	9	10	8.33
Averages for A		8.00	9.67	11.67	

Figure 12-16. The *AB* inter-action plot for the data in Table 12-21.

and then bring the mean on target. Taguchi claims he found empirically that the use of the SN_T ratio coupled with a two-stage optimization procedure would lead to a combination of factor levels where the standard deviation is minimized and the mean is on target. The optimization procedure consists of (1) finding the set of controllable factors that affect SN_T, called the *control factors,* and setting them to levels that maximize SN_T and then (2) finding the set of factors that have significant effects on the mean but do not influence the SN_T ratio, called the *signal factors,* and using these factors to bring the mean on target.

Given that this partitioning of factors is possible, SN_T is an example of a performance measure independent of adjustment (PERMIA) [see Leon et al. (1987)]. The signal factors would be the adjustment factors. The motivation behind the signal-to-noise ratio is to uncouple location and dispersion effects. It can be shown that the use of SN_T is equivalent to an analysis of the standard deviation of the logarithm of the original data. Thus, using SN_T implies that a log transformation will *always* uncouple location and dispersion effects. There is no assurance that this will happen. A much safer approach is to investigate what type of transformation is appropriate.

Note that we can write the SN_T ratio as

$$SN_T = 10 \log \left(\frac{\bar{y}^2}{S^2} \right)$$

$$= 10 \log (\bar{y}^2) - 10 \log (S^2)$$

If the mean is fixed at a target value (estimated by $\bar{y}$), then maximizing the SN_T ratio is equivalent to minimizing $\log (S^2)$. Using $\log (S^2)$ would require fewer calculations, is more intuitively appealing, and would provide a clearer understanding of the factor relationships that influence process variability—in other words, it would provide better process understanding. Furthermore, if we minimize $\log (S^2)$ directly, we eliminate the risk of obtaining wrong answers from the maximization of SN_T if some of the manipulated factors drive the mean $\bar{y}$ upward instead of driving S^2 downward. In general, if the response variable can be expressed in terms of the model

$$y = \mu(x_d, x_a)\epsilon(x_d)$$

where x_d is the subset of factors that drive the dispersion effects and x_a is the subset of adjustment factors that do not affect variability, then maximizing SN_T will be equivalent to minimizing the standard deviation. Considering the other potential problems surrounding SN_T, it is likely to be safer to work directly with the standard deviation (or its logarithm) as a response variable.

The ratios SN_L and SN_S are even more troublesome. Schmidt and Boudot (1989) have conducted a simulation study that addresses the effectiveness of these signal-to-noise ratios in detecting dispersion or variability effects. They show that SN_L and SN_S are completely *ineffective* in identifying dispersion effects, although they may serve to identify *location* effects, that is, factors that drive the mean. The reason for this is relatively easy to see. Consider the SN_S (smaller-the-better) ratio:

$$SN_S = -10 \log \left(\frac{1}{n} \sum_{i=1}^{n} y_i^2 \right)$$

The ratio is motivated by the assumption of a quadratic loss function with y nonnegative. The loss function for such a case would be

$$L = C \frac{1}{n} \sum_{i=1}^{n} \bar{y}_i^2$$

where C is a constant. Now

$$\log L = \log C + \log \left(\frac{1}{n} \sum_{i=1}^{n} \bar{y}_i^2 \right)$$

and

$$SN_S = 10 \log C - 10 \log L$$

so maximizing SN_S will minimize L. However, it is easy to show that

$$\frac{1}{n} \sum_{i=1}^{n} y_i^2 = \bar{y}^2 + \frac{1}{n} \left(\sum_{i=1}^{n} y_i^2 - n\bar{y}^2 \right)$$

$$= \bar{y}^2 + \left(\frac{n-1}{n} \right) s^2$$

Therefore, the use of SN_S as a response variable confounds location and dispersion effects.

The confounding of location and dispersion effects was observed in the analysis of the SN_L ratio in the pull-off force example. In Figures 12-10 and 12-11 notice that the plots of $\bar{y}$ and SN_L versus each factor have approximately the same shape, implying that both responses measure location. In fact, it can be shown that *none* of the factors A through D in this example have any major effect on process variability. Furthermore, since the SN_S and SN_L ratios involve y^2 and

$1/y^2$, they will be very sensitive to outliers or values near zero, and they are not invariant to linear transformation of the original response. We strongly recommend that these signal-to-noise ratios not be used.

A better approach for isolating location and dispersion effects is to analyze $\bar{y}$ and $\log(S)$ as separate responses. If no replication is available to estimate variability at each run in the design, methods for analyzing residuals can be used.

Finally, we turn to some of the applications of the analysis of variance recommended by Taguchi. As an example for discussion, consider the experiment reported by Quinlan (1985) at a symposium on Taguchi methods sponsored by the American Supplier Institute. The experiment concerned the quality improvement of speedometer cables. Specifically, the objective was to reduce the shrinkage in the plastic casing material. (Excessive shrinkage causes the cables to be noisy.) The experiment used an L_{16} orthogonal array (the 2_{III}^{15-11} design). The shrinkage values for four samples taken from 3000-foot lengths of the product manufactured at each set of test conditions are shown in Table 12-22, along with $\bar{y}$ and SN_S.

Quinlan, following the Taguchi approach to data analysis, used SN_S as the response variable in an analysis of variance. The error mean square was formed by pooling the mean squares associated with the seven effects that had the smallest absolute magnitude. This resulted in all eight remaining factors having significant effects (in order of magnitude: E, G, K, A, C, F, D, H). The author did note that E and G were the most important.

The pooling of mean squares as in this example is a procedure that has long been known to produce considerable bias in the test results. To illustrate the problem, consider the 15 NID(0, 1) random numbers shown in column 1 of Table 12-23. The square of each of these numbers, shown in column 2 of the table, is a single-degree-of-freedom mean square corresponding to the observed random number. The seven smallest random numbers are marked with an asterisk in column 1 of Table 12-23. The corresponding mean squares are pooled to form a mean square for error with seven degrees of freedom. This quantity is

$$MS_E = \frac{0.5088}{7} = 0.0727$$

Finally, column 3 of Table 12-23 presents the F ratio formed by dividing each of the eight remaining mean squares by MS_E. Now $F_{.05,1,7} = 5.59$, and this implies that five of the eight effects would be judged significant at the .05 level. Recall that since the original data came from a normal distribution with mean zero, *none* of the effects are different from zero.

Analysis methods such as this virtually guarantee erroneous conclusions. The normal probability plotting of effects avoids this invalid pooling of mean squares and provides a simple, easy to interpret method of analysis. Box (1988) provides an alternate analysis of the Quinlan data that correctly reveals factors E and G to be important along with other interesting results not apparent in the original analysis.

It is important to note that the Taguchi analysis identified negligible factors

Table 12-22 The Speedometer Cable Experiment

Run	H	D	−L	B	−J	F	N	A	−I	−E	M	−C	K	G	−O	y_1	y_2	y_3	y_4	SN_S	$\bar{y}$
1	−	−	+	−	+	−	−	−	+	+	−	+	−	−	+	0.49	0.54	0.46	0.45	6.2626	0.4850
2	+	−	−	−	−	−	+	−	−	+	+	+	+	−	−	0.55	0.60	0.57	0.58	4.8024	0.5750
3	−	+	−	−	+	+	+	−	+	−	+	+	−	−	−	0.07	0.09	0.11	0.08	21.0375	0.0875
4	+	+	−	−	+	+	−	−	−	−	−	+	+	+	+	0.16	0.16	0.19	0.19	15.1074	0.1750
5	−	+	+	+	+	+	+	−	+	+	−	−	+	+	−	0.13	0.22	0.20	0.23	14.0285	0.1950
6	+	−	+	+	−	−	−	−	−	+	+	−	−	+	+	0.16	0.17	0.13	0.12	16.6857	0.1450
7	−	−	−	+	+	−	−	−	+	−	−	−	+	−	+	0.24	0.22	0.19	0.25	12.9115	0.2250
8	+	+	−	+	−	−	+	−	−	−	+	−	−	−	+	0.13	0.19	0.19	0.19	15.0446	0.1750
9	−	−	−	−	+	−	−	+	−	−	−	−	+	+	−	0.08	0.10	0.14	0.18	17.6700	0.1250
10	+	+	−	−	+	−	+	+	+	+	−	−	−	+	−	0.07	0.04	0.19	0.18	17.2700	0.1200
11	−	−	−	−	−	+	+	+	−	+	+	+	+	−	+	0.48	0.49	0.44	0.41	6.8183	0.4550
12	+	+	+	−	−	+	−	+	+	−	−	+	−	−	+	0.54	0.53	0.53	0.54	5.4325	0.5350
13	−	−	+	+	−	+	+	+	−	−	+	+	+	−	−	0.13	0.17	0.21	0.17	15.2724	0.1700
14	+	−	−	+	+	+	−	+	+	+	−	+	−	−	−	0.28	0.26	0.26	0.30	11.1976	0.2750
15	−	+	−	+	−	−	−	+	−	+	−	+	−	+	−	0.34	0.32	0.30	0.41	9.2436	0.3425
16	+	+	+	+	+	−	+	+	+	+	+	+	−	+	+	0.58	0.62	0.59	0.54	4.6836	0.5825

Table 12-23 Bias Introduced by Pooling

NID (0, 1) Random Numbers	Mean Squares with One Degree of Freedom	F_0
-0.8607	0.7408	10.19
-0.8820	0.7779	10.70
0.3608*	0.1302	—
0.0227*	0.0005	—
0.1903*	0.0362	—
-0.3071*	0.0943	—
1.2075	1.4581	20.06
0.5641	0.3182	4.38
-0.3936*	0.1549	—
-0.6940	0.4816	6.63
-0.3028*	0.0917	—
0.5832	0.3401	4.68
0.0324*	0.0010	—
1.0202	1.0408	14.32
-0.6347	0.4028	5.54

as significant. This can have profound impact on our use of experimental design to enhance process knowledge. Experimental design methods should make gaining process knowledge easier, not harder.

Some Final Remarks ▪ In this section we have directed some major criticisms towards the specific methods of experimental design and data analysis used in the Taguchi approach to parameter design. Remember that these comments have focused on technical issues, and that the broad *philosophy* recommended by Taguchi, which we discussed in Section 12-5.1, is inherently sound.

On the other hand, many companies have reported success with the use of Taguchi's parameter design methods. If the methods are flawed, why do they produce successful results? Taguchi advocates often refute criticism with the remark that "they work." We must remember that the "best guess" and "one-factor-at-a-time" methods will also work—and occasionally they produce good results. This is no reason to claim that they are good methods. Most of the successful applications of Taguchi's technical methods have been in industries where there was no history of good experimental design practice. Designers and developers were using the best guess and one-factor-at-a-time methods (or other unstructured approaches), and since the Taguchi approach is based on the factorial design concept, it often produced better results than the methods it replaced. In other words, the factorial design is so powerful that, even when it is used inefficiently, it will work better than almost anything else.

As pointed out earlier, the Taguchi approach to parameter design often leads to *large, comprehensive* experiments with 70 or more runs. Many of the successful applications of this approach have been in industries characterized by a high-volume, low-cost manufacturing environment. In such situations, large designs may not be a real problem, as it is really no more difficult to make 72 runs than to make 16 or 32 runs. On the other hand, in industries characterized by low-volume, high-cost manufacturing (such as the aerospace industry, chemical and process industries, electronics and semiconductor manufacturing, and so forth); these methodological inefficiencies can be significant.

A final point concerns the learning process. If the Taguchi approach to parameter design works and yields good results, we may still not know what has caused the result because of the aliasing of critical interactions. In other words, we may have solved a problem (a short-term success), but we may not have gained *process knowledge,* which could be invaluable in future problems.

In summary, we should support Taguchi's philosophy of quality engineering. However, we must rely on simpler, more efficient methods that are easier to learn and apply to carry this philosophy into practice.

12-6 PROBLEMS

12-1 The effects of developer strength (A) and development time (B) on the density of photographic plate film are being studied. Three strengths and three times are used, and four replicates of a 3^2 factorial experiment are run. The data from this experiment follow. Analyze this data using the standard methods for factorial experiments.

Developer Strength	Development Time (Minutes)					
	10		14		18	
1	0	2	1	3	2	5
	5	4	4	2	4	6
2	4	6	6	8	9	10
	7	5	7	7	8	5
3	7	10	10	10	12	10
	8	7	8	7	9	8

12-2 Analyze the data in Problem 12-1 using Yates' algorithm.

12-3 Compute the *I* and *J* components of the two-factor interaction in Problem 12-1.

12-4 An experiment was performed to study the effect of three different types of 32 ounce bottles (A) and three different shelf types (B)—smooth permanent shelves, end-aisle displays with grilled shelves, and beverage coolers—on the time it takes to stock ten 12-bottle cases on the shelves. Three workers—(factor C) were employed in the experiment, and two replicates of a 3^3 factorial design were run. The observed time data are shown in the following table. Analyze these data and draw conclusions.

Worker	Bottle Type	Replicate I Permanent	End Aisle	Cooler	Replicate II Permanent	End Aisle	Cooler
1	Plastic	3.45	4.14	5.80	3.36	4.19	5.23
	28-mm glass	4.07	4.38	5.48	3.52	4.26	4.85
	38-mm glass	4.20	4.26	5.67	3.68	4.37	5.58
2	Plastic	4.80	5.22	6.21	4.40	4.70	5.88
	28-mm glass	4.52	5.15	6.25	4.44	4.65	6.20
	38-mm glass	4.96	5.17	6.03	4.39	4.75	6.38
3	Plastic	4.08	3.94	5.14	3.65	4.08	4.49
	28-mm glass	4.30	4.53	4.99	4.04	4.08	4.59
	38-mm glass	4.17	4.86	4.85	3.88	4.48	4.90

12-5 A medical researcher is studying the effect of lidocaine on the enzyme level in the heart muscle of beagle dogs. Three different commercial brands of lidocaine (A), three dosage levels (B), and three dogs (C) are used in the experiment, and two replicates of a 3^3 factorial design are run. The observed enzyme levels follow. Analyze the data from this experiment.

Lidocaine Brand	Dosage Strength	Replicate I Dog 1	2	3	Replicate II Dog 1	2	3
1	1	86	84	85	84	85	86
	2	94	99	98	95	97	90
	3	101	106	98	105	104	103
2	1	85	84	86	80	82	84
	2	95	98	97	93	99	95
	3	108	114	109	110	102	100
3	1	84	83	81	83	80	79
	2	95	97	93	92	96	93
	3	105	100	106	102	111	108

12-6 Compute the I and J components of the two-factor interactions for Example 12-1.

12-7 Confound a 3^3 design in three blocks using the ABC^2 component of the three-factor interaction. Compare your results with the design in Figure 12-6.

12-8 Confound a 3^4 design in three blocks using the AB^2CD component of the four-factor interaction.

12-9 Consider the data from the first replicate of Problem 12-4. Assuming that all 27 observations could not be run on the same day, set up a design for conducting the experiment over three days with AB^2C confounded with blocks. Analyze the data.

12-10 Outline the analysis of variance table for the 3^4 design in nine blocks. Is this a practical design?

12-11 Consider the data in Problem 12-4. If ABC is confounded in replicate I and ABC^2 is confounded in replicate II, perform the analysis of variance.

12-12 Consider the data from replicate I of Problem 12-4. Suppose that only a one-third fraction of this design with $I = ABC$ is run. Construct the design, determine the alias structure, and analyze the data.

12-13 Construct a 3_{IV}^{4-1} design with $I = ABCD$. Write out the alias structure for this design.

12-14 Verify that the design in Problem 12-13 is a resolution IV design.

12-15 Construct a 3^{5-2} design with $I = ABC$ and $I = CDE$. Write out the alias structure for this design. What is the resolution of this design?

12-16 Construct a 3^{9-6} design, and verify that it is a resolution III design.

12-17 Construct a 4×2^3 design confounded in two blocks of 16 observations each. How could Yates' algorithm be employed to analyze this design?

12-18 Outline the analysis of variance table for a $2^2 3^2$ factorial design. Discuss how this design may be confounded in blocks.

12-19 Starting with a 16-run 2^4 design, show how two three-level factors can be incorporated in this experiment. How many two-level factors can be included if we want some information on two-factor interactions?

12-20 Starting with a 16-run 2^4 design, show how one three-level factor and three two-level factors can be accommodated and still allow the estimation of two-factor interactions.

12-21 In Problem 11-26, you met Harry and Judy Peterson-Nedry, two friends of the author who have a winery and vineyard in Newberg, Oregon. That problem described the application of two-level fractional factorial designs to their 1985 Pinot Noir product. In 1987, they wanted to conduct another Pinot Noir experiment. The variables for this experiment were

Variable	Levels
Clone of Pinot Noir	Wadenswil, Pommard
Berry size	Small, Large
Fermentation temperature	80°F, 85°F, 90/80°F, 90°F
Whole Berry	None, 10%
Maceration time	10 days, 21 days
Yeast type	Assmanhau, Champagne
Oak type	Troncais, Allier

Harry and Judy decided to use a 16-run two-level fractional factorial design, treating the four levels of fermentation temperature as two two-level variables. As in Problem 11-26, they used the rankings from a taste-test panel as the response variable. The design and the resulting average ranks are shown below:

Run	Clone	Berry Size	Ferm. Temp.	Whole Berry	Macer. Time	Yeast Type	Oak Type	Average Rank
1	−	−	−	−	−	−	−	−	4
2	+	−	−	−	−	+	+	+	10
3	−	+	−	−	+	−	+	+	6
4	+	+	−	−	+	+	−	−	9
5	−	−	+	−	+	+	+	−	11
6	+	−	+	−	+	−	−	+	1
7	−	+	+	−	−	+	−	+	15
8	+	+	+	−	−	−	+	−	5
9	−	−	−	+	+	+	−	+	12
10	+	−	−	+	+	−	+	−	2
11	−	+	−	+	−	+	+	−	16
12	+	+	−	+	−	−	−	+	3
13	−	−	+	+	−	−	+	+	8
14	+	−	+	+	−	+	−	−	14
15	−	+	+	+	+	−	−	−	7
16	+	+	+	+	+	+	+	+	13

(a) Describe the aliasing in this design.

(b) Analyze the data and draw conclusions.

(c) What comparisons can you make between this experiment and the 1985 Pinot Noir experiment from Problem 11-26?

12-22   In a recent article ("Let's All Beware the Latin Square," *Quality Engineering,* Vol. 1, 1989, pp. 453–465) J. S. Hunter illustrates some of the problems associated with 3^{k-p} fractional factorial designs. Factor A is the amount of ethanol added to a standard fuel and factor B represents the air/fuel ratio. The response variable is carbon monoxide (CO) emission in g/m^3. The design is shown below:

Design				Observations	
A	B	x_1	x_2	y	
1	1	−1	−1	66	62
2	1	0	−1	78	81
3	1	+1	−1	90	94
1	2	−1	0	72	67
2	2	0	0	80	81
3	2	+1	0	75	78
1	3	−1	+1	68	66
2	3	0	+1	66	69
3	3	+1	+1	60	58

Notice that we have used the notation system of 1, 2, and 3 to represent the low, medium, and high levels for the factors. We have also used a "geometric nota-tion" of -1, 0, and $+1$. Each run in the design is replicated twice.

(a) Verify that the second-order model

$$\hat{y} = 78.5 + 4.5x_1 - 7.0x_2 - 4.5x_1^2 - 4.0x_2^2 - 9.0x_1x_2$$

is a reasonable model for this experiment. Sketch the CO concentration con-tours in the x_1, x_2 space.

(b) Now suppose that instead of only two factors, we had used *four* factors in a 3^{4-2} fractional factorial design and obtained *exactly* the same data as in part (a). The design would be as follows:

		Design						Observations	
A	B	C	D	x_1	x_2	x_3	x_4	y	
1	1	1	1	-1	-1	-1	-1	66	62
2	1	2	2	0	-1	0	0	78	81
3	1	3	3	$+1$	-1	$+1$	$+1$	90	94
1	2	3	2	-1	0	$+1$	0	72	67
2	2	1	3	0	0	-1	$+1$	80	81
3	2	2	1	$+1$	0	0	-1	75	78
1	3	2	3	-1	$+1$	0	$+1$	68	66
2	3	3	1	0	$+1$	$+1$	-1	66	69
3	3	1	2	$+1$	$+1$	-1	0	60	58

Confirm that this design is an L_9 orthogonal array.

(c) Calculate the marginal averages of the CO response at each level of the four factors A, B, C, and D. Construct plots of these marginal averages and inter-pret the results. Do factors C and D appear to have strong effects? Do these factors *really* have any effect on CO emission? Why is their apparent effect strong?

(d) The design in part (b) allows the model

$$y = \beta_0 + \sum_{i=1}^{4} \beta_i x_i + \sum_{i=1}^{4} \beta_{ii} x_i^2 + \epsilon$$

to be fitted. Suppose that the *true* model is

$$y = \beta_0 + \sum_{i=1}^{4} \beta_i x_i + \sum_{i=1}^{4} \beta_{ii} x_i^2 + \sum\sum_{i<j} \beta_{ij} x_i x_j + \epsilon$$

Show that if $\hat{\beta}_j$ represents the least squares estimates of the coefficients in the fitted model, then

$$E(\hat{\beta}_0) = \beta_0 - \beta_{23} - \beta_{24} - \beta_{34}$$

$$E(\hat{\beta}_1) = \beta_1 - (\beta_{23} - \beta_{24} - \beta_{34})/2$$

$$E(\hat{\beta}_2) = \beta_2 - (\beta_{13} + \beta_{14})/2$$

$$E(\hat{\beta}_3) = \beta_3 - (\beta_{12} + \beta_{14})/2$$

$$E(\hat{\beta}_4) = \beta_4 - (\beta_{12} + \beta_{13})/2$$

$$E(\hat{\beta}_{11}) = \beta_{11} + (\beta_{23} + \beta_{24} + \beta_{34})/2$$

$$E(\hat{\beta}_{22}) = \beta_{22} + (\beta_{13} - \beta_{14})/2 + \beta_{34}$$

$$E(\hat{\beta}_{33}) = \beta_{33} + (\beta_{14} - \beta_{12})/2 + \beta_{24}$$

$$E(\hat{\beta}_{44}) = \beta_{44} + (\beta_{12} - \beta_{13})/2 + \beta_{23}$$

Does this help explain the strong effects for factors C and D observed graphically in part (c)?

Chapter 13
Nested or Hierarchial Designs

13-1 INTRODUCTION

In certain multifactor experiments the levels of one factor (e.g., factor B) are similar but not identical for different levels of another factor (e.g., A). Such an arrangement is called a *nested* or *hierarchial* design, with the levels of factor B nested under the levels of factor A. For example, consider a company that purchases its raw material from three different suppliers. The company wishes to determine if the purity of the raw material is the same from each supplier. There are four batches of raw material available from each supplier, and three determinations of purity are to be taken from each batch. The situation is depicted in Figure 13-1.

This is a *two-stage nested* or *hierarchial design,* with batches nested under suppliers. At first glance, you may ask why the two factors, suppliers and batches, are not crossed. If the factors were crossed, then batch 1 would always refer to the same batch, batch 2 would always refer to the same batch, and so on. This is clearly not the case since the batches from each supplier are unique for that particular supplier. That is, batch 1 from supplier 1 has no connection with batch 1 from any other supplier, batch 2 from supplier 1 has no connection with batch 2 from any other supplier, and so forth. To emphasize the fact that the batches from each supplier are different batches, we may renumber the batches as 1, 2, 3, and 4 from supplier 1; 5, 6, 7, and 8 from supplier 2; and 9, 10, 11, and 12 from supplier 3, as shown in Figure 13-2.

Sometimes we may be uncertain as to whether a factor is crossed or nested. If the levels of the factor can be renumbered arbitrarily as in Figure 13-2, then the factor is nested.

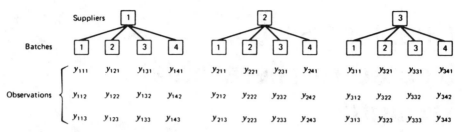

Figure 13-1. A two-stage nested design.

13-2 THE TWO-STAGE NESTED DESIGN

13-2.1 Statistical Analysis

The linear statistical model for the two-stage nested design is

$$y_{ijk} = \mu + \tau_i + \beta_{j(i)} + \epsilon_{(ij)k} \begin{cases} i = 1, 2, \ldots, a \\ j = 1, 2, \ldots, b \\ k = 1, 2, \ldots, n \end{cases} \tag{13-1}$$

That is, there are a levels of factor A, b levels of factor B nested under each level of A, and n replicates. The subscript $j(i)$ indicates that the jth level of factor B is nested under the ith level of factor A. It is convenient to think of the replicates as being nested within the combination of levels of A and B; thus, the subscript $(ij)k$ is used for the error term. This is a *balanced* nested design since there are an equal number of levels of B within each level of A and an equal number of replicates. Since every level of factor B does not appear with every level of factor A, there can be no interaction between A and B.

We may write the total corrected sum of squares as

$$\sum_{i=1}^{a} \sum_{j=1}^{b} \sum_{k=1}^{n} (y_{ijk} - \bar{y}_{...})^2$$

$$= \sum_{i=1}^{a} \sum_{j=1}^{b} \sum_{k=1}^{n} [(\bar{y}_{i..} - \bar{y}_{...}) + (\bar{y}_{ij.} - \bar{y}_{i..}) + (y_{ijk} - \bar{y}_{ij.})]^2 \tag{13-2}$$

Expanding the right-hand side of Equation 13-2 yields

$$\sum_{i=1}^{a} \sum_{j=1}^{b} \sum_{k=1}^{n} (y_{ijk} - \bar{y}_{...})^2 = bn \sum_{i=1}^{a} (\bar{y}_{i..} - \bar{y}_{...})^2 + n \sum_{i=1}^{a} \sum_{j=1}^{b} (\bar{y}_{ij.} - \bar{y}_{i..})^2$$

$$+ \sum_{i=1}^{a} \sum_{j=1}^{b} \sum_{k=1}^{n} (y_{ijk} - \bar{y}_{ij.})^2 \tag{13-3}$$

since the three cross-product terms are zero. Equation 13-3 indicates that the total sum of squares can be partitioned into a sum of squares due to factor A, a sum of squares due to factor B under the levels of A, and a sum of squares due

Figure 13-2. Alternate layout for the two-stage nested design.

to error. Symbolically, we may write Equation 13-3 as

$$SS_T = SS_A + SS_{B(A)} + SS_E \qquad (13\text{-}4)$$

There are $abn - 1$ degrees of freedom for SS_T, $a - 1$ degrees of freedom for SS_A, $a(b - 1)$ degrees of freedom for $SS_{B(A)}$, and $ab(n - 1)$ degrees of freedom for error. Note that $abn - 1 = (a - 1) + a(b - 1) + ab(n - 1)$. If the errors are NID($0, \sigma^2$), we may divide each sum of squares on the right of Equation 13-4 by its degrees of freedom to obtain independently distributed mean squares such that the ratio of any two mean squares is distributed as F.

The appropriate statistics for testing the effects of factors A and B depend on whether A and B are fixed or random. If factors A and B are fixed, we assume that $\sum_{i=1}^{a}\tau_i = 0$ and $\sum_{j=1}^{b}\beta_{j(i)} = 0$ ($i = 1, 2, \ldots, a$). That is, the A treatment effects sum to zero, and the B treatment effects sum to zero within each level of A. Alternatively, if A and B are random, then we assume that τ_i is NID($0, \sigma_\tau^2$) and $\beta_{j(i)}$ is NID($0, \sigma_\beta^2$). Mixed models with A fixed and B random are also widely encountered. The expected mean squares can be determined by a straightforward application of the rules in Chapter 8. Table 13-1 gives the expected mean squares for these situations.

Table 13-1 indicates that if the levels of A and B are fixed, then H_0: $\tau_i = 0$ is tested by MS_A/MS_E and H_0: $\beta_{j(i)} = 0$ is tested by $MS_{B(A)}/MS_E$. If A is a fixed factor and B is random, then H_0: $\tau_i = 0$ is tested by $MS_A/MS_{B(A)}$ and H_0: $\sigma_\beta^2 = 0$ is tested by $MS_{B(A)}/MS_E$. Finally, if both A and B are random factors, we test H_0: $\sigma_\tau^2 = 0$ by $MS_A/MS_{B(A)}$ and H_0: $\sigma_\beta^2 = 0$ by $MS_{B(A)}/MS_E$. The test procedure is summarized in an analysis of variance table as shown in Table 13-2. Computing formulas for the sums of squares may be obtained by expanding the quantities

Table 13-1 Expected Mean Squares in the Two-Stage Nested Design

$E(MS)$	A Fixed B Fixed	A Fixed B Random	A Random B Random
$E(MS_A)$	$\sigma^2 + \dfrac{bn\sum \tau_i^2}{a - 1}$	$\sigma^2 + n\sigma_\beta^2 + \dfrac{bn\sum \tau_i^2}{a - 1}$	$\sigma^2 + n\sigma_\beta^2 + bn\sigma_\tau^2$
$E(MS_{B(A)})$	$\sigma^2 + \dfrac{n\sum\sum \beta_{j(i)}^2}{a(b - 1)}$	$\sigma^2 + n\sigma_\beta^2$	$\sigma^2 + n\sigma_\beta^2$
$E(MS_E)$	σ^2	σ^2	σ^2

Table 13-2   Analysis of Variance Table for the Two-Stage Nested Design

Source of Variation	Sum of Squares	Degrees of Freedom	Mean Square
A	$\sum_i \dfrac{y_{i..}^2}{bn} - \dfrac{y_{...}^2}{abn}$	$a - 1$	MS_A
B within A	$\sum_i \sum_j \dfrac{y_{ij.}^2}{n} - \sum_i \dfrac{y_{i..}^2}{bn}$	$a(b - 1)$	$MS_{B(A)}$
Error	$\sum_i \sum_j \sum_k y_{ijk}^2 - \sum_i \sum_j \dfrac{y_{ij.}^2}{n}$	$ab(n - 1)$	MS_E
Total	$\sum_i \sum_j \sum_k y_{ijk}^2 - \dfrac{y_{...}^2}{abn}$	$abn - 1$	

in Equation 13-3 and simplifying. They are

$$SS_A = \sum_{i=1}^{a} \frac{y_{i..}^2}{bn} - \frac{y_{...}^2}{abn} \tag{13-5}$$

$$SS_{B(A)} = \sum_{i=1}^{a} \sum_{j=1}^{b} \frac{y_{ij.}^2}{n} - \sum_{i=1}^{a} \frac{y_{i..}^2}{bn} \tag{13-6}$$

$$SS_E = \sum_{i=1}^{a} \sum_{j=1}^{b} \sum_{k=1}^{n} y_{ijk}^2 - \sum_{i=1}^{a} \sum_{j=1}^{b} \frac{y_{ij.}^2}{n} \tag{13-7}$$

$$SS_T = \sum_{i=1}^{a} \sum_{j=1}^{b} \sum_{k=1}^{n} y_{ijk}^2 - \frac{y_{...}^2}{abn} \tag{13-8}$$

We see that Equation 13-6 for $SS_{B(A)}$ can be written as

$$SS_{B(A)} = \sum_{i=1}^{a} \left[\sum_{j=1}^{b} \frac{y_{ij.}^2}{n} - \frac{y_{i..}^2}{bn} \right]$$

This expresses the idea that $SS_{B(A)}$ is the sum of squares between levels of B for each level of A, summed over all the levels of A.

Example 13-1

Consider the company described in Section 13-1 that buys raw material in batches from three different suppliers. The purity of this raw material varies considerably, which causes problems in manufacturing the finished product. We wish to determine if the variability in purity is attributable to differences between the suppliers. Four batches of raw material are selected at random from each supplier, and three determinations of purity are made on each batch. This is, of course, a two-stage nested design. The data, after coding by subtracting 93, are shown in Table 13-3. The sums of squares are computed as follows:

$$SS_T = \sum_{i=1}^{a} \sum_{j=1}^{b} \sum_{k=1}^{n} y_{ijk}^2 - \frac{y_{...}^2}{abn}$$

$$= 153.00 - \frac{(13)^2}{36} = 148.31$$

$$SS_A = \sum_{i=1}^{a} \frac{y_{i..}^2}{bn} - \frac{y_{...}^2}{abn}$$

$$= \frac{(-5)^2 + (4)^2 + (14)^2}{(4)(3)} - \frac{(13)^2}{36}$$

$$= 19.75 - 4.69 = 15.06$$

$$SS_{B(A)} = \sum_{i=1}^{a} \sum_{j=1}^{b} \frac{y_{ij.}^2}{n} - \sum_{i=1}^{a} \frac{y_{i..}^2}{bn}$$

$$= \frac{(0)^2 + (-9)^2 + (-1)^2 + \cdots + (2)^2 + (6)^2}{3} - 19.75$$

$$= 89.67 - 19.75 = 69.92$$

and

$$SS_E = \sum_{i=1}^{a} \sum_{j=1}^{b} \sum_{k=1}^{n} y_{ijk}^2 - \sum_{i=1}^{a} \sum_{j=1}^{b} \frac{y_{ij.}^2}{n}$$

$$= 153.00 - 89.67 = 63.33$$

The analysis of variance is summarized in Table 13-4. Suppliers are fixed and batches are random, so the expected mean squares are obtained from the middle column of Table 13-1. They are repeated for convenience in Table 13-4. The analysis of variance indicates that at the 5 percent level, there is no significant effect on purity due to suppliers but the purity of batches of raw material from the same supplier does differ significantly.

∎

The practical implications of this experiment and the analysis are very important. The objective of the experimenter is to find the source of the variability

Table 13-3   Coded Purity Data for Example 13-1 (Code: y_{ijk} = purity − 93)

		Supplier 1				Supplier 2				Supplier 3			
	Batches	1	2	3	4	1	2	3	4	1	2	3	4
		1	−2	−2	1	1	0	−1	0	2	−2	1	3
		−1	−3	0	4	−2	4	0	3	4	0	−1	2
		0	−4	1	0	−3	2	−2	2	0	2	2	1
Batch totals	$y_{ij.}$	0	−9	−1	5	−4	6	−3	5	6	0	2	6
Supplier totals	$y_{i..}$		−5				4				14		

Table 13-4   Analysis of Variance for the Data in Example 13-1

Source of Variation	Sum of Squares	Degrees of Freedom	Mean Square	Expected Mean Square	F_0
Suppliers	15.06	2	7.53	$\sigma^2 + 3\sigma_\beta^2 + 6 \sum \tau_i^2$	0.97
Batches (within suppliers)	69.92	9	7.77	$\sigma^2 + 3\sigma_\beta^2$	2.94[a]
Error	63.33	24	2.64	σ^2	
Total	148.31	35			

[a]Significant at 5 percent.

in raw material purity. If it results from differences among suppliers, then we may be able to solve the problem by selecting the "best" supplier. However, that solution is not applicable here because the major source of variability is the batch-to-batch purity variation *within* suppliers. Therefore, we must attack the problem by working with the suppliers to reduce their batch-to-batch variability. This may involve modifications to the suppliers' production processes or their internal quality assurance system.

Notice what would have happened if we had incorrectly analyzed this design as a two-factor factorial experiment. If batches are considered to be crossed with suppliers, we obtain batch totals of 2, -3, -2, and 16, with each batch $\times$ suppliers cell containing three replicates. Thus, a sum of squares due to batches and an interaction sum of squares can be computed. The complete factorial analysis of variance is shown in Table 13-5.

This analysis indicates that batches differ significantly at the 5 percent level and that there is a significant interaction between batches and suppliers. However, it is difficult to give a practical interpretation of the batches $\times$ suppliers interaction. For example, does this significant interaction mean that the supplier effect is not constant from batch to batch? Furthermore, the significant interaction coupled with the nonsignificant supplier effect could lead the analyst to conclude that suppliers really differ but their effect is masked by the significant interaction. Comparing Tables 13-4 and 13-5, notice that

$$SS_B + SS_{S \times B} = 25.64 + 44.28 = 69.92 \equiv SS_{B(S)}$$

Table 13-5   Incorrect Analysis of the Two-Stage Nested Design in Example 13-1 as a Factorial (Suppliers Fixed, Batches Random)

Source of Variation	Sum of Squares	Degrees of Freedom	Mean Square	F_0
Suppliers (S)	15.06	2	7.53	1.02
Batches (B)	25.64	3	8.55	3.24[a]
S $\times$ B interaction	44.28	6	7.38	2.80[a]
Error	63.33	24	2.64	
Total	148.31	35		

[a] Significant at 5 percent.

That is, the sum of squares for batches within suppliers consists of the sum of squares of the batches and the sum of squares for the batches × suppliers interaction. The degrees of freedom have a similar property; that is,

$$\frac{\text{Batches}}{3} + \frac{\text{Batches} \times \text{suppliers}}{6} = \frac{\text{Batches within suppliers}}{9}$$

Therefore, a computer program for analyzing factorial designs could also be used for the analysis of nested designs by pooling the "main effect" of the nested factor and interactions of that factor with the factor under which it is nested. This is useful information, since a computer program specifically written for the analysis of nested designs may not be readily available whereas programs for analyzing factorial designs are widely available.

13-2.2 Diagnostic Checking

The major tool used in diagnostic checking is residual analysis. For the two-stage nested design, the residuals are

$$e_{ijk} = y_{ijk} - \hat{y}_{ijk}$$

However, the fitted value is

$$\hat{y}_{ijk} = \hat{\mu} + \hat{\tau}_i + \hat{\beta}_{j(i)}$$

and as we will subsequently show in Section 13-2.3,

$$\hat{y}_{ijk} = \bar{y}_{...} + (\bar{y}_{i..} - \bar{y}_{...}) + (\bar{y}_{ij.} - \bar{y}_{i..})$$
$$= \bar{y}_{ij.}$$

Thus, the residuals are

$$e_{ijk} = y_{ijk} - \bar{y}_{ij.}$$

where $\bar{y}_{ij.}$ are the individual batch averages.

The observations, fitted values, and residuals for the purity data in Example 13-1 follow:

Observed Value y_{ijk}	Fitted Value $\hat{y}_{ijk} = \bar{y}_{ij.}$	$e_{ijk} = y_{ijk} - \bar{y}_{ij.}$
1	0.00	1.00
−1	0.00	−1.00
0	0.00	0.00
−2	−3.00	1.00
−3	−3.00	0.00
−4	−3.00	−1.00
−2	−0.33	−1.67
0	−0.33	0.33
1	−0.33	1.33
1	1.67	−0.67
4	1.67	2.33
0	1.67	−1.67
1	−1.33	2.33

Observed Value y_{ijk}	Fitted Value $\hat{y}_{ijk} = \bar{y}_{ij.}$	$e_{ijk} = y_{ijk} - \bar{y}_{ij.}$
−2	−1.33	−0.67
−3	−1.33	−1.67
0	2.00	−2.00
4	2.00	2.00
2	2.00	0.00
−1	−1.00	0.00
0	−1.00	1.00
−2	−1.00	−1.00
0	1.67	−1.67
3	1.67	1.33
2	1.67	0.33
2	2.00	0.00
4	2.00	2.00
0	2.00	−2.00
−2	0.00	−2.00
0	0.00	0.00
2	0.00	2.00
1	0.67	0.33
−1	0.67	−1.67
2	0.67	1.33
3	2.00	1.00
2	2.00	0.00
1	2.00	−1.00

The usual diagnostic checks—including normal probability plots, checking for outliers, and plotting the residuals versus fitted values—may now be performed. As an illustration, the residuals are plotted versus the fitted values in Figure 13-3. This plot does not reveal any obvious model inadequacies.

13-2.3 Estimation of the Model Parameters

If both factors A and B are fixed, we may estimate the parameters in the balanced two-stage nested design model

$$y_{ijk} = \mu + \tau_i + \beta_{j(i)} + \epsilon_{(ij)k}$$

The least squares normal equations for this model are shown in Equations 13-9. For clarity in reading Equation 13-9, we have shown the parameter associated with each normal equation immediately to the left of that equation. For example, the first normal equation refers to μ, the second to τ_1, the third to τ_2, and so on.

The normal equations appear rather formidable; however, they may be easily derived using the rules of Section 4-4. As an illustration, consider the normal equation for τ_1, which is the second normal equation in Equations 13-9. The right-hand side of ths equation is $y_{1..}$ since $y_{1..}$ is the total of all observations

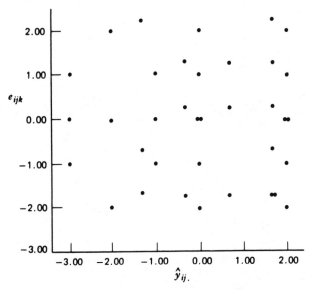

Figure 13-3. Plot of residuals versus fitted values for Example 13-1.

containing τ_1. Furthermore, in $y_{1..}$ the parameter μ appears bn times, τ_1 appears bn times, and each $\beta_{j(1)}$ ($j = 1, 2, \ldots, b$) appears n appears times. Hence, the normal eqution for τ_1 should be $bn\hat{\mu} + bn\hat{\tau}_1 + n\Sigma_{j=1}^{b}\hat{\beta}_{j(1)} = y_{1..}$, as shown in Equations 13-9.

There are $1 + a + ab$ normal equations in $1 + a + ab$ unknowns ($\hat{\mu}$, $\hat{\tau}_i$, and $\hat{\beta}_{j(i)}$ for $i = 1, 2, \ldots, a$ and $j = 1, 2, \ldots, b$). However, adding the equations associated with the τ_i yields the first normal equation, and adding the β equations in each level of τ_i yields the normal equation for that τ_i. Therefore, there are $1 + a$ linear dependencies in the normal equations. The usual restrictions imposed on the estimates of the parameters are

$$\sum_{i=1}^{a} \hat{\tau}_i = 0$$

and

$$\sum_{j=1}^{b} \hat{\beta}_{j(i)} = 0 \qquad i = 1, 2, \ldots, a$$

These constraints greatly simplify the normal equations, resulting in the solution $\hat{\mu} = \bar{y}_{...}$ and

$$\hat{\tau}_i = \bar{y}_{i..} - \bar{y}_{...} \qquad i = 1, 2, \ldots, a \qquad (13\text{-}10)$$

$$\hat{\beta}_{j(i)} = \bar{y}_{ij.} - \bar{y}_{i..} \qquad i = 1, 2, \ldots, a \qquad j = 1, 2, \ldots, b \qquad (13\text{-}11)$$

Notice that this solution to the normal equations has considerable intuitive ap-

$$
\begin{aligned}
\mu:\quad & abn\hat{\mu} + bn\hat{\tau}_1 + bn\hat{\tau}_2 + \cdots + bn\hat{\tau}_a + n\hat{\beta}_{1(1)} + n\hat{\beta}_{2(1)} + \cdots + n\hat{\beta}_{b(1)} + n\hat{\beta}_{1(2)} + n\hat{\beta}_{2(2)} + \cdots + n\hat{\beta}_{b(2)} + \cdots + n\hat{\beta}_{1(a)} + n\hat{\beta}_{2(a)} + \cdots + n\hat{\beta}_{b(a)} = y_{...} \\[4pt]
\tau_1:\quad & bn\hat{\mu} + bn\hat{\tau}_1 + n\hat{\beta}_{1(1)} + n\hat{\beta}_{2(1)} + \cdots + n\hat{\beta}_{b(1)} = y_{1..} \\[4pt]
\tau_2:\quad & bn\hat{\mu} + bn\hat{\tau}_2 + n\hat{\beta}_{1(2)} + n\hat{\beta}_{2(2)} + \cdots + n\hat{\beta}_{b(2)} = y_{2..} \\
& \qquad\qquad\ddots \\
\tau_a:\quad & bn\hat{\mu} + bn\hat{\tau}_a + n\hat{\beta}_{1(a)} + n\hat{\beta}_{2(a)} + \cdots + n\hat{\beta}_{b(a)} = y_{a..} \\[4pt]
\beta_{1(1)}:\quad & n\hat{\mu} + n\hat{\tau}_1 + n\hat{\beta}_{1(1)} = y_{11.} \\[4pt]
\beta_{2(1)}:\quad & n\hat{\mu} + n\hat{\tau}_1 + n\hat{\beta}_{2(1)} = y_{21.} \\
& \qquad\qquad\ddots \\
\beta_{b(1)}:\quad & n\hat{\mu} + n\hat{\tau}_1 + n\hat{\beta}_{b(1)} = y_{b1.} \\[4pt]
\beta_{1(2)}:\quad & n\hat{\mu} + n\hat{\tau}_2 + n\hat{\beta}_{1(2)} = y_{12.} \\[4pt]
\beta_{2(2)}:\quad & n\hat{\mu} + n\hat{\tau}_2 + n\hat{\beta}_{2(2)} = y_{22.} \\
& \qquad\qquad\ddots \\
\beta_{b(2)}:\quad & n\hat{\mu} + n\hat{\tau}_2 + n\hat{\beta}_{b(2)} = y_{b2.} \\
& \qquad\qquad\ddots \\
\beta_{1(a)}:\quad & n\hat{\mu} + n\hat{\tau}_a + n\hat{\beta}_{1(a)} = y_{1a.} \\[4pt]
\beta_{2(a)}:\quad & n\hat{\mu} + n\hat{\tau}_a + n\hat{\beta}_{2(a)} = y_{2a.} \\
& \qquad\qquad\ddots \\
\beta_{b(a)}:\quad & n\hat{\mu} + n\hat{\tau}_a + n\hat{\beta}_{b(a)} = y_{ba.}
\end{aligned}
\tag{13-9}
$$

peal; the A treatment effects are estimated by the average of all observations under each level of A minus the grand average, and the B treatment effects in each level of A are estimated by the corresponding cell average minus the average under that level of A. Using this solution, the general regression significance test can be applied to develop the analysis of variance for the balanced two-stage nested design (see Problem 13-5).

The solution to the normal equations was used in Section 13-2.2 to obtain the residuals for a nested design. Recall that we required $\hat{y}_{ijk}$ in order to compute the residuals e_{ijk}. Using Equations 13-10 and 13-11 we see that

$$\hat{y}_{ijk} = \hat{\mu} + \hat{\tau}_i + \hat{\beta}_{j(i)}$$
$$= \bar{y}_{...} + (\bar{y}_{i..} - \bar{y}_{...}) + (\bar{y}_{ij.} - \bar{y}_{i..})$$
$$= \bar{y}_{ij.}$$

which is the result used previously in Section 13-2.2.

Variance Components ▪ For the random effects case, the analysis of variance method can be used to estimate the variance components σ^2, σ_β^2, and σ_τ^2. From the expected mean squares in the last column of Table 13-1, we obtain

$$\hat{\sigma}^2 = MS_E \tag{13-12}$$

$$\hat{\sigma}_\beta^2 = \frac{MS_{B(A)} - MS_E}{n} \tag{13-13}$$

and

$$\hat{\sigma}_\tau^2 = \frac{MS_A - MS_{B(A)}}{bn} \tag{13-14}$$

Many applications of nested designs involve a mixed model, with the main factor (A) fixed and the nested factor (B) random. This is the case for the problem described in Example 13-1; suppliers (factor A) are fixed, whereas batches of raw material (factor B) are random. The effects of the suppliers may be estimated by

$$\hat{\tau}_1 = \bar{y}_{1..} - \bar{y}_{...} = \frac{-5}{12} - \frac{13}{36} = \frac{-28}{36}$$

$$\hat{\tau}_2 = \bar{y}_{2..} - \bar{y}_{...} = \frac{4}{12} - \frac{13}{36} = \frac{-1}{36}$$

$$\hat{\tau}_3 = \bar{y}_{3..} - \bar{y}_{...} = \frac{14}{12} - \frac{13}{36} = \frac{29}{36}$$

To estimate the variance components σ^2 and σ_β^2, we eliminate the line in the analysis of variance table pertaining to suppliers and apply the analysis of variance estimation method to the next two lines. This yields

$$\hat{\sigma}^2 = MS_E = 2.64$$

and

$$\hat{\sigma}_\beta^2 = \frac{MS_{B(A)} - MS_E}{n} = \frac{7.77 - 2.64}{3} = 1.71$$

From the analysis in Example 13-1, we know that the τ_i do not differ significantly from zero, whereas the variance component σ_β^2 is greater than zero.

13-3 THE GENERAL *m*-STAGE NESTED DESIGN

The results of Section 13-2 can be easily extended to the case of *m* completely nested factors. Such a design would be called an *m*-stage nested design. As an example, suppose a foundry wishes to investigate the hardness of two different formulations of a metal alloy. Three heats of each alloy formulation are prepared, two ingots are selected at random from each heat for testing, and two hardness measurements are made on each ingot. The situation is illustrated in Figure 13-4.

In this experiment, heats are nested under the levels of the factor alloy formulation, and ingots are nested under the levels of the factor heats. Thus, this is a three-stage nested design with two replicates.

The model for the general three-stage nested design is

$$y_{ijkl} = \mu + \tau_i + \beta_{j(i)} + \gamma_{k(ij)} + \epsilon_{(ijk)l} \begin{cases} i = 1,2, \ldots, a \\ j = 1,2, \ldots, b \\ k = 1,2, \ldots, c \\ l = 1,2, \ldots, n \end{cases} \qquad (13\text{-}15)$$

For our example, τ_i is the effect of the *i*th alloy formulation, $\beta_{j(i)}$ is the effect of the *j*th heat within the *i*th alloy, $\gamma_{k(ij)}$ is the effect of the *k*th ingot wihtin the *j*th heat and *i*th alloy, and $\epsilon_{(ijk)l}$ is the usual NID(0, σ^2) error term. Extension of this model to *m* factors is straightforward.

Notice that in the above example the overall variability in hardness consists

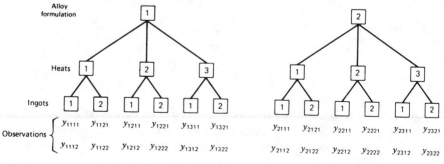

Figure 13-4. A three-stage nested design.

of three components: one that results from alloy formulations, one that results from heats, and one that results from analytical test error. These components of the variability in overall hardness are illustrated in Figure 13-5.

This example demonstrates how the nested design is often used in analyzing processes to identify the major sources of variablity in the output. For instance, if the alloy formulation variance component is large, then this implies that overall hardness variability could be reduced by using only one alloy formulation.

The calculation of the sums of squares and the analysis of variance for the *m*-stage nested design are similar to the analysis presented in Section 13-2. For example, the analysis of variance for the three-stage nested design is summarized in Table 13-6. Computational formulas for the sums of squares are also shown in this table. Notice that they are a simple extension of the computational formulas for the two-stage nested design.

To determine the proper test statistics we must find the expected mean squares using the methods of Chapter 8. For example, if factors *A* and *B* are fixed and factor *C* is random, then we may derive the expected mean squares as shown in Table 13-7. This table indicates the proper test statistics for this situation.

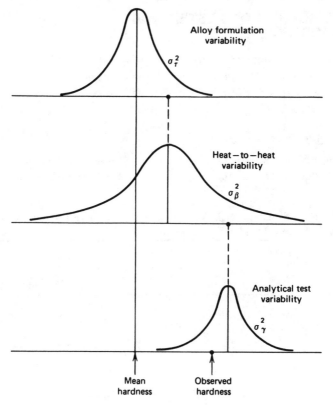

Figure 13-5. Sources of variation in the 3-stage nested design example.

Table 13-6 Analysis of Variance for the Three-Stage Nested Design

Source of Variation	Sum of Squares	Degrees of Freedom	Mean Square
A	$\displaystyle\sum_i \frac{y_{i...}^2}{bcn} - \frac{y_{....}^2}{abcn}$	$a - 1$	MS_A
B (within A)	$\displaystyle\sum_i \sum_j \frac{y_{ij.}^2}{cn} - \sum_i \frac{y_{i...}^2}{bcn}$	$a(b - 1)$	$MS_{B(A)}$
C (within B)	$\displaystyle\sum_i \sum_j \sum_k \frac{y_{ijk.}^2}{n} - \sum_i \sum_j \frac{y_{ij.}^2}{cn}$	$ab(c - 1)$	$MS_{C(B)}$
Error	$\displaystyle\sum_i \sum_j \sum_k \sum_l y_{ijkl}^2 - \sum_i \sum_j \sum_k \frac{y_{ijk.}^2}{n}$	$abc(n - 1)$	MS_E
Total	$\displaystyle\sum_i \sum_j \sum_k \sum_l y_{ijkl}^2 - \frac{y_{....}^2}{abcn}$	$abcn - 1$	

13-4 DESIGNS WITH NESTED AND CROSSED FACTORS

Occasionally in a multifactor experiment, some factors will be crossed and other factors will be nested. We sometimes call these designs *nested-factorial designs*. The statistical analysis of one such design with three factors is illustrated in the following example.

Example 13-2

An industrial engineer is studying the hand insertion of electronic components on printed circuit boards to improve the speed of the assembly operation. He has de-

Table 13-7 Expected Mean Square Derivation for a Three-Stage Nested Design with A and B Fixed and C Random

Factor	F a i	F b j	R c k	R n l	Expected Mean Square
τ_i	0	b	c	n	$\sigma^2 + n\sigma_\gamma^2 + \dfrac{bcn \sum \tau_i^2}{a - 1}$
$\beta_{j(i)}$	1	0	c	n	$\sigma^2 + n\sigma_\gamma^2 + \dfrac{cn \sum \sum \beta_{j(i)}^2}{a(b - 1)}$
$\gamma_{k(ij)}$	1	1	1	n	$\sigma^2 + n\sigma_\gamma^2$
$\epsilon_{l(ijk)}$	1	1	1	1	σ^2

signed three assembly fixtures and two workplace layouts that seem promising. Operators are required to perform the assembly, and it is decided to randomly select four operators for each fixture–layout combination. However, because the workplaces are in different locations within the plant, it is difficult to use the *same* four operators for each layout. Therefore, the four operators chosen for layout 1 are different individuals from the four operators chosen for layout 2. The treatment combinations in this design are run in random order and two replicates are obtained. The assembly times are measured in seconds and are shown in Table 13-8.

In this experiment, operators are nested within the levels of layouts, whereas fixtures and layouts are crossed. Thus, this design has both nested and crossed factors. The linear statistical model for this design is

$$y_{ijkl} = \mu + \tau_i + \beta_j + \gamma_{k(j)} + (\tau\beta)_{ij} + (\tau\gamma)_{ik(j)} + \epsilon_{(ijk)l} \begin{cases} i = 1,2,3 \\ j = 1,2 \\ k = 1,2,3,4 \\ l = 1,2 \end{cases}$$

where τ_i is the effect of the ith fixture, β_j is the effect of the jth layout, $\gamma_{k(j)}$ is the effect of the kth operator within the jth level of layout, $(\tau\beta)_{ij}$ is the fixture × layout interaction, $(\tau\gamma)_{ik(j)}$ is the fixture × operators within layouts interaction, and $\epsilon_{(ijk)l}$ is the usual error term. Notice that no layout × operator interaction can exist because all the operators do not use all the layouts. Similarly, there can be no three-way fixture × layout × operator interaction. Furthermore, fixtures and layouts are fixed factors, whereas the operator is a random factor. The expected mean squares are derived in Table 13-9. The proper test statistic for any effect or interaction can be found from inspection of this table.

The total sum of squares is

$$SS_T = \sum_{i=1}^{a}\sum_{j=1}^{b}\sum_{k=1}^{c}\sum_{l=1}^{n} y_{ijkl}^2 - \frac{y_{....}^2}{abcn}$$

$$= 32,956.00 - \frac{(1252)^2}{48} = 299.67$$

Table 13-8 Assembly Time Data For Example 13-2

		Layout 1				Layout 2				$y_{i...}$
Operator	1	2	3	4	1	2	3	4	
Fixture 1	22	23	28	25	26	27	28	24	404
	24	24	29	23	28	25	25	23	
Fixture 2	30	29	30	27	29	30	24	28	447
	27	28	32	25	28	27	23	30	
Fixture 3	25	24	27	26	27	26	24	28	401
	21	22	25	23	25	24	27	27	
Operator totals, $y_{.jk.}$	149	150	171	149	163	159	151	160	
Layout totals, $y_{.j..}$		619				633			$1252 = y_{....}$

Table 13-9   Expected Mean Square Derivation for Example 13-2

Factor	F 3 i	F 2 j	R 4 k	R 2 l	Expected Mean Square
τ_i	0	2	4	2	$\sigma^2 + 2\sigma_{\tau\gamma}^2 + 8 \sum \tau_i^2$
β_j	3	0	4	2	$\sigma^2 + 6\sigma_\gamma^2 + 24 \sum \beta_j^2$
$\gamma_{k(j)}$	3	1	1	2	$\sigma^2 + 6\sigma_\gamma^2$
$(\tau\beta)_{ij}$	0	0	4	2	$\sigma^2 + 2\sigma_{\tau\gamma}^2 + 4 \sum \sum (\tau\beta)_{ij}^2$
$(\tau\gamma)_{ik(j)}$	0	1	1	2	$\sigma^2 + 2\sigma_{\tau\gamma}^2$
$\epsilon_{(ijk)l}$	1	1	1	1	σ^2

The sums of squares for fixtures and layouts are computed in the usual manner; that is,

$$SS_{\text{Fixtures}} = \sum_{i=1}^{a} \frac{y_{i...}^2}{bcn} - \frac{y_{....}^2}{abcn}$$

$$= 32,739.13 - \frac{(1252)^2}{48} = 82.80$$

and

$$SS_{\text{Layouts}} = \sum_{j=1}^{b} \frac{y_{.j..}^2}{acn} - \frac{y_{....}^2}{abcn}$$

$$= 32,660.42 - \frac{(1252)^2}{48} = 4.08$$

The fixtures × layouts or *FL* interaction is computed from the fixtures × layouts cell totals (which are 198, 206, 228, 219, 193, and 208) as follows.

$$SS_{FL} = \sum_{i=1}^{a} \sum_{j=1}^{b} \frac{y_{ij..}^2}{cn} - \frac{y_{....}^2}{abcn} - SS_{\text{Fixtures}} - SS_{\text{Layouts}}$$

$$= 32,762.25 - \frac{(1252)^2}{48} - 82.80 - 4.08 = 19.04$$

To find the sum of squares for operators within layouts, we use the operator totals. The sum of squares is

$$SS_{O(L)} = \sum_{j=1}^{a} \sum_{k=1}^{b} \frac{y_{.jk.}^2}{an} - \sum_{j=1}^{b} \frac{y_{.j..}^2}{acn}$$

$$= 32,732.33 - 32,660.42 = 71.91$$

Notice that apart from the additional factor fixtures this is the same computational formula used to obtain the sum of squares for any nested factor. $SS_{O(L)}$ has $b(c - 1) = 2(4 - 1) = 6$ degrees of freedom.

We do not have available the computational formula or the number of degrees of freedom for the fixtures × operators within layouts interaction. To determine the number of degrees of freedom and a computational formula for the sum of squares due to this effect, we use the rules of Chapter 8. The term in the model for this effect is $(\tau\gamma)_{ik(j)}$, and using Rule 4 from Chapter 8, we obtain $b(a - 1)(c - 1) = 2(3 - 1)(4 - 1) = 12$ degrees of freedom. The sum of squares is obtained by using Rule 5 in Chapter 8, starting with the symbolic form $b(a - 1)(c - 1) = abc - bc - ab + b$. Then

$$
SS_{FO(L)} = \overset{abc}{\sum_{i=1}^{a}\sum_{j=1}^{b}\sum_{k=1}^{c}\frac{y_{ijk.}^{2}}{n}} - \overset{-bc}{\sum_{j=1}^{b}\sum_{k=1}^{c}\frac{y_{.jk.}^{2}}{an}} - \overset{-ab}{\sum_{i=1}^{a}\sum_{j=1}^{b}\frac{y_{ij..}^{2}}{cn}} + \overset{+b}{\sum_{j=1}^{b}\frac{y_{.j..}^{2}}{acn}}
$$

$$
= 32{,}900.00 - 32{,}732.33 - 32{,}762.25 + 32{,}660.42
$$

$$
= 65.84
$$

Finally, the error sum of sqaures is obtained by subtraction as

$$
SS_{E} = SS_{T} - SS_{\text{Fixtures}} - SS_{\text{Layouts}} - SS_{FL} - SS_{O(L)} - SS_{FO(L)}
$$

$$
= 299.67 - 82.80 - 4.08 - 19.04 - 71.91 - 65.84
$$

$$
= 56.00
$$

The complete analysis of variance is shown in Table 13-10. We see that assembly fixtures are significant and that operators within layouts also differ significantly. There is also a significant interaction between fixtures and operators within layouts, indicating that the effects of the different fixtures are not the same for all operators. The workplace layouts seem to have little effect on the assembly time. Therefore, to minimize assembly time, we should concentrate on fixture types 1 and 3. (Note that the fixture totals in Table 13-8 are smaller for fixture types 1 and 3 than for type 2. This difference in fixture type means could be formally tested using multiple comparisons.) Furthermore, the interaction between operators and fixtures implies that some operators are more effective than others using the same fixtures. Perhaps these operator–fixture effects could be isolated and the operators performance improved by retraining them.

Table 13-10 Analysis of Variance for Example 13-2

Source of Variation	Sum of Squares	Degrees of Freedom	Mean Square	F_0
Fixtures (F)	82.80	2	41.40	7.54[b]
Layouts (L)	4.08	1	4.09	0.34
Operators (within layouts), $O(L)$	71.91	6	11.99	5.15[b]
FL	19.04	2	9.52	1.73
$FO(L)$	65.84	12	5.49	2.36[a]
Error	56.00	24	2.33	
Total	299.67	47		

[a] Significant at 5 percent.
[b] Significant at 1 percent.

Computer programs for the anlaysis of factorial experiments may also be used to analyze designs with nested and crossed factors. For instance, the experiment in Example 13-2 could be considered as a three-factor factorial, with fixtures (F), operators (O), and layouts (L) as the factors. Then certain sums of squares and degrees of freedom from the factorial analysis would be pooled to form the appropriate quantities required for the design with nested and crossed factors as follows:

Factorial Analysis		Nested-Factorial Analysis	
Sum of Squares	Degrees of Freedom	Sum of Squares	Degrees of Freedom
SS_F	2	SS_F	2
SS_L	1	SS_L	1
SS_{FL}	2	SS_{FL}	2
SS_O	3	$SS_{O(L)} = SS_O + SS_{LO}$	6
SS_{LO}	3		
SS_{FO}	6	$SS_{FO(L)} = SS_{FO} + SS_{FOL}$	12
SS_{FOL}	6		
SS_E	24	SS_E	24
SS_T	47	SS_T	47

13-5   PROBLEMS

13-1   An ammunition manufacturer is studying the burning rate of powder from three production processes. Four batches of powder are randomly selected from the output of each process and three determinations of burning rate are made on each batch. The results follow. Analyze the data and draw conclusions.

	Process 1				Process 2				Process 3			
Batch	1	2	3	4	1	2	3	4	1	2	3	4
	25	19	15	15	19	23	18	35	14	35	38	25
	30	28	17	16	17	24	21	27	15	21	54	29
	26	20	14	13	14	21	17	25	20	24	50	33

13-2   The surface finish of metal parts made on four machines is being studied. An experiment is conducted in which each machine is run by three different opertors and two specimens from each operator are collected and tested. Because of the location of the machines, different operators are used on each machine, and the operators are chosen at random. The data are shown in the following table. Analyze the data and draw conclusions.

Operator	Machine 1			Machine 2			Machine 3			Machine 4		
	1	2	3	1	2	3	1	2	3	1	2	3
	79	94	46	92	85	76	88	53	46	36	40	62
	62	74	57	99	79	68	75	56	57	53	56	47

13-3 A manufacturing engineer is studying the dimensional variability of a particular component that is produced on three machines. Each machine has two spindles, and four components are randomly selected from each spindle. The results follow. Analyze these data, assuming that machines and spindles are fixed factors.

Spindle	Machine 1		Machine 2		Machine 3	
	1	2	1	2	1	2
	12	8	14	12	14	16
	9	9	15	10	10	15
	11	10	13	11	12	15
	12	8	14	13	11	14

13-4 In order to simplify production scheduling, an industrial engineer is studying the possibility of assigning one time standard to a particular class of jobs, believing that differences between jobs are negligible. To see if this simplification is possible, six jobs are randomly selected. Each job is given to a different group of three operators. Each operator completes the job twice at different times during the week, and the following results were obtained. What are your conclusions about the use of a common time standard for all jobs in this class? What value would you use for the standard?

Job	Operator 1		Operator 2		Operator 3	
1	158.3	159.4	159.2	159.6	158.9	157.8
2	154.6	154.9	157.7	156.8	154.8	156.3
3	162.5	162.6	161.0	158.9	160.5	159.5
4	160.0	158.7	157.5	158.9	161.1	158.5
5	156.3	158.1	158.3	156.9	157.7	156.9
6	163.7	161.0	162.3	160.3	162.6	161.8

13-5   Use the general regression significance test to develop the analysis of variance for the balanced two-stage nested design.

13-6   Consider the three-stage nested design shown in Figure 13-4 to investigate alloy hardness. Using the data that follow, analyze the design, assuming that alloy chemistry and heats are fixed whereas ingots are random.

	Alloy Chemistry											
	1						2					
Heats	1		2		3		1		2		3	
Ingots	1	2	1	2	1	2	1	2	1	2	1	2
	40	27	95	69	65	78	22	23	83	75	61	35
	63	30	67	47	54	45	10	39	62	64	77	42

13-7   Derive the expected mean squares for a balanced three-stage nested design, assuming that A is fixed and that B and C are random. Obtain formulas for estimating the variance components.

13-8   Derive the expected mean squares for a balanced three-stage nested design if all three factors are random. Obtain formulas for estimating the variance components.

13-9   Verify the expected mean squares given in Table 13-1.

13-10   *Unbalanced Nested Designs.* Consider an unbalanced two-stage nested design with b_i levels of B under the ith level of A and n_{ij} replicates in the (ij)th cell.

(a) Write down the least squares normal equations for this situation. Solve the normal equations.

(b) Construct the analysis of variance table for the unbalanced two-stage nested design.

(c) Analyze the following data, using the results in part (b).

Factor A	1		2		
Factor B	1	2	1	2	3
	6	-3	5	2	1
	4	1	7	4	0
	8		9	3	-3
			6		

13-11 *Variance Components in the Unbalanced Two-Stage Nested Design.* Consider the model

$$Y_{ijk} = \mu + \tau_i + \beta_{j(i)} + \epsilon_{k(ij)} \begin{cases} i = 1,2, \ldots, a \\ j = 1,2, \ldots, b_i \\ k = 1,2, \ldots, n_{ij} \end{cases}$$

where A and B are random factors. Show that

$$E(MS_A) = \sigma^2 + c_1 \sigma_\beta^2 + c_2 \sigma_\tau^2$$
$$E(MS_{B(A)}) = \sigma^2 + c_0 \sigma_\beta^2$$
$$E(MS_E) = \sigma^2$$

where

$$c_0 = \frac{N - \sum\limits_{i=1}^{a} \left(\sum\limits_{j=1}^{b_i} n_{ij}^2 / n_{i.} \right)}{b - a}$$

$$c_1 = \frac{\sum\limits_{i=1}^{a} \left(\sum\limits_{j=1}^{b_i} n_{ij}^2 / n_{i.} \right) - \sum\limits_{i=1}^{a} \sum\limits_{j=1}^{b_i} n_{ij}^2 / N}{a - 1}$$

$$c_2 = \frac{N - \dfrac{\sum\limits_{i=1}^{a} n_{i.}^2}{N}}{a - 1}$$

13-12 A process engineer is testing the yield of a product manufactured on three machines. Each machine can be operated at two power settings. Furthermore, a machine has three stations on which the product is formed. An experiment is conducted in which each machine is tested at both power settings, and three observations on yield are taken from each station. The runs are made in random order, and the results follow. Analyze this experiment, assuming that all three factors are fixed.

	Machine 1			Machine 2			Machine 3		
Station	1	2	3	1	2	3	1	2	3
	34.1	33.7	36.2	32.1	33.1	32.8	32.9	33.8	33.6
Power setting 1	30.3	34.9	36.8	33.5	34.7	35.1	33.0	33.4	32.8
	31.6	35.0	37.1	34.0	33.9	34.3	33.1	32.8	31.7
	24.3	28.1	25.7	24.1	24.1	26.0	24.2	23.2	24.7
Power setting 2	26.3	29.3	26.1	25.0	25.1	27.1	26.1	27.4	22.0
	27.1	28.6	24.9	26.3	27.9	23.9	25.3	28.0	24.8

13-13 Suppose that in Problem 13-12 a large number of power settings could have been used, and that the two selected for the experiment were chosen randomly. Obtain the expected mean squares for this situation and modify the previous analysis appropriately.

13-14   A structural engineer is studying the strength of aluminum alloy purchased from three vendors. Each vendor submits the alloy in standard-sized bars of 1.0, 1.50, or 2.0 inches. The processing of different sizes of bar stock from a common ingot involves different forging techniques, and so this factor may be important. Furthermore, the bar stock is forged from ingots made in different heats. Each vendor submits two test specimens of each size bar stock from three heats. The resulting strength data follow. Analyze the data assuming that vendors and bar size are fixed and heats are random.

Heat	Vendor 1			Vendor 2			Vendor 3		
	1	2	3	1	2	3	1	2	3
Bar size: 1 inch	1.230	1.346	1.235	1.301	1.346	1.315	1.247	1.275	1.324
	1.259	1.400	1.206	1.263	1.392	1.320	1.296	1.268	1.315
$1\frac{1}{2}$ inch	1.316	1.329	1.250	1.274	1.384	1.346	1.273	1.260	1.392
	1.300	1.362	1.239	1.268	1.375	1.357	1.264	1.265	1.364
2 inch	1.287	1.346	1.273	1.247	1.362	1.336	1.301	1.280	1.319
	1.292	1.382	1.215	1.215	1.328	1.342	1.262	1.271	1.323

13-15   Suppose that in Problem 13-14 the bar stock may be purchased in many sizes and that the three sizes actually used in the experiment were selected randomly. Obtain the expected mean squares for this situation and modify the previous analysis appropriately.

Chapter 14
Multifactor Experiments with Randomization Restrictions

In Chapters 5 and 6 we presented designs for experiments with *randomization restrictions*—that is, experiments in which we cannot assign the treatment combinations to the experimental units in a completely random manner. The randomized block and Latin square designs are effective design techniques for dealing with one and two randomization restrictions, respectively. These designs were discussed in Chapters 5 and 6 relative to a single factor. However, there are many *multifactor* experimental situations in which it is necessary to restrict complete randomization. These design problems are the subject of this chapter. For further reading on these designs, see Ostle (1963), John (1971), Hicks (1973), and Anderson and McLean (1974).

14-1 RANDOMIZED BLOCKS AND LATIN SQUARES AS MULTIFACTOR DESIGNS

Consider a factorial experiment with two factors, say A and B. The linear statistical model for this design is

$$y_{ijk} = \mu + \tau_i + \beta_j + (\tau\beta)_{ij} + \epsilon_{ijk} \begin{cases} i = 1,2,\ldots,a \\ j = 1,2,\ldots,b \\ k = 1,2,\ldots,n \end{cases} \tag{14-1}$$

where τ_i, β_j, and $(\tau\beta)_{ij}$ represent the effects of factors A, B, and the AB interaction, respectively. The statistical analysis for this design is given in Chapter 7. Now suppose that to run this experiment a particular raw material is required. This raw material is available in batches that are not large enough to allow *all abn* treatment combinations to be run from the *same* batch. However, if a batch

461

contains enough material for ab observations, then an alternative design is to run each of the n replicates using a separate batch of raw material. Consequently, the batches of raw material represent a randomization restriction or a *block*, and a single replicate of a complete factorial experiment is run within each block. The model for this new design is

$$y_{ijk} = \mu + \tau_i + \beta_j + (\tau\beta)_{ij} + \delta_k + \epsilon_{ijk} \begin{cases} i = 1,2, \ldots , a \\ j = 1,2, \ldots , b \\ k = 1,2, \ldots , n \end{cases} \quad (14\text{-}2)$$

where δ_k is the effect of the kth block. Of course, within a block the order in which the treatment combinations are run is completely randomized.

The model (Equation 14-2) assumes that interaction between blocks and treatments is negligible. This was assumed previously in the analysis of randomized block designs. If these interactions do exist, they cannot be separated from the error component. In fact, the error term in this model really consists of the $(\tau\delta)_{ik}$, $(\beta\delta)_{jk}$, and $(\tau\beta\delta)_{ijk}$ interactions. The analysis of variance is outlined in Table 14-1. The layout closely resembles that of a factorial design, with the error sum of squares reduced by the sum of squares for blocks. Computationally, we find the sum of squares for blocks as the sum of squares between the n block totals $\{y_{..k}\}$.

The expected mean squares for this design are derived in Table 14-2, assuming that factor A is fixed and that factor B and the blocks are random. In the left-hand side of this table we have expanded the model to include the interactions of treatments with blocks so that one may more clearly see that experimental error is composed of these interactions. In using the expected mean square algorithm, we have denoted the observations by y_{ijkh}, where the subscript h represents replication. In this design we have only one replicate, so $h = 1$. Because

Table 14-1   Analysis of Variance for a Two-Factor Factorial in a Randomized Complete Block

Source of Variation	Sum of Squares	Degrees of Freedom
Blocks	$\sum_k \dfrac{y_{..k}^2}{ab} - \dfrac{y_{...}^2}{abn}$	$n - 1$
A	$\sum_i \dfrac{y_{i..}^2}{bn} - \dfrac{y_{...}^2}{abn}$	$a - 1$
B	$\sum_j \dfrac{y_{.j.}^2}{an} - \dfrac{y_{...}^2}{abn}$	$b - 1$
AB	$\sum_i \sum_j \dfrac{y_{ij.}^2}{n} - \dfrac{y_{...}^2}{abn} - SS_A - SS_B$	$(a - 1)(b - 1)$
Error	Subtraction	$(ab - 1)(n - 1)$
Total	$\sum_i \sum_j \sum_k y_{ijk}^2 - \dfrac{y_{...}^2}{abn}$	$abn - 1$

Table 14-2 Expected Mean Square Derivation for a Factorial Experiment in a Randomized Block

Expanded Model				Collapsed Model			
Factor	Degrees of Freedom	a b n 1 F R R R i j k h	Expected Mean Square	Factor	Degrees of Freedom	a b n 1 F R R R i j k h	Expected Mean Square
τ_i	$a - 1$	0 b n 1	$\sigma^2 + \sigma^2_{\tau\beta\delta} + b\sigma^2_{\tau\delta} + n\sigma^2_{\tau\beta} + \dfrac{bn\sum \tau_i^2}{a-1}$	τ_i	$a - 1$	0 b n 1	$\sigma^2 + n\sigma^2_{\tau\beta} + \dfrac{bn\sum \tau_i^2}{a-1}$
β_j	$b - 1$	a 1 n 1	$\sigma^2 + a\sigma^2_{\beta\delta} + an\sigma^2_{\beta}$	β_j	$b - 1$	a 1 n 1	$\sigma^2 + an\sigma^2_{\beta}$
$(\tau\beta)_{ij}$	$(a-1)(b-1)$	0 1 n 1	$\sigma^2 + \sigma^2_{\tau\beta\delta} + n\sigma^2_{\tau\beta}$	$(\tau\beta)_{ij}$	$(a-1)(b-1)$	0 1 n 1	$\sigma^2 + n\sigma^2_{\tau\beta}$
δ_k	$(n-1)$	a b 1 1	$\sigma^2 + a\sigma^2_{\tau\delta} + ab\sigma^2_{\delta}$	δ_k	$n - 1$	a b 1 1	$\sigma^2 + ab\sigma^2_{\delta}$
$(\tau\delta)_{ik}$	$(a-1)(n-1)$	0 b 1 1	$\sigma^2 + \sigma^2_{\tau\beta\delta} + b\sigma^2_{\tau\delta}$				
$(\beta\delta)_{jk}$	$(b-1)(n-1)$	a 1 1 1	$\sigma^2 + a\sigma^2_{\beta\delta}$				
$(\tau\beta\delta)_{ijk}$	$(a-1)(b-1)(n-1)$	0 1 1 1	$\sigma^2 + \sigma^2_{\tau\beta\delta}$	$\epsilon_{(ijk)h}$	$(ab-1)(n-1)$	1 1 1 1	σ^2
$\epsilon_{(ijk)h}$	0	1 1 1 1	σ^2 (not estimable)				

each block contains only one replicate, the error variance σ^2 is not estimable. If we could run r replicates in each block, an estimate of σ^2 could be computed.

As discussed above, it is relatively standard to assume that the interactions of blocks and treatments are negligible and to pool their mean squares to obtain an estimate of error. The right-hand side of Table 14-2 shows how the expanded model is collapsed to produce the expected mean squares.

In the previous example, the randomization was restricted to within a batch of raw material. In practice, a variety of phenomena may cause randomization restrictions, such as time, operators, and so on. For example, if we could not run the entire factorial experiment on one day, trhen the experimenter could run a complete replicate on day 1, a second replicate on day 2, and so on. Consequently, each day would be a block.

Example 14-1

An engineer is studying methods for improving the ability to detect targets on a radar scope. Two factors she considers to be important are the amount of background noise or "ground clutter," on the scope and the type of filter placed over the screen. An experiment is designed using three levels of ground clutter and two filter types. The experiment is performed by randomly selecting a treatment combination (ground clutter level and filter type) and then introducing a signal representing the target into the scope. The intensity of this target is increased until the operator observes it. The intensity level at detection is then measured as the response variable. Because of operator availability, it is convenient to select an operator and keep him or her at the scope until all the necessary runs have been made. Furthermore, operators differ in their skill and ability to use the scope. Consequently, it seems logical to use the operators as blocks. Four operators are randomly selected. Once an operator is chosen, the order in which the six treatment combinations are run is randomly determined. Thus, we have a 3 × 2 factorial experiment run in a randomized complete block. The data are shown in Table 14-3.

The linear model for this experiment is

$$y_{ijk} = \mu + \tau_i + \beta_j + (\tau\beta)_{ij} + \delta_k + \epsilon_{ijk} \begin{cases} i = 1,2,3 \\ j = 1,2 \\ k = 1,2,3,4 \end{cases}$$

where τ_i represents the ground clutter effect (fixed), β_j represents the filter type effect (fixed), $(\tau\beta)_{ij}$ is the interaction, δ_k is the block effect (random), and ϵ_{ijk} is the NID(0,

Table 14-3   Intensity Level at Target Detection

Operators (blocks)	1		2		3		4	
Filter Type	1	2	1	2	1	2	1	2
Ground clutter								
Low	90	86	96	84	100	92	92	81
Medium	102	87	106	90	105	97	96	80
High	114	93	112	91	108	95	98	83

σ^2) error component. The sums of squares for ground clutter, filter type, and their interaction are computed in the usual manner. The sum of squares due to blocks is found from the operator totals $\{y_{..k}\}$ as follows:

$$SS_{Blocks} = \sum_{k=1}^{n} \frac{y_{..k}^2}{ab} - \frac{y_{...}^2}{abn}$$

$$= \frac{(572)^2 + (579)^2 + (597)^2 + (530)^2}{(3)(2)} - \frac{(2278)^2}{(3)(2)(4)}$$

$$= 402.17$$

The complete analysis of variance for this experiment is summarized in Table 14-4. A column has been added to this table listing the expected mean squares. These expected mean squares may be derived by the algorithm in Chapter 8. Since both factors are fixed, all effects are tested by dividing their mean squares by the mean square error. Both ground clutter level and filter type are significant at the 1 percent level, whereas their interaction is significant only at the 10 percent level. Thus, we conclude that both ground clutter level and the type of scope filter used affect the operator's ability to detect the target, and there is some evidence of mild interaction between these factors.

∎

In the case of two randomization restrictions, each with p levels, if the number of treatment combinations in a k-factor factorial design exactly equals the number of restriction levels, that is, if $p = ab \ldots m$, then the factorial design may be run in a $p \times p$ Latin square. For example, consider a modification of the radar target detection experiment of Example 14-1. The factors in this experiment are filter type (two levels) and ground clutter (three levels), and operators are considered as blocks. Suppose now that because of the setup time required, only six runs can be made per day. Thus, days become a second randomization restriction, resulting in the 6×6 Latin square design, as shown in Table 14-5. In this table we have used the lowercase letters f_i and g_j to represent the ith and jth levels of filter type and ground clutter, respectively. That is, f_1g_2 represents filter type 1 and medium ground clutter. Note that now six operators are required,

Table 14-4   Analysis of Variance for Example 14-1

Source of Variation	Sum of Square	Degrees of Freedom	Mean Square	Expected Mean Square	F_0
Ground Clutter (G)	335.58	2	167.79	$\sigma^2 + 8 \sum \tau_i^2/2$	15.13[a]
Filter Type (F)	1066.67	1	1066.67	$\sigma^2 + 12 \sum \beta_j^2$	96.19[a]
GF	77.08	2	38.54	$\sigma^2 + 4 \sum\sum (\tau\beta)_{ij}^2/6$	3.48
Blocks	402.17	3	134.06	$\sigma^2 + 6\sigma_\delta^2$	
Error	166.33	15	11.09	σ^2	
Total	2047.83	23			

[a] Significant at 1 percent.

Table 14-5 Radar Detection Experiment Run in a 6 × 6 Latin Square

Day	Operator					
	1	2	3	4	5	6
1	$A(f_1 g_1 = 90)$	$B(f_1 g_2 = 106)$	$C(f_1 g_3 = 108)$	$D(f_2 g_1 = 81)$	$F(f_2 g_3 = 90)$	$E(f_2 g_2 = 88)$
2	$C(f_1 g_3 = 114)$	$A(f_1 g_1 = 96)$	$B(f_1 g_2 = 105)$	$F(f_2 g_3 = 83)$	$E(f_2 g_2 = 86)$	$D(f_2 g_1 = 84)$
3	$B(f_1 g_2 = 102)$	$E(f_2 g_2 = 90)$	$F(f_2 g_3 = 95)$	$A(f_1 g_1 = 92)$	$D(f_2 g_1 = 85)$	$C(f_1 g_3 = 104)$
4	$E(f_2 g_2 = 87)$	$D(f_2 g_1 = 84)$	$A(f_1 g_1 = 100)$	$B(f_1 g_2 = 96)$	$C(f_1 g_3 = 110)$	$F(f_2 g_3 = 91)$
5	$F(f_2 g_3 = 93)$	$C(f_1 g_3 = 112)$	$D(f_2 g_1 = 92)$	$E(f_2 g_2 = 80)$	$A(f_1 g_1 = 90)$	$B(f_1 g_2 = 98)$
6	$D(f_2 g_1 = 86)$	$F(f_2 g_3 = 91)$	$E(f_2 g_2 = 97)$	$C(f_1 g_3 = 98)$	$B(f_1 g_2 = 100)$	$A(f_1 g_1 = 92)$

rather than four as in the original experiment, so the number of treatment combinations in the 3×2 factorial design exactly equals the number of restriction levels. Furthermore, in this design, each operator would be used only once on each day. The Latin letters A, B, C, D, E, and F represent the $3 \times 2 = 6$ factorial treatment combinations as follows: $A = f_1 g_1$, $B = f_1 g_2$, $C = f_1 g_3$, $D = f_2 g_1$, $E = f_2 g_2$, and $f = f_2 g_3$.

The five degrees of freedom between the six Latin letters corespond to the main effects of filter type (one degree of freedom), ground clutter (two degrees of freedom), and their interaction (two degrees of freedom). The linear statistical model for this design is

$$y_{ijkl} = \mu + \alpha_i + \tau_j + \beta_k + (\tau\beta)_{jk} + \theta_l + \epsilon_{ijkl} \begin{cases} i = 1,2, \ldots, 6 \\ j = 1,2,3 \\ k = 1,2, \\ l = 1,2, \ldots, 6 \end{cases} \tag{14-3}$$

where τ_j and β_k are effects of ground clutter and filter type, respectively, and α_i and θ_l represent the randomization restrictions of days and operators, respectively. To compute the sums of squares, the following two-way table of treatment totals is helpful:

Ground Clutter	Filter Type 1	Filter Type 2	$y_{.j..}$
Low	560	512	1072
Medium	607	528	1135
High	646	543	1189
$y_{..k.}$	1813	1583	$3396 = y_{....}$

Furthermore, the row and column totals are

Rows ($y_{.jkl}$): 563   568   568   568   565   564
Columns ($y_{ijk.}$): 572   579   597   530   561   557

The analysis of variance is summarized in Table 14-6. We have added a column to this table indicating how the number of degrees of freedom for each sum of squares is determined.

We have illustrated running factorial experiments in randomized blocks or Latin squares. However, other multifactor designs could be treated similarly. The statistical analysis of such designs is straightforward and follows the general methods outlined above. One may also use incomplete block designs as the basis of multifactor experiments. An interesting example of this is discussed by John (1971, p. 240).

Table 14-6 Analysis of Variance for the Radar Detection Experiment Run as a 3 × 2 Factorial in a Latin Square

Source of Variation	Sum of Squares	Degrees of Freedom	General Formula for Degrees of Freedom	Mean Square	F_0
Ground Clutter, G	571.50	2	$a - 1$	285.75	28.86[a]
Filter Type, F	1469.44	1	$b - 1$	1469.44	148.43[a]
GF	126.73	2	$(a - 1)(b - 1)$	63.37	6.40[b]
Days (rows)	4.33	5	$ab - 1$	0.87	
Operators (columns)	428.00	5	$ab - 1$	85.60	
Error	198.00	20	$(ab - 1)(ab - 2)$	9.90	
Total	2798.00	35	$(ab)^2 - 1$		

[a] Significant at 1 percent.
[b] Significant at 5 percent.

14-2 THE SPLIT-PLOT DESIGN

In some multifactor designs involving randomized blocks, we may be unable to randomize completely the order of the runs within the block. This often results in a generalization of the randomized block design called a *split-plot design*. As an example, consider a paper manufacturer who is interested in three different pulp preparation methods and four different cooking temperatures for the pulp and who wishes to study the effect of these two factors on the tensile strength of the paper. Each replicate of a factorial experiment requires 12 observations, and the experimeter has decided to run three replicates. However, the pilot plant is only capable of making 12 runs per day, so the experimenter decides to run one replicate on each of the three days and to consider the days or replicates as blocks. On any day, he conducts the experiment as follows. A batch of pulp is produced by one of the three methods under study. Then this batch is divided into four samples, and each sample is cooked at one of the four temperatures. Then a second batch of pulp is made up using another of the three methods. This second batch is also divided into four samples that are tested at the four temperatures. The process is then repeated, using a batch of pulp produced by the third method. The data are shown in Table 14-7.

Table 14-7 The Experiment on the Tensile Strength of Paper

Pulp Preparation Method	Block 1			Block 2			Block 3		
	1	2	3	1	2	3	1	2	3
Temperature:									
200	30	34	29	28	31	31	31	35	32
225	35	41	26	32	36	30	37	40	34
250	37	38	33	40	42	32	41	39	39
275	36	42	36	41	40	40	40	44	45

Initially, we might consider this to be a factorial experiment with three levels of preparation method (factor A) and four levels of temperature (factor B) in a randomized block. If this is the case, then the order of experimentation within the block should be completely randomized. That is, within a block, we should randomly select a treatment combination (a preparation method and a temperature) and obtain an observation, then we should randomly select another treatment combination and obtain a second observation, and so on, until the 12 observations in the block have been taken. However, the experimenter did not collect the data this way. He made up a batch of pulp and obtained observations for all four temperatures from that batch. Because of the economics of preparing the batches and the size of the batches, this is the only feasible way to collect the data.

Each block in the design is divided into three parts called *whole plots,* and the preparation methods are called the whole plot or *main* treatments. Each whole plot is divided into four parts called subplots (or split-plots), and one temperature is assigned to each. Temperature is called the *subplot* treatment. Note that if there are other uncontrolled or undesigned factors present, and if these uncontrolled factors vary as the pulp preparation methods are changed, then any effect of the undersigned factors on the response will be completely confounded with the effect of the pulp preparation methods. Since the whole plot treatments in a split-plot design are confounded with the whole plots and the subplot treatments are not confounded, it is best to assign the factor we are most interested in to the subplots, if possible.

The linear model for the split-plot design is

$$y_{ijk} = \mu + \tau_i + \beta_j + (\tau\beta)_{ij} + \gamma_k + (\tau\gamma)_{ik}$$

$$+ (\tau\beta\gamma)_{ijk} \begin{cases} i = 1, 2, \ldots, a \\ j = 1, 2, \ldots, b \\ k = 1, 2, \ldots, c \end{cases} \quad (14\text{-}4)$$

where τ_i, β_j, and $(\tau\beta)_{ij}$ represent the whole plot and correspond respectively to blocks (factor A), main treatments (factor B), and whole plot error (AB); and γ_k, $(\tau\gamma)_{ik}$, $(\beta\gamma)_{jk}$, and $(\tau\beta\gamma)_{ijk}$ represent the subplot and correspond respectively to the subplot treatment (factor C), the AC and BC interactions, and the subplot error. Note that the whole plot error is the AB interaction and the subplot error is the three-factor interaction ABC. The sums of squares for these factors are computed as in the three-way analysis of variance without replication.

The expected mean squares for the split-plot design, with blocks random and main treatments and subplot treatments fixed, are derived in Table 14-8. Note that the main factor (B) in the whole plot is tested against the whole plot error, whereas the subtreatment (C) is tested against the block × subtreatment (AC) interaction. The BC interaction is tested against the subplot error. Notice that there are no tests for the block effect (A) or the block × subtreatment (AC) interaction. Some authors believe that the subplot error should be found by pooling the ABC and AC interactions and then testing both the subtreatment and the BC interaction by this pooled estimate. If the experimenter has previous experience in the area and is reasonably sure that blocks do not interact with the various

Table 14-8   Expected Mean Square Derivation for Split-Plot Design

	Factor	$\begin{array}{c}a\\R\\i\end{array}$	$\begin{array}{c}b\\F\\j\end{array}$	$\begin{array}{c}c\\F\\k\end{array}$	$\begin{array}{c}1\\R\\h\end{array}$	Expected Mean Square
Whole plot	τ_i	1	b	c	1	$\sigma^2 + bc\sigma_\tau^2$
	β_j	a	0	c	1	$\sigma^2 + c\sigma_{\tau\beta}^2 + \dfrac{ac\sum \beta_j^2}{b-1}$
	$(\tau\beta)_{ij}$	1	0	c	1	$\sigma^2 + c\sigma_{\tau\beta}^2$
Subplot	γ_k	a	b	0	1	$\sigma^2 + b\sigma_{\tau\gamma}^2 + \dfrac{ab\sum \gamma_k^2}{(c-1)}$
	$(\tau\gamma)_{ik}$	1	b	0	1	$\sigma^2 + b\sigma_{\tau\gamma}^2$
	$(\beta\gamma)_{jk}$	a	0	0	1	$\sigma^2 + \sigma_{\tau\beta\gamma}^2 + \dfrac{a\sum\sum (\beta\gamma)_{jk}^2}{(b-1)(c-1)}$
	$(\tau\beta\gamma)_{ijk}$	1	0	0	1	$\sigma^2 + \sigma_{\tau\beta\gamma}^2$
	$\epsilon_{(ijk)h}$	1	1	1	1	σ^2 (not estimable)

treatments, this may be appropriate. On the other hand, if one cannot be reasonably sure that the interactions are negligible, then pooling is inadvisable.

The analysis of variance for the tensile strength data in Table 14-7 is summarized in Table 14-9. Since both preparation methods and temperatures are fixed and blocks are random, the expected mean squares in Table 14-8 apply. The mean square for preparation methods is compared to the whole plot error mean square, and the mean square for temperatures is compared to the block × temperature AC mean square. Finally, the preparation method × temperature mean square is tested against the subplot error. Both preparation methods and temperature have a significant effect on strength.

Table 14-9   Analysis of Variance for the Split-Plot Design Using Tensile Strength Data from Table 14-7

Source of Variation	Sum of Squares	Degrees of Freedom	Mean Square	F_0
Blocks (A)	77.55	2	38.78	
Preparation method (B)	128.39	2	64.20	7.08[a]
AB (whole plot error)	36.28	4	9.07	
Temperature (C)	434.08	3	144.69	41.94[b]
AC	20.67	6	3.45	
BC	75.17	6	12.53	2.96
ABC (subplot error)	50.83	12	4.24	
Total	822.97	35		

[a] Significant at 5 percent.
[b] Significant at 1 percent.

Note from Table 14-9 that the subplot error (4.24) is less than the whole plot error (9.07). This is the usual case in split-plot designs since the subplots are genrally more homogeneous than the whole plots. Because the subplot treatments are compared with greater precision, it is preferable to assign the treatment we are most interested in to the subplots, if possible.

The split-plot design has an agricultural heritage, with the whole plots usually being large areas of land and the subplots being small areas of land. For example, several varieties of a crop could be planted in different fields (whole plots), one variety to a field. Then each field could be divided into, say, four subplots, and each subplot could be treated with a different type of fertilizer. Here the crop varieties are the main treatments and the different fertilizers are the subtreatments.

Despite its agricultural basis, the split-plot design is useful in many scientific and industrial experiments. In these experimental settings, it is not unusual to find that some factors require large experimental units whereas other factors require small experimental units, such as in the tensile strength problem described above. Alternatively, we sometimes find that complete randomization is not feasible because it is more difficult to change the levels of some factors than others. The hard-to-vary factors form the whole plots whereas the easy-to-vary factors are run in the subplots.

In principle, we must carefully consider how the data will be collected and incorporate all restrictions on randomization into the analysis. We illustrate this point using a modification of the eye focus time experiment in Chapter 11. Suppose there are only two factors, visual acuity (A) and illumination level (B). A factorial experiment with a levels of acuity, b levels of illumination, and n replicates would require that all abn observations be taken in random order. However, in the test apparatus, it is fairly difficult to adjust these two factors to different levels, so the experimenter decides to obtain the n replicates by adjusting the device to one of the a acuities and one of the b illumination levels and running all n observations at once. In the factorial design, the error actually represents the scatter or noise in the system plus the ability of the subject to reproduce the same focus time. The model for the factorial design could be written as

$$y_{ijk} = \mu + \tau_i + \beta_j + (\tau\beta)_{ij} + \phi_{ijk} + \theta_{ijk} \begin{cases} i = 1,2, \ldots, a \\ j = 1,2, \ldots, b \\ k = 1,2, \ldots, n \end{cases} \quad (14\text{-}5)$$

where ϕ_{ijk} represents the scatter or noise in the sytem that results from "experimental error" (that is our failure to duplicate exactly the same levels of acuity and illumination on different runs, variability in environmental conditions, and the like), and θ_{ijk} represents the "reproducibility error" of the subject. Usually, we combine these components into one overall error term, say $\epsilon_{ijk} = \phi_{ijk} + \theta_{ijk}$. Assume that $V(\epsilon_{ijk}) = \sigma^2 = \sigma_\phi^2 + \sigma_\theta^2$. Now, in the factorial design, the error mean square has expectation $\sigma^2 = \sigma_\phi^2 + \sigma_\theta^2$, with $ab(n-1)$ degrees of freedom.

If we restrict the randomization as in the second design above, then the "error" mean square in the analysis of variance provides an estimate of the "reproducibility error" σ_θ^2 with $ab(n-1)$ degrees of freedom, but it yields no information on the "experimental error" σ_θ^2. Thus, the mean square for error in this second design is too small; consequently, we will wrongly reject the null hypothesis very frequently. As pointed out by John (1971), this design is similar to a split-plot design with ab whole plots, each divided into n subplots, and no subtreatment. The situation is also similar to *subsampling*, as described by Ostle (1963). Assuming that A and B are fixed, the expected mean squares in this case are

$$E(MS_A) = \sigma_\theta^2 + n\sigma_\phi^2 + \frac{bn \sum \tau_i^2}{a - 1}$$

$$E(MS_B) = \sigma_\theta^2 + n\sigma_\phi^2 + \frac{an \sum \beta_j^2}{b - 1}$$

$$E(MS_{AB}) = \sigma_\theta^2 + n\sigma_\phi^2 + \frac{n \sum\sum (\tau\beta)_{ij}^2}{(a - 1)(b - 1)}$$ (14-6)

$$E(MS_E) = \sigma_\theta^2$$

Thus, there are no tests on the main effects unless interaction is negligible. The situation is exactly that of a two-way analysis of variance with one observation per cell. If both factors are random, then the main effects may be tested against the AB interaction. If only one factor is random, then the fixed factor can be tested against the AB interaction.

John (1971) also states that, in general, if one analyzes a factorial design and all the main effects and interactions are significant, then one should examine carefully *how* the data were collected. There may be randomization restrictions in the model not accounted for in the analysis, and consequently, the data should not be analyzed as a factorial. Anderson and McLean (1974, p. 191) present an example where the experimenter has confused the factorial and split-plot layouts.

14-3 THE SPLIT–SPLIT-PLOT DESIGN

The concept of split-plot designs can be extended to situations in which randomization restrictions may occur at any number of levels within a block. If there are two levels of randomization restrictions within a block, then the layout is called a split–split plot-design. The following example illustrates such a design.

Example 14-2

A medical researcher is studying the absorption times of a particular type of antibiotic capsule. There are three technicians, three dosage strengths, and four capsule wall thicknesses of interest to the researcher. Each replicate of a factorial experiment would require 36 observations. The experimenter has decided on four replicates, and it is necessary to run each replicate on a different day. Thus, the days are *blocks*. Within a block (day), the experiment is performed by assigning a unit of antibiotic to a technician who conducts the experiment on the three dosage strengths and the four wall thicknesses. Once a particular dosage strength is formulated, all four wall thicknesses are tested at that strength. Then another dosage strength is selected and all four wall thicknesses are tested. Finally, the third dosage strength and the four wall thicknesses are tested. Meanwhile, two other laboratory technicians also follow this plan, each starting with a unit of antibiotic.

Note that there are *two* randomization restrictions within a block: technician

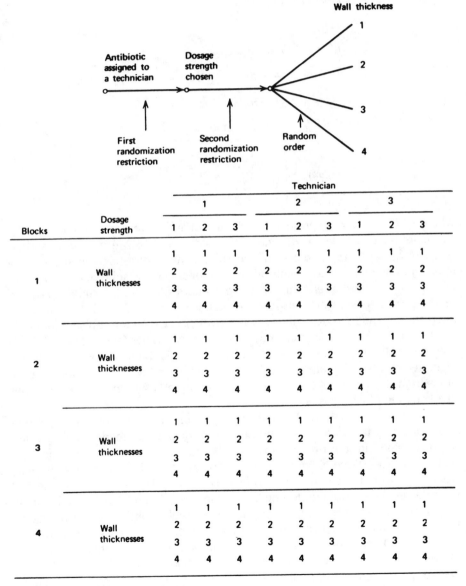

Figure 14-1. A split-split plot design.

and dosage strength. The whole plots correspond to the technician. The order in which the technicians are assigned the units of antibiotic is randomly determined. The dosage strengths form three subplots. Dosage strength may be randomly assigned to a subplot. Finally, within a particular dosage strength, the four capsule wall thicknesses are tested in random order, forming four sub-subplots. The wall thicknesses are usually called sub-subtreatments. Because there are *two* randomization restrictions within a block (some authors say two "splits" in the design) the design is called a split–split-plot design. Figure 14-1 illustrates the randomization restrictions and experimental layout in this design.

■

The linear statistical model for the split–split-plot design is

$$
y_{ijkh} = \mu + \tau_i + \beta_j + (\tau\beta)_{ij} + \gamma_k + (\tau\gamma)_{ik} + (\beta\gamma)_{jk} + (\tau\beta\gamma)_{ijk}
$$
$$
+ \delta_h + (\tau\delta)_{ih} + (\beta\delta)_{jh}
$$
$$
+ (\tau\beta\delta)_{ijh} + (\gamma\delta)_{kh} + (\tau\gamma\delta)_{ikh} + (\beta\gamma\delta)_{jkh} \qquad (14\text{-}7)
$$
$$
+ (\tau\beta\gamma\delta)_{ijkh} \begin{cases} i = 1,2,\ldots,a \\ j = 1,2,\ldots,b \\ k = 1,2,\ldots,c \\ h = 1,2,\ldots,d \end{cases}
$$

where τ_i, β_j, and $(\tau\beta)_{ij}$ represent the whole plot and correspond to blocks (factor A), main treatments (factor B), and whole plot error (AB), respectively; and γ_k, $(\tau\gamma)_{jk}$, $(\beta\gamma)_{jk}$, and $(\tau\beta\gamma)_{ijk}$ represent the subplot and correspond to the subplot treatment (factor C), the AC and BC interactions, and the subplot error, respectively; and δ_h and the remaining parameters correspond to the sub-subplot and represent respectively the sub-subplot treatment (factor D) and the remaining interactions. The four-factor interaction $(\gamma\beta\gamma\delta)_{ijkh}$ is called the sub-subplot error.

Assuming that blocks are random and that the other factors are fixed, we may derive the expected mean squares as shown in Table 14-10. Tests on the main treatments, subtreatments, sub-subtreatments and their interactions are obvious from inspection of this table. Note that no tests on blocks or interactions involving the blocks exist.

The statistical analysis of a split–split-plot design is like that of a single replicate of a four-factor factorial. The number of degrees of freedom for each test are determined in the usual manner. To illustrate, in Example 14-2, where we had four blocks, three technicians, three dosage strengths, and four wall thicknesses, we would have only $(a - 1)(b - 1) = (4 - 1)(3 - 1) = 6$ whole plot error degrees of freedom for testing technicians. This is a relatively small number of degrees of freedom, and the experimenter might consider using additional blocks to increase the precision of the test. If there are a blocks, then we will have $2(a - 1)$ degrees of freedom for whole plot error. Thus, five blocks yield $2(5 - 1) = 8$ error degrees of freedom, six blocks yield $2(6 - 1) = 10$ error degrees of freedom, seven blocks yield $2(7 - 1) = 12$ error degrees of freedom, and so on. Consequently, we would probably not want to run fewer than four

Table 14-10   Expected Mean Square Derivation for the Split–Split-Plot Design

	Factor	$a\ b\ c\ d\ 1$ $R\ F\ F\ F\ R$ $i\ j\ k\ h\ l$	Expected Mean Square
Whole plot	τ_i	$1\ b\ c\ d\ 1$	$\sigma^2 + bcd\sigma_\tau^2$
	β_j	$a\ 0\ c\ d\ 1$	$\sigma^2 + cd\sigma_{\tau\beta}^2 + \dfrac{acd \sum \beta_j^2}{(b-1)}$
	$(\tau\beta)_{ij}$	$1\ 0\ c\ d\ 1$	$\sigma^2 + cd\sigma_{\tau\beta}^2$
Subplot	γ_k	$a\ b\ 0\ d\ 1$	$\sigma^2 + bd\sigma_{\tau\gamma}^2 + \dfrac{abd \sum \gamma_k^2}{(c-1)}$
	$(\tau\gamma)_{ik}$	$1\ b\ 0\ d\ 1$	$\sigma^2 + bd\sigma_{\tau\gamma}^2$
	$(\beta\gamma)_{jk}$	$a\ 0\ 0\ d\ 1$	$\sigma^2 + d\sigma_{\tau\beta\gamma}^2 + \dfrac{ad \sum\sum (\beta\gamma)_{jh}^2}{(b-1)(c-1)}$
	$(\tau\beta\gamma)_{ijk}$	$1\ 0\ 0\ d\ 1$	$\sigma^2 + d\sigma_{\tau\beta\gamma}^2$
Sub-subplot	δ_h	$a\ b\ c\ 0\ 1$	$\sigma^2 + bc\sigma_{\tau\delta}^2 + \dfrac{abc \sum \gamma_k^2}{(c-1)}$
	$(\tau\delta)_{ih}$	$1\ b\ c\ 0\ 1$	$\sigma^2 + bc\sigma_{\tau\delta}^2$
	$(\beta\delta)_{jh}$	$a\ 0\ c\ 0\ 1$	$\sigma^2 + c\sigma_{\tau\beta\delta}^2 + \dfrac{ac \sum\sum (\beta\delta)_{jh}^2}{(b-1)(d-1)}$
	$(\tau\beta\delta)_{ijh}$	$1\ 0\ c\ 0\ 1$	$\sigma^2 + c\sigma_{\tau\beta\delta}^2$
	$(\gamma\delta)_{kh}$	$a\ b\ 0\ 0\ 1$	$\sigma^2 + b\sigma_{\tau\gamma\delta}^2 + \dfrac{ab \sum\sum (\gamma\delta)_{kh}^2}{(c-1)(d-1)}$
	$(\tau\gamma\delta)_{ikh}$	$1\ b\ 0\ 0\ 1$	$\sigma^2 + b\sigma_{\tau\gamma\delta}^2$
	$(\beta\gamma\delta)_{jkh}$	$a\ 0\ 0\ 0\ 1$	$\sigma^2 + \sigma_{\tau\beta\gamma\delta}^2 + \dfrac{a \sum\sum\sum (\beta\gamma\delta)_{ijk}^2}{(b-1)(c-1)(d-1)}$
	$(\tau\beta\gamma\delta)_{ijkh}$	$1\ 0\ 0\ 0\ 1$	$\sigma^2 + \sigma_{\tau\beta\gamma\delta}^2$
	$\epsilon_{l(ijkh)}$	$1\ 1\ 1\ 1\ 1$	σ^2 (not estimable)

blocks since this would yield only 4 error degrees of freedom. Each additional block allows us to gain 2 degrees of freedom for error. If we could afford to run five blocks, we could increase the precision of the test by one-third (from 6 to 8 degrees of freedom). Also, in going from five to six blocks, there is an additional 25 percent gain in precision. If resources permit, the experimenter should run five or six blocks.

14-4   PROBLEMS

14-1   The yield of a chemical process is being studied. The two factors under study are the temperature and the pressure. Three levels of each factor are selected, and only nine observations can be run in one day. The experimenter runs a complete replicate of the design on each day. The data are shown in the following table. Analyze the data, assuming that the days are blocks.

	Day 1 Pressure			Day 2 Pressure		
Temperature	250	260	270	250	260	270
Low	86.3	84.0	85.8	86.1	85.2	87.3
Medium	88.5	87.3	89.0	89.4	89.9	90.3
High	89.1	90.2	91.3	91.7	93.2	93.7

14-2   Consider the data in Problem 9-5. Analyze the data assuming that replicates are blocks. Assume that all factors are fixed.

14-3   Consider the data in Problem 9-1. Analyze the data assuming that replicates are blocks. Assume that the levels of tool geometry are randomly selected and that speed and angle are fixed.

14-4   Rework Problem 14-3 computing all the interactions of the three factors with blocks.

14-5   Rework Example 13-1 assuming that each replicate represents a block. Derive the expected mean squares.

14-6   Rework Example 13-2 assuming that each replicate represents a block. Derive the expected mean squares.

14-7   Outline the analysis of variance for a 2^3 factorial experiment run in a Latin square.

14-8   Steel is normalized by heating above the critical temperature, soaking, and then air cooling. This process increases the strength of the steel, refines the grain, and homogenizes the structure. An experiment is performed to determine the effect of temperature and heat treatment time on the strength of normalized steel. Two temperatures and three times are selected. The experiment is performed by heating the oven to a randomly selected temperature and inserting three specimens. After 10 minutes one specimen is removed, after 20 minutes the second is removed, and after 30 minutes the final specimen is removed. Then the temperature is changed to the other level and the process is repeated. Four shifts are required to collect the data, which follow. Analyze the data shown in Table 14.11 and draw conclusions, assuming both factors are fixed.

14-9   Repeat Problem 14-8 assuming that the temperature and time are randomly selected.

14-10   An experiment is designed to study pigment dispersion in paint. Four different mixes of a particular pigment are studied. The procedure consists of preparing a particular mix and then applying that mix to a panel by three application methods (brushing, spraying, and rolling). The response measured is the percentage reflectance of pigment. Three days are required to run the experiment, and the data obtained follow. Analyze the data and draw conclusions, assuming that mixes and application methods are fixed.

Day	Application Method	Mix			
		1	2	3	4
1	1	64.5	66.3	74.1	66.5
	2	68.3	69.5	73.8	70.0
	3	70.3	73.1	78.0	72.3
2	1	65.2	65.0	73.8	64.8
	2	69.2	70.3	74.5	68.3
	3	71.2	72.8	79.1	71.5
3	1	66.2	66.5	72.3	67.7
	2	69.0	69.0	75.4	68.6
	3	70.8	74.2	80.1	72.4

Table 14-11

Shift	Time (minutes)	Temperature (°F)	
		1500	1600
1	10	63	89
	20	54	91
	30	61	62
2	10	50	80
	20	52	72
	30	59	69
3	10	48	73
	20	74	81
	30	71	69
4	10	54	88
	20	48	92
	30	59	64

14-11 Repeat Problem 14-10 assuming that the mixes are random and the application methods are fixed.

14-12 Consider the split–split-plot design described in Example 14-2. Suppose that this experiment is conducted as described and that the data that follow in Table 14-12 are obtained. Analyze this data and draw conclusions.

14-13 Rework Problem 14-12 assuming that the dosage strengths are chosen at random.

14-14 Suppose that in Problem 14-12 four technicians had been used. Assuming that all the factors are fixed, how many blocks should be run to obtain an adequate number of degrees of freedom on the test for differences among technicians?

Table 14-12

Blocks	Dosage Strengths	Technician								
		1			2			3		
		1	2	3	1	2	3	1	2	3
	Wall Thickness									
1	1	95	71	108	96	70	108	95	70	100
	2	104	82	115	99	84	100	102	81	106
	3	101	85	117	95	83	105	105	84	113
	4	108	85	116	97	85	109	107	87	115
2	1	95	78	110	100	72	104	92	69	101
	2	106	84	109	101	79	102	100	76	104
	3	103	86	116	99	80	108	101	80	109
	4	109	84	110	112	86	109	108	86	113
3	1	96	70	107	94	66	100	90	73	98
	2	105	81	106	100	84	101	97	75	100
	3	106	88	112	104	87	109	100	82	104
	4	113	90	117	121	90	117	110	91	112
4	1	90	68	109	98	68	106	98	72	101
	2	100	84	112	102	81	103	102	78	105
	3	102	85	115	100	85	110	105	80	110
	4	114	88	118	118	85	116	110	95	120

14-15 Consider the experiment described in Example 14-2. Demonstrate how the order in which the treatment combinations are run would be determined if this experiment were run as (a) a split–split plot, (b) a split plot, (c) a factorial design in a randomized block, and (d) a completely randomized factorial design.

Chapter 15
Regression Analysis

15-1 INTRODUCTION

In many problems there are two or more variables that are related, and it is important to model and explore this relationship. For example, in a chemical process the yield of product is related to the operating temperature. It may be of interest to build a model relating yield to temperature and then use the model for prediction, process optimization, or process control.

In general, suppose that there is a single *dependent variable* or response y that depends on k *independent* or *regressor variables,* for example, $x_1, x_2, \ldots,$ x_k. The relationship between these variables is characterized by a mathematical model called a *regression equation.* The regression model is fit to a set of sample data. In some instances, the experimenter knows the exact form of the true functional relationship between y and $x_1, x_2, \ldots, x_k$, say $y = \phi(x_1, x_2, \ldots, x_k)$. However, in most cases, the true functional relationship is unknown, and the experimenter chooses an appropriate function to approximate ϕ. Polynomial models are widely used as approximating functions, and for that reason we discuss the fitting of polynomials to data in this chapter. More extensive presentations of regression analysis are in Montgomery and Peck (1982) and Draper and Smith (1981).

Regression methods are frequently used to analyze data from *unplanned experiments,* such as might arise from observations of uncontrolled phenomena or historical records. Regression analysis is also highly useful in designed experiments. Generally, the analysis of variance in a designed experiment helps to identify *which* factors are important, and regression is used to build a quantitative model relating the important factors to the response.

15-2 SIMPLE LINEAR REGRESSION

We wish to determine the relationship between a single regressor variable x and a response variable y. The regressor variable x is usually assumed to be a continuous variable that is controllable by the experimenter. Then if the experiment is *designed*, we choose the values of x and observe the corresponding value of y.

Suppose the true relationship between y and x is a straight line, and that the observation y at each level of x is a random variable. Now, the expected value of y for each value of x is

$$E(y|x) = \beta_0 + \beta_1 x \qquad (15\text{-}1)$$

where the parameters of the straight line β_1 and β_1 are unknown constants. We assume that each observation y can be described by the model

$$y = \beta_0 + \beta_1 x + \epsilon \qquad (15\text{-}2)$$

where ϵ is a random error with mean zero and variance σ^2. The $\{\epsilon\}$ are also assumed to be uncorrelated random variables. The regression model (Equation 15-2) involving only a single regressor variable x is often called the *simple linear regression model*.

If we have n pairs of data $(y_1, x_1), (y_2, x_2), \ldots, (y_n, x_n)$, we may estimate the model parameters β_0 and β_1 by least squares. By Equation 15-2, we may write

$$y_j = \beta_0 + \beta_1 x_j + \epsilon_j \qquad j = 1, 2, \ldots, n$$

and the least squares function is

$$L = \sum_{j=1}^{n} \epsilon_j^2 = \sum_{j=1}^{n} (y_j - \beta_0 - \beta_1 x_j)^2 \qquad (15\text{-}3)$$

Minimizing the least squares function is simplifed if we rewrite the model, Equation 15-2, as

$$y = \beta_0' + \beta_1(x - \bar{x}) + \epsilon \qquad (15\text{-}4)$$

where $\bar{x} = (1/n)\sum_{j=1}^{n} x_j$ and $\beta_0' = \beta_0 + \beta_1 \bar{x}$. In Equation 15-4 we have simply corrected the regressor variable for its average, resulting in a transformation on the intercept. Equation 15-4 is frequently called the *transformed simple linear regression model* or, more simply, the *transformed model*. In addition to simplifying the estimation problem, use of the transformed model allows other inference tasks to be performed more easily.

By employing the transformed model, the least squares function becomes

$$L = \sum_{j=1}^{n} [y_j - \beta_0' - \beta_1(x_j - \bar{x})]^2 \qquad (15\text{-}5)$$

The least squares estimators of β_0' and β_1, say $\hat{\beta}_0'$ and $\hat{\beta}_1$, must satisfy

$$\left.\frac{\partial L}{\partial \beta_0'}\right|_{\hat{\beta}_0', \hat{\beta}_1} = -2 \sum_{j=1}^{n} [y_j - \hat{\beta}_0' - \hat{\beta}_1(x_j - \bar{x})] = 0$$

$$\left.\frac{\partial L}{\partial \beta_1}\right|_{\hat{\beta}_0', \hat{\beta}_1} = -2 \sum_{j=1}^{n} [y_j - \hat{\beta}_0' - \hat{\beta}_1(x_j - \bar{x})](x_j - \bar{x}) = 0$$

Simplifying these two equations yields

$$n\hat{\beta}_0' = \sum_{j=1}^{n} y_j \tag{15-6}$$

$$\hat{\beta}_1 \sum_{j=1}^{n} (x_j - \bar{x})^2 = \sum_{j=1}^{n} y_j(x_j - \bar{x})$$

Equations 15-6 are called the *least squares normal equations*. Their solutions are

$$\hat{\beta}_0' = \frac{1}{n} \sum_{j=1}^{n} y_j = \bar{y} \tag{15-7}$$

$$\hat{\beta}_1 = \frac{\displaystyle\sum_{j=1}^{n} y_j(x_j - \bar{x})}{\displaystyle\sum_{j=1}^{n} (x_j - \bar{x})^2} \tag{15-8}$$

Thus, $\hat{\beta}_0'$ and $\hat{\beta}_1$ are the least squares estimators of the intercept and slope, respectively.

The fitted simple linear regression model is

$$\hat{y} = \hat{\beta}_0' + \hat{\beta}_1(x - \bar{x}) \tag{15-9}$$

If we wish to present our results in terms of the *original* intercept, β_0, then

$$\hat{\beta}_0 = \hat{\beta}_0' - \hat{\beta}_1\bar{x}$$

and the fitted model is

$$\hat{y} = \hat{\beta}_0 + \hat{\beta}_1 x \tag{15-10}$$

Notationally, it is convenient to give special symbols to the numerator and denominator of Equation 15-8. That is, let

$$S_{xx} = \sum_{j=1}^{n} (x_j - \bar{x})^2 = \sum_{j=1}^{n} x_j^2 - \frac{\left(\displaystyle\sum_{j=1}^{n} x_j\right)^2}{n} \tag{15-11}$$

and

$$S_{xy} = \sum_{j=1}^{n} y_j(x_j - \bar{x}) = \sum_{j=1}^{n} x_j y_j - \frac{\left(\displaystyle\sum_{j=1}^{n} x_j\right)\left(\displaystyle\sum_{j=1}^{n} y_j\right)}{n} \tag{15-12}$$

We call S_{xx} the corrected sum of squares of x and S_{xy} the corrected sum of the cross-products of x and y. The extreme right-hand sides of Equations 15-11 and 15-12 are the usual computational formulas. Using this new notation, the least squares estimator of the slope is

$$\hat{\beta}_1 = \frac{S_{xy}}{S_{xx}} \tag{15-13}$$

Example 15-1

A study was made to determine the effect of stirring rate on the amount of impurity in paint produced by a chemical process. The study yielded the following data.

Stirring rate, rpm (x)	20	22	24	26	28	30	32	34	36	38	40	42
Impurity, %(y)	8.4	9.5	11.8	10.4	13.3	14.8	13.2	14.7	16.4	16.5	18.9	18.5

A plot of percent impurity versus stirring rate is shown in Figure 15-1. This plot is called a *scatter diagram*, and it is very useful in identifying the relationship between two variables. This *scatter diagram* suggests that a straight-line relationship may be appropriate, so the model $y = \beta_0 + \beta_1 x + \epsilon$ is proposed, and the following quantities are computed:

$$n = 12 \quad \sum_{j=1}^{12} x_j = 372 \quad \sum_{j=1}^{12} y_j = 166.4 \quad \sum_{j=1}^{12} x_j^2 = 12{,}104$$

$$\bar{y} = 13.87 \quad \bar{x} = 31 \quad \sum_{j=1}^{12} y_j^2 = 2435.14 \quad \sum_{j=1}^{12} y_j x_j = 5419.60$$

From Equations 15-11 and 15-12, we find

$$S_{xx} = 12{,}104 - \frac{(372)^2}{12} = 572.00$$

$$S_{xy} = 5419.60 - \frac{(372)(166.4)}{12} = 261.20$$

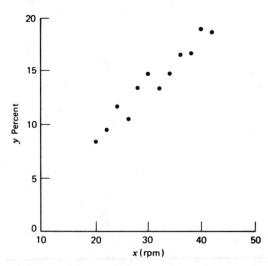

Figure 15-1. Scatter diagram of percent impurity (y) versus stirring rate (x) for Example 15-1.

Thus,

$$\hat{\beta}_1 = \frac{S_{xy}}{S_{xx}} = \frac{261.20}{572.00} = 0.4566$$

$$\hat{\beta}_0' = \bar{y} = 13.8667$$

and the fitted model is

$$\hat{y} = \hat{\beta}_0' + \hat{\beta}_1(x - \bar{x})$$

$$= 13.8667 + 0.4566(x - 31)$$

If we wish to express the model in terms of the original intercept, then

$$\hat{\beta}_0 = \hat{\beta}_0' - \hat{\beta}_1\bar{x}$$

$$= 13.8667 - (0.4566)(31) = -0.2879$$

and since $\hat{y} = \hat{\beta}_0 + \hat{\beta}_1 x$, we have

$$\hat{y} = -0.2879 + 0.4566x$$

Table 15-1 shows the observations y_i the fitted values $\hat{y}_i$, and the residuals $e_i = y_i - \hat{y}_i$ from this model. The residuals are useful in examining the adequacy of the least squares fit.

■

The statistical properties of the least squares estimators are of considerable value in judging the adequacy of the fitted model. The estimators $\hat{\beta}_0'$ and $\hat{\beta}_1$ are random variables since they are just linear combinations of the y_i and the y_i are random variables. We now investigate the bias and variance properties of these

Table 15-1 Observations y_i, Fitted Values $\hat{y}_i$, and Residuals $e_i = y_i - \hat{y}_i$ for Example 15-1

y_i	$\hat{y}_i$	$e_i = y_i - \hat{y}_i$
8.4	8.8441	-0.4441
9.5	7.7573	-0.2573
11.8	10.6705	1.1295
10.4	11.5837	-1.1837
13.3	12.4969	0.8031
14.8	13.4101	1.3899
13.2	14.3233	-1.1233
14.7	15.2365	-0.5365
16.4	16.1497	0.2503
16.5	17.0629	-0.5629
18.9	17.9761	0.9239
18.5	18.8893	-0.3893

estimators. Consider first $\hat{\beta}_1$. The expected value of $\hat{\beta}_1$ is

$$
\begin{aligned}
E(\hat{\beta}_1) &= E\left(\frac{S_{xy}}{S_{xx}}\right) \\
&= \frac{1}{S_{xx}} E\left[\sum_{j=1}^{n} y_j(x_j - \bar{x})\right] \\
&= \frac{1}{S_{xx}} E\left[\sum_{j=1}^{n} (\beta_0' + \beta_1(x_j - \bar{x}) + \epsilon_j)(x_j - \bar{x})\right] \\
&= \frac{1}{S_{xx}} \left\{ E\left[\beta_0' \sum_{j=1}^{n} (x_j - \bar{x})\right] + E\left[\beta_1 \sum_{j=1}^{n} (x_j - \bar{x})^2\right] + E\left[\sum_{j=1}^{n} \epsilon_j(x_j - \bar{x})\right] \right\} \\
&= \frac{1}{S_{xx}} \beta_1 S_{xx} \\
&= \beta_1
\end{aligned}
$$

since $\sum_{j=1}^{n}(x_j - \bar{x}) = 0$, and by assumption $E(\epsilon_j) = 0$. Thus, $\hat{\beta}_1$ is an *unbiased* estimator of the true slope β_1. Now consider the variance of $\hat{\beta}_1$. Since we have assumed that $V(\epsilon_j) = \sigma^2$, it follows that $V(y_j) = \sigma^2$, and

$$
\begin{aligned}
V(\hat{\beta}_1) &= V\left(\frac{S_{xy}}{S_{xx}}\right) \\
&= \frac{1}{S_{xx}^2} V \sum_{j=1}^{n} y_j(x_j - \bar{x})
\end{aligned}
\tag{15-14}
$$

The random variables y_j are uncorrelated because the ϵ_j are uncorrelated. Therefore, the variance of the sum in Equation 15-14 is just the sum of the variances, and the variance of each term in the sum, say $V[y_j(x_j - \bar{x})]$, is $\sigma^2(x_j - \bar{x})^2$. Thus,

$$
\begin{aligned}
V(\hat{\beta}_1) &= \frac{1}{S_{xx}^2} \sigma^2 \sum_{j=1}^{n} (x_j - \bar{x})^2 \\
&= \frac{\sigma^2}{S_{xx}}
\end{aligned}
\tag{15-15}
$$

By using a similar approach, we can show that

$$
E(\hat{\beta}_0') = \beta_0' \qquad V(\hat{\beta}_0') = \frac{\sigma^2}{n}
\tag{15-16}
$$

and

$$
E(\hat{\beta}_0) = \beta_0 \qquad V(\hat{\beta}_0) = \sigma^2\left[\frac{1}{n} + \frac{\bar{x}^2}{S_{xx}}\right]
\tag{15-17}
$$

To find $V(\hat{\beta}_0)$, we must make use of the result $\text{Cov}(\hat{\beta}_0', \hat{\beta}_1) = 0$. However, the covariance of $\hat{\beta}_0$ and $\hat{\beta}_1$ is not zero; in fact, $\text{Cov}(\hat{\beta}_0, \hat{\beta}_1) = -\sigma^2 \bar{x}/S_{xx}$. (Refer to Problems 15-14 and 15-15.) Note that $\hat{\beta}_0'$ and $\hat{\beta}_0$ are unbiased estimators of β_0' and β_0, respectively.

It is necessary to obtain an estimate of σ^2. This estimate may be obtained from the residuals $e_i = y_i - \hat{y}_i$. The sum of the squares of the residuals, or the error sum of squares, would be

$$SS_E = \sum_{j=1}^{n} e_j^2$$

$$= \sum_{j=1}^{n} (y_j - \hat{y}_j)^2$$

(15-18)

A more convenient computing formula for SS_E may be found by substituting the estimated model $\hat{y}_i = \bar{y} + \hat{\beta}_1(x_i - \bar{x})$ into Equation 15-18 and simplifying as follows:

$$SS_E = \sum_{j=1}^{n} [y_j - \bar{y} - \hat{\beta}_1(x_j - \bar{x})]^2$$

$$= \sum_{j=1}^{n} [y_j^2 + \bar{y}^2 + \hat{\beta}_1^2(x_j - \bar{x})^2 - 2\bar{y}y_j - 2\hat{\beta}_1 y_j(x_j - \bar{x}) - 2\hat{\beta}_1 \bar{y}(x_j - \bar{x})]$$

$$= \sum_{j=1}^{n} y_j^2 + n\bar{y}^2 + \hat{\beta}_1^2 S_{xx} - 2\bar{y} \sum_{j=1}^{n} y_j - 2\hat{\beta}_1 S_{xy} - 2\hat{\beta}_1 \bar{y} \sum_{j=1}^{n} (x_j - \bar{x})$$

(15-19)

The last term in Equation 15-19 is zero, $2\bar{y}\sum_{j=1}^{n}y_j = 2n\bar{y}^2$, and $\hat{\beta}_1^2 S_{xx} = \hat{\beta}_1(S_{xy}/S_{xx})S_{xx} = \hat{\beta}_1 S_{xy}$. Therefore Equation 15-19 becomes

$$SS_E = \sum_{j=1}^{n} y_j^2 - n\bar{y}^2 - \hat{\beta}_1 S_{xy}$$

But $\sum_{j=1}^{n} y_j^2 - n\bar{y}^2 = \sum_{j=1}^{n}(y_j - \bar{y})^2 \equiv S_{yy}$, say, which is just the corrected sum of squares of the y's, so we may write SS_E as

$$SS_E = S_{yy} - \hat{\beta}_1 S_{xy}$$

(15-20)

By taking expectation of SS_E, we may show that $E(SS_E) = (n - 2)\sigma^2$. Therefore,

$$\hat{\sigma}^2 = \frac{SS_E}{n - 2} \equiv MS_E$$

(15-21)

is an unbiased estimator of σ^2. Note that MS_E is the *error* or *residual mean square*.

Regression analysis is widely used and frequently *misused*. There are several common abuses of regression analysis that should be briefly mentioned. Care should be taken in selecting the variables with which regression models are to be constructed and in determining the form of the approximating function. It is quite possible to develop relationships among variables that are completely meaningless in a practical sense. For example, one might attempt to relate the shear strength of spot welds with the number of boxes of paper used by the data processing department. A straight line may even appear to provide a good fit to the data, but the relationship is unreasonable and cannot be relied on. The analyst familiar with the process under study must be the final judge of such functional relationships.

Regression relationships are valid only for values of the regressor variable within the range of the original data. The linear relationship that we have tentatively assumed may be valid over the original range of x, but it is unlikely to remain so as we encounter x values beyond that range. In other words, as we move beyond the original range of x, we become less certain about the validity of the assumed model. Regression models should never be used for extrapolation.

Finally, one occasionally feels that the model $y = \beta x + \epsilon$ is appropriate. The omission of the intercept from this model implies, of course, that $y = 0$ when $x = 0$. This is a very strong assumption that often is unjustified. Even when two variables, such as the height and weight of persons, would seem to qualify for the use of this model, we would usually obtain a better fit by including the intercept because of the limited range of data concerning the regressor variable.

15-3   HYPOTHESIS TESTING IN SIMPLE LINEAR REGRESSION

To test hypotheses about the slope and intercept of the regression model, we must make an additional assumption about the error term, namely, that the ϵ_i are normally distributed. Thus, we are assuming that the errors ϵ_i are NID$(0, \sigma^2)$. Later, we discuss how these asumptions can be checked through *residual analysis*.

Suppose the experimenter wishes to test the hypothesis that the slope equals some value, for example, $\beta_{1,0}$. The appropriate hypotheses are

$$H_0: \beta_1 = \beta_{1,0} \qquad (15\text{-}22)$$
$$H_1: \beta_1 \neq \beta_{1,0}$$

where we have specified a two-sided alternative. Now since the ϵ_i are NID$(0, \sigma^2)$, it follows directly that the observations y_i are NID$(\beta_0 + \beta_1 x_i, \sigma^2)$. Thus, $\hat{\beta}_1$ is a linear combination of independent normal random variables; consequently, $\hat{\beta}_1$ is $N(\beta_1, \sigma^2/S_{xx})$, using the bias and variance properties of $\hat{\beta}_1$ from Section 15-2. Furthermore, $\hat{\beta}_1$ is independent of MS_E. Then as a result of the normality assumption, the statistic

$$t_0 = \frac{\hat{\beta}_1 - \hat{\beta}_{1,0}}{\sqrt{MS_E/S_{xx}}} \qquad (15\text{-}23)$$

follows the t distribution with $n - 2$ degrees of freedom under $H_0: \beta_1 = \beta_{1,0}$. We would reject $H_0: \beta_1 = \beta_{1,0}$ if

$$|t_0| > t_{\alpha/2, n-2} \qquad (15\text{-}24)$$

where t_0 is computed from Equation 15-23.

A similar procedure can be used to test hypotheses about the intercept. To test

$$H_0: \beta_0 = \beta_{0,0} \qquad (15\text{-}25)$$
$$H_1: \beta_0 \neq \beta_{0,0}$$

we would use the statistic

$$t_0 = \frac{\hat{\beta}_0 - \beta_{0,0}}{\sqrt{MS_E \left(\dfrac{1}{n} + \dfrac{\bar{x}^2}{S_{xx}} \right)}} \qquad (15\text{-}26)$$

and reject the null hypothesis if $|t_0| > t_{\alpha/2, n-2}$.

A very important special case of the hypotheses in Equation 15-22 is

$$H_0: \beta_1 = 0 \qquad (15\text{-}27)$$
$$H_1: \beta_1 \neq 0$$

The hypothesis $H_0: \beta_1 = 0$ relates to the *significance* of regression. Failing to reject $H_0: \beta_1 = 0$ is equivalent to concluding that there is no *linear* relationship between x and y; that is, the best estimator of y_i for any x_i is $\hat{y}_i = \bar{y}$. In many instances this may mean that there is no *causal* relationship between x and y or that the true relationship is not linear.

The test procedure for $H_0: \beta_1 = 0$ may be developed from two approaches. The first approach starts with the following partitioning of the total corrected sum of squares for y:

$$S_{yy} \equiv \sum_{j=1}^{n} (y_j - \bar{y})^2 = \sum_{j=1}^{n} (\hat{y}_j - \bar{y})^2 + \sum_{j=1}^{n} (y_j - \hat{y}_j)^2 \qquad (15\text{-}28)$$

The two components of S_{yy} measure, respectively, the amount of variability in the y_j accounted for by the regression line and the residual variation left unexplained by the regression line. We recognize $SS_E = \sum_{j=1}^{n}(y_j - \hat{y}_j)^2$ as the *error* or *residual* sum of squares and $SS_R = \sum_{j=1}^{n}(\hat{y}_j - \bar{y})^2$ as the *regression* sum of squares. Thus, Equation 15-28 may be written as

$$S_{yy} = SS_R + SS_E \qquad (15\text{-}29)$$

From Equation 15-20, we obtain the computing formula for SS_R as

$$SS_R = \hat{\beta}_1 S_{xy} \qquad (15\text{-}30)$$

S_{yy} has $n - 1$ degrees of freedom, and SS_R and SS_E have 1 and $n - 2$ degrees of freedom, respectively.

We may show that $E[SS_E/(n-2)] = \sigma^2$, that $E[SS_R/1] = \sigma^2 + \beta_1^2 S_{xx}$, and that SS_E and SS_R are independent. Thus, if $H_0: \beta_1 = 0$ is true, the statistic

$$F_0 = \frac{SS_R/1}{SS_E/(n-2)} = \frac{MS_R}{MS_E} \qquad (15\text{-}31)$$

follows the $F_{1, n-2}$ distribution, and we would reject H_0 if $F_0 > F_{\alpha, 1, n-2}$. The test procedure is usually arranged in an analysis of variance table such as Table 15-2.

The test for significance of regression may also be developed from Equation 15-23 with $\beta_{1,0} = 0$, say

$$t_0 = \frac{\hat{\beta}_1}{\sqrt{MS_E/S_{xx}}} \tag{15-32}$$

By squaring both sides of Equation 15-32, we obtain

$$t_0^2 = \frac{\hat{\beta}_1^2 S_{xx}}{MS_E} = \frac{\hat{\beta}_1 S_{xy}}{MS_E} = \frac{MS_R}{MS_E} \tag{15-33}$$

Note that t_0^2 in Equation 15-33 is identical to F_0 in Equation 15-31. It is true in general that the square of a t random variable with f degrees of freedom is an F random variable with one and f degrees of freedom in the numerator and denominator, respectively. Thus, the test using t_0 is equivalent to the test based on F_0.

Example 15-2

For the data given in Example 15-1, we test for significance of regression. The fitted model is $\hat{y} = -0.2879 + 0.4566x$, and S_{yy} is computed as

$$S_{yy} = \sum_{j=1}^{n} y_j^2 - \frac{\left(\sum_{j=1}^{n} y_j\right)^2}{12}$$

$$= 2435.14 - \frac{(166.4)^2}{12} = 127.73$$

The regression sum of squares is

$$SS_R = \hat{\beta}_1 S_{xy} = (0.4566)(261.20) = 119.26$$

Thus, the error sum of squares is

$$SS_E = S_{yy} - SS_R$$

$$= 127.73 - 119.26 = 8.47$$

The analysis of variance for testing $H_0: \beta_1 = 0$ is summarized in Table 15-3. Since $F_{.01,1,10} = 10.0$, we reject H_0 and conclude that $\beta_1 \neq 0$. Note that the error mean square in Table 15-3 is the estimate of σ^2 from Equation 15-21.

∎

Table 15-2   Analysis of Variance for Testing Significance of Regression

Source of Variation	Sum of Squares	Degrees of Freedom	Mean Square	F_0
Regression	$SS_R = \hat{\beta}_1 S_{xy}$	1	MS_R	MS_R/MS_E
Error or residual	$SS_E = S_{yy} - \hat{\beta}_1 S_{xy}$	$n - 2$	MS_E	
Total	Sy_y	$n - 1$		

Table 15-3 Analysis of Variance for Example 15-2

Source of Variation	Sum of Squares	Degrees of Freedom	Mean Square	F_0
Regression	119.26	1	119.26	140.80
Error	8.47	10	0.847	
Total	127.73	11		

15-4 INTERVAL ESTIMATION IN SIMPLE LINEAR REGRESSION

In addition to point estimates of the slope and intercept, it is possible to obtain interval estimates of these parameters. If the ϵ_i are normally and independently distributed, then

$$(\hat{\beta}_1 - \beta_1)/\sqrt{MS_E/S_{xx}} \quad \text{and} \quad (\hat{\beta}_0 - \beta_0)/\sqrt{MS_E\left(\frac{1}{n} + \frac{\bar{x}^2}{S_{xx}}\right)}$$

are both distributed as t with $n - 2$ degrees of freedom. Therefore, a $100(1 - \alpha)$ percent confidence interval on β_1 is given by

$$\left[\hat{\beta}_1 \pm t_{\alpha/2,n-2}\sqrt{\frac{MS_E}{S_{xx}}}\right] \tag{15-34}$$

Similarly, a $100(1 - \alpha)$ percent confidence interval on β_0 is given by

$$\left[\hat{\beta}_0 \pm t_{\alpha/2,n-2}\sqrt{MS_E\left(\frac{1}{n} + \frac{\bar{x}^2}{S_{xx}}\right)}\right] \tag{15-35}$$

As an illustration, a 95 percent confidence interval for β_1 for the data in Example 15-1 would be found from Equation 15-34 as

$$\left[0.4566 \pm (2.228)\sqrt{\frac{0.847}{572.00}}\right]$$

or

$$[0.4566 \pm 0.0857]$$

Thus, the 95 percent confidence interval for β_1 is $0.3709 \le \beta_1 \le 0.5423$.

A confidence interval may be constructed for the mean response at a specified x, for example, x_0. This is a confidence interval about $E(y|x_0)$ and is often called a confidence interval about the regression line. Since $E(y|x_0) = \beta_0' + \beta_1(x_0 - \bar{x})$, we may obtain a point estimate of $E(y|x_0)$ from the fitted model as

$$\widehat{E(y|x_0)} \equiv \hat{y}_0 = \hat{\beta}_0' + \hat{\beta}_1(x_0 - \bar{x})$$

Now it is clear that $E(\hat{y}_0) = \hat{\beta}'_0 + \hat{\beta}_1(x_0 - \bar{x})$, since $\hat{\beta}'_0$ and $\hat{\beta}_1$ are unbiased, and furthermore that

$$V(\hat{y}_0) = \sigma^2\left[\frac{1}{n} + \frac{(x_0 - \bar{x})^2}{S_{xx}}\right]$$

since $\text{Cov}(\hat{\beta}'_0, \hat{\beta}_1) = 0$. Also, $\hat{y}_0$ is normally distributed as $\hat{\beta}_0$ and $\hat{\beta}_1$ are normally distributed. Therefore, a $100(1 - \alpha)$ percent confidence interval about the true regression line at $x = x_0$ may be computed from

$$\left[\hat{y}_0 \pm t_{\alpha/2,n-2}\sqrt{MS_E\left(\frac{1}{n} + \frac{(x_0 - \bar{x})^2}{S_{xx}}\right)}\right] \tag{15-36}$$

Notice that the width of the confidence interval for $E(y|x_0)$ is a function of x_0. The interval width is a minimum for $x_0 = \bar{x}$ and widens as $|x_0 - \bar{x}|$ increases.

Example 15-3

We can construct a 95 percent confidence interval about the regression line for the data in Example 15-1. Since $\hat{y}_0 = -0.2879 + 0.4566x_0$, the 95 percent confidence interval is given by

$$\left[\hat{y}_0 \pm 2.228\sqrt{(0.847)\left(\frac{1}{12} + \frac{(x_0 - 31)^2}{572.00}\right)}\right]$$

The predicted values and the 95 percent confidence limits for $x_0 = x_i$, $i = 1, 2,$ ..., 12 are displayed in Table 15-4. To demonstrate the use of this table, we could find the 95 percent confidence interval on the true regression line at $x_0 = 26$ (say) as

$$11.5837 - 0.73 \leq E(y|x_0 = 26) \leq 11.5837 + 0.73$$

or

$$10.85 \leq E(y|x_0 = 26) \leq 12.31$$

The fitted model and the 95 percent confidence interval about the true regression line is shown in Figure 15-2.

    ■

Another useful concept in simple linear regression is the *prediction interval*, an interval estimate on the mean of k *future* observations at a particular value of x, say x_0. To illustrate the prediction interval, suppose that in Example 15-1 the analyst wishes to construct an interval estimate on the mean impurity on the next four batches of paint processed with stirring rate $x_0 = 34$. A confidence interval is inadequate because it refers the true mean impurity (an unknown constant) and not future observations of the random variable.

Let the jth future observation on the response at x_0 be denoted by y_{0j}, and let the mean of these values be denoted by

$$\bar{y}_0 = \frac{1}{k}\sum_{j=1}^{k} y_{0j}$$

Table 15-4　Confidence Limits for Example 15-3

x_0	20	22	24	26	28	30	32	34	36	38	40	42
$\hat{y}_0$	8.8441	9.7573	10.6705	11.5837	12.4967	13.4101	14.3233	15.2365	16.1497	17.0629	17.9761	18.8893
95 percent confidence limits	±1.11	±0.97	±0.84	±0.73,	±0.65	±0.60	±0.60	±0.65	±0.73	±0.84	±0.97	±1.11

Figure 15-2. Fitted model and the 95 percent confidence interval for Example 15-3.

Note that if $k = 1$, then we are predicting a single future observation. The best predictor of the future value $\bar{y}_0$ is $\hat{y}_0 = \hat{\beta}_0' + \hat{\beta}_1(x_0 - \bar{x})$. The random variable

$$\psi = \bar{y}_0 - \hat{y}_0$$

is normally distributed with mean zero and variance

$$V(\psi) = V(\bar{y}_0 - \hat{y}_0)$$
$$= \sigma^2 \left[\frac{1}{k} + \frac{1}{n} + \frac{(x_0 - \bar{x})^2}{S_{xx}} \right]$$

because $\bar{y}_0$ is independent of $\hat{y}_0$. Thus, the $100(1 - \alpha)$ percent prediction interval on the mean of k future observations at x_0 is

$$\left[\hat{y}_0 \pm t_{\alpha/2, n-2} \sqrt{MS_E \left(\frac{1}{k} + \frac{1}{n} + \frac{(x_0 - \bar{x})^2}{S_{xx}} \right)} \right] \qquad (15\text{-}37)$$

Notice that the prediction interval is of minimum width at $x_0 = \bar{x}$ and widens as $|x_0 - \bar{x}|$ increases. Furthermore, if $k = 1$, then Equation 15-37 yields a prediction interval on a single future observation at x_0. By comparing Equation 15-37 with Equation 15-36, we observe that the prediction interval at x_0 is always wider than the confidence interval at x_0. This results from the prediction interval depending on both the error from the fitted model and the error associated with future observations.

To illustrate the construction of a prediction interval, we may use the data in Example 15-1 and find a 95 percent prediction interval on the mean impurity of the next two batches of paint produced at $x_0 = 34$ from

$$\left[15.2365 \pm 2.228 \sqrt{(0.847)\left(\frac{1}{2} + \frac{1}{12} + \frac{(34-31)^2}{572.00}\right)} \right]$$

This calculation yields [15.2365 ± 1.5870]. Thus, the 95 percent prediction interval for $k = 2$ at $x_0 = 34$ is $13.6495 \le \bar{y}_0 \le 16.8235$.

15-5 MODEL ADEQUACY CHECKING

15-5.1 Residual Analysis

As in fitting any linear model, analysis of the residuals from a regression model is necessary to determine the adequacy of the least squares fit. It is helpful to examine a normal probability plot, a plot of residuals versus fitted values, and a plot of residuals versus each regressor variable. In addition, if there are variables not included in the model that are of potential interest, then the residuals should be plotted against these omitted factors. Any structure in such a plot would indicate that the model could be improved by the addition of that factor.

A normal probability plot of the residuals from the simple linear regression model in Example 15-1 is shown in Figure 15-3. This plot does not indicate any serious violation of the normality assumption. The residuals are plotted versus the fitted values $\hat{y}_i$ in Figure 15-4 and versus the level so the regressor stirring rate x_i in Figure 15-5. These plots do not reveal any major difficulty, so we conclude that the simple linear regression model is an adequate fit to the paint impurity data.

15-5.2 The Lack-of-Fit Test

Regression models are often fitted to data when the true functional relationship is unknown. Naturally, we would like to know whether the order of the model tentatively assumed is correct. This section describes a test for the validity of this assumption.

The danger of using a regression model that is a poor approximation of the true functional relationship is illustrated in Figure 15-6. Obviously, a polynomial of degree two or greater should have been used for this hypothetical situation. The result is that a very poor model has been obtained.

We present a test for the "goodness of fit" of a regression model. Although the case of only one independent variable is explicitly treated, the procedure will generalize to k regressor variables easily. The hypotheses we wish to test are

H_0: The model adequately fits the data

H_1: The model does not fit the data

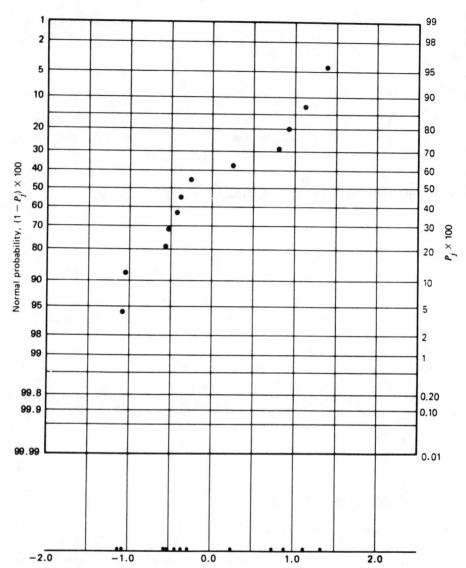

Figure 15-3.   Normal probability plot of residuals for Example 15-1.

The test involves partitioning the error or residual sum of squares into the following two components.

$$SS_E = SS_{PE} + SS_{LOF}$$

where SS_{PE} is the sum of squares attributable to "pure" experimental error, and SS_{LOF} is the sum of squares attributahble to the lack of fit of the model. To compute SS_{PE} we require observations on y for at least one level of x. Suppose that we

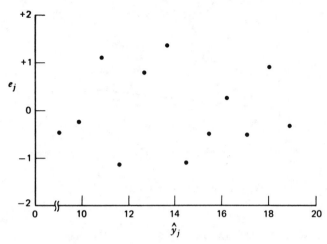

Figure 15-4. Plot of residuals versus $\hat{y}_j$ for Example 15-1.

have n observations such that

$$y_{11}, y_{12}, \ldots, y_{1n_1} = \text{Repeated observations at } x_1$$
$$y_{21}, y_{22}, \ldots, y_{2n_2} = \text{Repeated observations at } x_2$$
$$\vdots$$
$$y_{m1}, y_{m2}, \ldots, y_{mn_m} = \text{Repeated observations at } x_m$$

We see that there are m distinct levels of x. The contribution to the pure-error

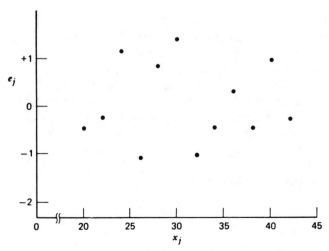

Figure 15-5. Plot of residuals versus x_j for Example 15-1.

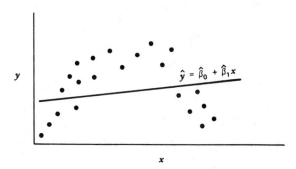

Figure 15-6. A regression model displaying lack of fit.

sum of squares at x_1, say, would be

$$\sum_{u=1}^{n_1} (y_{1u} - \bar{y}_1)^2 \tag{15-38}$$

The total sum of squares for pure error would be obtained by summing Equation 15-38 over all levels of x as

$$SS_{PE} = \sum_{i=1}^{m} \sum_{u=1}^{n_i} (y_{iu} - \bar{y}_i)^2$$

There are $n_e = \sum_{i=1}^{m}(n_i - 1) = n - m$ degrees of freedom associated with the pure-error sum of squares. The sum of squares for lack of fit is simply

$$SS_{LOF} = SS_E - SS_{PE}$$

with $n - 2 - n_e = m - 2$ degrees of freedom. The test statistic for lack of fit would then be

$$F_0 = \frac{SS_{LOF}/(m - 2)}{SS_{PE}/(n - m)} = \frac{MS_{LOF}}{MS_{PE}} \tag{15-39}$$

and we would reject the hypothesis of model adequacy if $F_0 > F_{\alpha, m-2, n-m}$.

This test procedure may be easily introduced into the analysis of variance conducted for the significance of regression. If the null hypothesis of model adequacy is rejected, then the model must be abandoned and attempts must be made to find a more appropriate model. If H_0 is not rejected, then there is no apparent reason to doubt the adequacy of the model, and MS_{PE} and MS_{LOF} are often combined to estimate σ^2.

Example 15-4

Suppose we have the following data.

x	1.0	1.0	2.0	3.3	3.3	4.0	4.0	4.0	4.7	5
y	2.3	1.8	2.8	1.8	3.7	2.6	2.6	2.2	3.2	2
x	5.6	5.6	5.6	6.0	6.0	6.5	6.9
y	3.5	2.8	2.1	3.4	3.2	3.4	5.0

We compute the quantities $S_{yy} = 10.97$, $S_{xy} = 13.62$, $S_{xx} = 52.53$, $\bar{y} = 2.847$, and $\bar{x} = 4.382$. The regression model is $\hat{y} = 1.708 + 0.260x$, and the regression sum of squares is $SS_R = \hat{\beta}_1 S_{xy} = (0.260)(13.62) = 3.541$. The pure-error sum of squares is computed as follows:

Level of x	$\Sigma(y_i - \bar{y})^2$	Degrees of Freedom
1.0	0.1250	1
3.3	1.8050	1
4.0	0.1066	2
5.6	0.9800	2
6.0	0.0200	1
Totals	3.0366	7

The analysis of variance is summarized in Table 15-5. Since $F_{.25,8,7} = 1.70$, we cannot reject the hypothesis that the tentative model adequately describes the data. We will pool lack-of-fit and pure-error mean squares to form the denominator mean square in the test for significance of regression. Also, since $F_{.05,1,15} = 4.54$, we must conclude that $\beta_1 \neq 0$.

∎

In fitting a regression model to experimental data, a good practice is to use the lowest-degree model that adequately describes the data. The lack-of-fit test may be useful in this respect. However, it is always possible to fit a polynomial of degree $n - 1$ to n data points, and the experimenter should not consider using a model that is "saturated," that is, which has very nearly as many regressor variables as observations on y.

15-5.3   The Coefficient of Determination

The quantity

$$R^2 = \frac{SS_R}{S_{yy}} = \frac{\sum_{j=1}^{n} (\hat{y}_j - \bar{y})^2}{\sum_{j=1}^{n} (y_j - \bar{y})^2} \tag{15-40}$$

Table 15-5   Analysis of Variance for Example 15-4

Source of Variation	Sum of Squares	Degrees of Freedom	Mean Square	F_0
Regression	3.541	1	3.541	7.15
Residual	7.429	15	0.4952	
(Lack of fit)	4.3924	8	0.5491	1.27
(Pure error)	3.0366	7	0.4338	
Total	10.970	16		

is called the coefficient of determination, and it is often used to judge the adequacy of a regression model. Clearly, $0 < R^2 \leq 1$. We often refer loosely to R^2 as the proportion of variability in the data explained or accounted for by the regression model. If the regressor x is a random variable so that y and x may be viewed as jointly distributed random variables, then R is just the simple correlation between y and x. However, if x is not a random variable, then the concept of correlation between y and x is undefined. For the data in Example 15-1, we have $R^2 = SS_R/S_{yy} = 119.26/127.73 = 0.9337$; that is, 93.37 percent of the variability in the data is accounted for the model.

The statistic R^2 should be used with caution since it is always possible to make R^2 unity by simply adding enough terms to the mode. For example, we can obtain a "perfect" fit to n data points with a polynomial of degree $n - 1$. Also, R^2 will always increase if we add a variable to the model, but this does not necessarily mean the new model is superior to the old one. Unless the error sum of squares in the new model is reduced by an amount equal to the original error mean square, the new model will have a *larger* error mean square than the old one because of the loss of one residual degree of freedom. Thus, the new model will actually be worse than the old one.

15-6 MULTIPLE LINEAR REGRESSION

Many regression problems involve more than one regressor variable. For example, the yield of a chemical process may depend on the temperature, the pressure, and the concentration of catalyst. In this case, at least three regressor variables will be necessary.

The general problem of fitting the model

$$y = \beta_0 + \beta_1 x_1 + \beta_2 x_2 + \cdots + \beta_k x_k + \epsilon \qquad (15\text{-}41)$$

is called the multiple linear regression problem. The unknown parameters $\{\beta_i\}$ are usually called regression coefficients. Note that the model (Equation 15-41) describes a hyperplane in the k-dimensional space of the regressor variables $\{x_i\}$.

The method of least squares is used to estimate the regression coefficients in Equation 15-41. Suppose that $n > k$ observations are available, and let x_{ij} denote the jth observation or level of variable x_i. The data will appear as in Table 15-6.

Table 15-6   Data For Multiple Linear Regression

y	x_1	x_2	$\cdots$	x_k
y_1	x_{11}	x_{21}	$\cdots$	x_{k1}
y_2	x_{12}	x_{22}	$\cdots$	x_{k2}
$\vdots$	$\vdots$	$\vdots$		$\vdots$
y_n	x_{1n}	x_{2n}	$\cdots$	x_{kn}

The estimation procedure requires that the random error component have $E(\epsilon) = 0$ and $V(\epsilon) = \sigma^2$ and that the $\{\epsilon\}$ are uncorrelated.

We may write the model in terms of the data as

$$
\begin{aligned}
y_j &= \beta_0 + \beta_1 x_{1j} + \beta_2 x_{2j} + \cdots + \beta_k x_{kj} + \epsilon_j \\
&= \beta_0 + \sum_{i=1}^{k} \beta_i x_{ij} + \epsilon_j \qquad j = 1, 2, \ldots, n
\end{aligned}
\tag{15-42}
$$

As in the simple linear regression case, the intercept is redefined as

$$
\beta_0' = \beta_0 + \beta_1 \bar{x}_1 + \beta_2 \bar{x}_2 + \cdots + \beta_k \bar{x}_k
\tag{15-43}
$$

where $\bar{x}_i = (1/n)\sum_{j=1}^{n} x_{ij}$ is the average level for the ith regressor variable. The model now becomes

$$
y_j = \beta_0' + \sum_{i=1}^{k} \beta_i (x_{ij} - \bar{x}_i) + \epsilon_j \qquad j = 1, 2, \ldots, n
\tag{15-44}
$$

and the least squares function is

$$
L = \sum_{j=1}^{n} \left[y_j - \beta_0' - \sum_{i=1}^{k} \beta_i (x_{ij} - \bar{x}_i) \right]^2
\tag{15-45}
$$

It will be convenient to define

$$
S_{ii} = \sum_{j=1}^{n} (x_{ij} - \bar{x}_i)^2 = \sum_{j=1}^{n} x_{ij}^2 - \frac{\left(\sum_{j=1}^{n} x_{ij} \right)^2}{n} \qquad i = 1, 2, \ldots, k
$$

$$
S_{rs} = S_{sr} = \sum_{j=1}^{n} (x_{rj} - \bar{x}_r)(x_{sj} - \bar{x}_s) = \sum_{j=1}^{n} x_{rj} x_{sj} - \frac{\left(\sum_{j=1}^{n} S_{rj} \right)\left(\sum_{j=1}^{n} x_{sj} \right)}{n} \qquad r \neq s
\tag{15-46}
$$

$$
S_{iy} = \sum_{j=1}^{n} y_j (x_{ij} - \bar{x}_i) = \sum_{j=1}^{n} y_j x_{ij} - \frac{\left(\sum_{j=1}^{n} y_j \right)\left(\sum_{j=1}^{n} x_{ij} \right)}{n} \qquad i = 1, 2, \ldots, k
$$

Note that S_{ii} is the corrected sum of squares of the ith regressor variable, S_{rs} is the corrected sum of cross-products between x_r and x_s, and S_{iy} is the corrected sum of cross-products between x_i and y.

The least squares estimators of $\beta_0', \beta_1, \ldots, \beta_k$ must satisfy

$$
\left. \frac{\partial L}{\partial \beta_0'} \right|_{\beta_0', \beta_1, \ldots, \beta_k} = -2 \left[\sum_{j=1}^{n} y_j - \hat{\beta}_0' - \sum_{u=1}^{k} \hat{\beta}_u (x_{uj} - \bar{x}_u) \right] = 0
\tag{15-47}
$$

and

$$
\left. \frac{\partial L}{\partial \beta_i} \right|_{\beta_0', \beta_1, \ldots, \beta_k} = -2 \left[\sum_{j=1}^{n} y_j - \hat{\beta}_0' - \sum_{u=1}^{k} \hat{\beta}_u (x_{uj} - \bar{x}_u) \right] (x_{ij} - \bar{x}_i)
$$

$$
= 0 \qquad i = 1, 2, \ldots, k
\tag{15-48}
$$

Simplifying Equations 15-47 and 15-48, and invoking Equations 15-46, we obtain the least squares normal equations

$$n\hat{\beta}_0' = \sum_{j=1}^{n} y_j$$

$$\hat{\beta}_1 S_{i1} + \hat{\beta}_2 S_{i2} + \cdots + \hat{\beta}_k S_{ik} = S_{iy} \qquad i = 1, 2, \ldots, k \qquad (15\text{-}49)$$

Note that there are $p = k + 1$ normal equations, one for each of the unknown regression coefficients. The solution to the normal equations will be the least squares estimators, $\hat{\beta}_0', \hat{\beta}_1, \ldots, \hat{\beta}_k$.

It is simpler to solve the normal equations if they are first expressed in matrix notation. We present a matrix development of the normal equations that parallels the development of Equation 15-49. The model in terms of the observations, Equation 15-42, may be written in matrix notation as

$$\mathbf{y} = \mathbf{X}\boldsymbol{\beta} + \boldsymbol{\epsilon}$$

where

$$\mathbf{y} = \begin{bmatrix} y_1 \\ y_2 \\ \vdots \\ y_n \end{bmatrix} \qquad \mathbf{X} = \begin{bmatrix} 1 & (x_{11} - \bar{x}_1) & (x_{21} - \bar{x}_2) & \cdots & (x_{k1} - \bar{x}_k) \\ 1 & (x_{12} - \bar{x}_1) & (x_{22} - \bar{x}_2) & \cdots & (x_{k2} - \bar{x}_k) \\ \vdots & \vdots & \vdots & & \vdots \\ 1 & (x_{1n} - \bar{x}_1) & (x_{2n} - \bar{x}_2) & \cdots & (x_{kn} - \bar{x}_k) \end{bmatrix}$$

$$\boldsymbol{\beta} = \begin{bmatrix} \beta_0' \\ \beta_1 \\ \vdots \\ \beta_k \end{bmatrix} \qquad \text{and} \qquad \boldsymbol{\epsilon} = \begin{bmatrix} \epsilon_1 \\ \epsilon_2 \\ \vdots \\ \epsilon_n \end{bmatrix}$$

In general, $\mathbf{y}$ is an $(n \times 1)$ vector of the responses, $\mathbf{X}$ is an $(n \times p)$ matrix of the levels of the regressor variables, $\boldsymbol{\beta}$ is a $(p \times 1)$ vector of the regression coefficients, and $\boldsymbol{\epsilon}$ is an $(n \times 1)$ vector of random errors.

We wish to find the vector of least squares estimators $\hat{\boldsymbol{\beta}}$ that minimizes

$$L = \sum_{j=1}^{n} \epsilon_j^2 = \boldsymbol{\epsilon}'\boldsymbol{\epsilon} = (\mathbf{y} - \mathbf{X}\boldsymbol{\beta})'(\mathbf{y} - \mathbf{X}\boldsymbol{\beta})$$

Note that L may be expressed as

$$L = \mathbf{y}'\mathbf{y} - \boldsymbol{\beta}'\mathbf{X}'\mathbf{y} - \mathbf{y}'\mathbf{X}\boldsymbol{\beta} + \boldsymbol{\beta}'\mathbf{X}'\mathbf{X}\boldsymbol{\beta} \qquad (15\text{-}50)$$
$$= \mathbf{y}'\mathbf{y} - 2\boldsymbol{\beta}'\mathbf{X}'\mathbf{y} + \boldsymbol{\beta}'\mathbf{X}'\mathbf{X}\boldsymbol{\beta}$$

since $\boldsymbol{\beta}'\mathbf{X}'\mathbf{y}$ is a (1×1) matrix, or a scalar, and its transpose $(\boldsymbol{\beta}'\mathbf{X}'\mathbf{y})' = \mathbf{y}'\mathbf{X}\boldsymbol{\beta}$ is the same scalar. The least squares estimators must satisfy

$$\frac{\partial L}{\partial \boldsymbol{\beta}}\bigg| = -2\mathbf{X}'\mathbf{y} + 2\mathbf{X}'\mathbf{X}\hat{\boldsymbol{\beta}} = \mathbf{0}$$

which simplifies to

$$\mathbf{X'X\hat{\beta}} = \mathbf{X'y} \tag{15-51}$$

Equations 15-51 are the least squares normal equations. They are identical to Equations 15-49. To solve the normal equations, multiply both sides of Equation 15-51 by the inverse of $\mathbf{X'X}$. Thus, the least squares estimator of β is

$$\hat{\beta} = (\mathbf{X'X})^{-1}\mathbf{X'y} \tag{15-52}$$

It is easy to see that the matrix form of the normal equations is identical to the scalar form. Writing out Equation 15-51 in detail we obtain, in general,

$$\begin{bmatrix} n & 0 & 0 & \cdots & 0 \\ 0 & S_{11} & S_{12} & \cdots & S_{1k} \\ 0 & S_{12} & S_{22} & \cdots & S_{2k} \\ \vdots & \vdots & \vdots & & \vdots \\ 0 & S_{1k} & S_{2k} & \cdots & S_{kk} \end{bmatrix} \begin{bmatrix} \hat{\beta}_0' \\ \hat{\beta}_1 \\ \hat{\beta}_2 \\ \vdots \\ \hat{\beta}_k \end{bmatrix} \begin{bmatrix} \sum_{j=1}^{n} y_i \\ S_{1y} \\ S_{2y} \\ \vdots \\ S_{ky} \end{bmatrix}$$

If the indicated matrix muluplication is carried out, the scalar form of the normal equations will result. In this form it is easy to see that $\mathbf{X'X}$ is a $(p \times p)$ symmetric matrix and $\mathbf{X'y}$ is a $(p \times 1)$ column vector. Note the special structure of $\mathbf{X'X}$. The diagonal elements $\mathbf{X'X}$ are the sums of squares of the columns of $\mathbf{X}$, and the off-diagonal elements are the sums of cross-products of columns of $\mathbf{X}$.

The statistical properties of least squares estimator $\hat{\beta}$ may be easily investigated. Consider first the bias.

$$\begin{aligned} E(\hat{\beta}) &= E[(\mathbf{X'X})^{-1}\mathbf{X'y}] \\ &= E[(\mathbf{X'X})^{-1}\mathbf{X'}(\mathbf{X\beta} + \epsilon)] \\ &= E[(\mathbf{X'X})^{-1}\mathbf{X'X\beta} + (\mathbf{X'X})^{-1}\mathbf{X'\epsilon}] \\ &= \beta \end{aligned}$$

since $E(\epsilon) = \mathbf{0}$ and $(\mathbf{X'X})^{-1}\mathbf{X'X} = \mathbf{I}$. Thus, $\hat{\beta}$ is an unbiased estimator of β.

The variance property of $\hat{\beta}$ is expressed in the *covariance matrix:*

$$\text{Cov}(\hat{\beta}) \equiv E\{[\hat{\beta} - E(\hat{\beta})][\hat{\beta} - E(\hat{\beta})]'\} \tag{15-53}$$

which is just a symmetric matrix whose ith main diagonal element is the variance of $\hat{\beta}_i$ and whose (ij)th element is the covariance between $\hat{\beta}_i$ and $\hat{\beta}_j$. The covariance matrix of $\hat{\beta}$ is

$$\text{Cov}(\hat{\beta}) = \sigma^2(\mathbf{X'X})^{-1} \tag{15-54}$$

Example 15-5

A beer distributor is analyzing the delivery system for the product. Specifically, the distributor is interested in predicting the amount of time required to service a retail outlet. The industrial engineer in charge of the study has suggested that the two most important factors influencing the delivery time are the number of cases of beer delivered and the maximum distance the deliveryman must travel. The engineer has collected a sample of delivery time data, which are shown in Table 15-7.

We fit the multiple linear regression model

$$y = \beta_0' + \beta_1(x_1 - \bar{x}_1) + \beta_2(x_2 - \bar{x}_2) + \epsilon$$

to these data. Since $\bar{x}_1 = 18$ and $\bar{x}_2 = 28$, we may write the model in matrix terms as

$$
\begin{array}{ccc}
\beta_0' & \beta_1 & \beta_2
\end{array}
$$

$$
\begin{bmatrix}
24 \\ 27 \\ 29 \\ 31 \\ 25 \\ 33 \\ 26 \\ 28 \\ 31 \\ 39 \\ 33 \\ 30 \\ 25 \\ 42 \\ 40
\end{bmatrix}
=
\begin{bmatrix}
1 & -8 & 2 \\
1 & -3 & -3 \\
1 & -8 & 12 \\
1 & 2 & -10 \\
1 & 7 & -6 \\
1 & 0 & 3 \\
1 & -6 & -2 \\
1 & -4 & 6 \\
1 & -2 & 1 \\
1 & 4 & 9 \\
1 & 6 & -8 \\
1 & -1 & -3 \\
1 & -5 & -1 \\
1 & 12 & -5 \\
1 & 6 & 5
\end{bmatrix}
\begin{bmatrix}
\beta_0' \\ \beta_1 \\ \beta_2
\end{bmatrix}
+
\begin{bmatrix}
\epsilon_1 \\ \epsilon_2 \\ \epsilon_3 \\ \epsilon_4 \\ \epsilon_5 \\ \epsilon_6 \\ \epsilon_7 \\ \epsilon_8 \\ \epsilon_9 \\ \epsilon_{10} \\ \epsilon_{11} \\ \epsilon_{12} \\ \epsilon_{13} \\ \epsilon_{14} \\ \epsilon_{15}
\end{bmatrix}
$$

Table 15-7 Delivery Time Data for Example 15-5

Number of Cases x_1	Distance x_2	Time (minutes) y
10	30	24
15	25	27
10	40	29
20	18	31
25	22	25
18	31	33
12	26	26
14	34	28
16	29	31
22	37	39
24	20	33
17	25	30
13	27	25
30	23	42
24	33	40

where for clarity we have labeled each column of the **X** matrix to indicate the regressor variable associated with that column. It is easy to verify that

$$\mathbf{X'X} = \begin{bmatrix} 15 & 0 & 0 \\ 0 & 504 & -213 \\ 0 & -213 & 548 \end{bmatrix} \quad \mathbf{X'y} = \begin{bmatrix} 463 \\ 345 \\ 63 \end{bmatrix}$$

and

$$(\mathbf{X'X})^{-1} = \begin{bmatrix} (1/15) & 0 & 0 \\ 0 & 0.00237411 & 0.00092278 \\ 0 & 0.00092278 & 0.00218349 \end{bmatrix}$$

Thus, the vector of the least squares estimators is

$$\hat{\boldsymbol{\beta}} = \begin{bmatrix} \hat{\beta}_0' \\ \hat{\beta}_1 \\ \hat{\beta}_2 \end{bmatrix} = (\mathbf{X'X})^{-1}\mathbf{X'y} = \begin{bmatrix} 30.866667 \\ 0.877203 \\ 0.455919 \end{bmatrix}$$

The final regression model is

$$\hat{y} = 30.86667 + 0.877203(x_1 - 18) + 0.455918(x_2 - 28)$$

Table 15-8 presents the observations, fitted values, and residuals from this model. A normal probability plot of these residuals is shown in Figure 15-7. The residuals are plotted versus the fitted values $\hat{y}_i$ in Figure 15-8 and versus the regressors x_1 and x_2 in Figures 15-9 and 15-10, respectively. The primary feature of these plots is the large residual associated with run number 5, $e_5 = -9.2716$. The error estimate for

Table 15-8 Observations, Fitted Values, and Residuals for Example 15-5

y_i	$\hat{y}_i$	$e_i = y_i - \hat{y}_i$
24	24.7609	−0.7609
27	26.8673	0.1327
29	29.3201	−0.3201
31	28.0619	2.9381
25	34.2716	−9.2716
33	32.2344	0.7656
26	24.6916	1.3084
28	30.0934	−2.0934
31	29.5682	1.4318
39	38.4788	0.5212
38	32.4825	0.5175
30	28.6217	1.3783
25	26.0247	−1.0247
42	39.1135	2.8865
40	38.4095	1.5905

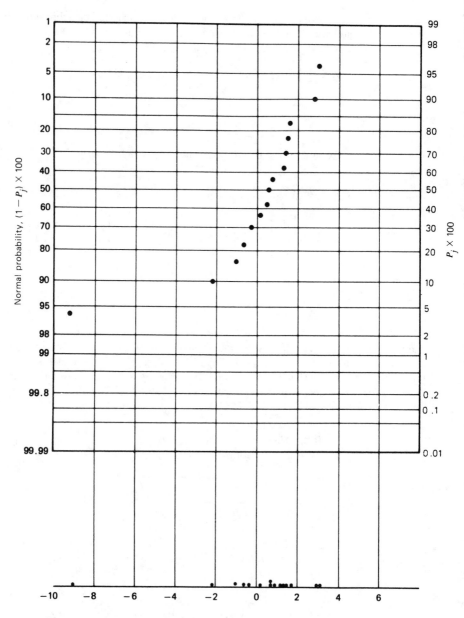

Figure 15-7. Normal probability plot of residuals for Example 15-5.

this model is $\hat{\sigma}^2 = MS_E = 9.86$ (see Example 15-7). Since the standardized value of e_5 is larger than 3, we suspect that run 5 is a potential outlier. (The residual e_5 actually contributes about 80 percent of the residual sum of squares for this model.) It is possible that a data-recording error has been made. (Perhaps $y_5 = 35$ instead of 25.) However, since there is no way to verify this and since there is no other nonstatistical basis for deleting the run, we decide to leave run 5 in the model.

∎

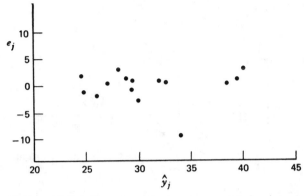

Figure 15-8. Plot of residuals e_j versus fitted values $\hat{y}_j$ for Example 15-5.

Example 15-6

Regression Analysis of a Designed Experiment

A chemical engineer is investigating the yield of a process. Three process variables are of interest: temperature, pressure, and percent concentration. Each variable can be run at a low and a high level, and the engineer decides to run a 2^3 design involving these three variables. The experiment and the resulting yields are shown in Table 15-9, where we have used the $+1$, -1 coded variable notation normally employed in 2^k factorial designs to represent the factor levels.

Suppose that the engineer decides to fit a "main effects" only model, say

$$y = \beta_0 + \beta_1 x_1 + \beta_2 x_2 + \beta_3 x_3 + \epsilon$$

For this model we have

$$\mathbf{X'X} = \begin{bmatrix} 8 & 0 & 0 & 0 \\ 0 & 8 & 0 & 0 \\ 0 & 0 & 8 & 0 \\ 0 & 0 & 0 & 8 \end{bmatrix} = 8\mathbf{I}_4 \qquad \mathbf{X'y} = \begin{bmatrix} 409 \\ 45 \\ 85 \\ 9 \end{bmatrix}$$

Since $\mathbf{X'X}$ is *diagonal*, the required inverse is just $(\mathbf{X'X})^{-1} = (1/8)\mathbf{I}_4$, and the least

Figure 15-9. Plot of residuals e_j versus number of cases x_1 for Example 15-5.

Figure 15-10. Plot of residuals e_j versus distance x_2 for Example 15-5.

squares estimates are

$$\hat{\boldsymbol{\beta}} = \begin{bmatrix} \hat{\beta}_0 \\ \hat{\beta}_1 \\ \hat{\beta}_2 \\ \hat{\beta}_3 \end{bmatrix} = \begin{bmatrix} 51\frac{1}{8} \\ 5\frac{5}{8} \\ 10\frac{5}{8} \\ 1\frac{1}{8} \end{bmatrix}$$

Thus, the fitted model is

$$\hat{y} = 51\tfrac{1}{8} + 5\tfrac{5}{8}x_1 + 10\tfrac{5}{8}x_2 + 1\tfrac{1}{8}x_3$$

These regression coefficients are closely connected to the effect estimates that would be obtained from the usual analysis of a 2^3 design. For example, the effect of temperature (x_1 in the foregoing model) is

$$T = \frac{[-32 - 36 - 57 + 46 + 65 - 57 + 48 + 68]}{4}$$

$$= \frac{45}{4}$$

$$= 11\tfrac{1}{4}$$

Notice that the regression coefficient for x_1 is

$$(11\tfrac{1}{4})/2 = 5\tfrac{5}{8}$$

That is, the regression coefficient is exactly one-half the usual effect estimate. This

Table 15-9 Data for Example 15-6

Run	x_1 (temp)	x_2 (pressure)	x_3 (% concentration)	y (% yield)
1	-1	-1	-1	32
2	-1	-1	1	36
3	-1	1	-1	57
4	1	-1	-1	46
5	1	1	-1	65
6	-1	1	1	57
7	1	-1	1	48
8	1	1	1	68

will always be true for a 2^k design in fact, we used this result in Chapters 9 through 11 to produce regression models, fitted values, and residuals for several two-level experiments. This example demonstrates that the effect estimates from a 2^k design are (apart from a constant) least squares estimators.

∎

In Example 15-6, the inverse matrix is easy to obtain because $\mathbf{X'X}$ is diagonal. Intuitively, this seems to be advantageous, not only because of the computational simplicity but also because the estimators of all the regression coefficients are uncorrelated, that is, $\text{Cov}(\hat{\beta}_i, \hat{\beta}_j) = 0$. If we can choose the levels of x_{ij} before the data are collected, we might wish to design the experiment so that a diagonal $\mathbf{X'X}$ will result.

In practice, it may be relatively easy to do this. We know that the off-diagonal elements in $\mathbf{X'X}$ are the sums of cross-products of the columns in $\mathbf{X}$. Therefore, we must make the dot product of the columns of $\mathbf{X}$ equal to zero; that is, these columns must be *orthogonal*. Experimental designs that have this property for fitting a regression model, such as Equation 15-41, are called orthogonal designs. In general, the 2^k factorial design is an orthogonal design for fittting the multiple linear regression model. Other designs for fitting polynomials are discussed in Chapter 16.

15-7 HYPOTHESIS TESTING IN MULTIPLE LINEAR REGRESSION

In multiple linear regression the experimenter frequently wishes to test hypotheses about the model parameters. This requires the additional assumption that the errors are NID$(0, \sigma^2)$. As a direct result of this assumption, the observations y_i are NID$(\beta_0 + \sum_{i=1}^{k} \beta_i x_{ij}, \sigma^2)$.

Consider testing for significance of regression. In multiple linear regression, this is accomplished by testing

$$H_0: \beta_1 = \beta_2 = \cdots = \beta_k = 0$$
$$H_1: \beta_i \neq 0 \quad \text{at least one } i \quad\quad (15\text{-}55)$$

Rejection of H_0 in Equation 15-55 implies that at least one variable in the model contributes significantly to the fit. The test procedure for Equation 15-55 is a generalization of the procedure used in simple linear regression. The total sum of squares S_{yy} is partitioned into regression and error sums of squares,

$$S_{yy} = SS_R + SS_E$$

and if $H_0: \beta_i = 0$ is true, then $SS_R/\sigma^2 \sim \chi_k^2$, where the number of degrees of freedom for χ^2 are equal to the number of regressor variables in the model. Also, we can show that $SS_E/\sigma^2 \sim \chi_{n-k-1}^2$, and SS_E and SS_R are independent. Therefore, the test procedure for $H_0: \beta_i = 0$ is to compute

$$F_0 = \frac{SS_R/k}{SS_E/(n - k - 1)} = \frac{MS_R}{MS_E} \tag{15-56}$$

and to reject H_0 if $F_0 > F_{\alpha,k,n-k-1}$. The procedure is usually summarized in an analysis of variance table such as Table 15-10.

A computational formula for the regression sum of squares SS_R may be obtained as follows:

$$SS_E = \sum_{j=1}^{n} (y_j - \hat{y}_j)^2$$

$$= (\mathbf{y} - \mathbf{X}\hat{\boldsymbol{\beta}})'(\mathbf{y} - \mathbf{X}\hat{\boldsymbol{\beta}})$$

$$= \mathbf{y'y} - \hat{\boldsymbol{\beta}}'\mathbf{X'y}$$

$$= \sum_{j=1}^{n} y_j^2 - [\hat{\beta}_0', \hat{\beta}_1, \ldots, \hat{\beta}_k] \begin{bmatrix} \sum_{j=1}^{n} y_j \\ S_{1y} \\ \vdots \\ S_{ky} \end{bmatrix}$$

$$= \sum_{j=1}^{n} y_j^2 - \frac{\left(\sum_{j=1}^{n} y_j\right)^2}{n} - \sum_{i=1}^{k} \hat{\beta}_i S_{iy}$$

$$= S_{yy} - \sum_{i=1}^{k} \hat{\beta}_i S_{iy}$$

Thus, since $S_{yy} = SS_E + SS_R$, we see that the regression sum of squares is

$$SS_R = \sum_{i=1}^{k} \hat{\beta}_i S_{iy} \tag{15-57}$$

Example 15-7

Consider the data in Example 15-5. The fitted model is $\hat{y} = 30.866667 + 0.877203(x_1 - 18) + 0.455918(x_2 - 28)$, and from $\mathbf{X'y}$ we see that $S_{1y} = 345$ and

Table 15-10 Analysis of Variance for Significance of Regression in Multiple Linear Regression

Source of Variation	Sum of Squares	Degrees of Freedom	Mean Square	F_0
Regression	SS_R	k	MS_R	MS_R/MS_E
Error or residual	SS_E	$n - k - 1$	MS_E	
Total	S_{yy}	$n - 1$		

$S_{2y} = 63$. The total sum of squares is

$$S_{yy} = \sum_{j=1}^{15} y_j^2 - \frac{\left(\sum_{j=1}^{15} y_j\right)^2}{15} = 14{,}741 - \frac{(463)^2}{15} = 449.73$$

and from Equation 15-57 the regression sum of squares is

$$SS_R = \sum_{i=1}^{2} \hat{\beta}_i S_{iy} = (0.877203)(345) + (0.455918)(63) = 331.36$$

The analysis of variance is summarized in Table 15-11. Since $F_{.01,2,12} = 6.93$, we conclude that at least one variable contributes significantly to the regression. ∎

We are frequently interested in testing hypotheses on the individual regression coefficients. Such tests would be useful in determining the value of each of the regressor variables in the model. For example, the model might be more effective with the inclusion of additional variables or perhaps with the deletion of one or more of the variables already in the model.

Adding a variable to a regression model always causes the sum of squares for regression to increase and the error sum of squares to decrease. We must decide whether the increase in the regression sum of squares is sufficient to warrant using the additional variable in the model. Furthermore, by adding an unimportant variable to the model, we can actually *increase* the mean square error, thereby decreasing the usefulness of the model.

The least squares estimator $\hat{\boldsymbol{\beta}}$ is a random variable. Also, since $\hat{\boldsymbol{\beta}}$ is a linear combination of the observations y_j, the distribution of $\hat{\boldsymbol{\beta}}$ is

$$\hat{\boldsymbol{\beta}} \sim N[\boldsymbol{\beta}, \sigma^2(\mathbf{X'X})^{-1}]$$

Thus, the variance of the regression coefficient $\hat{\beta}_i$ is σ^2 times the $(i + 1)$st diagonal element of $(\mathbf{X'X})^{-1}$, say C_{ii}. Therefore, each regression coefficient has the distributional property

$$\hat{\beta}_i \sim N(\beta_i, \sigma^2 C_{ii})$$

The hypotheses for testing the significance of any individual coefficient, for example, β_i, are

$$H_0: \beta_i = 0$$

$$H_1: \beta_i \neq 0 \qquad (15\text{-}58)$$

Table 15-11 Analysis of Variance for Significance of Regression in Example 15-7

Source of Variation	Sum of Squares	Degrees of Freedom	Mean Square	F_0
Regression	331.36	2	165.68	16.80
Error	118.37	12	9.86	
Total	449.73	14		

The appropriate test statistic for Equation 15-58 is

$$t_0 = \frac{\hat{\beta}_i}{\sqrt{MS_E C_{ii}}} \qquad (15\text{-}59)$$

and $H_0: \beta_i = 0$ is rejected if $|t_0| > t_{\alpha/2, n-k-1}$. This test procedure can be misleading, however, because the $\hat{\beta}_i$ are not usually independent. That is, there will usually be elements C_{ij} that are not zero. This implies that the $\hat{\beta}_i$ are not independent and that, consequently, the t tests in Equation 15-59 will not be independent. The results of this may be that β_i *appears* to be significant only because its estimator $\hat{\beta}_i$ is not independent of $\hat{\beta}_j$, and β_i really *is* significant.

We need a procedure whereby the contribution to the regression sum of squares for a parameter (e.g., $\hat{\beta}_i$) can be determined *given* that other parameters $\hat{\beta}_j$ $(j \neq i)$ are already in the model. That is, we wish to measure the value of adding a regressor x_i to a model that originally did not include such a term. The general regression significance test can be used to do this.

Although we have discussed the general regression significance test before, we now give a statement of the procedure in the present context. Suppose the model is $\mathbf{y} = \mathbf{X}\boldsymbol{\beta} + \boldsymbol{\epsilon}$, and the $\boldsymbol{\beta}$ vector is partitioned as follows:

$$\boldsymbol{\beta} = \begin{bmatrix} \boldsymbol{\beta}_1 \\ \boldsymbol{\beta}_2 \end{bmatrix}$$

where $\boldsymbol{\beta}_1$ is $(r \times 1)$ and $\boldsymbol{\beta}_2$ is $[(p - r) \times 1]$. We wish to test the hypotheses

$$H_0: \boldsymbol{\beta}_1 = \mathbf{0}$$
$$H_1: \boldsymbol{\beta}_1 \neq \mathbf{0} \qquad (15\text{-}60)$$

The model may be written as

$$\mathbf{y} = \mathbf{X}\boldsymbol{\beta} + \boldsymbol{\epsilon} = \mathbf{X}_1\boldsymbol{\beta}_1 + \mathbf{X}_2\boldsymbol{\beta}_2 + \boldsymbol{\epsilon} \qquad (15\text{-}61)$$

where $\mathbf{X}_1$ represents the columns of $\mathbf{X}$ associated with $\boldsymbol{\beta}_1$ and $\mathbf{X}_2$ represents the columns of $\mathbf{X}$ associated with $\boldsymbol{\beta}_2$.

For the *full* model (including both $\boldsymbol{\beta}_1$ and $\boldsymbol{\beta}_2$), we know that $\hat{\boldsymbol{\beta}} = (\mathbf{X}'\mathbf{X})^{-1}\mathbf{X}'\mathbf{y}$. Also,

$$SS_R(\boldsymbol{\beta}) = \hat{\boldsymbol{\beta}}'\mathbf{X}'\mathbf{y} \qquad (p \text{ degrees of freedom})$$

and

$$MS_E = \frac{\sum_{j=1}^{n} y_j^2 - \hat{\boldsymbol{\beta}}'\mathbf{X}'\mathbf{y}}{n - p}$$

$SS_R(\boldsymbol{\beta})$ is called the regression sum of squares *due to* $\boldsymbol{\beta}$. To find the contribution of the terms in $\boldsymbol{\beta}_1$ to the regression, we fit the model assuming the null hypothesis $H_0: \boldsymbol{\beta}_1 = \mathbf{0}$ to be true. The *reduced* model is found from Equation 15-61 as

$$\mathbf{y} = \mathbf{X}_2\boldsymbol{\beta}_2 + \boldsymbol{\epsilon}. \qquad (15\text{-}62)$$

The least squares estimator of $\boldsymbol{\beta}_2$ is $\hat{\boldsymbol{\beta}}_2 = (\mathbf{X}_2'\mathbf{X}_2)^{-1}\mathbf{X}_2'\mathbf{y}$, and

$$SS_R(\boldsymbol{\beta}_2) = \hat{\boldsymbol{\beta}}_2'\mathbf{X}_2'\mathbf{y} \quad (p - r \text{ degrees of freedom}) \tag{15-63}$$

The regression sum of squares due to $\boldsymbol{\beta}_1$ *adjusted for the presence of* $\boldsymbol{\beta}_2$ *already in the model* is

$$SS_R(\boldsymbol{\beta}_1|\boldsymbol{\beta}_2) = SS_R(\boldsymbol{\beta}) - SS_R(\boldsymbol{\beta}_2) \tag{15-64}$$

This sum of squares has r degrees of freedom. It is sometimes called the "extra sum of squares" due to $\boldsymbol{\beta}_1$. Now $SS_R(\boldsymbol{\beta}_1|\boldsymbol{\beta}_2)$ is independent of MS_E and the null hypothesis $\boldsymbol{\beta}_1 = \mathbf{0}$ may be tested by the statistic

$$F_0 = \frac{SS_R(\boldsymbol{\beta}_1|\boldsymbol{\beta}_2)/r}{MS_E} \tag{15-65}$$

If $F_0 > F_{\alpha,r,n-p}$, we reject H_0 and conlcude that at least one of the parameters in $\boldsymbol{\beta}_1$ is not zero.

This procedure is extremely useful. If we have several variables in a regression model, we can measure the value of β_i as though it were added to the model last by computing

$$SS_R(\beta_i|\beta_0,\beta_1, \ldots, \beta_{i-1},\beta_{i+1}, \ldots, \beta_k) \quad i = 1, 2, \ldots, k$$

Example 15-8

We use the data in Example 15-5 to illustrate the general regression significance test. The regression model is

$$\hat{y} = 30.86667 + 0.877203(x_1 - 18) + 0.455918(x_2 - 28)$$

We would like to test the hypothesis $H_0: \beta_1 = 0$ against $H_1: \beta_1 \neq 0$. In the general regression significance test notation, this implies that

$$\boldsymbol{\beta} = \begin{bmatrix} \boldsymbol{\beta}_1 \\ \boldsymbol{\beta}_2 \end{bmatrix} = \begin{bmatrix} \beta_1 \\ \beta_0' \\ \beta_2 \end{bmatrix}$$

Note that $\boldsymbol{\beta}_1 = \beta_1$ is a (1×1) vector (thus $r = 1$) and $\boldsymbol{\beta}_2$ is a (2×1) vector. For the full model, the regression sum of squares is

$$SS_R(\boldsymbol{\beta}) = SS_R(\beta_0', \beta_1, \beta_2) = \hat{\boldsymbol{\beta}}'\mathbf{X}'\mathbf{y}$$

$$= \hat{\beta}_0' \sum_{j=1}^{15} y_j + \sum_{i=1}^{2} \hat{\beta}_i S_{iy}$$

$$= 14,622.63$$

Note that this is not the same regression sum of squares computed in Example 15-7 because $SS_R(\boldsymbol{\beta})$ includes a contribution to regression due to the intercept. Also, $SS_R(\boldsymbol{\beta})$ has $p = 3$ degrees of freedom. The error mean square, with $n - p = 15 - 3 = 12$ degrees of freedom, is

$$MS_E = \frac{\sum_{j=1}^{15} y_j^2 - \hat{\boldsymbol{\beta}}\mathbf{X}'\mathbf{y}}{n - p} = \frac{14,741 - 14,622.63}{15 - 3} = 9.86$$

To test H_0: $\beta_1 = 0$, we must find $SS_R(\beta_1|\beta_2) = SS_R(\beta_1|\beta_0', \beta_2)$. This is obtained by fitting the reduced model $y = \beta_0' + \beta_2(x_2 - \bar{x}) + \epsilon$. We find that $\hat{y} = 30.866667 + 0.114964(x_2 - 28)$ and, as a result,

$$SS_R(\beta_2) = \hat{\beta}_2' X_2' y$$

$$= \hat{\beta}_0' \sum_{j=1}^{15} y_j + \hat{\beta}_2 S_{2y}$$

$$= 14{,}298.51$$

with $p - r = 2$ degrees of freedom. Thus, the regression sum of squares for β_1 *adjusted* for β_2 is

$$SS_R(\beta_1|\beta_2) = SS_R(\beta_1|\beta_0', \beta_2)$$

$$= SS_R(\beta) - SS_R(\beta_2)$$

$$= 14{,}622.63 - 14{,}298.51 = 324.12$$

with $r = 1$ degree of freedom. From Equation 15-65, the test statistic is

$$F_0 = \frac{SS_R(\beta_1|\beta_0', \beta_2)/1}{MS_E} = \frac{324.12}{9.86} = 32.87$$

and since $F_{.01,1,12} = 9.33$, we conclude that $\beta_1 \neq 0$. Therefore, x_1 contributes significantly to the model.

∎

15-8   OTHER LINEAR REGRESSION MODELS

The linear model $y = X\beta + \epsilon$ is a general model. It can be used to fit any relationship that is *linear* in the unknown parameters β. An example would be the kth degree polynomial in one variable

$$y = \beta_0 + \beta_1 x + \beta_2 x^2 + \cdots + \beta_k x^k + \epsilon$$

which we used previously in estimating the polynomial effects of a quantitative factor. Other examples include the second-degree polynomial in two variables

$$y = \beta_0 + \beta_1 x_1 + \beta_2 x_2 + \beta_{11} x_1^2 + \beta_{22} x_2^2 + \beta_{12} x_1 x_2 + \epsilon$$

and the trignometric model

$$y = \beta_0 + \beta_1 \sin x + \beta_2 \cos x + \epsilon$$

Use of the *general linear model* $y = X\beta + \epsilon$ to fit models such as these is illustrated in the following example.

Example 15-9

A Quadratic Polynomial

Suppose we have the following data and we wish to fit the model $y = \beta_0 + \beta_1 x + \beta_2 x^2 + \epsilon$.

x	1.0	1.2	1.4	1.6	1.8	2.0
y	6.15	7.90	9.40	10.50	11.00	14.00

In matrix notation the model is

$$
y = \begin{bmatrix} 6.15 \\ 7.90 \\ 9.40 \\ 10.50 \\ 11.00 \\ 14.00 \end{bmatrix} \qquad
\begin{matrix} \beta_0 & \beta_1 & \beta_2 \end{matrix} \\
X = \begin{bmatrix} 1 & 1.0 & 1.00 \\ 1 & 1.2 & 1.44 \\ 1 & 1.4 & 1.96 \\ 1 & 1.6 & 2.56 \\ 1 & 1.8 & 3.24 \\ 1 & 2.0 & 4.00 \end{bmatrix} \qquad
\beta = \begin{bmatrix} \beta_0 \\ \beta_1 \\ \beta_2 \end{bmatrix}
$$

The least squares estimator of β is

$$
\hat{\beta} = (X'X)^{-1}X'y = \begin{bmatrix} 6.0000 & 9.000 & 14.2000 \\ 9.0000 & 14.2000 & 23.4000 \\ 14.2000 & 23.4000 & 39.9664 \end{bmatrix}^{-1} \begin{bmatrix} 58.95 \\ 93.39 \\ 154.47 \end{bmatrix}
$$

or

$$
\hat{\beta} = \begin{bmatrix} 79.5702 & -109.2841 & 35.7137 \\ -109.2841 & 152.0959 & -50.2225 \\ 35.7137 & -50.2225 & 16.7408 \end{bmatrix} \begin{bmatrix} 58.95 \\ 93.39 \\ 154.47 \end{bmatrix} = \begin{bmatrix} 1.3284 \\ 4.0798 \\ 1.0044 \end{bmatrix}
$$

Therefore, the regression model would be

$$
\hat{y} = 1.3284 + 4.0798x + 1.0044x^2
$$

∎

This general method can be employed if one wishes to fit a response curve to data arising from a designed experiment with unequal spacing of the design variables. If the levels of the design variables are equally spaced, then the use of orthogonal polynomials greatly simplifies the computational effort, as illustrated in Sections 4-3 and 7-6. A derivation of the formulas used in those sections is now given.

Orthogonal Polynomials ▪ In Chapters 4 and 7 we illustrated the use of orthogonal polynomials to compute the polynomial effects of quantitative factors in an experimental design. The procedure involves fitting the model

$$
y_j = \alpha_0 + \alpha_1 P_1(x_j) + \alpha_2 P_2(x_j) + \cdots + \alpha_k P_k(x_j) + \epsilon_j
$$

where $P_u(x_j)$ is a uth-order orthogonal polynomial; that is, $P_0(x_j) = 1$, $\sum_{j=1}^{n} P_u(x_j) = 0$, and $\sum_{j=1}^{n} P_u(x_j)P_s(x_j) = 0$ for $u \neq s$. The first five orthogonal polynomials are given in Section 4-3. The least-squares normal equations for this model are

$$
\begin{bmatrix}
n & 0 & 0 & \cdots & 0 \\
0 & \sum_{j=1}^{n} \{P_1(x_j)\}^2 & 0 & \cdots & 0 \\
0 & 0 & \sum_{j=1}^{n} \{P_2(x_j)\}^2 & & \\
\vdots & & & \ddots & \vdots \\
0 & & & & \sum_{j=1}^{n} \{P_k(x_j)\}^2
\end{bmatrix}
\begin{bmatrix}
\hat{\alpha}_0 \\
\hat{\alpha}_1 \\
\vdots \\
\hat{\alpha}_k
\end{bmatrix}
$$

$$
=
\begin{bmatrix}
\sum_{j=1}^{n} y_i \\
\sum_{j=1}^{n} y_i P_1(x_i) \\
\vdots \\
\sum_{j=1}^{n} y_i P_k(x_i)
\end{bmatrix}
$$

Note that all off-diagonal elements in the **X'X** matrix are zero because of the orthogonality property of the polynomials $P_u(x_j)$. Therefore, the least squares estimators of the model parameters are

$$\hat{\alpha}_0 = \bar{y}$$

$$\hat{\alpha}_i = \frac{\sum_{j=1}^{n} y_j P_i(x_j)}{\sum_{j=1}^{n} \{P_i(x_j)\}^2} \qquad i = 1, 2, \ldots, k$$

and the regression sum of squares for any model parameter adjusted for the presence of other model parameters is

$$SS_R(\alpha_i) = \hat{\alpha}_i \sum_{j=1}^{n} y_j P_i(x_j) = \frac{\left\{ \sum_{j=1}^{n} y_j P_i(x_j) \right\}^2}{\sum_{j=1}^{n} \{P_i(x_j)\}^2} \qquad i = 1, 2, \ldots, k$$

with $SS_R(\alpha_0) = (\sum_{j=1}^{n} y_j)^2/n$. These results were used previously in Section 4-3.

15-9 SAMPLE COMPUTER PRINTOUT

Computer programs are used extensively in the application of regression analysis. The basic output from one such program, the SAS procedure PROC REG, is shown in Figure 15-11 for the delivery time data in Example 15-5. The numerical results agree closely with those presented previously, with discrepancies attributable to round-off errors. In running this program, we used the centered x's [$(x_{1j} - \bar{x}_1)$ and $(x_{2j} - \bar{x}_2)$] as input data so that the program would calculate the transformed intercept $\hat{\beta}_0' = 30.866667 = \bar{y}$. If the original x_1 and x_2 values had been entered, the program would have calculated the original intercept $\hat{\beta}_0$.

Most of the information displayed in the computer output is self-explanatory. The R^2 statistic is computed as in simple linear regression; that is,

$$R^2 = \frac{SS_R}{S_{yy}} = \frac{331.359}{449.733} = 0.7368$$

Roughly speaking, about 73.68 percent of the variability in delivery time is explained by the linear model in the two regressors x_1 = cases and x_2 = distance. The adjusted R^2 is defined as

$$R^2_{Adj} = 1 - \left(\frac{n-1}{n-p}\right)(1 - R^2)$$
$$= 1 - \left(\frac{15-1}{15-3}\right)(1 - 0.7368)$$
$$= 0.6929$$

The advantage of the adjusted R^2 is that it does not automatically increase as new regressors are inserted into the model. For further reading, see Montgomery and Peck (1982, p. 251).

Modern regression computer programs such as PROC REG will also produce the residual plots displayed earlier and will compute many other valuable model diagnostics. The reader is strongly encouraged to become familiar with and use these procedures when fitting regression models.

15-10 PROBLEMS

15-1 The tensile strength of a paper product is related to the amount of hardwood in the pulp. Ten samples are produced in the pilot plant, and the data obtained are shown in the following table. Fit a simple linear regression model expressing strength as a function of hardwood concentration using these data.

Strength	Percent Hardwood	Strength	Percent Hardwood
160	10	181	20
171	15	188	25
175	15	193	25
182	20	195	28
184	20	200	30

DEP VARIABLE: TIME

SOURCE	DF	SUM OF SQUARES	MEAN SQUARE	F VALUE	PROB > F
MODEL	2	331.359	165.679	16.795	0.0001
ERROR	12	118.375	9.864561		
C TOTAL	14	119.733			

ROOT MSE	3.140790	R-SQUARE	0.7368	
DEP MEAN	30.866667	ADJ R-SQ	0.6929	
C.V.	10.17535			

VARIABLE	DF	PARAMETER ESTIMATE	STANDARD ERROR	T FOR H0: PARAMETER = 0	PROB > |T|
INTERCEP	1	30.866667	0.810948	38.062	0.0001
CASES	1	0.877205	0.153035	5.732	0.0001
DIST	1	0.455921	0.146762	3.107	0.0091

OBS	ACTUAL	PREDICT VALUE	RESIDUAL
1	24.000	24.761	− .760871
2	27.000	26.867	0.132709
3	29.000	29.320	− .320079
4	31.000	28.062	2.938
5	25.000	34.272	− 9.272
6	33.000	32.234	0.765571
7	26.000	24.692	1.308
8	28.000	30.093	− 2.093
9	31.000	29.568	1.432
10	39.000	38.479	0.521228
11	33.000	32.483	0.517472
12	30.000	28.622	1.378
13	25.000	26.025	− 1.025
14	42.000	39.114	2.886
15	40.000	38.409	1.591

SUM OF RESIDUALS 9.59233E-14
SUM OF SQUARED RESIDUALS 118.3747

Figure 15-11. Sample computer output for Example 15-5.

15-2 Using the results of Problem 15-1, test the regression model for lack of fit and significance of regression.

15-3 Using the results of Problem 15-1, construct 95 percent confidence intervals on the slope and intercept.

15-4 A plant distills liquid air to produce oxygen, nitrogen, and argon. The percentage of impurity in the oxygen is thought to be linearly related to the amount of impurities in the air as measured by the "pollution count" in parts per million (ppm). Fit a linear regression model to the following data:

Purity (%)	93.3	92.0	92.4	91.7	94.0	94.6	93.6
Pollution count (ppm)	1.10	1.45	1.36	1.59	1.08	0.75	1.20
	93.1	93.2	92.9	92.2	91.3	90.1	91.6	91.9
	0.99	0.83	1.22	1.47	1.81	2.03	1.75	1.68

15-5 Using the results of Problem 15-4, does a linear relationship between purity and the pollution count seem reasonable?

15-6 Construct 95 percent confidence intervals on the slope and intercept of the linear regression model in Problem 15-4.

15-7 The yield of a chemical process is related to the concentration of the reactant and the operating temperature. Fit a multiple linear regression model to the following data:

Yield	Concentration	Temperature
81	1.00	150
89	1.00	180
83	2.00	150
91	2.00	180
79	1.00	150
87	1.00	180
84	2.00	150
90	2.00	180

15-8 Test the multiple linear regression model in Problem 15-7 for significance of regression.

15-9 Suppose that in Problem 15-7 we had the additional data points shown in the following table. How do these additional observation change the estimates of the regression coefficients found in Problem 15-7. Test the new model for lack of fit and significance of regression.

Yield	Concentration	Temperature
86	1.50	165
85	1.50	165
87	1.50	165
86	1.50	165
85	1.50	165
85	1.50	165

15-10   The brake horsepower developed by an automobile engine on a dynomometer is thought to be a function of the engine speed in revolutions per minute (rpm), the road octane number of the fuel, and the engine compression. An experiment is run in the laboratory and the data that follow are collected. Fit a multiple linear regression model to these data.

Brake Horsepower	rpm	Road Octane Number	Compression
225	2000	90	100
212	1800	94	95
229	2400	88	110
222	1900	91	96
219	1600	86	100
278	2500	96	110
246	3000	94	98
237	3200	90	100
233	2800	88	105
224	3400	86	97
223	1800	90	100
230	2500	89	104

15-11   Test the model in Problem 15-10 for significance of regression. Use the general regression significance test to determine if octane number is an important variable in this model.

15-12   Given the following data, fit the second-order polynomial regression model

$$\hat{y} = \hat{\beta}_0 + \hat{\beta}_1 x_1 + \hat{\beta}_2 x_2 + \hat{\beta}_{11} x_1^2 + \hat{\beta}_{22} x_2^2 + \hat{\beta}_{12} x_1 x_2$$

y	x_1	x_2
26	1.0	1.0
24	1.0	1.0
175	1.5	4.0
160	1.5	4.0
163	1.5	4.0
55	0.5	2.0

62	1.5	2.0
100	0.5	3.0
26	1.0	1.5
30	0.5	1.5
70	1.0	2.5
71	0.5	2.5

15-13 Fit a quadratic model to the data that follow using orthogonal polynomials.

y	10.0	9.7	8.5	6.8	4.5	3.6	−1.7	−5.4	−9.8
x	1.0	1.5	2.0	2.5	3.0	3.5	4.0	4.5	5.0

15-14 Prove that in simple linear regression $Cov(\hat{\beta}_0', \hat{\beta}_1) = 0$.

15-15 Verify the bias and variance properties of $\hat{\beta}_0'$ and $\hat{\beta}_0$ given in Equations 15-16 and 15-17. Show that $Cov(\hat{\beta}_0, \hat{\beta}_1) = -\sigma^2 \bar{x}/S_{xx}$.

15-16 Derive the expected value of the mean square regression for the simple linear regression model.

15-17 Find the residuals for Problem 15-1. Analyze them by the methods discussed in Section 15-5.

15-18 *Relationship Between Analysis of Variance and Regression.* Any analysis of variance model can be expressed in terms of the general linear model $\mathbf{y} = \mathbf{X}\boldsymbol{\beta} + \boldsymbol{\epsilon}$, where the $\mathbf{X}$ matrix consists of zeros and ones. Show that the single-factor model $y_{ij} = \mu + \tau_i + \varepsilon_{ij}$, $i = 1, 2, 3$, $j = 1, 2, 3, 4$ can be written in general linear model form. Then

(a) Write the normal equations $(\mathbf{X'X})\hat{\boldsymbol{\beta}} = \mathbf{X'y}$, and compare them with the normal equations found for this model in Chapter 3, Section 3-3.3.

(b) Find the rank of $\mathbf{X'X}$. Can $(\mathbf{X'X})^{-1}$ be obtained?

(c) Suppose the first normal equation is deleted and the restriction $\sum_{i=1}^{3} n\hat{\tau}_i = 0$ is added. Can the resulting system of equations be solved? If so, find the solution. Find the regression sum of squares $\hat{\boldsymbol{\beta}}'\mathbf{X'y}$, and compare it to the treatment sum of squares in the single-factor model.

15-19 Suppose that we are fitting a straight line and we desire to make the variance of $\hat{\beta}_1$ as small as possible. Restricting ourselves to an even number of experimental points, where should we place these points so as to minimize $V(\hat{\beta}_1)$? (Note: Use the design called for in this exercise with *great* caution because, even though it minimizes $V(\hat{\beta}_1)$, it has some undesirable properties; for example, see Myers (1971). Only if you are *very sure* the true functional relationship is linear should you consider using this design.)

15-20 *Weighted Least Squares.* Suppose that we are fitting the straight line $y = \beta_0 + \beta_1 x + \varepsilon$, but the variance of the y's now depends on the level of x; that is,

$$V(y|x_i) = \sigma_i^2 = \frac{\sigma^2}{w_i} \qquad i = 1, 2, \ldots, n$$

where the w_i are known constants, often called weights. Show that the resulting least squares normal equations are

$$\hat{\beta}_0 \sum_{i=1}^{n} w_i + \hat{\beta}_1 \sum_{i=1}^{n} w_i x_i = \sum_{i=1}^{n} w_i y_i$$

$$\hat{\beta}_0 \sum_{i=1}^{n} w_i x_i + \hat{\beta}_1 \sum_{i=1}^{n} w_i x_i^2 = \sum_{i=1}^{n} w_i x_i y_i$$

15-21   ***Nonlinear Regression.*** This chapter showed that polynomial models could be fit by multiple linear regression techniques. Actually, any model that is linear in the unknown parameters can be treated in such a fashion. Regression models that are not linear in the parameters are called *nonlinear* models and cannot be handled by the methods discussed in this chapter. Sometimes models that appear nonlinear can be made linear by the use of some appropriate transformation. Could you fit the model

$$y = \beta_0 e^{\beta_1 x} \varepsilon$$

by the methods discussed in this chapter? Suppose the model were

$$y = \beta_0 e^{\beta_1 x} + \varepsilon$$

instead. Obtain the least squares normal equations for this latter case and try to solve them.

Chapter 16
Response Surface
Methods and Designs

16-1 INTRODUCTION TO RESPONSE SURFACE METHODOLOGY

Response surface methodology, or RSM, is a collection of mathematical and statistical techniques that are useful for the modeling and analysis of problems in which a response of interest is influenced by several variables and the objective is to optimize this response. For example, suppose that a chemical engineer wishes to find the levels of temperature (x_1) and pressure (x_2) that maximize the yield (y) of a process. The process yield is a function of the levels of temperature and pressure, say

$$y = f(x_1, x_2) + \epsilon$$

where ϵ represents the noise or error observed in the response y. If we denote the expected response by $E(y) = f(x_1, x_2) = \eta$, then the surface represented by

$$\eta = f(x_1, x_2)$$

is called a *response surface*.

We may represent the response surface graphically as shown in Figure 16-1, where η is plotted versus the levels of x_1 and x_2. Notice that the response is represented as a solid surface in a three-dimensional space. To help visualize the shape of a response surface, we often plot the contours of the response surface as shown in Figure 16-2. In the contour plot, lines of constant response are drawn in the x_1, x_2 plane. Each contour corresponds to a particular height of the response surface. Such a plot is helpful in studying the levels of x_1 and x_2 that result in changes in the shape or height of the response surface.

In most RSM problems, the form of the relationship between the response and the independent variables is unknown. Thus, the first step in RSM is to find a suitable approximation for the true functional relationship between y and the set of independent variables. Usually, a low-order polynomial in some region of the independent variables is employed. If the response is well modeled by a

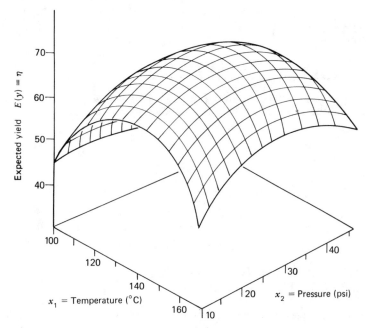

Figure 16-1.   A three-dimensional response surface showing the expected yield (η) as a function of temperature (x_1) and pressure (x_2).

Figure 16-2.   A contour plot of a response surface.

linear function of the independent variables, then the approximating function is the first-order model

$$y = \beta_0 + \beta_1 x_1 + \beta_2 x_2 + \cdots + \beta_k x_k + \epsilon \qquad (16\text{-}1)$$

If there is curvature in the system, then a polynomial of higher degree must be used, such as the second-order model

$$y = \beta_0 + \sum_{i=1}^{k} \beta_i x_i + \sum_{i=1}^{k} \beta_{ii} x_i^2 + \sum_{i}\sum_{j}\beta_{ij} x_i x_j + \epsilon \qquad (16\text{-}2)$$
$$\scriptstyle i<j$$

Almost all RSM problems utilize one or both of these approximating polynomials. Of course, it is unlikely that a polynomial model will be a reasonable approximation of the true functional relationship over the entire space of the independent variables, but for a relatively small region they usually work quite well.

The method of least squares, discussed in Chapter 15, is used to estimate the parameters in the approximating polynomials. The response surface analysis is then done in terms of the fitted surface. If the fitted surface is an adequate approximation of the true response function, then analysis of the fitted surface will be approximately equivalent to analysis of the actual system. The model parameters can be estimated most effectively if proper experimental designs are used to collect the data. Designs for fitting response surfaces are called *response surface designs*. These designs are dicsussed in Section 16-4.

RSM is a *sequential* procedure. Often, when we are at a point on the response surface that is remote from the optimum, such as the current operating conditions in Figure 16-2, there is little curvature in the system and the first-order model will be appropriate. Our objective here is to lead the experimenter rapidly and efficiently to the general vicinity of the optimum. Once the region of the optimum has been found, a more elaborate model such, as the second-order model, may be employed, and an analysis may be performed to locate the optimum. From Figure 16-2, we see that the analysis of a response surface can be thought of as "climbing a hill," where the top of the hill represents the point of maximum response. If the true optimum is a point of minimum response, then we may think of "descending into a valley."

The eventual objective of RSM is to determine the optimum operating conditions for the system or to determine a region of the factor space in which operating specifications are satisfied. RSM is not used primarily to gain understanding of the *physical mechanism* of the system, although RSM may assist in the gaining of such knowledge. Furthermore, note that "optimum" in RSM is used in a special sense. The "hill climbing" procedures of RSM guarantee convergence to a local optimum only.

16-2 THE METHOD OF STEEPEST ASCENT

Frequently, the initial estimate of the optimum operating conditions for the system will be far from the actual optimum. In such circumstances, the objective of the experimenter is to move rapidly to the general vicinity of the optimum. We wish to use a simple and economically efficient experimental procedure. When we are remote from the optimum, we usually assume that a first-order model is an adequate approximation to the true surface in a small region of the x's.

The *method of steepest ascent* is a procedure for moving sequentially along the path of steepest ascent, that is, in the direction of the maximum increase in the response. Of course, if minimization is desired, then we are talking about the method of steepest *descent*. The fitted first-order model is

$$\hat{y} = \hat{\beta}_0 + \sum_{i=1}^{k} \hat{\beta}_i x_i \qquad (16\text{-}3)$$

and the first-order response surface, that is, the contours of $\hat{y}$, is a series of parallel lines such as that shown in Figure 16-3. The direction of steepest ascent is the direction in which $\hat{y}$ increases most rapidly. This direction is parallel to the normal to the fitted response surface. We usually take as the *path of steepest ascent* the line through the center of the region of interest and normal to the fitted surface. Thus, the steps along the path are proportional to the regression coefficients $\{\hat{\beta}_i\}$. The actual step size is determined by the experimenter based on process knowledge or other practical considerations.

Experiments are conducted along the path of steepest ascent until no further increase in response is observed. Then a new first-order model may be fit, a new

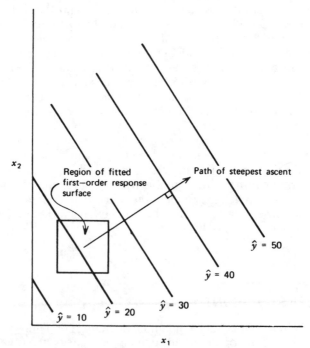

Figure 16-3. First-order response surface and path of steepest ascent.

path of steepest ascent determined, and the procedure continued. Eventually, the experimenter will arrive in the vicinity of the optimum. This is usually indicated by lack of fit of a first-order model. At that time additional experiments, which will be described in Section 16-3, are conducted to obtain a more precise estimate of the optimum.

Example 16-1

A chemical engineer is interested in determining the operating conditions that maximize the yield of a process. Two controllable variables influence process yield: reaction time and reaction temperature. The engineer is currently operating the process with a reaction time of 35 minutes and a temperature of 155°F, which result in yields of around 40 percent. Since it is unlikely that this region contains the optimum, she fits a first-order model and applies the method of steepest ascent.

The engineer decides that the region of exploration for fitting the first-order model should be (30, 40) minutes of reaction time and (150, 160)°F. To simplify the calculations, the independent variables will be coded to a $(-1, 1)$ interval. Thus, if ξ_1 denotes the *natural* variable time and ξ_2 denotes the *natural* variable temperature, then the *coded* variables are

$$x_1 = \frac{\xi_1 - 35}{5} \quad \text{and} \quad x_2 = \frac{\xi_2 - 155}{5}$$

The data are shown in Table 16-1. Note that the design used to collect this data is a 2^2 factorial augmented by five center points. Repeat observations at the center are used to estimate the experimental error and to allow for checking the adequacy of the first-order model. Also, the design is centered about the current operating conditions for the process.

A first-order model may be fit to these data by least squares. Following the methods of Chapter 15, we obtain the following model in the coded variables.

$$\hat{y} = 40.44 + 0.775x_1 + 0.325x_2$$

Before exploring along the path of steepest ascent, the adequacy of the first-order

Table 16-1 Process Data for Fitting the First-Order Model

Natural Variables		Coded Variables		Response
ξ_1	ξ_2	x_1	x_2	y
30	150	-1	-1	39.3
30	160	-1	1	40.0
40	150	1	-1	40.9
40	160	1	1	41.5
35	155	0	0	40.3
35	155	0	0	40.5
35	155	0	0	40.7
35	155	0	0	40.2
35	155	0	0	40.6

model should be investigated. The 2^2 design with center points allows the experimenter to

1. Obtain an estimate of error.
2. Check for interactions (cross-product terms) in the model.
3. Check for quadratic effects (curvature).

The repeat observations at the center can be used to calculate an estimate of error as follows:

$$\hat{\sigma}^2 = \frac{(40.3)^2 + (40.5)^2 + (40.7)^2 + (40.2)^2 + (40.6)^2 - (202.3)^2/5}{4}$$

$$= 0.0430$$

The first-order model assumes that the variables x_1 and x_2 have an additive effect on the response. Interaction between the variables would be measured by the coefficient β_{12} of a cross-product term x_1x_2 added to the model. The least squares estimate of this coefficient is just one-half the interaction effect calculated as in an ordinary 2^2 factorial design or

$$\hat{\beta}_{12} = \tfrac{1}{4} [(1 \times 39.3) + (1 \times 41.5) + (-1 \times 40.0) + (-1 \times 40.9)]$$

$$= \tfrac{1}{4}(-0.1)$$

$$= -0.025$$

The single degree of freedom sum of squares for interaction is

$$SS_{\text{Interaction}} = \frac{(-0.1)^2}{4}$$

$$= 0.0025$$

Comparing $SS_{\text{Interaction}}$ to $\hat{\sigma}^2$ gives the following lack-of-fit statistic:

$$F = \frac{SS_{\text{Interaction}}}{\hat{\sigma}^2}$$

$$= \frac{0.0025}{0.0430}$$

$$= 0.058$$

which would be compared to $F_{\alpha,1,4}$. Clearly interaction is nonsignificant.

Another check of the adequacy of the straight-line model is obtained by comparing the average response at the four points in the factorial portion of the design, say $\bar{y}_f = 40.425$, with the average response at the design center, say $\bar{y}_c = 40.46$. If the design is sitting on a curved surface, then $\bar{y}_f - \bar{y}_c$ is a measure of the overall curvature in the surface. If β_{11} and β_{22} are the coefficients of the "pure quadratic" terms x_1^2 and x_2^2, then $\bar{y}_f - \bar{y}_c$ is an estimate of $\beta_{11} + \beta_{22}$. In our example, an estimate of the pure quadratic term is

$$\hat{\beta}_{11} + \hat{\beta}_{22} = \bar{y}_f - \bar{y}_c$$

$$= 40.425 - 40.46$$

$$= -0.035$$

The single-degree-of-freedom sum of squares associated with the null hypothesis, H_0: $\beta_{11} + \beta_{22} = 0$, is

$$SS_{\text{Pure Quadratic}} = \frac{n_f n_c (\bar{y}_f - \bar{y}_c)^2}{n_f + n_c}$$

$$= \frac{(4)(5)(-0.035)^2}{4 + 5}$$

$$= 0.0027$$

where n_f and n_c are the number of points in the factorial portion and the number of center points, respectively. Since

$$F = \frac{SS_{\text{Pure Quadratic}}}{\hat{\sigma}^2}$$

$$= \frac{0.0027}{0.0430}$$

$$= 0.063$$

which would be compared to $F_{\alpha,1,4}$. There is no indication of a pure quadratic affect.

The analysis of variance for this model is summarized in Table 16-2. Both the interaction and curvature checks are not significant, whereas the F test for the overall regression is significant. Furthermore, the standard error of $\hat{\beta}_1$ and $\hat{\beta}_2$ is

$$s.e.(\hat{\beta}_i) = \sqrt{\frac{MS_E}{4}} = \sqrt{\frac{\hat{\sigma}^2}{4}} = \sqrt{\frac{0.0430}{4}} = 0.10 \qquad i = 1, 2$$

Both regression coefficients $\hat{\beta}_1$ and $\hat{\beta}_2$ are large relative to their standard errors. At this point, we have no reason to question the adequacy of the first-order model.

To move away from the design center—the point ($x_1 = 0$, $x_2 = 0$)—along the path of steepest ascent, we would move 0.775 units in the x_1 direction for every 0.325 units in the x_2 direction. Thus, the path of steepest ascent passes through the point ($x_1 = 0$, $x_2 = 0$) and has a slope 0.325/0.775. The engineer decides to use 5 minutes of reaction time as the basic step size. Using the relationship between ξ_1 and x_1, we see that 5 minutes of reaction time is equivalent to a step in the *coded* variable x_1 of $\Delta x_1 = 1$. Therefore, the steps along the path of steepest ascent are $\Delta x_1 = 1.0000$ and $\Delta x_2 = (0.325/0.775)\Delta x_1 = 0.42$.

Table 16-2   Analysis of Variance for the First-Order Model

Source of Variation	Sum of Squares	Degrees of Freedom	Mean Square	F_0
Regression (β_1, β_2)	2.8250	2	1.4125	47.83[a]
Residual	0.1772	6		
(Interaction)	(0.0025)	1	0.0025	0.058
(Pure quadratic)	(0.0027)	1	0.0027	0.063
(Pure error)	(0.1720)	4	0.0430	
Total	3.0022	8		

[a] Significant at 1 percent.

Table 16-3 Steepest Ascent Experiment for Example 16-1

Steps	Coded Variables		Natural Variables		Response
	x_1	x_2	ξ_1	ξ_2	y
Origin	0	0	35	155	
Δ	1.00	0.42	5	2	
Origin + Δ	1.00	0.42	40	157	41.0
Origin + 2Δ	2.00	0.84	45	159	42.9
Origin + 3Δ	3.00	1.26	50	161	47.1
Origin + 4Δ	4.00	1.68	55	163	49.7
Origin + 5Δ	5.00	2.10	60	165	53.8
Origin + 6Δ	6.00	2.52	65	167	59.9
Origin + 7Δ	7.00	2.94	70	169	65.0
Origin + 8Δ	8.00	3.36	75	171	70.4
Origin + 9Δ	9.00	3.78	80	173	77.6
Origin + 10Δ	10.00	4.20	85	175	80.3
Origin + 11Δ	11.00	4.62	90	179	76.2
Origin + 12Δ	12.00	5.04	95	181	75.1

The engineer computes points along this path and observes the yields at these points until a decrease in response is noted. The results are shown in Table 16-3 in both coded and natural variables. Although the coded variables are easier to manipulate mathematically, the natural variables must be used in running the process. Figure 16-4 plots the yield at each step along the path of steepest ascent. Increases in response are observed through the tenth step; however, all steps beyond this point

Figure 16-4. Yield versus steps along the path of steepest ascent for Example 16-1.

Table 16-4   Data for Second First-Order Model

Natural Variables		Coded Variables		Response
ξ_1	ξ_2	x_1	x_2	y
80	170	−1	−1	76.5
80	180	−1	1	77.0
90	170	1	−1	78.0
90	180	1	1	79.5
85	175	0	0	79.9
85	175	0	0	80.3
85	175	0	0	80.0
85	175	0	0	79.7
85	175	0	0	79.8

result in a decrease in yield. Therefore, another first-order model should be fit in the general vicinity of the point ($\xi_1 = 85$, $\xi_2 = 175$).

A new first-order model is fit around the point ($\xi_1 = 85$, $\xi_2 = 175$). The region of exploration for ξ_1 is [80, 90], and for ξ_2 it is [170, 180]. Thus, the coded variables are

$$x_1 = \frac{\xi_1 - 85}{5} \qquad x_2 = \frac{\xi_2 - 175}{5}$$

Once again, a 2^2 design with five center points is used. The data are shown in Table 16-4.

The first-order model fit to the coded data in Table 16-4 is

$$\hat{y} = 78.97 + 1.00x_1 + 0.50x_2$$

The analysis of variance for this model, including the interaction and pure quadratic term checks, is shown in Table 16-5. The interaction and pure quadratic checks imply that the first-order model is not an adequate approximation. This curvature in the true

Table 16-5   Analysis of Variance for the Second First-Order Model

Source of Variation	Sum of Squares	Degree of Freedom	Mean Square	F_0
Regression	5.00	2		
Residual	11.1200	6		
(Interaction)	(0.2500)	1	0.2500	4.72[a]
(Pure quadratic)	(10.6580)	1	10.6580	201.09[b]
(Pure error)	(0.2120)	4	0.0530	
Total	16.1200	8		

[a] Significant at 10 percent.
[b] Significant at 1 percent.

surface may indicate that we are near the optimum. At this point, additional analysis must be done to locate the optimum more precisely.

■

From Example 16-1 we notice that the path of steepest ascent is proportional to the signs and magnitudes of the regression coefficient in the fitted first-order model

$$\hat{y} = \hat{\beta}_0 + \sum_{i=1}^{k} \hat{\beta}_i x_i$$

It is easy to give a general algorithm for determining the coordinates of a point on the path of steepest ascent. Assume that the point $x_1 = x_2 = \cdots = x_k = 0$ is the base or origin point. Then

1.   Choose a step size in one of the process variables, say Δx_j. Usually, we would select the variable we know the most about, or we would select the variable that has the largest absolute regression coefficient $|\hat{\beta}_j|$.

2.   The step size in the other variables is

$$\Delta x_i = \frac{\hat{\beta}_i}{\hat{\beta}_j / \Delta x_j} \qquad i = 1, 2, \ldots, k; \, i \neq j$$

3.   Convert the Δx_i from coded variables to the natural variables.

To illustrate, consider the path of steepest ascent computed in Example 16-1. Since x_1 has the largest regression coefficient, we select reaction time as the variable in step 1 of the above procedure. Five minutes of reaction time is the step size (based on process knowledge). In terms of the coded variables, this is $\Delta x_1 = 1.0$. Therefore, from guideline 2, the step size in temperature is

$$\Delta x_2 = \frac{\hat{\beta}_2}{\hat{\beta}_1 / \Delta x_1} = \frac{0.325}{(0.775/1.0)} = 0.42$$

To convert the coded step sizes ($\Delta x_1 = 1.0$ and $\Delta x_2 = 0.42$) to the natural units of time and temperature, we use the relationships

$$\Delta x_1 = \frac{\Delta \xi_1}{5} \qquad \text{and} \qquad \Delta x_2 = \frac{\Delta \xi_2}{5}$$

which results in

$$\Delta \xi_1 = \Delta x_1(5) = 1.0(5) = 5 \text{ min}$$

and

$$\Delta \xi_2 = \Delta x_2(5) = 0.42(5) = 2°F$$

16-3 ANALYSIS OF A SECOND-ORDER MODEL

When the experimenter is relatively close to the optimum, a model of degree 2 or higher is usually required to approximate the response because of curvature in the true response surface. In most cases, the second-order model

$$\hat{y} = \hat{\beta}_0 + \sum_{i=1}^{k} \hat{\beta}_i x_i + \sum_{i=1}^{k} \hat{\beta}_{ii} x_i^2 + \sum_i \sum_j_{i<j} \hat{\beta}_{ij} x_i x_j \tag{16-4}$$

is adequate. In this section we will show how to use this fitted model to find the optimum set of operating conditions for the x's and to characterize the nature of the response surface.

16-3.1 Location of the Stationary Point

Suppose we wish to find the levels of $x_1, x_2, \ldots, x_k$ that optimize the predicted response. This point, if it exists, will be the set of $x_1, x_2, \ldots, x_k$ for which the partial derivatives $\partial \hat{y}/\partial x_1 = \partial \hat{y}/\partial x_2 = \cdots = \partial \hat{y}/\partial x_k = 0$. This point, say $x_{1,0}, x_{2,0}, \ldots, x_{k,0}$, is called the *stationary point*. The stationary point could represent (1) a point of maximum response, (2) a point of minimum response, or (3) a *saddle point*. These three possibilities are shown in Figure 16-5.

We may obtain a general solution for the stationary point. Writing the second-order model in matrix notation, we have

$$\hat{y} = \hat{\beta}_0 + \mathbf{x'b} + \mathbf{x'Bx} \tag{16-5}$$

where

$$\mathbf{x} = \begin{bmatrix} x_1 \\ x_2 \\ \vdots \\ x_k \end{bmatrix} \quad \mathbf{b} = \begin{bmatrix} \hat{\beta}_1 \\ \hat{\beta}_2 \\ \vdots \\ \hat{\beta}_k \end{bmatrix} \quad \text{and} \quad \mathbf{B} = \begin{bmatrix} \hat{\beta}_{11}, \hat{\beta}_{12}/2, \ldots, \hat{\beta}_{1k}/2 \\ \hat{\beta}_{22}, \ldots, \hat{\beta}_{2k}/2 \\ \ddots \\ \text{sym.} \qquad \hat{\beta}_{kk} \end{bmatrix}$$

That is, $\mathbf{b}$ is a $(k \times 1)$ vector of the first-order regression coefficients and $\mathbf{B}$ is a $(k \times k)$ symmetric matrix whose main diagonal elements are the *pure* quadratic coefficients $(\hat{\beta}_{ii})$ and whose off-diagonal elements are one-half the *mixed* quadratic coefficients $(\hat{\beta}_{ij}, i \neq j)$. The derivative of $\hat{y}$ with respect to the elements of the vector $\mathbf{x}$ equated to $\mathbf{0}$ is

$$\frac{\partial \hat{y}}{\partial \mathbf{x}} = \mathbf{b} + 2\mathbf{Bx} = \mathbf{0} \tag{16-6}$$

The stationary point is the solution to Equation 16-6, or

$$\mathbf{x}_0 = -\tfrac{1}{2} \mathbf{B}^{-1} \mathbf{b} \tag{16-7}$$

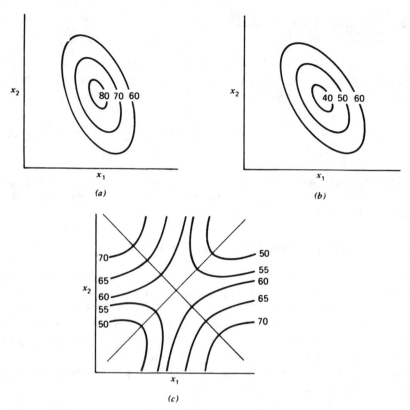

Figure 16-5. Stationary points in a fitted second-order response surface. (a) Maximum response (b) Minimum response. (c) Saddle point.

Furthermore, by substituting Equation 16-7 into Equation 16-5, we can find the predicted response at the stationary point as

$$\hat{y}_0 = \hat{\beta}_0 + \tfrac{1}{2}\,\mathbf{x}_0'\mathbf{b} \qquad (16\text{-}8)$$

16-3.2 Characterizing the Response Surface

Once we have found the stationary point, it is usually necessary to characterize the response surface in the immediate vicinity of this point. By characterize, we mean determine whether the stationary point is a point of maximum or minimum response or a saddle point and the relative sensitivity of the response to the variables $x_1, x_2, \ldots, x_k$.

The most straightforward way to do this is to examine a contour plot of the fitted model. If there are only two or three process variables (the x's), the construction and interpretation of this contour plot is relatively easy. However, even when there are relatively few variables, a more formal analysis can be useful.

It is helpful first to transform the model into a new coordinate system with the origin at the stationary point x_0 and then to rotate the axes of this system until they are parallel to the principal axes of the fitted response surface. This transformation is shown in Figure 16-6. We can show that this results in the fitted model

$$\hat{y} = \hat{y}_0 + \lambda_1 w_1^2 + \lambda_2 w_2^2 + \cdots + \lambda_k w_k^2 \qquad (16\text{-}9)$$

where the $\{w_i\}$ are the transformed independent variables and the $\{\lambda_i\}$ are constants. Equation 16-9 is called the *canonical form* of the model. Furthermore, the $\{\lambda_i\}$ are just the *eigenvalues* or *characteristic roots* of the matrix B.

The nature of the response surface can be determined from the stationary point and the *sign* and *magnitude* of the $\{\lambda_i\}$. First, suppose that the stationary point is within the region of exploration for fitting the second-order model. If the $\{\lambda_i\}$ are all positive, then x_0 is a point of minimum response; if the $\{\lambda_i\}$ are all negative, then x_0 is a pont of maximum response; and if the $\{\lambda_i\}$ have different signs, x_0 is a saddle point. Furthermore, the surface is steepest in the w_i direction for which $|\lambda_i|$ is the greatest. For example, Figure 16-6, depicts a system for which x_0 is a maximum (λ_1 and λ_2 are negative) with $|\lambda_1| > |\lambda_2|$.

Example 16-2

We will continue the analysis of the chemical process in Example 16-1. A second-order model in the variables x_1 and x_2 cannot be fit using the data in Table 16-4. The experimenter decides to augment these data with enough points to fit a second-order model.[1] She obtains four observations at ($x_1 = 0$, $x_2 = \pm1.414$) and ($x_1 = \pm1.414$, $x_2 = 0$). The complete data set is shown in Table 16-6, and the design is displayed in Figure 16-7. This design is called a *central composite design*. It will be discussed in more detail in Section 16-4.2.

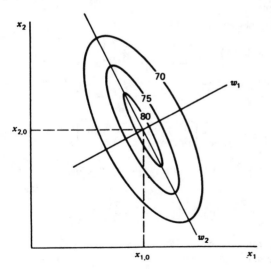

Figure 16-6. Canonical form of the second-order model.

---

[1] The engineer ran the additional four observations at about the same time she ran the original nine observations. If substantial time had elapsed between the two sets of runs, then blocking would have been necessary. Blocking in response surface designs is discussed in Section 16-4.3.

Table 16-6   Central Composite Design for Example 15-2

Natural Variables		Coded Variables		Response
ξ_1	ξ_2	x_1	x_2	y
80	170	−1	−1	76.5
80	180	−1	1	77.0
90	170	1	−1	78.0
90	180	1	1	79.5
85	175	0	0	79.9
85	175	0	0	80.3
85	175	0	0	80.0
85	175	0	0	79.7
85	175	0	0	79.8
92.07	175	1.414	0	78.4
77.93	175	−1.414	0	75.6
85	182.07	0	1.414	78.5
85	167.93	0	−1.414	77.0

A second-order model is fit to the coded data by least squares, yielding

$$\hat{y} = 79.9408 + 0.9949x_1 + 0.5151x_2 - 1.3770x_1^2 - 1.0018x_2^2 + 0.2500x_1x_2$$

The analysis of variance for this model is shown in Table 16-7. Lack of fit is not significant and regression is significant, so we conclude that the second-order model adequately approximates the true surface. It is important to be sure that the second-order model is adequate since a common abuse of RSM is to interpret a response surface that is inadequately estimated.

We now perform the canonical analysis. Note that

$$\mathbf{b} = \begin{bmatrix} 0.9949 \\ 0.5151 \end{bmatrix} \qquad \mathbf{B} = \begin{bmatrix} -1.3770 & 0.1250 \\ 0.1250 & -1.0018 \end{bmatrix}$$

and from Equation 16-7 the stationary point is

$$\mathbf{x}_0 = -\tfrac{1}{2}\mathbf{B}^{-1}\mathbf{b}$$

$$= -\tfrac{1}{2}\begin{bmatrix} -0.7345 & -0.0917 \\ -0.0917 & -1.0096 \end{bmatrix}\begin{bmatrix} 0.9949 \\ 0.5151 \end{bmatrix} = \begin{bmatrix} 0.3890 \\ 0.3056 \end{bmatrix}$$

That is, $x_{1,0} = 0.3890$ and $x_{2,0} = 0.3056$. In terms of the natural variables, the stationary point is

$$0.3890 = \frac{\xi_1 - 85}{5} \qquad 0.3056 = \frac{\xi_2 - 175}{5}$$

which yields $\xi_1 = 86.9450 \approx 87$ minutes of reaction time and $\xi_2 = 176.5280 \approx 176.5°F$. Using Equation 16-8, we may find the predicted response at the stationary point as $\hat{y}_0 = 80.21$.

The simplest way to characterize the response surface is to construct a contour plot of the yield as a function of time and temperature. Figure 16-8 is the three-

Figure 16-7.   Central composite design for Example 16-2.

dimensional response surface generated from the fitted model, and the corresponding two-dimensional contour plot is shown in Figure 16-9. It is very clear from examining both figures that the optimum is very near 175°F and 85 minutes of reaction time (compare this with the mathematically determined stationary point) and that the response is at a maximum at this point. From examination of the contour plot in Figure 16-9, we note that the process may be slightly more sensitive to changes in reaction time than to changes in temperature.

We may also use the canonical analysis described in this section to characterize the response surface. First, it is necessary to express the fitted model in canonical form (Equation 16-9). The eigenvalues λ_1 and λ_2 are the roots of the determinantal

Table 16-7   Analysis of Variance for Second-Order Model

Source of Variation	Sum of Squares	Degrees of Freedom	Mean Square	F_0
Regression	28.2560	5	5.6512	81.20[a]
Error	0.4871	7	0.0696	
(Lack of fit)	(0.2751)	3	0.0917	1.73
(Pure error)	(0.2120)	4	0.0530	
Total	28.7431	12		

[a] Significant at 1 percent.

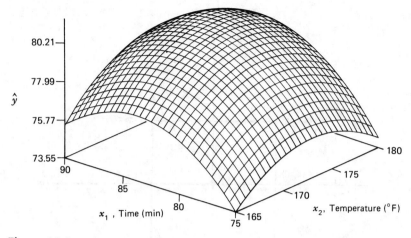

Figure 16-8. The response surface for the yield as a function of time (x_1) and temperature (x_2), where the predicted yield $\hat{y} = 79.9408 + 0.9949x_1 + 0.5151x_2 - 1.3770x_1^2 - 1.0018x_2^2 + 0.2500x_1x_2$ with time $= 85 + 5x_1$ and temperature $= 175 + 5x_2$.

equation

$$|\mathbf{B} - \lambda \mathbf{I}| = 0$$

$$\begin{vmatrix} -1.3770 - \lambda & 0.1250 \\ 0.1250 & -1.0018 - \lambda \end{vmatrix} = 0$$

which reduces to

$$\lambda^2 + 2.3788\lambda + 1.3639 = 0$$

The roots of this quadratic equation are $\lambda_1 = -0.9641$ and $\lambda_2 = -1.4147$. Thus,

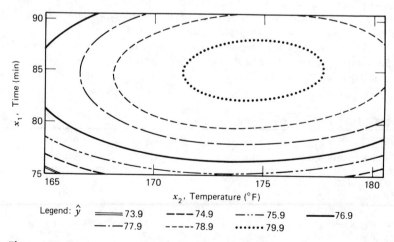

Legend: $\hat{y}$ ──── 73.9   ─ ─ ─ 74.9   ─ ··─ 75.9   ━━━ 76.9
        ─ · ─ 77.9   ─ ─ ─ ─ 78.9   •••••• 79.9

Figure 16-9. Contour plot of the yield as a function of time and temperature.

the canonical form of the fitted model is

$$\hat{y} = 80.21 - 0.9641w_1^2 - 1.4147w_2^2$$

Since both λ_1 and λ_2 are negative and the stationary point is within the region of exploration, we conclude that the stationary point is a maximum.

■

In some RSM problems it may be necessary to find the relationship between the *canonical* variables $\{w_i\}$ and the *design* variables $\{x_i\}$. This is particularly true if it is impossible to operate the process at the stationary point. As an illustration, suppose that in Example 16-2 we could not operate the process at $\xi_1 = 87$ minutes and $\xi_2 = 176.5$ of because this combination of factors results in excessive cost. We now wish to "back away" from the stationary point to a point of lower cost without incurring large losses in yield. The canonical form of the model indicates that the surface is less sensitive to yield loss in the w_1 direction. Exploration of the canonical form requires converting points in the (w_1, w_2) space to points in the (x_1, x_2) space.

In general, the variables $\mathbf{x}$ are related to the canonical variables $\mathbf{w}$ by

$$\mathbf{w} = \mathbf{M}'(\mathbf{x} - \mathbf{x}_0)$$

where $\mathbf{M}$ is a $(k \times k)$ orthogonal matrix. The columns of $\mathbf{M}$ are the normalized eigenvectors associated with the $\{\lambda_i\}$. That is, if $\mathbf{m}_i$ is the ith column of $\mathbf{M}$, then $\mathbf{m}_i$ is the solution to

$$(\mathbf{B} - \lambda_i \mathbf{I})\mathbf{m}_i = \mathbf{0} \tag{16-10}$$

for which $\Sigma_{j=1}^{k} m_{ji}^2 = 1$.

We illustrate the procedure using the fitted second-order model in Example 16-2. For $\lambda_1 = -0.9641$, Equation 16-10 becomes

$$\begin{bmatrix} (-1.3770 + 0.9641) & 0.1250 \\ 0.1250 & (-1.0018 + 0.9641) \end{bmatrix} \begin{bmatrix} m_{11} \\ m_{21} \end{bmatrix} = \begin{bmatrix} 0 \\ 0 \end{bmatrix}$$

or

$$-0.4129m + 0.1250m = 0$$

$$0.1250M_{11} - 0.0377m_{21} = 0$$

We wish to obtain the normalized solution to these equations, that is, the one for which $m_{11}^2 + m_{21}^2 = 1$. There is no unique solution to these equations, so it is most convenient to assign an arbitrary value to one unknown, solve the system, and normalize the solution. Letting $m_{21}^* = 1$, we find $m_{11}^* = 0.3027$. To normalize this solution, we divide m_{11}^* and m_{21}^* by

$$\sqrt{(m_{11}^*)^2 + (m_{21}^*)^2} = \sqrt{(0.3027)^2 + (1)^2} = 1.0448$$

This yields the normalized solution

$$m_{11} = \frac{m_{11}^*}{1.0448} = \frac{0.3027}{1.0448} = 0.2897$$

and

$$m_{21} = \frac{m_{21}^*}{1.0448} = \frac{1}{1.0448} = 0.9571$$

which is the first column of the **M** matrix.

Using $\lambda_2 = -1.4147$, we can repeat the above procedure, obtaining $m_{12} = -0.9574$ and $m_{22} = 0.2888$ as the second column of **M**. Thus, we have

$$\mathbf{M} = \begin{bmatrix} 0.2897 & -0.9574 \\ 0.9571 & 0.2888 \end{bmatrix}$$

The relationship between the **w** and **x** variables is

$$\begin{bmatrix} w_1 \\ w_2 \end{bmatrix} = \begin{bmatrix} 0.2897 & 0.9571 \\ -0.9574 & 0.2888 \end{bmatrix} \begin{bmatrix} x_1 - 0.3890 \\ x_2 - 0.3056 \end{bmatrix}$$

or

$$w_1 = 0.2897(x_1 - 0.3890) + 0.9571(x_2 - 0.3056)$$

$$w_2 = -0.9574(x_1 - 0.3890) + 0.2888(x_2 - 0.3056)$$

If we wished to explore the response surface in the vicinity of the stationary point, we could determine appropriate points at which to take observations in the (w_1, w_2) space and then use the above relationship to convert these points into the (x_1, x_2) space so that the runs may be made.

16-3.3 Ridge Systems

It is not unusual to encounter variations of the pure maximum, minimum, or saddle point response surfaces discussed in the previous section. Ridge systems, in particular, are fairly common. Consider the canonical form of the second-order model given previously in Equation 16-9:

$$\hat{y} = \hat{y}_0 + \lambda_1 w_1^2 + \lambda_2 w_2^2 + \ldots + \lambda_k w_k^2$$

Now suppose that the stationary point $\mathbf{x}_0$ is within the region of experimentation; furthermore, let one or more of the λ_i be very small (e.g., $\lambda_i \approx 0$). The response variable is then very insensitive to the variables w_i multiplied by the small λ_i.

An example is shown in Figure 16-10 for $k = 2$ variables with $\lambda_1 = 0$. (In practice, λ_1 would be close to zero.) The canonical model for this response surface is theoretically

$$\hat{y} = \hat{y}_0 + \lambda_2 w_2^2$$

with λ_2 negative. Notice that the severe elongation in the w_1 direction has resulted in a line of centers at $\hat{y} = 70$ and the optimum may be taken anywhere along that line. This type of response surface is called a *stationary ridge system*.

If the stationary point is far outside the region of exploration for fitting the second-order model and one (or more) λ_i is near zero, then the surface may be

Figure 16-10. A stationary ridge system.

a *rising ridge*. Figure 16-11 illustrates a rising ridge for $k = 2$ variables with λ_1 near zero and λ_2 negative. In this type of ridge system, we cannot draw inferences about the true surface or the stationary point since $\mathbf{x}_0$ is outside the region where we have fit the model. However, further exploration is warranted in the w_1 direction. If λ_2 had been positive, we would call this system a falling ridge.

The distance of the stationary point from the design center is

$$d = \left(\sum_{i=1}^{k} x_{i0}^2 \right)^{1/2} \tag{16-11}$$

where x_{i0}, $i = 1, 2, \ldots, k$, are the coordinates of the stationary point. When interpreting rising (or falling) ridge systems, d will usually be greater than unity; consequently, attempting to draw conclusions about the behavior of the response surface at $\mathbf{x}_0$ is risky. As noted above, the best approach is to continue exploration along the ridge in the direction of the optimum. In these cases, another canonical form may be helpful, say

$$\hat{y} = \hat{\beta}_0 + \theta_1 w_1 + \theta_2 w_2 + \cdots + \theta_k w_k$$
$$+ \lambda_1 w_1^2 + \lambda_2 w_2^2 + \cdots + \lambda_k w_k^2 \tag{16-12}$$
$$= \hat{\beta}_0 + \mathbf{w}'\boldsymbol{\theta} + \mathbf{w}'\boldsymbol{\Lambda}\mathbf{w}$$

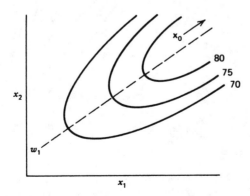

Figure 16-11. A rising ridge system.

where $\boldsymbol{\theta} = \mathbf{M}'\mathbf{b}$ and $\boldsymbol{\Lambda} = \text{diag}\,(\lambda_1, \lambda_2, \ldots, \lambda_k)$. In this canonical form, the λ's (as before) determine the type of fitted surface and the θ's measure the slopes of the surface at the original origin $\mathbf{x} = \mathbf{0}$ in the directions of the rotated axes w_1, $w_2, \ldots, w_k$.

For example, consider the rising ridge system shown in Figure 16-12. The equation for this surface is

$$\hat{y} = 82.71 + 8.80x_2 + 8.19x_2 - 6.95x_1^2 - 2.07x_2^2 + 7.59x_1x_2$$

and it can be shown that the canonical form corresponding to Equation 16-12 is

$$\hat{y} = 82.71 - 9.02w_1^2 + 2.97w_2$$

Since $\lambda_1 = 0$ and $\lambda_2 = -9.02$, the system is a rising ridge ($\lambda_2 < 0$ implies a maximum at the stationary point), and the linear term $2.97w_2$ is the contribution

Figure 16-12. A stationary ridge system. (Adapted from *Statistics for Experimenters*, by G. E. P. Box, W. G. Hunter, and J. S. Hunter, Wiley, New York, 1978, with permission.)

from the tilt of the ridge. Further exploration along the canonical direction w_2 in the direction of the optimum is warranted.

16-4 EXPERIMENTAL DESIGNS FOR FITTING RESPONSE SURFACES

Fitting and analyzing response surfaces is greatly facilitated by the proper choice of an experimental design. In this section, we discuss some aspects of selecting appropriate designs for fitting response surfaces.

When selecting a response surface design, some of the features of a desirable design are as follows:

1. Provides a reasonable distribution of data points (and hence information) throughout the region of interest.

2. Allows model adequacy, including lack of fit, to be investigated.

3. Allows experiments to be performed in blocks.

4. Allows designs of higher-order to be built up sequentially.

5. Provides an internal estimate of error.

6. Does not require a large number of runs.

7. Does not require too many levels of the independent variables.

8. Ensures simplicity of calculation of the model parameters.

These features are sometimes conflicting, so judgment must often be applied in design selection. For more information on the choice of a response surface design, refer to Box and Draper (1987) and Khuri and Cornell (1987).

16-4.1 Designs for Fitting the First-order Model

Suppose we wish to fit the first-order model in k variables

$$y = \beta_0 + \sum_{i=1}^{k} \beta_i x_i + \epsilon \qquad (16-13)$$

There is a unique class of designs that minimize the variance of the regression coefficients $\{\hat{\beta}_i\}$. These are the *orthogonal* first-order designs. A first-order design is orthogonal if the off-diagonal elements of the $(\mathbf{X}'\mathbf{X})$ matrix are all zero. This implies that the cross-products of the columns of the $\mathbf{X}$ matrix sum to zero.

The class of orthogonal first-order designs includes the 2^k factorial and fractions of the 2^k series in which main effects are not aliased with each other. In using these designs, we assume that the k factors are coded to the standardized levels ± 1. As an example, suppose we use a 2^3 design to fit the first-order model

$$y = \beta_0 + \beta_1 x_1 + \beta_2 x_2 + \beta_3 x_3 + \epsilon$$

The **X** matrix for fitting this model is

$$
\mathbf{X} =
\begin{array}{cccc}
\beta_0 & \beta_1 & \beta_2 & \beta_3 \\
\end{array}
\left[
\begin{array}{rrrr}
1 & -1 & -1 & -1 \\
1 & 1 & -1 & -1 \\
1 & -1 & 1 & -1 \\
1 & 1 & 1 & -1 \\
1 & -1 & -1 & 1 \\
1 & 1 & -1 & 1 \\
1 & -1 & 1 & 1 \\
1 & 1 & 1 & 1 \\
\end{array}
\right]
$$

It is easy to verify that the off-diagonal elements of $(\mathbf{X'X})$ are zero for this design.

The 2^k design does not afford an estimate of the experimental error unless some runs are repeated. A common method of including replication in the 2^k design is to augment the design with several observations at the center (the point $x_i = 0$, $i = 1, 2, \ldots, k$). The addition of center points to the 2^k design does not influence the $\{\hat{\beta}_i\}$ for $i \geq 1$, but the estimate of β_0 becomes the grand average of all observations. Furthermore, the addition of center points does not alter the orthogonality property of the design. Example 16-1 illustrates the use of a 2^2 design augmented with five center points to fit a first-order model.

Another orthogonal first-order design is the *simplex*. The simplex is a regularly sided figure with $k + 1$ vertices in k dimensions. Thus, for $k = 2$ the simplex design is an equilateral triangle, and for $k = 3$ it is a regular tetrahedron. Simplex designs in two and three dimensions are shown in Figure 16-13.

16-4.2 Designs for Fitting the Second-Order Model

An experimental design for fitting a second-order model must have at least three levels of each factor. There are many designs that could be used for fitting a second-order model, so to select an appropriate design it is helpful to have a design criterion. For the first-order model, *orthogonality* is the optimal design property as it minimizes the variance of the regression coefficients. Orthogonality is also desirable in the first-order case because it results in a very nice property for the variance of the predicted response $[V(\hat{y})]$. We will now discuss this property.

Figure 16-14a shows the information surface and contours for a 2^2 design with four center points used to fit a first-order model, where we define the information function as the reciprocal of the variance; that is,

$$I_x = V(\hat{y})^{-1}$$

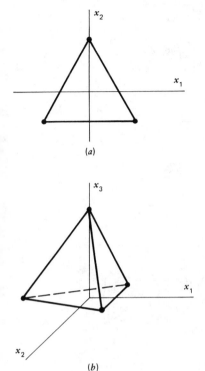

x_2

x_1

(a)

x_3

x_1

x_2

(b)

Figure 16-13. The simplex design for (a) $k = 2$ variables and (b) $k = 3$ variables.

Notice that the information contours (or equivalently, the variance contours) are concentric circles. An experimental design is said to be *rotatable* if the variance of the predicted response $\hat{y}$ at some point **x** is a function only of the *distance* of the point from the design center and is not a function of direction. Furthermore, a design with this property will leave the variance of $\hat{y}$ unchanged when the design is rotated about the center $(0, 0, \ldots, 0)$; hence, the name *rotatable* design.

Rotability is a very important property in the selection of a response surface design. Since the purpose of RSM is optimization and the location of the optimum is unknown prior to running the experiment, it makes sense to use a design that provides equal precision of estimation in all direction. Note that any first-order orthogonal design is rotatable.

Figure 16-14b shows the information surface and contours for a 3^2 design used to fit a second-order model. Notice that the 3^2 design is not rotatable. For this reason, 3^k designs and their fractions are not good choices for second-order response surface designs.

Figure 16-14c shows the information surface and contours for a central composite design similar to the one used in Example 16-2 to fit a second-order model. (This design has only four center points.) Notice that this design is rotatable. Recall the central composite design consists of a 2^k factorial or fractional factorial

Figure 16-14. Information surfaces and contours for various designs. (a) Information surface and contours for a 2^2 factorial used as a first-order design. (b) Information surface and contours for 3^2 factorial used as a second-order design. (c) Information surface and contours for a second-order rotatable design consisting of eight point on a circle plus four center points. [Adapted from Box and Draper (1987) with permission of the publisher.]

(coded to the usual ± 1 notation) augmented by $2k$ *axial* points $(\pm\alpha, 0, 0, \ldots,$ $0)$, $(0, \pm\alpha, 0, \ldots, 0)$, $(0, 0, \pm\alpha, \ldots, 0)$, $\ldots$, $(0, 0, 0, \ldots, \pm\alpha)$ and n_c center points $(0, 0, \ldots, 0)$. Central composite designs for $k = 2$ and $k = 3$ are shown in Figure 16-15. The central composite design is probably the most widely-used experimental design for fitting a second-order response surface.

A central composite design is made rotatable by the choice of α. The value of α for rotatability depends on the number of points in the factorial portion of the design; in fact, $\alpha = (n_f)^{1/4}$ yields a rotatable central composite design where

Figure 16-15. Central composite designs for $k = 2$ and $k = 3$.

n_f is the number of points used in the factorial portion of the design. Example 16-2 illustrates the use of a rotatable central composite design. That example has two factors, and the factorial portion contains $n_f = 2^2 = 4$ points. Thus, the value of α for rotatability is $\alpha = (4)^{1/4} = 1.414$. Another useful property of the central composite design is that it may be "built up" from the first-order design (the 2^k design) by adding the axial points and perhaps several center points. This sequential assembly property was demonstrated in Examples 16-1 and 16-2.

Other properties of the central composite design may be controlled by choice of the number of center points, n_c. With proper choice of n_c the central composite design may be made *orthogonal,* or it can be made a *uniform-precision* design. In a uniform-precision design, the variance of $\hat{y}$ at the origin is equal to the variance of $\hat{y}$ at unit distance from the origin. A uniform-precision design affords more protection against bias in the regression coefficients than does an orthogonal design, because of the presence of third-order and higher terms in the true surface. Table 16-8 provides the design parameters for both orthogonal and uniform-precision rotatable central composite designs for various values of k.

A variation of the central composite design is the *face-centered central composite design,* in which $\alpha = 1$. This design locates the star or axial points on the centers of the faces of the cube, as shown in Figure 16-16 for the $k = 3$. This variation of the central composite design is sometimes used because it requires only three levels of each factor, and in practice it is frequently difficult to change factor levels. However, face-centered central composite designs are not rotatable, and this is considered a serious disadvantage.

Box and Behnken (1960) have proposed some three-level designs for fitting response surfaces. These designs are formed by combining 2^k factorials with incomplete block designs. The resulting designs are usually very efficient in terms of the number of required runs, and they are rotatable (or nearly rotatable).

Table 16-9 shows a three-variable Box–Behnken design. The design is shown geometrically in Figure 16-17. Notice that the Box–Behnken design does not contain any points at the vertices of the cubic region created by the upper and lower limits for each variable. This could be advantageous when the points on

Table 16-8 Orthogonal and Uniform-Precision Rotatable Central Composite Designs

k	2	3	4	5	5 $\frac{1}{2}$ rep.	6	6 $\frac{1}{2}$ rep.	7 $\frac{1}{2}$ rep.	8 $\frac{1}{2}$ rep
n_f	4	8	16	32	16	64	32	64	128
Axial points, n_a	4	6	8	10	10	12	12	14	16
n_c (up)	5	6	7	10	6	15	9	14	20
n_c (orth)	8	9	12	17	10	24	15	22	33
N (up)	13	20	31	52	32	91	53	92	164
N (orth)	16	23	36	59	36	100	59	100	177
α	1.414	1.682	2.000	2.378	2.000	2.828	2.378	2.828	3.3

SOURCE: Adapted from "Multifactor Experimental Designs for Exploring Response Surfaces" by G. E. P. Box and J. S. Hunter, *Annals of Mathemal Statistics,* Vol. 28, pp. 195–242, 1957, with permission of the publisher.

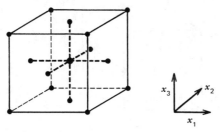

Figure 16-16. A face-centered central composite design for $k = 3$.

the corners of the cube represent factor-level combinations that are prohibitively expensive or impossible to test because of physical process constraints.

There are several other rotatable designs that are occasionally useful for problems involving two or three variables. These designs consist of points that are equally spaced on a circle ($k = 2$) or a sphere ($k = 3$) and form regular polygons or polyhedrons. Because the design points are equidistant from the origin, these arrangements are often called *equiradial designs*.

For $k = 2$, a rotatable equiradial design is obtained by combining $n_2 \geq 5$ points equally spaced on a circle with $n_1 \geq 1$ points at the center of the circle. Particularly useful designs for $k = 2$ are the pentagon and hexagon. These designs are shown in Figure 16-18 a and b. For $k = 3$ the only equiradial arrangements that contain enough points to allow all parameters in the second-order model to be estimated are the icosahedron (20 points) and the dodecahedron (12 points).

Table 16-9   A Three-Variable Box–Behnken Design

Run	x_1	x_2	x_3
1	-1	-1	0
2	-1	1	0
3	1	-1	0
4	1	1	0
5	-1	0	-1
6	-1	0	1
7	1	0	-1
8	1	0	1
9	0	-1	-1
10	0	-1	1
11	0	1	-1
12	0	1	1
13	0	0	0
14	0	0	0
15	0	0	0

Figure 16-17. A Box-Behnken design for three factors.

16-4.3 Blocking in Response Surface Designs

When using response surface designs, it is often necessary to consider blocking to eliminate nuisance variables. For example, this problem may occur when a second-order design is assembled sequentially from a first-order design, as was illustrated in Examples 16-1 and 16-2. Considerable time may elapse between the running of the first-order design and the running of the supplemental experiments required to build up a second-order design, and during this time test conditions may change, thus necessitating blocking.

A response surface design is said to block orthogonally if it is divided into blocks such that block effects do not affect the parameter estimates of the response surface model. If a 2^k or 2^{k-p} design is used as a first-order response surface design, the methods of Chapter 10 may be used to arrange the runs in 2^r blocks. The center points in these designs should be allocated equally among the blocks.

For a second-order design to block orthogonally, two conditions must be satisfied. If there are n_b observations in the bth block, then these conditions are

1. Each block must be a first-order orthogonal design; that is,

$$\sum_{u}^{n_b} x_{iu} x_{ju} = 0 \qquad i \neq j = 0, 1, \ldots, k \text{ for all } b$$

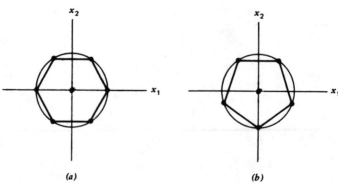

(a)   (b)

Figure 16-18. Equiradial designs for two variables. (a) Hexagon. (b) Pentagon.

where x_{iu} and x_{ju} are the levels of ith and jth variables in the uth run of the experiment with $x_{0u} = 1$ for all u.

2. The fraction of the total sum of squares for each variable contributed by every block must be equal to the fraction of the total observations that occur in the block; that is,

$$\frac{\displaystyle\sum_{u}^{n_b} x_{iu}^2}{\displaystyle\sum_{u=1}^{N} x_{iu}^2} = \frac{n_b}{N} \qquad i = 1, 2, \ldots, k \text{ for all } b$$

where N is the number of runs in the design.

As an example of applying these conditions, consider a rotatable central composite design in $k = 2$ variables with $N = 12$ runs. We may write the levels of x_1 and x_2 for this design in the design matrix

$$
\mathbf{D} =
\begin{bmatrix}
-1 & -1 \\
1 & -1 \\
-1 & 1 \\
1 & 1 \\
0 & 0 \\
0 & 0 \\
1.414 & 0 \\
-1.414 & 0 \\
0 & 1.414 \\
0 & -1.414 \\
0 & 0 \\
0 & 0
\end{bmatrix}
\begin{array}{l}
\left.\vphantom{\begin{matrix}-1\\1\\-1\\1\\0\\0\end{matrix}}\right\} \text{Block 1} \\[2.2em]
\left.\vphantom{\begin{matrix}1.414\\-1.414\\0\\0\\0\\0\end{matrix}}\right\} \text{Block 2}
\end{array}
$$

Notice that the design has been arranged in two blocks, with the first block consisting of the factorial portion of the design plus two center points and the second block consisting of the axial points plus two additional center points. It is clear that condition 1 is met; that is, both blocks are first-order orthogonal designs. To investigate condition two, consider first block 1 and note that

$$\sum_{u}^{n_1} x_{1u}^2 = \sum_{u}^{n_1} x_{2u}^2 = 4$$

$$\sum_{u=1}^{N} x_{1u}^2 = \sum_{u=1}^{N} x_{2u}^2 = 8 \qquad \text{and} \qquad n_1 = 6$$

Therefore,

$$\frac{\displaystyle\sum_{u}^{n_1} x_{iu}^2}{\displaystyle\sum_{u=1}^{N} x_{iu}^2} = \frac{n_1}{N}$$

$$\frac{4}{8} = \frac{6}{12}$$

Thus, condition 2 is satisfied in block 1. For block 2, we have

$$\sum_u^{n_2} x_{1u}^2 = \sum_u^{n_2} x_{2u}^2 = 4 \quad \text{and} \quad n_2 = 6$$

Therefore,

$$\frac{\sum\limits_u^{n_2} x_{iu}^2}{\sum\limits_{u=1}^{N} x_{iu}^2} = \frac{n_2}{N}$$

$$\frac{4}{8} = \frac{6}{12}$$

Since condition 2 is also satisfied in block 2, this design blocks orthogonally.

In general, the central composite design can always be constructed to block orthogonally in two blocks with the first block consisting of n_f factorial points plus n_{cf} center points and the second block consisting of $n_a = 2k$ axial points plus n_{ca} center points. The first condition for orthogonal blocking will always hold regardless of the value used for α in the design. For the second condition to hold,

$$\frac{\sum\limits_u^{n_2} x_{iu}^2}{\sum\limits_u^{n_1} x_{iu}^2} = \frac{n_a + n_{ca}}{n_f + n_{cf}} \tag{16-14}$$

The left-hand side of Equation 16-14 is $2\alpha^2/n_f$, and after substituting in this quantity, we may solve the equation for the value of α that will result in orthogonal blocking as

$$\alpha = \left[\frac{n_f(n_a + n_{ca})}{2(n_f + n_{cf})} \right]^{1/2} \tag{16-15}$$

This value of α does not, in general, result in a rotatable design. If the design is also required to be rotatable, then $\alpha = (n_f)^{1/4}$ and

$$(n_f)^{1/2} = \frac{n_f(n_a + n_{ca})}{2(n_f + n_{cf})} \tag{16-16}$$

It is not always possible to find a design that exactly satisfies Equation 16-16. For example if $k = 3$ then $n_f = 8$ and $n_a = 6$, and Equation 16-16 reduces to

$$(8)^{1/2} = \frac{8(6 + n_{ca})}{2(8 + n_{cf})}$$

$$2.83 = \frac{48 + 8n_{ca}}{16 + 2n_{cf}}$$

It is impossible to find values of n_{ca} and n_{cf} that exactly satisfy this last equation. However, note that if $n_{cf} = 3$ and $n_{ca} = 2$, then the right-hand side is

$$\frac{48 + 8(2)}{16 + 2(3)} = 2.91$$

so the design nearly blocks orthogonally. In practice, one could relax somewhat the requirement of either rotatability or orthogonal blocking without any major loss of information.

The central composite design is very versatile in its ability to accommodate blocking. If k is large enough, the factorial portion of the design can be divided into two or more blocks. (The number of blocks must be a power of 2, with the axial portion forming a single block. Table 16-10 presents several useful blocking arrangements for the central composite design.

There are two important points about the analysis of variance when the response surface design has been run in blocks. The first concerns the use of center points to calculate an estimate of pure error. Only center points that are run in the same block can be considered to be replicates, so the pure error term can only be calculated within each block. If the variability is consistent across blocks, then these pure error estimates could be pooled. The second point concerns the block effect. If the design blocks orthogonally in m blocks, the sum of squares for blocks is

$$SS_{\text{Blocks}} = \sum_{b=1}^{m} \frac{B_b^2}{n_b} - \frac{G^2}{N} \tag{16-17}$$

where B_b is the total of the n_b observations in the bth block and G is the grand total of all N observations in all m blocks. When blocks are not exactly orthogonal, the general regression significance test (the "extra" sum of squares method) described in Chapter 15 can be used.

16-5 MIXTURE EXPERIMENTS

In previous sections, we have presented response surface designs for those situations in which the levels of each factor are independent of the levels of other factors. In mixture experiments, the factors are the components or ingredients of a mixture, and consequently, their levels are not independent. For example, if $x_1, x_2, \ldots, x_p$ denote the proportions of p components of a mixture, then

$$0 \leq x_i \leq 1 \qquad i = 1, 2, \ldots, p$$

and

$$x_1 + x_2 + \ldots + x_p = 1 \qquad (\text{i.e., } 100\%)$$

These restrictions are illustrated graphically in Figure 16-19 for $p = 2$ and $p = 3$ components. For two components, the factor space for the design includes all

Table 16-10   Some Rotatable and Near-Rotatable Central Composite Designs that Block Orthogonally

k	2	3	4	5	5 $\frac{1}{2}$ rep.	6	6 $\frac{1}{2}$ rep.	7	7 $\frac{1}{2}$ rep.
Factorial Block(s)									
n_f	4	8	16	32	16	64	32	128	64
Number of blocks	1	2	2	4	1	8	2	16	8
Number of points in each block	4	4	8	8	16	8	16	8	8
Number of center points in each block	3	2	2	2	6	1	4	1	1
Total number of points in each block	7	6	10	10	22	9	20	9	9
Axial Block									
n_a	4	6	8	10	10	12	12	14	14
n_{ca}	3	2	2	4	1	6	2	11	4
Total number of points in the axial block	7	8	10	14	11	18	14	25	18
Total number of points N in the Design	14	20	30	54	33	90	54	169	80
Values of α									
Orthogonal Blocking	1.4142	1.6330	2.0000	2.3664	2.0000	2.8284	2.3664	3.3636	2.8284
Rotatability	1.4142	1.6818	2.0000	2.3784	2.0000	2.8284	2.3784	3.3333	2.8284

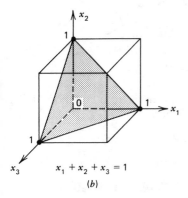

Figure 16-19. Constrained factor space for mixtures with (a) $p = 2$ components and (b) $p = 3$ components.

values of the two components that lie on the line segment $x_1 + x_2 = 1$, with each component being bounded by 0 and 1. With three components, the mixture space is a triangle with vertices corresponding to formulations that are *pure blends* (mixtures that are 100 percent of a single component).

When there are three components of the mixture, the constrained experimental region can be conveniently represented on *trilinear* coordinate paper as shown in Figure 16-20. Each of the three sides of the graph in Figure 16-20 represents a mixture that has none of one of the three components (the component labeled on the opposite vertex). The nine grid lines in each direction mark off 10 percent increments in the respective components.

Simplex designs are used to study the effects of mixture components on the response variable. A $\{p, m\}$ *simplex lattice design* for p components consists of points defined by the following coordinate settings: the proportions assumed by each component take the $m + 1$ equally spaced values from 0 to 1,

$$x_i = 0, \frac{1}{m}, \frac{2}{m}, \ldots, 1 \qquad i = 1, 2, \ldots, p \qquad (16\text{-}18)$$

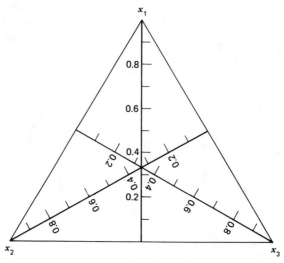

Figure 16-20.   Trilinear coordinate system.

and all possible combinations (mixtures) of the proportions from Equation 16-18 are used. As an example, let $p = 3$ and $m = 2$. Then

$$x_i = 0, \tfrac{1}{2}, 1 \qquad i = 1, 2, 3$$

and the simplex lattice consists of the following six runs:

$$(x_1, x_2, x_3) = (1, 0, 0), (0, 1, 0), (0, 0, 1), (\tfrac{1}{2}, \tfrac{1}{2}, 0), (\tfrac{1}{2}, 0, \tfrac{1}{2}), (0, \tfrac{1}{2}, \tfrac{1}{2})$$

This design is shown in Figure 16-21. The three vertices $(1, 0, 0)$, $(0, 1, 0)$, and

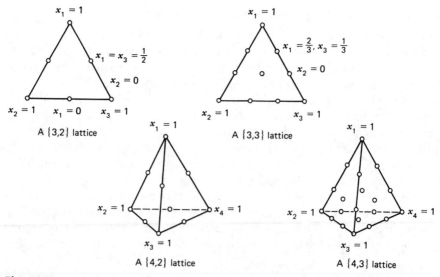

Figure 16-21.   Some simplex lattice designs for $p = 3$ and $p = 4$ components.

0, 0, 1) are the pure blends, whereas the points $(\frac{1}{2}, \frac{1}{2}, 0)$, $(\frac{1}{2}, 0, \frac{1}{2})$, and $(0, \frac{1}{2}, \frac{1}{2})$ are binary blends or two-component mixtures located at the midpoints of the three sides of the triangle. Figure 16-21 also shows the {3, 3}, the {4, 2}, and the {4, 3} simplex lattice designs. In general, the number of points in a {p, m} simplex lattice design is

$$N = \frac{(p + m - 1)!}{m!(p - 1)!}$$

An alternative to the simplex lattice design is the *simplex centroid design*. In a p-component simplex centroid design, there are $2^p - 1$ points, corresponding to the p permutations of $(1, 0, 0, \ldots, 0)$, the $\binom{p}{2}$ permutations of $(\frac{1}{2}, \frac{1}{2}, 0, \ldots, 0)$, the $\binom{p}{3}$ permutation of $(\frac{1}{3}, \frac{1}{3}, \frac{1}{3}, 0, \ldots, 0)$, $\ldots$, and the overall centroid $(\frac{1}{p}, \frac{1}{p}, \cdots \frac{1}{p})$. Figure 16-22 shows some simplex centroid designs.

A criticism of the simplex designs described above is that most of the experimental runs occur on the boundary of the region and, consequently, include only $p - 1$ of the p components. It is usually desirable to augment the simplex lattice or simplex centroid with additional points in the interior of the region where the blends will consist of all p mixture components. For more discussion, see Cornell (1981).

Mixture models differ from the usual polynomials employed in response surface work, because of the constraint $\Sigma x_i = 1$. The standard forms of the mixture models that are in widespread use are

Linear:

$$E(y) = \sum_{i=1}^{p} \beta_i x_i \tag{16-19}$$

Quadratic:

$$E(y) = \sum_{i=1}^{p} \beta_i x_i + \sum \sum_{i<j} \beta_{ij} x_i x_j \tag{16-20}$$

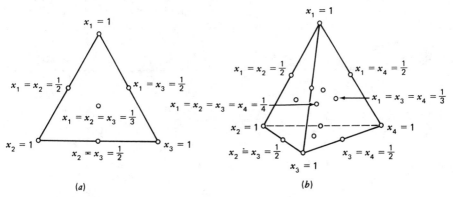

Figure 16-22. Simplex centroid designs with three and four components. (a) $p = 3$. (b) $p = 4$.

Full Cubic:

$$E(y) = \sum_{i=1}^{p} \beta_i x_i + \sum_{i<j}^{p} \beta_{ij} x_i x_j$$

$$+ \sum_{i<j}^{p} \delta_{ij} x_i x_j (x_i - x_j) \qquad (16\text{-}21)$$

$$+ \sum_{i<j<k} \beta_{ijk} x_i x_j x_k$$

Special Cubic:

$$E(y) = \sum_{i=1}^{p} \beta_i x_i + \sum_{i<j}^{p} \beta_{ij} x_i x_j$$

$$+ \sum_{i<j<k} \beta_{ij} x_i x_j x_k \qquad (16\text{-}22)$$

The terms in these models have relatively simple interpretations. In Equations 16-19 through 16-22, the parameter β_i represents the expected response to the pure blend $x_i = 1$ and $x_j = 0$ when $j \neq i$. The portion $\sum_{i=1}^{p} \beta_i x_i$ is called the *linear blending* portion. When there is curvature arising from nonlinear blending between component pairs, the parameters β_{ij} represent either synergistic or antagonistic blending. Higher-order terms are frequently necessary in mixture models because (1) the phenomena studied may be complex and (2) the experimental region is frequently the entire operability region and is therefore large, requiring an elaborate model.

Example 16-3

A Three-Component Mixture

Cornell (1981) describes a mixture experiment in which three components—polyethylene (x_1), polystyrene (x_2), and polypropylene (x_3)—were blended to form fiber that will be spun into yarn for draperies. The response variable of interest is yarn elongation in kilograms of force applied. A {3,2} simplex lattice design is used to study the product. The design and the observed responses are shown in Table 16-11. Notice

Table 16-11   The {3,2} Simplex Lattice Design for the Yarn Elongation Problem

Design Point	Component Proportions			Observed Elongation Values	Average Elongation Value ($\bar{y}$)
	x_1	x_2	x_3		
1	1	0	0	11.0, 12.4	11.7
2	$\frac{1}{2}$	$\frac{1}{2}$	0	15.0, 14.8, 16.1	15.3
3	0	1	0	8.8, 10.0	9.4
4	0	$\frac{1}{2}$	$\frac{1}{2}$	10.0, 9.7, 11.8	10.5
5	0	0	1	16.8, 16.0	16.4
6	$\frac{1}{2}$	0	$\frac{1}{2}$	17.7, 16.4, 16.6	16.9

that all of the design points involve either *pure* or *binary* blends; that is, at most only two of the three components are used in any formulation of the product. Replicate observations are also run, with two replicates at each of the pure blends and three replicates at each of the binary blends. The error standard deviation can be estimated from these replicate observations as $\hat{\sigma} = 0.85$. Cornell fits the second-order mixture polynomial to the data, resulting in

$$\hat{y} = 11.7x_1 + 9.4x_2 + 16.4x_3 + 19.0x_1x_2 + 11.4x_1x_3 - 9.6x_2x_3$$

This model can be shown to be an adequate representation of the response. Note that since $\hat{\beta}_3 > \hat{\beta}_1 > \hat{\beta}_2$, we would conclude that component 3 (polypropylene) produces yarn with the highest elongation. Furthermore, since $\hat{\beta}_{12}$ and $\hat{\beta}_{13}$ are positive, blending components 1 and 2 or components 1 and 3 produce higher elongation values than would be expected just by averaging the elongations of the pure blends. This is an example of "synergistic" bending effects. Components 2 and 3 have antagonistic blending effects since $\hat{\beta}_{23}$ is negative.

Figure 16-23 plots the contours of elongation, and this may be helpful in interpreting the results. From examining the figure, we note that if maximum elongation is desired, a blend of components 1 and 3 should be chosen consisting of about 80 percent component 3 and 20 percent component 1.

■

In some mixture problems, constraints on the individual components arise. Lower bound constraints of the form

$$l_i \leq x_i \leq 1 \qquad i = 1, 2, \ldots, p$$

are fairly common. This situation may be treated by the introduction of pseudocomponents, defined as

$$x_i' = \frac{x_i - l_i}{\left(1 - \sum_{j=1}^{p} l_j\right)} \qquad (16\text{-}23)$$

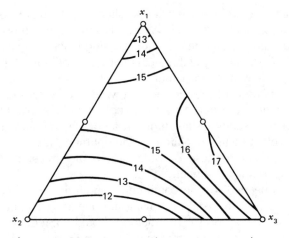

Figure 16-23. Contours of constant estimated yarn elongation from the second-order mixture model for Example 16-3.

with $\sum_{j=1}^{p} l_j < 1$. Now

$$x_1' + x_2' + \ldots + x_p' = 1$$

so the use of pseudocomponents allows the use of simplex-type designs when lower bounds are a part of the experimental situation. The formulations specified by the simplex design for the pseudocomponents are transformed into formulations for the original components by reversing the transformation Equation 16-23. That is, if x_i' is the value assigned to the ith pseudocomponent on one of the runs in the experiment, then the ith original mixture component is

$$x_i = l_i + \left(1 - \sum_{j=1}^{p} l_j \right) x_i' \tag{16-24}$$

Pseudocomponents can also be used with upper bound constraints of the form $x_i \leq u_i$, $i = 1, 2, \ldots, k$. The pseudocomponents are

$$x_i' = \frac{u_i - x_i}{\left(\sum_{j=1}^{p} u_j - 1 \right)} \tag{16-25}$$

where $\sum_{j=1}^{p} u_j - \min (u_j) < 1$. If the components have both upper and lower bound constraints, this method cannot be used. See Cornell (1981) for more information on designs for constrained mixture spaces.

16-6 EVOLUTIONARY OPERATION

Response surface methodology is often applied to pilot plant operations by research and development personnel. When it is applied to a full-scale production process, it is usually only done once (or very infrequently) because the experimental procedure is relatively elaborate. However, conditions that were optimum for the pilot plant may not be optimum for the full-scale process. The pilot plant may produce 2 pounds of product per day whereas the full-scale process may produce 2000 pounds per day. This "scale-up" of the pilot plant to the full-scale production process usually results in distortion of the optimum conditions. Even if the full-scale plant begins operation at the optimum, it will eventually "drift" away from that point because of variations in raw materials, environmental changes, and operating personnel.

A method is needed for the continuous monitoring and improvement of a full-scale process with the goal of moving the operating conditions toward the optimum or following a "drift." The method should not require large or sudden changes in operating conditions that might disrupt production. Evolutionary operation (EVOP) was proposed by Box (1957) as such an operating procedure. It is designed as a method of routine plant operation that is carried out by manufacturing personnel with minimum assistance from the research and development staff.

EVOP consists of systematically introducing small changes in the levels of the operating variables under consideration. Usually, a 2^k design is employed to do this. The changes in the variables are assumed to be small enough that serious disturbances in yield, quality, or quantity will not occur, yet large enough that potential improvements in process performance will eventually be discovered. Data are collected on the response variables of interest at each point of the 2^k design. When one observation has been taken at each design point, a *cycle* is said to have been completed. The effects and interactions of the process variables may then be computed. Eventually, after several cycles, the effect of one or more process variables or their interactions may appear to have a significant effect on the response. At this point, a decision may be made to change the basic operating conditions to improve the response. When improved conditions have been detected, a *phase* is said to have been completed.

In testing the significance of process variables and interactions, an estimate of experimental error is required. This is calculated from the cycle data. Also, the 2^k design is usually centered about the current best operating conditions. By comparing the response at this point with the 2^k points in the factorial portion, we may check on curvature or *change in mean* (CIM); that is, if the process is really centered at the maximum, say, then the response at the center should be significantly greater than the responses at the 2^k-peripheral points.

In theory, EVOP can be applied to k process variables. In practice, only two or three variables are usually considered. We will give an example of the procedure for two variables. Box and Draper (1969) give a detailed discussion of the three-variable case, including necessary forms and worksheets.

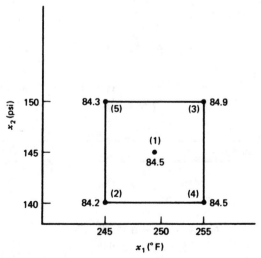

Figure 16-24. A 2^2 design for EVOP.

Example 16-4

Consider a chemical process whose yield is a function of temperature (x_1) and pressure (x_2). The current operating conditions are $x_1 = 250°F$ and $x_2 = 145$ psi. The EVOP procedure uses the 2^2 design plus the center point shown in Figure 16-24. The cycle is completed by running each design point in numerical order (1, 2, 3, 4, 5). The yields in the first cycle are shown in Figure 16-24.

The yields from the first cycle are entered in the EVOP calculation sheet, shown in Table 16-12. At the end of the first cycle, no estimate of the standard deviation can be made. The effects and interaction for temperature and pressure are calculated in the usual manner for a 2^2 design.

A second cycle is then run and the yield data entered in another EVOP calculation sheet, shown in Table 16-13. At the end of the second cycle, the experimental error can be estimated and the estimates of the effects compared to approximate 95 percent (two standard deviation) limits. Note that the range refers to the range of the differences in row (iv); thus the range is $+1.0 - (-1.0) = 2.0$. Since none of the effects in Table 16-13 exceed their error limits, the true effect is probably zero, and no changes in operating conditions are contemplated.

The results of a third cycle are shown in Table 16-14. The effect of pressure now exceeds its error limit and the temperature effect is equal to the error limit. A change in operating conditions is now probably justified.

In light of the results, it seems reasonable to begin a new EVOP phase about

Table 16-12   EVOP Calculation Sheet for Example 16-4, $n = 1$

		Cycle: $n = 1$			Phase: 1
		Response: Yield			Date: 1/11/90

		Calculation of Averages				Calculation of Standard Deviatic
Operating Conditions	(1)	(2)	(3)	(4)	(5)	
(i) Previous cycle sum						Previous sum $S =$
(ii) Previous cycle average						Previous average S
(iii) New observations	84.5	84.2	84.9	84.5	84.3	New S = range $\times$ $f_{s,n} =$
(iv) Differences [(ii) − (iii)]						Range of (iv) =
(v) New sums [(i) + (iii)]	84.5	84.2	84.9	84.5	84.3	New sum $S =$
(vi) New averages [$\bar{y}_i$ = (v)/n]	84.5	84.2	84.9	84.5	84.3	New average $S =$ $\dfrac{\text{New sum } S}{n-1}$

Calculation of Effects	Calculation of Error Limit
Temperature effect $= \frac{1}{2}(\bar{y}_3 + \bar{y}_4 - \bar{y}_2 - \bar{y}_5) = 0.45$	For new average $\dfrac{2}{\sqrt{n}} S =$
Pressure effect $= \frac{1}{2}(\bar{y}_3 + \bar{y}_5 - \bar{y}_2 - \bar{y}_4) = 0.25$	For new effects $\dfrac{2}{\sqrt{n}} S =$
$T \times P$ interaction effect $= \frac{1}{2}(\bar{y}_2 + \bar{y}_3 - \bar{y}_4 - \bar{y}_5) = 0.15$	
Change-in-mean effect $\frac{1}{5}(\bar{y}_2 + \bar{y}_3 + \bar{y}_4 + \bar{y}_5 - 4\bar{y}_1) = 0.02$	For change in mean $\dfrac{1.78}{\sqrt{n}} S$

ble 16-13   EVOP Calculation Sheet for Example 16-4, $n = 2$

```
5 ┌─┐ 3
  │1│
2 └─┘ 4
```

| | | Cycle: $n = 2$ Response: Yield | | | | Phase: 1 Date: 1/11/90 |

		Calculation of Averages				Calculation of Standard Deviation
Operation Conditions	(1)	(2)	(3)	(4)	(5)	
Previous cycle sum	84.5	84.2	84.9	84.5	84.3	Previous sum $S =$
Previous cycle average	84.5	84.2	84.9	84.5	84.3	Previous average $S =$
New observations	84.9	84.6	85.9	83.5	84.0	New S = range $\times$ $f_{5,n} = 0.60$
Differences [(ii) − (iii)]	−0.4	−0.4	−1.0	+1.0	0.3	Range of (iv) = 2.0
New sums [(i) + (iii)]	169.4	168.8	170.8	168.0	168.3	New sum S = 0.60
New Averages	84.70	84.40	85.40	84.00	84.15	New average $S =$
[$\bar{y}_i = $ (v)/n]						$\dfrac{\text{New sum } S}{n-1} = 0.60$

Calculation of Effects	Calculation of Error Limits
nperature effect = $\frac{1}{2}(\bar{y}_3 + \bar{y}_4 - \bar{y}_2 - \bar{y}_5) = 0.43$	For new average $\dfrac{2}{\sqrt{n}} S = 0.85$
ssure effect = $\frac{1}{2}(\bar{y}_3 + \bar{y}_5 - \bar{y}_2 - \bar{y}_4) = 0.58$	For new effects $\dfrac{2}{\sqrt{n}} S = 0.85$
$\times$ P interaction effect = $\frac{1}{2}(\bar{y}_2 + \bar{y}_3 - \bar{y}_4 - \bar{y}_5) = 0.83$	
ınge-in-mean effect = $\frac{1}{5}(\bar{y}_2 + \bar{y}_3 + \bar{y}_4 + \bar{y}_5 - 4\bar{y}_1) = -0.17$	For change in mean $\dfrac{1.78}{\sqrt{n}} S = 0.76$

point (3). Thus, $x_1 = 225°F$ and $x_2 = 150$ psi would become the center of the 2^2 design in the second phase.

An important aspect of EVOP is feeding the information generated back to the process operators and supervisors. This is accomplished by a prominently displayed EVOP information board. The information board for this example at the end of cycle three is shown in Table 16-15.

∎

Most of the quantities on the EVOP calculation sheet follow directly from the analysis of the 2^k factorial design. For example, the variance of any effect, such as $\frac{1}{2}(\bar{y}_3 + \bar{y}_5 - \bar{y}_2 - \bar{y}_4)$, is simply

$$V\left[\frac{1}{2}(\bar{y}_3 + \bar{y}_5 - \bar{y}_2 - \bar{y}_4)\right] = \frac{1}{4}(\sigma_{\bar{y}_3}^2 + \sigma_{\bar{y}_5}^2 + \sigma_{\bar{y}_2}^2 + \sigma_{\bar{y}_4}^2)$$

$$= \frac{1}{4}(4\sigma_{\bar{y}}^2) = \frac{\sigma^2}{n}$$

where σ^2 is the variance of the observations (y). Thus, two standard deviation (corresponding to 95 percent) error limits on any effect would be $\pm 2\sigma/\sqrt{n}$. The

Table 16-14   EVOP Calculation Sheet for Example 16-4, $n = 3$

```
5 ┌──┐ 3
  │ I│
2 └──┘ 4
```

| | | Cycle: $n = 3$
Response: Yield | | | Phase: 1
Date: 1/11/90 |

			Calculation of Averages			Calculation of Standard Deviati
Operating Conditions		(1)	(2)	(3)	(4)	(5)	
(i)	Previous cycle sum	169.4	168.8	170.8	168.0	168.3	Previous sum S = 0.60
(ii)	Previous cycle average	84.70	84.40	85.40	84.00	84.15	Previous average . 0.60
(iii)	New observations	85.0	84.0	86.6	84.9	85.2	New S = range × $f_{5,n} = 0.56$
(iv)	Differences [(ii) − (iii)]	−0.30	+0.40	−1.20	−0.90	−1.05	Range of (iv) = 1
(v)	New sums [(i) + (iii)]	254.4	252.8	257.4	252.9	253.5	New sum S = 1.1
(vi)	New averages $[\bar{y}_i = (v)/n]$	84.80	84.27	85.80	84.30	84.50	New average S = $\dfrac{\text{New sum } S}{n-1} = ($

Calculation of Effects	Calculations of Error Limits
Temperature effect = $\frac{1}{2}(\bar{y}_3 + \bar{y}_4 - \bar{y}_2 - \bar{y}_5) = 0.67$	For new average $\dfrac{2}{\sqrt{n}} S = 0.67$
Pressure effect = $\frac{1}{2}(\bar{y}_3 + \bar{y}_5 - \bar{y}_2 - \bar{y}_4) = 0.87$	For new effects $\dfrac{2}{\sqrt{n}} S = 0.67$
$T \times P$ interaction effect = $\frac{1}{2}(\bar{y}_2 + \bar{y}_3 - \bar{y}_4 - \bar{y}_5) = 0.64$ Change-in-mean effect = $\frac{1}{5}(\bar{y}_2 + \bar{y}_3 + \bar{y}_4 + \bar{y}_5 - 4\bar{y}_1) = -0.07$	For change in mean $\dfrac{1.78}{\sqrt{n}} S = ($

Table 16-15   EVOP Information Board-Cycle Three

Response: Percent Yield

Requirement: Maximize

Error Limits for Averages: ± 0.67

Effects with	Temperature	0.67	±0.67
95% error	Pressure	0.87	±0.67
Limits:	$T \times P$	0.64	±0.67
	Change in mean	0.07	±0.60
Standard deviation:	0.58		

Table 16-16 Values of $f_{k,n}$

$n =$	2	3	4	5	6	7	8	9	10
$k = 5$	0.30	0.35	0.37	0.38	0.39	0.40	0.40	0.40	0.41
9	0.24	0.27	0.29	0.30	0.31	0.31	0.31	0.32	0.32
10	0.23	0.26	0.28	0.29	0.30	0.30	0.30	0.31	0.31

variance of the change in mean is

$$V(\text{CIM}) = V\left[\frac{1}{5}(\bar{y}_2 + \bar{y}_3 + \bar{y}_4 + \bar{y}_5 - 4\bar{y}_1)\right]$$

$$= \frac{1}{25}(4\sigma_{\bar{y}}^2 + 16\sigma_{\bar{y}}^2) = \left(\frac{20}{25}\right)\frac{\sigma^2}{n}$$

Thus, two standard deviation error limits on the CIM are $\pm(2\sqrt{20/25})\sigma/\sqrt{n} = \pm1.78\sigma/\sqrt{n}$.

The standard deviation σ is estimated by the range method. Let $y_i(n)$ denote the observation at the ith design point in cycle n and $\bar{y}_i(n)$ denote the corresponding average of $y_i(n)$ after n cycles. The quantities in row (iv) of the EVOP calculation sheet are the differences $y_i(n) - \bar{y}_i(n - 1)$. The variance of these differences is

$$V[y_i(n) - \bar{y}_i(n - 1)] \equiv \sigma_D^2 = \sigma^2\left(1 + \frac{1}{(n - 1)}\right) = \sigma^2\frac{n}{(n - 1)}$$

The range of the differences, say R_D, is related to the estimate of the standard deviation of the differences by $\hat{\sigma}_D = R_D/d_2$. The factor d_2 depends on the number of observations used in computing R_D. Now $R_D/d_2 = \hat{\sigma}\sqrt{n/(n - 1)}$, so

$$\hat{\sigma} = \sqrt{\frac{(n - 1)}{n}}\frac{R_D}{d_2} = (f_{k,n})R_D \equiv S$$

can be used to estimate the standard deviation of the observations, where k denotes the number of points used in the design. For a 2^2 design with one center point we have $k = 5$, and for a 2^3 design with one center point we have $k = 9$. Values of $f_{k,n}$ are given in Table 16-16.

16-7 PROBLEMS

16-1 A chemical plant produces oxygen by liquifying air and separating it into its component gases by fractional distillation. The purity of the oxygen is a function of the main condenser temperature and the pressure ratio between the upper and lower columns. Current operating conditions are temperature (ξ_1) $= -220°C$ and pressure ratio (ξ_2) $= 1.2$. Using the following data, find the path of steepest ascent.

Temperature (ξ_1)	Pressure Ratio (ξ_2)	Purity
−225	1.1	82.8
−225	1.3	83.5
−215	1.1	84.7
−215	1.3	85.0
−220	1.2	84.1
−220	1.2	84.5
−220	1.2	83.9
−220	1.2	84.3

16-2 An industrial engineer has developed a computer simulation model of a two-item inventory system. The decision variables are the order quantity and the reorder point for each item. The response to be minimized is total inventory cost. The simulation model is used to produce the data shown in the following table. Identify the experimental design. Find the path of steepest decent.

Item 1		Item 2		
Order Quantity (ξ_1)	Reorder Point (ξ_2)	Order Quantity (ξ_3)	Reorder Point (ξ_4)	Total Cost
100	25	250	40	625
140	45	250	40	670
140	25	300	40	663
140	25	250	80	654
100	45	300	40	648
100	45	250	80	634
100	25	300	80	692
140	45	300	80	686
120	35	275	60	680
120	35	275	60	674
120	35	275	60	681

16-3 Verify that the following design is a simplex. Fit the first-order model and find the path of steepest ascent.

x_1	x_2	x_3	y
0	$\sqrt{2}$	−1	18.5
−$\sqrt{2}$	0	1	19.8
0	−$\sqrt{2}$	−1	17.4
$\sqrt{2}$	0	1	22.5

16-4 For the first-order model

$$\hat{y} = 60 + 1.5x_1 - 0.8x_2 + 2.0x_3$$

find the path of steepest ascent. The variables are coded as $-1 \leq x_i \leq 1$.

16-5 The data shown in the following table were collected in an experiment to optimize crystal growth as a function of three variables x_1, x_2, and x_3. Large values of y (yield in grams) are desirable. Fit a second-order model and analyze the fitted surface. Under what set of conditions is optimum growth achieved?

x_1	x_2	x_3	y
-1	-1	-1	66
-1	-1	1	70
-1	1	-1	78
-1	1	1	60
1	-1	-1	80
1	-1	1	70
1	1	-1	100
1	1	1	75
-1.682	0	0	100
1.682	0	0	80
0	-1.682	0	68
0	1.682	0	63
0	0	-1.682	65
0	0	1.682	82
0	0	0	113
0	0	0	100
0	0	0	118
0	0	0	88
0	0	0	100
0	0	0	85

16-6 The following data were collected by a chemical engineer. The response y is filtration time, x_1 is temperature, and x_2 is pressure. Fit a second-order model and analyze the fitted surface.

x_1	x_2	y
-1	-1	54
-1	1	45
1	-1	32
1	1	47
-1.414	0	50
1.414	0	53
0	-1.414	47

x_1	x_2	y
0	1.414	51
0	0	41
0	0	39
0	0	44
0	0	42
0	0	40

16-7 A manufacturer of cutting tools has developed two empirical equations for tool life in hours (y_1) and for tool cost in dollars (y_2). Both models are linear functions of steel hardness (x_1) and manufacturing time (x_2). The two equations are

$$\hat{y}_1 = 10 + 5x_1 + 2x_2$$

$$\hat{y}_2 = 23 + 3x_1 + 4x_2$$

and both equations are valid over the range $-1.5 \leq x_i \leq 1.5$. Unit tool cost must be below \$27.5 and life must exceed 12 hours for the product to be competitive. Is there a feasible set of operating conditions for this process? Where would you recommend that the process be run?

16-8 The hexagon design that follows is used to collect data for fitting a second-order model. Fit the second-order model and perform a canonical analysis.

x_1	x_2	y
1	0	68
0.5	$\sqrt{0.75}$	74
-0.5	$\sqrt{0.75}$	65
-1	0	60
-0.5	$-\sqrt{0.75}$	63
0.5	$-\sqrt{0.75}$	70
0	0	58
0	0	60
0	0	57
0	0	55
0	0	69

16-9 Verify that an orthogonal first-order design is also first-order rotatable.

16-10 Show that augmenting a 2^k design with n_c center ponts does not affect the estimates of the $\beta_i(i = 1, 2, \ldots, k)$, but that the estimate of the intercept β_0 is the average of all $2^k + n_c$ observations.

16-11 *An Orthogonal Central Composite Design.* Suppose a central composite design is used to fit the second-order model

$$y = \beta_0 + \sum_{i=1}^{k} \beta_i x_i + \sum_{i=1}^{k} \beta_{ii}(x_i^2 - \overline{x_i^2}) + \sum_i \sum_j \beta_{ij} x_i x_j + \epsilon$$

Show that the design can be made orthogonal by the appropriate choice of α.

16-12 **The Rotatable Central Composite Design.** It can be shown that a second-order design is rotatable if $\sum_{u=1}^{n} x_{iu}^a x_{ju}^b = 0$ if a or b (or both) are odd and if $3\sum_{u=1}^{n} x_{iu}^4 = 3\sum_{u=1}^{n} x_{iu}^2 x_{ju}^2$. Show that for the central composite design these conditions lead to $\alpha = (n_f)^{1/4}$ for rotatability, where n_f is the number of points in the factorial portion.

16-13 Verify that the central composite design shown below blocks orthogonally.

Block 1			Block 2			Block 3		
x_1	x_2	x_3	x_1	x_2	x_3	x_1	x_2	x_3
0	0	0	0	0	0	-1.633	0	0
0	0	0	0	0	0	1.633	0	0
1	1	1	1	1	-1	0	-1.633	0
1	-1	-1	1	-1	1	0	1.633	0
-1	-1	1	-1	1	1	0	0	-1.633
-1	1	-1	-1	-1	-1	0	0	1.633
						0	0	0
						0	0	0

16-14 **Blocking in the Central Composite Design.** Consider a central composite design for $k = 4$ variables in two blocks. Can a rotatable design always be found that blocks orthogonally?

16-15 How could a hexagon design be run in two orthogonal blocks?

16-16 Yield during the first four cycles of a chemical process is shown in the following table. The variables are percent concentration (x_1) at levels 30, 31, and 32 and temperature (x_2) at 140, 142, and 144°F. Analyze by EVOP methods.

Cycle	Conditions				
	(1)	(2)	(3)	(4)	(5)
1	60.7	59.8	60.2	64.2	57.5
2	59.1	62.8	62.5	64.6	58.3
2	56.6	59.1	59.0	62.3	61.1
4	60.5	59.8	64.5	61.0	60.1

16-17 Design an EVOP calculation sheet for three variables. Derive the two standard error limits for the effects and for the change in mean.

16-18 Suppose that we approximate a response surface with a model of order d_1, such as $y = X_1\beta_1 + \epsilon$, when the true surface is described by a model of order $d_2 > d_1$; that is, $E(y) = X_1\beta_1 + X_1\beta_2$.

(a) Show that the regression coefficients are biased, that is, that $E(\hat{\beta}_1) = \beta_1 + A\beta_2$, where $A = (X_1'X_1)^{-1}X_1'X_2$. A is usually called the alias matrix.

(b) If $d_1 = 1$ and $d_2 = 2$, and a full 2^k is used to fit the model, use the result in part (a) to determine the alias structure.

(c) If $d_1 = 1$, $d_2 = 2$, and $k = 3$, find the alias structure assuming that a 2^{3-1} design is used to fit the model.

(d) If $d_1 = 1$, $d_2 = 2$, $k = 3$, and the simplex design in Problem 16-3 is used to fit the model, determine the alias structure and compare the results with part (c).

Chapter 17
Analysis of Covariance

17-1 INTRODUCTION

In previous chapters we have discussed the use of randomized block and Latin square designs to improve the precision with which comparisons between treatments are made. The analysis of covariance is another technique that is occasionally useful for improving the precision of an experiment. Suppose that in an experiment with a response variable y there is another variable, such as x, and that y is linearly related to x. Furthermore, suppose that x cannot be controlled by the experimenter but can be observed along with y. The variable x is called a *covariate* or *concomitant variable*. The analysis of covariance involves adjusting the observed response variable for the effect of the concomitant variable. If such an adjustment is not performed, the concomitant variable could inflate the error mean square and make true differences in the response due to treatments harder to detect. Thus, the analysis of covariance is a method of adjusting for the effects of an uncontrollable nuisance variable. As we will see, the procedure is a combination of analysis of variance and regression analysis.

As an example of an experiment in which the analysis of covariance may be employed, consider a study performed to determine if there is a difference in the strength of a monofilament fiber produced by three different machines. However, the strength of the fiber is also affected by its denure or thickness; consequently, a thicker fiber will generally be stronger than a thinner one. The analysis of covariance could be used to remove the effect of thickness (x) on strength (y) when testing for differences in strength between machines.

17-2 A SINGLE-FACTOR DESIGN WITH ONE COVARIATE

The basic procedure for the analysis of covariance is now illustrated for a single-factor experimental design with one covariate. Assuming that there is a linear relationship between the response and the covariate, an appropriate statistical model is

$$y_{ij} = \mu + \tau_i + \beta(x_{ij} - \bar{x}_{..}) + \varepsilon_{ij} \begin{cases} i = 1,2, \ldots, a \\ j = 1,2, \ldots, n \end{cases} \tag{17-1}$$

where y_{ij} is the jth observation on the response variable taken under the ith treatment or level of the single factor, x_{ij} is the measurement made on the co-variate or concomitant variable corresponding to y_{ij} (i.e., the ijth run), $\bar{x}_{..}$ is the mean of the x_{ij} values, μ is an overall mean, τ_i is the effect of the ith treatment, β is a linear regression coefficient indicating the dependency of y_{ij} on x_{ij}, and ε_{ij} is a random error component. We assume that the errors ε_{ij} are NID(0, σ^2), that the slope $\beta \neq 0$ and the true relationship between y_{ij} and x_{ij} is linear, that the regression coefficients for each treatment are identical, that the treatment effects sum to zero ($\Sigma_{i=1}^{a} \tau_i = 0$), and that the concomitant variable x_{ij} is not affected by the treatments.

Notice immediately from Equation 17-1 that the analysis of covariance model is a combination of the linear models employed in analysis of variance and regression. That is, we have treatment effects $\{\tau_i\}$ as in a single-factor analysis of variance and a regression coefficient β as in regression analysis. The concomitant variable in Equation 17-1 is expressed as $(x_{ij} - \bar{x}_{..})$ instead of x_{ij} so that the parameter μ is preserved as the overall mean. The model could have been written as

$$y_{ij} = \mu' + \tau_i + \beta x_{ij} + \varepsilon_{ij} \begin{cases} i = 1,2, \ldots, a \\ j = 1,2, \ldots, n \end{cases} \tag{17-2}$$

where μ' is a constant not equal to the overall mean, which for this model is $\mu' + \beta\bar{x}_{..}$. Equation 17-1 is more widely found in the literature.

To describe the analysis, we introduce the following notation:

$$S_{yy} = \sum_{i=1}^{a} \sum_{j=1}^{n} (y_{ij} - \bar{y}_{..})^2 = \sum_{i=1}^{a} \sum_{j=1}^{n} y_{ij}^2 - \frac{y_{..}^2}{an} \tag{17-3}$$

$$S_{xx} = \sum_{i=1}^{a} \sum_{j=1}^{n} (x_{ij} - \bar{x}_{..})^2 = \sum_{i=1}^{a} \sum_{j=1}^{n} x_{ij}^2 - \frac{x_{..}^2}{an} \tag{17-4}$$

$$S_{xy} = \sum_{i=1}^{a} \sum_{j=1}^{n} (x_{ij} - \bar{x}_{..})(y_{ij} - \bar{y}_{..}) = \sum_{i=1}^{a} \sum_{j=1}^{n} x_{ij}y_{ij} - \frac{(x_{..})(y_{..})}{an} \tag{17-5}$$

$$T_{yy} = \sum_{i=1}^{a} (\bar{y}_{i.} - \bar{y}_{..})^2 = \sum_{i=1}^{a} \frac{y_{i.}^2}{n} - \frac{y_{..}^2}{an} \tag{17-6}$$

$$T_{xx} = \sum_{i=1}^{a} (\bar{x}_{i.} - \bar{x}_{..})^2 = \sum_{i=1}^{a} \frac{x_{i.}^2}{n} - \frac{x_{..}^2}{an} \tag{17-7}$$

$$T_{xy} = \sum_{i=1}^{a} (\bar{x}_{i.} - \bar{x}_{..})(\bar{y}_{i.} - \bar{y}_{..}) = \sum_{i=1}^{a} \frac{(x_{i.})(y_{i.})}{n} - \frac{(x_{..})(y_{..})}{an} \tag{17-8}$$

$$E_{yy} = \sum_{i=1}^{a} \sum_{j=1}^{n} (y_{ij} - \bar{y}_{i.})^2 = S_{yy} - T_{yy} \tag{17-9}$$

$$E_{xx} = \sum_{i=1}^{a} \sum_{j=1}^{n} (x_{ij} - \bar{x}_{i.})^2 = S_{xx} - T_{xx} \qquad (17\text{-}10)$$

$$E_{xy} = \sum_{i=1}^{a} \sum_{j=1}^{n} (x_{ij} - \bar{x}_{i.})(y_{ij} - \bar{y}_{i.}) = S_{xy} - T_{xy} \qquad (17\text{-}11)$$

Note that, in general, $S = T + E$, where the symbols S, T, and E are used to denote sums of squares and cross-products for total, treatments, and error, respectively. The sums of squares for x and y must be nonnegative; however, the sums of cross-products (xy) may be negative.

We now show how the analysis of covariance adjusts the response variable for the effect of the covariate. Consider the full model (Equation 17-1). The least-squares estimators of μ, τ_i, and β are $\hat{\mu} = \bar{y}_{..}$, $\hat{\tau}_i = \bar{y}_{i.} - \bar{y}_{..} - \hat{\beta}(\bar{x}_{i.} - \bar{x}_{..})$, and

$$\hat{\beta} = \frac{E_{xy}}{E_{xx}} \qquad (17\text{-}12)$$

The error sum of squares in this model is

$$SS_E = E_{yy} - (E_{xy})^2/E_{xx} \qquad (17\text{-}13)$$

with $a(n - 1) - 1$ degrees of freedom. The experimental error variance is estimated by

$$MS_E = \frac{SS_E}{a(n - 1) - 1}$$

Now suppose that there is no treatment effect. The model (Equation 17-1) would then be

$$y_{ij} = \mu + \beta(x_{ij} - \bar{x}_{..}) + \varepsilon_{ij} \qquad (17\text{-}14)$$

and the least-squares estimators of μ and β are $\hat{\mu} = \bar{y}_{..}$ and $\hat{\beta} = S_{xy}/S_{xx}$. The sum of squares for error in this reduced model is

$$SS'_E = S_{yy} - (S_{xy})^2/S_{xx} \qquad (17\text{-}15)$$

with $an - 2$ degrees of freedom. In Equation 17-15, the quantity $(S_{xy})^2/S_{xx}$ is the reduction in the sum of squares of y obtained through the linear regression of y on x. Furthermore, note that SS_E is smaller than SS'_E [because the model (Equation 17-1) contains additional parameters $\{\tau_i\}$] and that the quantity $SS'_E - SS_E$ is a reduction in sum of squares due to the $\{\tau_i\}$. Therefore, the difference between SS'_E and SS_E, that is, $SS'_E - SS_E$, provides a sum of squares with $a - 1$ degrees of freedom for testing the hypothesis of no treatment effects. Consequently, to test $H_0: \tau_i = 0$, compute[1]

$$F_0 = \frac{(SS'_E - SS_E)/(a - 1)}{SS_E/[a(n - 1) - 1]} \qquad (17\text{-}16)$$

---

[1] This test statistic is a type III F ratio in SAS.

which, if the null hypothesis is true, is distributed as $F_{a-1,a(n-1)-1}$. Thus, we reject H_0: $\tau_i = 0$ if $F_0 > F_{\alpha,a-1,a(n-1)-1}$.

It is instructive to examine the display in Table 17-1. In this table we have presented the analysis of covariance as an "adjusted" analysis of variance. In the source of variation column, the total variability is measured by S_{yy} with an -1 degrees of freedom. The source of variation "regression" has the sum of squares $(S_{xy})^2/S_{xx}$ with one degree of freedom. If there were no concomitant variable, we would have $S_{xy} = S_{xx} = E_{xy} = E_{xx} = 0$. Then the sum of squares for error would be simply E_{yy} and the sum of squares for treatments would be $S_{yy} - E_{yy} = T_{yy}$. However, because of the presence of the concomitant variable, we must "adjust" S_{yy} and E_{yy} for the regression of y on x as shown in Table 16-1. The adjusted error sum of squares has $a(n-1) - 1$ degrees of freedom instead of $a(n-1)$ degrees of freedom because an additional parameter (the slope β) is fitted to the data.

The computations are usually displayed in an analysis of covariance table such as Table 17-2. This layout is employed because it conveniently summarizes all the required sums of squares and cross-products as well as the sums of squares for testing hypotheses about treatment effects. In addition to testing the hypothesis that there are no differences in the treatment effects, we frequently find it useful in interpreting the data to present the adjusted treatment means. These adjusted means are computed according to

$$\text{Adjusted } \bar{y}_{i.} = \bar{y}_{i.} - \hat{\beta}(\bar{x}_{i.} - \bar{x}_{..}) \qquad i = 1,2,\ldots,a \qquad (17\text{-}17)$$

where $\hat{\beta} = E_{xy}/E_{xx}$. This adjusted treatment mean is the least-squares estimator of $\mu + \tau_i$, $i = 1,2,\ldots,a$, in the model (Equation 17-1). The standard error of any adjusted treatment mean is

$$S_{\text{adj}.\bar{y}_{i.}} = \left[MS_E\left(\frac{1}{n} + \frac{(\bar{x}_{i.} - \bar{x}_{..})^2}{E_{xx}}\right) \right]^{1/2} \qquad (17\text{-}18)$$

Finally, we recall that the regression coefficient β in the model (Equation 17-1) has been assumed to be nonzero. We may test the hypothesis H_0: $\beta = 0$ by using the test statistic[2]

$$F_0 = \frac{(E_{xy})^2/E_{xx}}{MS_E} \qquad (17\text{-}19)$$

which under the null hypothesis is distributed as $F_{1,a(n-1)-1}$. Thus, we reject H_0: $\beta = 0$ if $F_0 > F_{\alpha,1,a(n-1)-1}$.

Example 17-1

Consider the experiment described in Section 17-1. Three different machines produce a monofilament fiber for a textile company. The process engineer is interested in determining if there is a difference in the breaking strength of the fiber produced by the three machines. However, the strength of a fiber is related to its diameter, with

---

[2] This test statistic is a type III F ratio in SAS.

Table 17-1   Analysis of Covariance as an "Adjusted" Analysis of Variance

Source of Variation	Sum of Squares	Degrees of Freedom	Mean Square	F_0
Regression	$(S_{xy})^2/S_{xx}$	1		
Treatments	$SS'_E - SS_E = $ $S_{yy} - (S_{xy})^2/S_{xx} -$ $[E_{yy} - (E_{xy})^2/E_{xx}]$	$a-1$	$\dfrac{SS'_E - SS_E}{a-1}$	$\dfrac{(SS'_E - SS_E)/(a-1)}{MS_E}$
Error	$SS_E = E_{yy} - (E_{xy})^2/E_{xx}$	$a(n-1)-1$	$MS_E = \dfrac{SS_E}{a(n-1)-1}$	
Total	S_{yy}	$an-1$		

Table 17-2   Analysis of Covariance for a Single Factor Experiment with One Covariate

Source of Variation	Degrees of Freedom	Sums of Squares and Products			Adjusted for Regression		
		x	xy	y		Degrees of Freedom	Mean Square
Treatments	$a-1$	T_{xx}	T_{xy}	T_{yy}			
Error	$a(n-1)$	E_{xx}	E_{xy}	E_{yy}	$SS_E = E_{yy} - (E_{xy})^2/E_{xx}$	$a(n-1)-1$	$MS_E = \dfrac{SS_E}{a(n-1)-1}$
Total	$an-1$	S_{xx}	S_{xy}	S_{yy}	$SS'_E = S_{yy} - (S_{xy})^2/S_{xx}$	$an-2$	
Adjusted treatments					$SS'_E - SS_E$	$a-1$	$\dfrac{SS'_E - SS_E}{a-1}$

Table 17-3 Breaking Strength Data for Example 17-1
(y = strength in pounds and x = diameter in 10^{-3} inches)

Machine 1		Machine 2		Machine 3	
y	x	y	x	y	x
36	20	40	22	35	21
41	25	48	28	37	23
39	24	39	22	42	26
42	25	45	30	34	21
49	32	44	28	32	15
207	126	216	130	180	106

thicker fibers being generally stronger than thinner ones. A random sample of five fiber specimens is selected from each machine. The fiber strength (y) and the corresponding diameter (x) for each specimen are shown in Table 17-3.

A graph of the breaking strength versus the fiber diameter is shown in Figure 17-1. There is a strong suggestion of a linear relationship between breaking strength and diameter, and it seems appropriate to remove the effect of diameter on strength by an analysis of covariance. Assuming that a linear relationship between breaking strength and diameter is appropriate, the model is

$$y_{ij} = \mu + \tau_i + \beta(x_{ij} - \bar{x}_{..}) + \varepsilon_{ij} \begin{cases} i = 1,2,3 \\ j = 1,2, \ldots, 5 \end{cases}$$

Using Equations 17-3 through 17-11, we may compute

$$S_{yy} = \sum_{i=1}^{3} \sum_{j=1}^{5} y_{ij}^2 - \frac{y_{..}^2}{an} = (36)^2 + (41)^2 + \cdots + (32)^2 - \frac{(603)^2}{(3)(5)} = 346.40$$

Figure 17-1. Breaking strength (y) versus fiber diameter (x) for Example 17-1.

$$S_{xx} = \sum_{i=1}^{3} \sum_{j=1}^{5} x_{ij} - \frac{x_{..}^2}{an} = (20)^2 + (25)^2 + \cdots + (15)^2 - \frac{(362)^2}{(3)(5)} = 261.73$$

$$S_{xy} = \sum_{i=1}^{3} \sum_{j=1}^{5} x_{ij} y_{ij} - \frac{(x_{..})(y_{..})}{an} = (20)(36) + (25)(41) + \cdots + (15)(32)$$

$$- \frac{(362)(603)}{(3)(5)} = 282.60$$

$$T_{yy} = \sum_{i=1}^{3} \frac{y_{i.}^2}{n} - \frac{y_{..}^2}{an} = \frac{(207)^2 + (216)^2 + (180)^2}{5} - \frac{(603)^2}{(3)(5)} = 140.40$$

$$T_{xx} = \sum_{i=1}^{3} \frac{x_{i.}^2}{n} - \frac{x_{..}^2}{an} = \frac{(126)^2 + (130)^2 + (106)^2}{5} - \frac{(362)^2}{(3)(5)} = 66.13$$

$$T_{xy} = \sum_{i=1}^{3} \frac{x_{i.} y_{i.}}{n} - \frac{(x_{..})(y_{..})}{an} = \frac{(126)(207) + (130)(216) + (106)(184)}{5}$$

$$- \frac{(362)(603)}{(3)(5)} = 96.00$$

$$E_{yy} = S_{yy} - T_{yy} = 346.40 - 140.40 = 206.00$$

$$E_{xx} = S_{xx} - T_{xx} = 261.73 - 66.13 = 195.60$$

$$E_{xy} = S_{xy} - T_{xy} = 282.60 - 96.00 = 186.60$$

From Equation 17-15, we find

$$SS_E' = S_{yy} - (S_{xy})^2 / S_{xx}$$

$$= 346.40 - (282.60)^2 / 261.73$$

$$= 41.27$$

with $an - 2 = (3)(5) - 2 = 13$ degrees of freedom; and from Equation 17-13, we find

$$SS_E = E_{yy} - (E_{xy})^2 / E_{xx}$$

$$= 206.00 - (186.60)^2 / 195.60$$

$$= 27.99$$

with $a(n - 1) - 1 = 3(5 - 1) - 1 = 11$ degrees of freedom.

The sum of squares for testing $H_0: \tau_i = 0$ is

$$SS_E' - SS_E = 41.27 - 27.99$$

$$= 13.28$$

with $a - 1 = 3 - 1 = 2$ degrees of freedom. These calculations are summarized in Table 17-4.

To test the hypothesis that machines differ in the breaking strength of fiber produced, that is, $H_0: \tau_i = 0$, we compute the test statistic from Equation 17-16 as

$$F_0 = \frac{(SS_E' - SS_E)/(a - 1)}{SS_E/[a(n - 1) - 1]}$$

$$= \frac{13.28/2}{27.99/11} = \frac{6.64}{2.54} = 2.61$$

Table 17-4   Analysis of Covariance for the Breaking Strength Data

Source of Variation	Degrees of Freedom	Sums of Squares and Products			Adjusted for Regression		
		x	xy	y	y	Degrees of Freedom	Mean Square
Machines	2	66.13	96.00	140.40			
Error	12	195.60	186.60	206.00	27.99	11	2.54
Total	14	261.73	282.60	346.40	41.27	13	
Adjusted machines					13.28	2	6.64

Comparing this to $F_{.10,2,11} = 2.86$, we find that the null hypothesis cannot be rejected. That is, there is no strong evidence that the fibers produced by the three machines differ in breaking strength.

The estimate of the regression coefficient is computed from Equation 17-12 as

$$\hat{\beta} = \frac{E_{xy}}{E_{xx}} = \frac{186.60}{195.60} = 0.9540$$

We may test the hypothesis $H_0: \beta = 0$ by using Equation 17-19. The test statistic is

$$F_0 = \frac{(E_{xy})^2/E_{xx}}{MS_E} = \frac{(186.60)^2/195.60}{2.54} = 70.08$$

and since $F_{.01,1,11} = 9.65$, we reject the hypothesis that $\beta = 0$. Therefore, there is a linear relationship between breaking strength and diameter, and the adjustment provided by the analysis of covariance was necessary.

The adjusted treatment means may be computed from Equation 17-17. These adjusted means are

$$\text{Adjusted } \bar{y}_{1.} = \bar{y}_{1.} - \hat{\beta}(\bar{x}_{1.} - \bar{x}_{..})$$
$$= 41.40 - (0.9540)(25.20 - 24.13) = 40.38$$
$$\text{Adjusted } \bar{y}_{2.} = \bar{y}_{2.} - \hat{\beta}(\bar{x}_{2.} - \bar{x}_{..})$$
$$= 43.20 - (0.9540)(26.00 - 24.13) = 41.42$$

and

$$\text{Adjusted } \bar{y}_{3.} = \bar{y}_{3.} - \hat{\beta}(\bar{x}_{3.} - \bar{x}_{..})$$
$$= 36.00 - (0.9540)(21.20 - 24.13) = 38.80$$

Comparing the adjusted treatment means with the unadjusted treatment means (the $\bar{y}_{i.}$), we note that the adjusted means are much closer together, another indication that the covariance analysis was necessary.

A basic assumption in the analysis of covariance is that the treatments do not influence the covariate x because the technique removes the effect of variations in the $\bar{x}_{i.}$. However, if the variability in the $\bar{x}_{i.}$ is due in part to the treatments, then analysis of covariance removes part of the treatment effect. Thus, we must be reasonably sure that the treatments do not affect the values x_{ij}. In some experiments this may be obvious from the nature of the covariate, whereas in others it may be more doubtful. In our example, there may be a difference in fiber diameter (x_{ij}) between the three machines. In such cases, Cochran and Cox (1957) suggest that an analysis of variance on the x_{ij} values may be helpful in determining the validity of this assumption. For our problem, this procedure yields

$$F_0 = \frac{66.13/2}{195.60/12} = \frac{33.07}{16.30} = 2.03$$

which is less than $F_{.10,2,12} = 2.81$, so there is no reason to believe that machines produce fibers of different diameters.

■

Diagnostic checking of the covariance model is based on residual analysis. For the covariance model, the residuals are

$$e_{ij} = y_{ij} - \hat{y}_{ij}$$

where the fitted values are

$$\hat{y}_{ij} = \hat{\mu} + \hat{\tau}_i + \hat{\beta}(x_{ij} - \bar{x}_{..}) = \bar{y}_{..} + [\bar{y}_{i.} - \bar{y}_{..} - \hat{\beta}(\bar{x}_{i.} - \bar{x}_{..})]$$
$$+ \hat{\beta}(x_{ij} - \bar{x}_{..}) = \bar{y}_{i.} + \hat{\beta}(x_{ij} - \bar{x}_{i.})$$

Thus,

$$e_{ij} = y_{ij} - \bar{y}_{i.} - \hat{\beta}(x_{ij} - \bar{x}_{i.}) \qquad (17\text{-}20)$$

To illustrate the use of Equation 17-20, the residual for the first observation from the first machine in Example 17-1 is

$$e_{11} = y_{11} - \bar{y}_{1.} - \hat{\beta}(x_{11} - \bar{x}_{1.}) = 36 - 41.4 - (0.9540)(20 - 25.2)$$
$$= 36 - 36.4392 = -0.4392$$

A complete listing of observations, fitted values, and residuals is given in the following table:

Observed Value y_{ij}	Fitted Value $\hat{y}_{ij}$	Residual $e_{ij} = y_{ij} - \hat{y}_{ij}$
36	36.4392	−0.4392
41	41.2092	−0.2092
39	40.2552	−1.2552
42	41.2092	0.7908
49	47.8871	1.1129
40	39.3840	0.6160
48	45.1079	2.8921
39	39.3840	−0.3840
45	47.0159	−2.0159
44	45.1079	−1.1079
35	35.8092	−0.8092
37	37.7171	−0.7171
42	40.5791	1.4209
34	35.8092	−1.8092
32	30.0852	1.9148

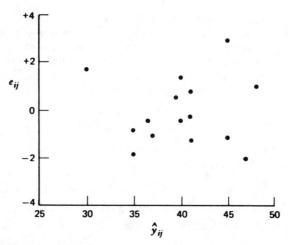

Figure 17-2. Plot of residuals versus fitted values for Example 17-1.

The residuals are plotted versus the fitted values $\hat{y}_{ij}$ in Figure 17-2, versus the covariate x_{ij} in Figure 17-3, and versus the machines in Figure 17-4. A normal probability plot of the residuals is shown in Figure 17-5. These plots do not reveal any major departures from the assumptions, so we conclude that the covariance model (Equation 17-1) is appropriate for the breaking strength data.

It is interesting to note what would have happened in this experiment if an analysis of covariance had not been performed, that is, if the breaking strength data (y) had been analyzed as a single-factor experiment in which the covariate x was ignored. The analysis of variance of the breaking strength data is shown in Table 17-5. Since $F_{.05,2,12} = 3.89$, we would conclude that machines differ significantly in the strength of fiber produced. This is exactly opposite *the conclusion* reached by the covariance analysis. If we suspected that the machines

Figure 17-3. Plot of residuals versus fiber diameter x for Example 17-1.

Figure 17-4. Plot of residuals versus machine.

differed significantly in their effect on fiber strength, then we would try to equalize the strength output of the three machines. However, in this problem the machines do not differ in the strength of fiber produced after the linear effect of fiber diameter is removed. It would be helpful to reduce the within-machine fiber diameter variability since this would probably reduce the strength variability in the fiber.

Sample Computer Output ▪ The computer output from the SAS General Linear Models procedure for the data in Example 17-1 is shown in Figure 17-6. This output is very similar to those presented previously. The model sum of squares contains both the sum of squares due to the machines and the sum of squares due to fiber diameter (x). The type I sums of squares correspond to a "sequential" partitioning of the model sum of squares, say

$$SS \text{ (Model)} = SS \text{ (Machine)} + SS \text{ (x|Machine)}$$
$$= 140.40 + 178.01$$
$$= 318.41$$

whereas the type III sums of squares correspond to the "extra" sum of squares for each factor, that is,

$$SS \text{ (Machine|x)} = 13.28$$

and

$$SS \text{ (x|Machine)} = 178.01$$

Note that SS (Machine|x) is the correct sum of squares to use for testing for no machine effect.

The program also computes the adjusted treatment means from Equation 17-17 (SAS refers to those as least squares means or "LSMEAN" on the sample

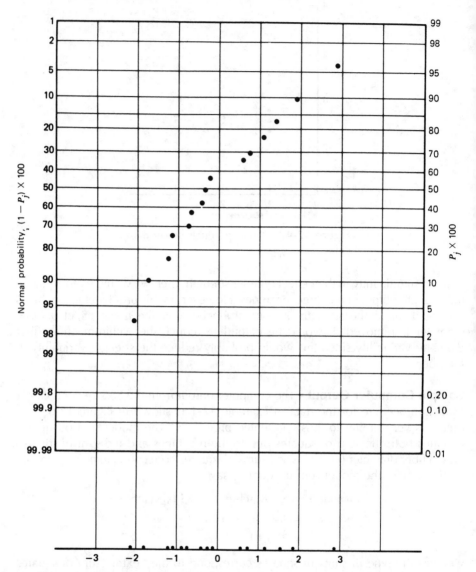

Figure 17-5.   Normal probability plot of residuals for Example 17-1.

output) and the standard errors, and it constructs the test statistic for the hypothesis that each adjusted treatment mean equals zero. In our example, all these means differ from zero. The program also compares all pairs of treatment means and displays the significance level at which the pair of means would be different. We observe that means 2 and 3 differ at the .0433 level, whereas means 1 and 3 differ at the .1803 level, and means 1 and 2 differ at the .3290 level. It seems reasonable to conclude that only means 2 and 3 differ significantly.

Table 17-5   Incorrect Analysis of the Breaking Strength Data as a
Single-Factor Experiment

Source of Variation	Sum of Squares	Degrees of Freedom	Mean Square	F_0
Machines	140.40	2	70.20	4.09[a]
Error	206.00	12	17.17	
Total	346.40	14		

[a] Significant at 5 percent.

17-3 DEVELOPMENT BY THE GENERAL REGRESSION SIGNIFICANCE TEST

It is possible to develop formally the procedure for testing $H_0: \tau_i = 0$ in the covariance model

$$y_{ij} = \mu + \tau_i + \beta(x_{ij} - \bar{x}_{..}) + \varepsilon_{ij} \begin{cases} i = 1,2, \ldots, a \\ j = 1,2, \ldots, n \end{cases} \quad (17\text{-}21)$$

using the general regression significance test. Consider estimating the parameters in the model (Equation 17-21) by least squares. The least squares function is

$$L = \sum_{i=1}^{a} \sum_{j=1}^{n} [y_{ij} - \mu - \tau_i - \beta(x_{ij} - \bar{x}_{..})]^2 \quad (17\text{-}22)$$

and from $\partial L/\partial \mu = \partial L/\partial \tau_i = \partial L/\partial \beta = 0$, we obtain the normal equations

$$\mu: \quad an\hat{\mu} + n \sum_{i=1}^{a} \hat{\tau}_i = y_{..} \quad (17\text{-}23a)$$

$$\tau_i: \quad n\hat{\mu} + n\hat{\tau}_i + \hat{\beta} \sum_{j=1}^{n} (x_{ij} - \bar{x}_{..}) = y_{i.} \quad i = 1,2, \ldots, a \quad (17\text{-}23b)$$

$$\beta: \quad \sum_{i=1}^{a} \hat{\tau}_i \sum_{j=1}^{n} (x_{ij} - \bar{x}_{..}) + \hat{\beta} S_{xx} = S_{xy} \quad (17\text{-}23c)$$

Adding the a equations in Equation 17-23b, we obtain Equation 17-23a since $\Sigma_{i=1}^{a} \Sigma_{j=1}^{n} (x_{ij} - \bar{x}_{..}) = 0$, so there is one linear dependency in the normal equations. Therefore, it is necessary to augment Equations 17-23 with a linearly independent equation in order to obtain a solution. A logical side condition is $\Sigma_{i=1}^{a} \hat{\tau}_i = 0$. Using this condition, we obtain from Equation 17-23a

$$\hat{\mu} = \bar{y}_{..} \quad (17\text{-}24a)$$

and from Equation 17-23b

$$\hat{\tau}_i = \bar{y}_{i.} - \bar{y}_{..} - \hat{\beta}(\bar{x}_{i.} - \bar{x}_{..}) \quad (17\text{-}24b)$$

13:16 WEDNESDAY, JANUARY 17, 1990

S T A T I S T I C A L   A N A L Y S I S   S Y S T E M

GENERAL LINEAR MODELS PROCEDURE

DEPENDENT VARIABLE: Y

SOURCE	DF	SUM OF SQUARES	MEAN SQUARE	F VALUE	PR > F	R-SQUARE	C.V.
MODEL	3	318.41411043	106.13803681	41.72	0.0001	0.919209	03.9678
ERROR	11	27.98588957	2.54417178			STD DEV	Y MEAN
CORRECTED TOTAL	14	346.40000000				1.59504601	40.20000000

SOURCE	DF	TYPE I SS	F VALUE	PR > F	DF	TYPE III SS	F VALUE	PR > F
MACHINE	2	140.40000000	27.59	0.0001	2	13.28385062	2.61	0.1181
X	1	178.01411043	69.97	0.0001	1	178.01411043	69.97	0.0001

OBSERVATION	OBSERVED VALUE	PREDICTED VALUE	RESIDUAL
1	36.00000000	36.43926380	-0.43926380
2	41.00000000	41.20920245	-0.20920245
3	39.00000000	40.25521472	-1.25521472
4	42.00000000	41.20920245	0.79079755
5	49.00000000	47.88711656	1.11288344
6	40.00000000	39.38404908	0.61595092
7	48.00000000	45.10797546	2.89202454
8	39.00000000	39.38404908	-0.38404908
9	45.00000000	47.01595092	-2.01595092
10	44.00000000	45.10797546	-1.10797546
11	35.00000000	35.80920245	-0.80920245
12	37.00000000	37.71717791	-0.71717791
13	42.00000000	40.57914110	1.42085890
14	34.00000000	35.80920245	-1.80920245
15	32.00000000	30.08527607	1.91472393

SUM OF RESIDUALS	0.00000000
SUM OF SQUARED RESIDUALS	27.98588957
SUM OF SQUARED RESIDUALS - ERROR SS	-0.00000000
FIRST ORDER AUTOCORRELATION	-0.03469267
DURBIN-WATSON D	1.93149012

LEAST SQUARES MEANS

MACHINE	Y LSMEAN	STD ERR LSMEAN	PROB > $|T|$ HO:LSMEAN=0	I/J	1	2	3
1	40.3824131	0.7236252	0.0001	1		0.3280	0.1803
2	41.4192229	0.7444169	0.0001	2	0.3280		0.0433
3	38.7983640	0.7878785	0.0001	3	0.1803	0.0433	

HO: LSMEAN(I)=LSMEAN(J)

NOTE: TO ENSURE OVERALL PROTECTION LEVEL, ONLY PROBABILITIES ASSOCIATED WITH PRE-PLANNED COMPARISONS SHOULD BE USED.

Equation 17-23c may be rewritten as

$$\sum_{i=1}^{a} (\bar{y}_{i.} - \bar{y}_{..}) \sum_{j=1}^{n} (x_{ij} - \bar{x}_{..}) - \hat{\beta} \sum_{i=1}^{a} (x_{i.} - \bar{x}_{..}) \sum_{j=1}^{n} (x_{ij} - \bar{x}_{..}) + \hat{\beta} S_{xx} = S_{xy}$$

after substituting for $\hat{\tau}_i$. But we see that

$$\sum_{i=1}^{a} (\bar{y}_{i.} - \bar{y}_{..}) \sum_{j=1}^{n} (x_{ij} - \bar{x}_{..}) = T_{xy}$$

and

$$\sum_{i=1}^{a} (\bar{x}_{i.} - \bar{x}_{..}) \sum_{j=1}^{n} (x_{ij} - \bar{x}_{..}) = T_{xx}$$

Therefore, the solution to Equation 17-23c is

$$\hat{\beta} = \frac{S_{xy} - T_{xy}}{S_{xx} - T_{xx}} = \frac{E_{xy}}{E_{xx}}$$

which was the result given previously in Section 17-2, Equation 17-12.

We may express the reduction in the total sum of squares due to fitting the model (Equation 17-21) as

$$R(\mu, \tau, \beta) = \hat{\mu} y_{..} + \sum_{i=1}^{a} \hat{\tau}_i y_{i.} + \hat{\beta} S_{xy}$$

$$= (\bar{y}_{..}) y_{..} + \sum_{i=1}^{a} [\bar{y}_{i.} - \bar{y}_{..} - (E_{xy}/E_{xx})(\bar{x}_{i.} - \bar{x}_{..})] y_{i.} + (E_{xy}/E_{xx}) S_{xy}$$

$$= y_{..}^2/an + \sum_{i=1}^{a} (\bar{y}_{i.} - \bar{y}_{..}) y_{i.} - (E_{xy}/E_{xx}) \sum_{i=1}^{a} (\bar{x}_{i.} - \bar{x}_{..}) y_{i.} + (E_{xy}/E_{xx}) S_{xy}$$

$$= y_{..}^2/an + T_{yy} - (E_{xy}/E_{xx})(T_{xy} - S_{xy})$$

$$= y_{..}^2/an + T_{yy} + (E_{xy})^2/E_{xx}$$

This sum of squares has $a + 1$ degrees of freedom since the rank of the normal equations is $a + 1$. The error sum of squares for this model is

$$SS_E = \sum_{i=1}^{a} \sum_{j=1}^{n} y_{ij}^2 - R(\mu, \tau, \beta)$$

$$= \sum_{i=1}^{a} \sum_{j=1}^{n} y_{ij}^2 - y_{..}^2/an - T_{yy} - (E_{xy})^2/E_{xx} \qquad (17\text{-}25)$$

$$= S_{yy} - T_{yy} - (E_{xy})^2/E_{xx}$$

$$= E_{yy} - (E_{xy})^2/E_{xx}$$

with $an - (a + 1) = a(n - 1) - 1$ degrees of freedom. This quantity was obtained previously as Equation 17-13.

Now consider the model restricted to the null hypothesis, that is, to $H_0: \tau_i = 0$. This reduced model is

$$y_{ij} = \mu + \beta(x_{ij} - \bar{x}_{..}) + \varepsilon_{ij} \begin{cases} i = 1, 2, \ldots, a \\ j = 1, 2, \ldots, n \end{cases} \quad (17\text{-}26)$$

We recognize this as a simple linear regression model, and by analogy with Equation 14-6, the normal equations for this model are

$$an\hat{\mu} = y_{..} \quad (17\text{-}27a)$$

$$\hat{\beta}S_{xx} = S_{xy} \quad (17\text{-}27b)$$

The solutions to these equations are $\hat{\mu} = \bar{y}_{..}$ and $\hat{\beta} = S_{xy}/S_{xx}$, and the reduction in the total sum of squares due to fitting the reduced model is

$$\begin{aligned} R(\mu, \beta) &= \hat{\mu}y_{..} + \hat{\beta}S_{xy} \\ &= (\bar{y}_{..})y_{..} + (S_{xy}/S_{xx})S_{xy} \quad (17\text{-}28) \\ &= y_{..}^2/an + (S_{xy})^2/S_{xx} \end{aligned}$$

This sum of squares has 2 degrees of freedom.

We may find the appropriate sum of squares for testing $H_0: \tau_i = 0$ as

$$\begin{aligned} R(\tau|\mu, \beta) &= R(\mu, \tau, \beta) - R(\mu, \beta) \\ &= y_{..}^2/an + T_{yy} + (E_{xy})^2/E_{xx} - y_{..}^2/an - (S_{xy})^2/S_{xx} \quad (17\text{-}29) \\ &= S_{yy} - (S_{xy})^2/S_{xx} - [E_{yy} - (E_{xy})^2/E_{xx}] \end{aligned}$$

using $T_{yy} = S_{yy} - E_{yy}$. Note that $R(\tau|\mu, \beta)$ has $a + 1 - 2 = a - 1$ degrees of freedom and is identical to the sum of squares given by $SS'_E - SS_E$ in Section 17-2. Thus, the test statistic for $H_0: \tau_i = 0$ is

$$F_0 = \frac{R(\tau|\mu, \beta)/(a - 1)}{SS_E/[a(n - 1) - 1]} = \frac{(SS'_E - SS_E)/(a - 1)}{SS_E/[a(n - 1) - 1]} \quad (17\text{-}30)$$

which we gave previously as Equation 17-16. Therefore, by using the general regression significance test, we have justified the heuristic development of the analysis of covariance in Section 17-2.

17-4 OTHER COVARIANCE MODELS

The general procedure for the analysis of covariance can be embedded into any experimental design model. For example, consider the randomized complete block design with a single covariate

$$y_{ij} = \mu + \tau_i + \gamma_j + \beta(x_{ij} - \bar{x}_{..}) + \varepsilon_{ij} \begin{cases} i = 1, 2, \ldots, a \\ j = 1, 2, \ldots, b \end{cases} \quad (17\text{-}31)$$

where y_{ij} is the observation on treatment i in block j, μ is an overall mean, τ_i is the effect of the ith treatment, γ_j is the effect of the jth block, β is a regression coefficient, x_{ij} is the ijth observation on the covariate, and ε_{ij} is the NID(0, σ^2) random error component.

The computational details of the analysis of covariance for the randomized complete block design are displayed in Table 17-6. The definitions of the quantities in this table are

$$R_{xx} = \sum \frac{x_{.j}^2}{a} - \frac{x_{..}^2}{ab} \qquad R_{yy} = \sum \frac{y_{.j}^2}{a} - \frac{y_{..}^2}{ab}$$

$$R_{xy} = \sum \frac{x_{.j}y_{.j}}{a} - \frac{x_{..}y_{..}}{ab}$$

$$T_{xx} = \sum \frac{x_{i.}^2}{b} - \frac{x_{..}^2}{ab} \qquad T_{yy} = \sum \frac{y_{i.}^2}{b} - \frac{y_{..}^2}{ab}$$

$$T_{xy} = \sum \frac{x_{i.}y_{i.}}{b} - \frac{x_{..}y_{..}}{ab}$$

$$S_{xx} = \sum\sum x_{ij}^2 - \frac{x_{..}^2}{ab} \qquad S_{yy} = \sum\sum y_{ij}^2 - \frac{y_{..}^2}{ab}$$

$$S_{xy} = \sum\sum x_{ij}y_{ij} - \frac{x_{..}y_{..}}{ab}$$

$$E_{xx} = S_{xx} - R_{xx} - T_{xx} \qquad E_{yy} = S_{yy} - R_{yy} - T_{yy}$$

$$E_{xy} = S_{xy} - R_{xy} - T_{xy}$$

Notice that the procedure is similar to the usual analysis of variance, except for the adjustment afforded by regression. Comparing Tables 17-6 and 17-2, we see that in the randomized complete block design we have reduced the residuals E_{yy}, E_{xx}, and E_{xy} by a sum of squares or products due to blocks. Also, the sum of squares for the adjusted treatments is computed as the difference between the adjusted error sum of squares and an adjusted sum of squares for "treatments plus error" so that block effects may be eliminated. The regression coefficient β is estimated by

$$\hat{\beta} = \frac{E_{xy}}{E_{xx}} \tag{17-32}$$

and the appropriate statistic for testing $H_0: \beta = 0$ is

$$F_0 = \frac{(E_{xy})^2/E_{xx}}{MS_E} \tag{17-33}$$

where the definition of MS_E is found in Table 16-6. Under the null hypothesis, F_0 is distributed as $F_{1,ab-a-b}$, so $H_0: \beta = 0$ would be rejected if $F_0 > F_{\alpha,1,ab-a-b}$.

The analysis of covariance may be extended to Latin squares, incomplete blocks, factorials, nested designs, and so on. It may also be used to compare the slopes of several regression lines. Finally, situations may arise in which a multiple covariance structure is necessary; that is, either y is related to x in a nonlinear fashion (such as $y = \beta_1 x + \beta_2 x^2$), or there are two or more covariates. These

Table 17-6 Analysis of Covariance for the Randomized Complete Block Design

Source of Variation	Degrees of Freedom	Sums of Squares and Products			Adjusted for Regression			
		x	xy	y	y	Degrees of Freedom	Mean Square	F_0
Blocks	$b - 1$	R_{xx}	R_{xy}	R_{yy}	—	—	—	
Treatments	$a - 1$	T_{xx}	T_{xy}	T_{yy}	—	—	—	
Error	$(a - 1)(b - 1)$	E_{xx}	E_{xy}	E_{yy}	$SS_E = E_y$ $-(E_{xy})^2/E_{xx}$	$ab - a - b$	$MS_E = \dfrac{SS_E}{ab - a - b}$	
Total	$ab - 1$	S_{xx}	S_{xy}	S_{yy}		$ab - 2$		
Treatments plus error	$b(a - 1)$	$S'_{xx} = T_{xx}$ $+ E_{xx}$	$S'_{xy} = T_{xy}$ $+ E_{xy}$	$S'_{yy} = T_{yy}$ $+ E_{yy}$	$SS'_E = S'_{yy}$ $- (S'_{xy})^2/S'_{xx}$	$b(a - 1) - 1$		
Adjusted treatments					$SS'_E - SS_E$	$a - 1$	$MS_A = \dfrac{SS'_E - SS_E}{a - 1}$	$\dfrac{MS_A}{MS_E}$

and other related topics are discussed in Cochran (1957), Duncan (1974), Federer (1957), and Ostle (1963). The latter two references are found in the third volume of *Biometrics* (1957), which is devoted almost exclusively to the analysis of covariance.

17-5 PROBLEMS

17-1 A soft drink distributor is studying the effectiveness of delivery methods. Three different types of hand trucks have been developed, and an experiment is performed in the company's methods engineering laboratory. The variable of interest is the delivery time in minutes (y); however, delivery time is also strongly related to the case volume delivered (x). Each hand truck is used four times and the data that follow are obtained. Analyze these data and draw appropriate conclusions.

	Hand Truck Type					
	1		2		3	
	y	x	y	x	y	x
	27	24	25	26	40	38
	44	40	35	32	22	26
	33	35	46	42	53	50
	41	40	26	25	18	20

17-2 Compute the adjusted treatment means and the standard errors of the adjusted treatment means for the data in Problem 17-1.

17-3 The sums of squares and products for a single-factor analysis of covariance follow. Complete the analysis and draw appropriate conclusions.

Source of Variation	Degrees of Freedom	Sums of Squares and Products		
		x	xy	y
Treatment	3	1500	1000	650
Error	12	6000	1200	550
Total	15	7500	2200	1200

17-4 Find the standard errors of the adjusted treatment means in Example 17-1.

17-5 Four different formulations of an industrial glue are being tested. The tensile strength of the glue is also related to the thickness. Five observations on strength

(y) in pounds and thickness (x) in 0.01 inches are obtained for each formulation. The data are shown in the following table. Analyze these data and draw appropriate conclusions.

				Glue Formulation				
1		2		3		4		
y	x	y	x	y	x	y	x	
46.5	13	48.7	12	46.3	15	44.7	16	
45.9	14	49.0	10	47.1	14	43.0	15	
49.8	12	50.1	11	48.9	11	51.0	10	
46.1	12	48.5	12	48.2	11	48.1	12	
44.3	14	45.2	14	50.3	10	48.6	11	

17-6 Compute the adjusted treatment means and their standard errors using the data in Problem 17-5.

17-7 An engineer is studying the effect of cutting speed on the rate of metal removal in a machining operation. However, the rate of metal removal is also related to the hardness of the test specimen. Five observations are taken at each cutting speed. The amount of metal removed (y) and the hardness of the specimen (x) are shown in the following table. Analyze the data using an analysis of covariance.

Cutting Speed (rpm)					
1000		1200		1400	
y	x	y	x	y	x
68	120	112	165	118	175
90	140	94	140	82	132
98	150	65	120	73	124
77	125	74	125	92	141
88	136	85	133	80	130

17-8 Show that in a single-factor analysis of covariance with a single covariate a 100 $(1 - \alpha)$ percent confidence interval on the ith adjusted treatment mean is

$$\bar{y}_{i.} - \hat{\beta}(\bar{x}_{i.} - \bar{x}_{..}) \pm t_{\alpha/2, a(n-1)-1} \left[MS_E \left(\frac{1}{n} + \frac{(\bar{x}_{i.} - \bar{x}_{..})^2}{E_{xx}} \right) \right]^{1/2}$$

Using this formula, calculate a 95 percent confidence interval on the adjusted mean of machine 1 in Example 17-1.

17-9 Show that in a single-factor analysis of covariance with a single covariate, the standard error of the difference between any two adjusted treatment means is

$$S_{\text{Adj}.\bar{y}_{i.} - \text{Adj}.\bar{y}_{j.}} = \left[MS_E \left(\frac{2}{n} + \frac{(\bar{x}_{i.} - \bar{x}_{j.})^2}{E_{xx}} \right) \right]^{1/2}$$

17-10 A randomized complete block design with one covariate produced the following table of corrected sums of squares and products. Complete the analysis of covariance and draw appropriate conclusions.

Source of Variation	Degrees of Freedom	Sums of Squares and Products		
		x	xy	y
Blocks	8	200	600	1200
Treatments	4	100	300	800
Error	32	600	700	1400

17-11 Write down the normal equations for a randomized complete block design with a single covariate. Solve the normal equations for the estimators of the model parameters.

17-12 Discuss how the operating characteristic curves for the analysis of variance can be used in the analysis of covariance.

Bibliography

1. Addelman, S. (1961). "Irregular Functions of the 2^n Factorial Experiments." *Technometrics,* Vol. 3, pp. 479–496.

2. Addleman, S. (1962). "Orthogonal Main Effect Plans for Asymmetric Factorial Experiments." *Technometrics,* Vol. 4, pp. 21–46.

3. Addelman, S. (1963). "Techniques for Constructing Fractional Replicate Plans." *Journal of the American Statistical Association,* Vol. 58, pp. 45–71.

4. Anderson, V. L. and R. A. McLean (1974). *Design of Experiments: A Realistic Approach.* Marcel Dekker Inc., New York.

5. Anscombe, F. J. (1960). "Rejection of Outliers." *Technometrics,* Vol. 2, pp. 123–147.

6. Anscombe, F. J. and J. W. Tukey (1963). "The Examination and Analysis of Residuals." *Technometrics,* Vol. 5, pp. 141–160.

7. Bancroft, T. A. (1968). *Topics in Intermediate Statistical Methods.* Iowa State University Press, Ames, Iowa.

8. Barnett, V. and T. Lewis (1978). *Outliers in Statistical Data.* Wiley, New York.

9. Bartlett, M. S. (1947). "The Use of Transformations." *Biometrics,* Vol. 3, pp. 39–52.

10. Bennett, C. A. and N. L. Franklin (1954). *Statistical Analysis in Chemistry and the Chemical Industry.* Wiley, New York.

11. Bose, R. C. and T. Shimamoto (1952). "Classification and Analysis of Partially Balanced Incomplete Block Designs with Two Associate Classes." *Journal of the American Statistical Association,* Vol. 47, pp. 151–184.

12. Bose, R. C., W. H. Clatworthy, and S. S. Shrikhande (1954). *Tables of Partially Balanced Designs with Two Associate Classes.* Technical Bulletin No. 107, North Carolina Agricultural Experiment Station.

13. Bowker, A. H. and G. J. Lieberman (1972). *Engineering Statistics.* 2nd edition. Prentice-Hall, Englewood Cliffs, N.J.

14. Box, G. E. P. and K. G. Wilson (1951). "On the Experimental Attainment of Optimum Conditions." *Journal of the Royal Statistical Society,* B, Vol. 13, pp. 1–45.

15. Box, G. E. P. (1954a). "Some Theorems on Quadratic Forms Applied in the Study of Analysis of Variance Problems: I. Effect of Inequality of Variance in the One-Way Classification." *Annals of Mathematical Statistics*, Vol. 25, pp. 290–302.

16. Box, G. E. P. (1954b). "Some Theorems on Quadratic Forms Applied in the Study of Analysis of Variance Problems: II. Effect of Inequality of Variance and of Correlation of Errors in the Two-Way Classification." *Annals of Mathematical Statistics*, Vol. 25, pp. 484–498.

17. Box, G. E. P. (1957). "Evolutionary Operation: A Method for Increasing Industrial Productivity." *Applied Statistics*, Vol. 6, pp. 81–101.

18. Box, G. E. P. and J. S. Hunter (1957). "Multifactor Experimental Designs for Exploring Response Surfaces." *Annals of Mathematical Statistics*, Vol. 28, pp. 195–242.

19. Box, G. E. P. and D. W. Behnken (1960). "Some New Three Level Designs for the Study of Quantitative Variables." *Technometrics*, Vol. 2, pp. 455–476.

20. Box, G. E. P. and J. S. Hunter (1961a). "The 2^{k-p} Fractional Factorial Designs, Part I." *Technometrics*, Vol. 3, pp. 311–352.

21. Box, G. E. P. and J. S. Hunter (1961b). "The 2^{k-p} Fractional Factorial Designs, Part II." *Technometrics*, Vol. 3, pp. 449–458.

22. Box, G. E. P. and D. R. Cox (1964). "An Analysis of Transformations." *Journal of the Royal Statistical Society*, B, Vol. 26, pp. 211–243.

23. Box, G. E. P. and N. R. Draper (1969). *Evolutionary Operation.* Wiley, New York.

24. Box, G. E. P., W. G. Hunter, and J. S. Hunter (1978). *Statistics for Experimenters.* Wiley, New York.

25. Box, G. E. P. and R. D. Meyer (1986). "An Analysis of Unreplicated Fractional Factorials." *Technometrics*, Vol. 28, pp. 11–18.

26. Box, G. E. P. and N. R. Draper (1987). *Empirical Model Building and Response Surfaces.* Wiley, New York.

27. Box, G. E. P. (1988). "Signal-to-Noise Ratios, Performance Criteria, and Transformation." *Technometrics*, Vol. 30, pp. 1–40.

28. Box, G. E. P., S. Bisgaard, and C. A. Fung (1988). "An Explanation and Critique of Taguchi's Contributions to Quality Engineering." *Quality and Reliability Engineering International*, Vol. 4, pp. 123–131.

29. Box, J. F. (1978). *R. A. Fisher: The Life of a Scientist.* Wiley, New York.

30. Carmer, S. G. and M. R. Swanson (1973). "Evaluation of Ten Pairwise Multiple Comparison Procedures by Monte Carlo Methods." *Journal of the American Statistical Association*, Vol. 68, No. 314, pp. 66–74.

31. Cochran, W. G. (1947). "Some Consequences When the Assumptions for the Analysis of Variance are not Satisfied." *Biometrics*, Vol. 3, pp. 22–38.

32. Cochran, W.G. (1957). "Analysis of Covariance: Its Nature and Uses." *Biometrics*, Vol. 13, No. 3, pp. 261–281.

33. Cochran, W. G. and G. M. Cox (1957). *Experimental Designs.* 2nd edition. Wiley, New York.

34. Connor, W. S. and M. Zelen (1959). *Fractional Factorial Experimental Designs for Factors at Three Levels.* National Bureau of Standards, Washington, D.C., Applied Mathematics Series, No. 54.

35. Conover, W. J. (1980). *Practical Nonparametric Statistics*. 2nd edition. Wiley, New York.

36. Conover, W. J. and R. L. Iman (1976). "On Some Alternative Procedures Using Ranks for the Analysis of Experimental Designs." *Communications in Statistics*, Vol. A5, pp. 1349–1368.

37. Conover, W. J. and R. L. Iman (1981). "Rank Transformations as a Bridge Between Parametric and Nonparametric Statistics" (with discussion). *The American Statistician*, Vol. 35, pp. 124–133.

38. Cornell, J. A. (1981). *Experiments with Mixtures: Designs, Models, and the Analysis of Mixture Data*. Wiley, New York.

39. Daniel, C. (1959). "Use of Half-Normal Plots in Interpreting Factorial Two Level Experiments." *Technometrics*, Vol. 1, pp. 311–342.

40. Daniel, C. (1976). *Applications of Statistics to Industrial Experimentation*. Wiley, New York.

41. Davies, O. L. (1956). *Design and Analysis of Industrial Experiments*. 2nd edition. Hafner Publishing Company, New York.

42. Dolby, J. L. (1963). "A Quick Method for Choosing a Transformation." *Technometrics*, Vol. 5, pp. 317–326.

43. Draper, N. R. and W. G. Hunter (1969). "Transformations: Some Examples Revisited." *Technometrics*, Vol. 11, pp. 23–40.

44. Draper, N. R. and H. Smith (1981). *Applied Regression Analysis*. 2nd edition. Wiley, New York.

45. Duncan, A. J. (1974). *Quality Control and Industrial Statistics*. 4th edition. Richard D. Irwin, Inc., Homewood, Ill.

46. Duncan, D. B. (1955). "Multiple Range and Multiple F Tests." *Biometrics*, Vol. 11, pp. 1–42.

47. Dunnett, C. W. (1964). "New Tables for Multiple Comparisons with a Control." *Biometrics*, Vol. 20, pp. 482–491.

48. Eisenhart, C. (1947). "The Assumptions Underlying the Analysis of Variance." *Biometrics*, Vol. 3, pp. 1–21.

49. Federer, W. T. (1957). "Variance and Covariance Analysis for Unbalanced Classifications." *Biometrics*, Vol. 13, No. 3, 333–362.

50. Fisher, R. A. and F. Yates (1953). *Statistical Tables for Biological, Agricultural, and Medical Research*. 4th edition. Oliver and Boyd, Edinburgh.

51. Fisher, R. A. (1958). *Statistical Methods for Research Workers*. 13th edition. Oliver and Boyd, Edinburgh.

52. Fisher, R. A. (1966). *The Design of Experiments*. 8th edition. Hafner Publishing Company, New York.

53. Gaylor, D. W. and T. D. Hartwell (1969). "Expected Mean Squares for Nested Classifications." *Biometrics*, Vol. 25, pp. 427–430.

54. Gaylor, D. W. and F. N. Hopper (1969). "Estimating the Degrees of Freedom for Linear Combinations of Mean Squares by Satterthwaite's Formula." *Technometrics*, Vol. 11, No. 4, pp. 699–706.

55. Good, I. J. (1955). "The Interaction Algorithm and Practical Fourier Analysis." *Journal of the Royal Statistical Society*, B, Vol. 20, pp. 361–372.

56. Good, I. J. (1958). Addendum to "The Interaction Algorithm and Practical Fourier Analysis." *Journal of the Royal Statistical Society*, B, Vol. 22, pp. 372–375.

57. Graybill, F. A. and D. L. Weeks (1959). "Combining Interblock and Intrablock Information in Balanced Incomplete Blocks." *Annals of Mathematical Statistics*, Vol. 30, pp. 799–805.

58. Graybill, F. A. (1961). *An Introduction to Linear Statistical Models*. Vol. 1. McGraw-Hill, New York.

59. Herzberg, A. M. and D. R. Cox (1969). "Recent Work on the Design of Experiments: A Bibliography and a Review," *Journal of the Royal Statistical Society*, A, 132, pp. 29–67.

60. Hicks, C. R. (1973). *Fundamental Concepts in the Design of Experiments*. 2nd edition. Holt, Rinehart and Winston, New York.

61. Hill, W. G. and W. G. Hunter (1966). "A Review of Response Surface Methodology: A Literature Survey." *Technometrics*, Vol. 8, pp. 571–590.

62. Hines, W. W. and D. C. Montgomery (1990). *Probability and Statistics in Engineering and Management Science*. 3rd edition. Wiley, New York.

63. Hocking, R. R. (1973). "A Discussion of the Two-Way Mixed Model." *The American Statistician*, Vol. 27, No. 4, pp. 148–152.

64. Hocking, R. R. and F. M. Speed (1975). "A Full Rank Analysis of Some Linear Model Problems." *Journal of the American Statistical Association*, Vol. 70, pp. 706–712.

65. Hocking, R. R., O. P. Hackney, and F. M. Speed (1978). "The Analysis of Linear Models with Unbalanced Data." In *Contributions to Survey Sampling and Applied Statistics*, H. A. David (ed.), Academic Press, New York.

66. Hunter, J. S. (1966). "The Inverse Yates Algorithm." *Technometrics*, Vol. 8, pp. 177–183.

67. John, J. A. and P. Prescott (1975). "Critical Values of a Test to Detect Outliers in Factorial Experiments." *Applied Statistics*, Vol. 24, pp. 56–59.

68. John, P. W. M. (1961). "The Three-Quarter Replicates of 2^4 and 2^5 Designs." *Biometrics*, Vol. 17, pp. 319–321.

69. John, P. W. M. (1962). "Three-Quarter Replicates of 2^n Designs." *Biometrics*, Vol. 18, pp. 171–184.

70. John, P. W. M. (1964). "Blocking a $3(2^{n-k})$ Designs." *Technometrics*, Vol. 6, pp. 371–376.

71. John, P. W. M. (1971). *Statistical Design and Analysis of Experiments*. The MacMillan Company, New York.

72. Kempthorne, O. (1952). *The Design and Analysis of Experiments*. Wiley, New York.

73. Keuls, M. (1952). "The Use of the Studentized Range in Connection with an Analysis of Variance." *Euphytica*, Vol. 1, pp. 112–122.

74. Khuri, A. I. and J. A. Cornell (1987). *Response Surfaces: Designs and Analyses*. Dekker, New York.

75. Kruskal, W. H. and W. A. Wallis (1952). "Use of Ranks on One Criterion Variance Analysis." *Journal of the American Statistical Association*, Vol. 47, pp. 583–621 (Corrections appear in Vol. 48, pp. 907–911).

76. Leon, R. V., A. C. Shoemaker, and R. N. Kackar (1987). "Performance Measures Independent of Adjustment." *Technometrics*, Vol. 29, pp. 253–265.

77. Margolin, B. H. (1967). "Systematic Methods of Analyzing $2^n 3^m$ Factorial Experiments with Applications." *Technometrics*, Vol. 9, pp. 245–260.

78. Margolin, B. H. (1969). "Results on Factorial Designs of Resolution IV for the 2^n and $2^n 3^m$ Series." *Technometrics*, Vol. 11, pp. 431–444.

79. Miller, R. G. (1966). *Simultaneous Statistical Inference*. McGraw-Hill, New York.

80. Miller, R. G., Jr. (1977). "Developments in Multiple Comparisons, 1966–1976." *Journal of the American Statistical Association*, Vol. 72, pp. 779–788.

81. Montgomerey, D. C. and E. A. Peck (1982). *Introduction to Linear Regression Analysis*. Wiley, New York.

82. Myers, R. H. (1971). *Response Surface Methodology*. Allyn and Bacon, Inc., Boston.

83. Neter, J. and W. Wasserman (1974). *Applied Linear Statistical Models*. Richard D. Irwin, Homewood, Ill.

84. Newman, D. (1939). "The Distribution of the Range in Samples from a Normal Population, Expressed in Terms of an Independent Estimate of Standard Deviation." *Biometrika*, Vol. 31, pp. 20–30.

85. O'Neill, R. and G. B. Wetherill (1971). "The Present State of Multiple Comparison Methods." *Journal of the Royal Statistical Society*, B, Vol. 33, pp. 218–241.

86. Ostle, B. (1963). *Statistics in Research*. 2nd edition. Iowa State Press, Ames, Iowa.

87. Pearson, E. S. and H. O. Hartley (1966). *Biometrika Tables for Statisticians*. Vol. 1, 3rd edition, Cambridge University Press, Cambridge.

88. Pearson, E. S. and H. O. Hartley (1972). *Biometrika Tables for Statisticians*. Vol. 2, Cambridge University Press, Cambridge.

89. Plackett, R. L. and J. P. Burman (1946). "The Design of Optimum Multifactorial Experiments." *Biometrika*, Vol. 33, pp. 305–325.

90. Quenouille, M. H. (1953). *The Design and Analysis of Experiments*. Charles Griffin and Company, London.

91. Quenouille, M. H. (1955). "Checks on the Calculation of Main Effects and Interactions in a 2^n Factorial Experiment." *Annals of Eugenics*, Vol. 19, pp. 151–152.

92. Quinlan, J. (1985). "Product Improvement by Application of Taguchi Methods." *Third Supplier Symposium on Taguchi Methods*, American Supplier Institute, Inc., Dearborne, MI.

93. Rayner, A. A. (1967). "The Square Summing Check on the Main Effects and Interactions in a 2^n Experiment as Calculated by Yates' Algorithm." *Biometrics*, Vol. 23, pp. 571–573.

94. Satterthwaite, F. E. (1946). "An Approximate Distribution of Estimates of Variance Components." *Biometrics Bull.*, Vol. 2, pp. 110–112.

95. Scheffé, H. (1953). "A Method for Judging All Contrasts in the Analysis of Variance," *Biometrika*, Vol. 40, pp. 87–104.

96. Scheffé, H. (1956a). "A 'Mixed Model' for the Analysis of Variance." *Annals of Mathematical Statistics*, Vol. 27, pp. 23–36.

97. Scheffé, H. (1956b). "Alternative Models for the Analysis of Variance." *Annals of Mathematical Statistics*, Vol. 27, pp. 251–271.

98. Scheffé, H. (1959). *The Analysis of Variance*. Wiley, New York.

99. Schmidt, S. R. and J. R. Boudot (1989). "A Monte Carlo Simulation Study Comparing Effectiveness of Signal-to-Noise Ratios and Other Methods for Identifying Dispersion Effects." *Rocky Mountain Quality Control Conference*.

100. Searle, S. R. and R. F. Fawcett (1970). "Expected Mean Squares in Variance Component Models Having Finite Populations." *Biometrics*, Vol. 26, pp. 243–254.

101. Searle, S. R. (1971a). *Linear Models*. Wiley, New York.

102. Searle, S. R. (1971b). "Topics in Variance Component Estimation." *Biometrics*, Vol. 27, pp. 1–76.

103. Searle, S. R., F. M. Speed, and H. V. Henderson (1981). "Some Computational and Model Equivalences in Analyses of Variance of Unequal-Subclass-Numbers Data." *The American Statistician*, Vol. 35, pp. 16–33.

104. Searle, S. R. (1987). *Linear Models for Unbalanced Data*. Wiley, New York.

105. Smith, C. A. B. and H. O. Hartley (1948). "Construction of Youden Squares." *Journal of the Royal Statistical Society*, B, Vol. 10, pp. 262–264.

106. Smith, H. F. (1957). "Interpretations of Adjusted Treatment Means and Regressions in Analysis of Covariance." *Biometrics*, Vol. 13, No. 3, pp. 282–308.

107. Speed, F. M. and R. R. Hocking (1976). "The Use of the $R(\ )$-Notation with Unbalanced Data." *The American Statistician*, Vol. 30, pp. 30–33.

108. Speed, F. M., R. R. Hocking, and O. P. Hackney (1978). "Methods of Analysis of Linear Models with Unbalanced Data." *Journal of the American Statistical Association*, Vol. 73, pp. 105–112.

109. Stefansky, W. (1972). "Rejecting Outliers in Factorial Designs." *Technometrics*, Vol. 14, pp. 469–479.

110. Taguchi, G. and Y. Wu (1980). *Introduction to Off-Line Quality Control*. Central Japan Quality Control Association, Nagoya, Japan.

111. Taguchi, G. (1986). *Introduction to Quality Engineering*. Asian Productivity Organization, UNIPUB, White Plains, New York.

112. Tukey, J. W. (1949a). "One Degree of Freedom for Non-Additivity." *Biometrics*, Vol. 5, pp. 232–242.

113. Tukey, J. W. (1949b). "Comparing Individual Means in the Analysis of Variance." *Biometrics*, Vol. 5, pp. 99–114.

114. Tukey, J. W. (1951). "Quick and Dirty Methods in Statistics, Part II, Simple Analysis for Standard Designs." *Proceedings of the Fifth Annual Convention, American Society for Quality Control*, pp. 189–197.

115. Tukey, J. W. (1953). "The Problem of Multiple Comparisons." Unpublished Notes, Princeton University.

116. Wine, R. L. (1964). *Statistics for Scientists and Engineers*, Prentice-Hall, Inc., Englewood Cliffs, N. J.

117. Winer, B. J. (1971). *Statistical Principles in Experimental Design*. 2nd edition. McGraw-Hill, New York.

118. Yates, F. (1934). "The Analysis of Multiple Classifications with Unequal Numbers in the Different Classes." *Journal of the American Statistical Association*, Vol. 29, pp. 52–66.

119. Yates, F. (1937). *Design and Analysis of Factorial Experiments*. Tech. Comm. No. 35, Imperial Bureau of Soil Sciences, London.

120. Yates, F. (1940). "The Recovery of Interblock Information in Balanced Incomplete Block Designs." *Annals of Eugenics*, Vol. 10, pp. 317–325.

Appendix

I. Cumulative Standard Normal Distribution[a]

$$\Phi(z) = \int_{-\infty}^{z} \frac{1}{\sqrt{2\pi}} e^{-u^2/2} \, du$$

z	.00	.01	.02	.03	.04	z
.0	.500 00	.503 99	.507 98	.511 97	.515 95	.0
.1	.539 83	.543 79	.547 76	.551 72	.555 67	.1
.2	.579 26	.583 17	.587 06	.590 95	.594 83	.2
.3	.617 91	.621 72	.625 51	.629 30	.633 07	.3
.4	.655 42	.659 10	.662 76	.666 40	.670 03	.4
.5	.691 46	.694 97	.698 47	.701 94	.705 40	.5
.6	.725 75	.729 07	.732 37	.735 65	.738 91	.6
.7	.758 03	.761 15	.764 24	.767 30	.770 35	.7
.8	.788 14	.791 03	.793 89	.796 73	.799 54	.8
.9	.815 94	.818 59	.821 21	.823 81	.826 39	.9
1.0	.841 34	.843 75	.846 13	.848 49	.850 83	1.0
1.1	.864 33	.866 50	.868 64	.870 76	.872 85	1.1
1.2	.884 93	.886 86	.888 77	.890 65	.892 51	1.2
1.3	.903 20	.904 90	.906 58	.908 24	.909 88	1.3
1.4	.919 24	.920 73	.922 19	.923 64	.925 06	1.4
1.5	.933 19	.934 48	.935 74	.936 99	.938 22	1.5
1.6	.945 20	.946 30	.947 38	.948 45	.949 50	1.6
1.7	.955 43	.956 37	.957 28	.958 18	.959 07	1.7
1.8	.964 07	.964 85	.965 62	.966 37	.967 11	1.8
1.9	.971 28	.971 93	.972 57	.973 20	.973 81	1.9
2.0	.977 25	.977 78	.978 31	.978 82	.979 32	2.0
2.1	.982 14	.982 57	.983 00	.983 41	.938 82	2.1
2.2	.986 10	.986 45	.986 79	.987 13	.987 45	2.2
2.3	.989 28	.989 56	.989 83	.990 10	.990 36	2.3
2.4	.991 80	.992 02	.992 24	.992 45	.992 66	2.4
2.5	.993 79	.993 96	.994 13	.994 30	.994 46	2.5
2.6	.995 34	.995 47	.995 60	.995 73	.995 85	2.6
2.7	.996 53	.996 64	.996 74	.996 83	.996 93	2.7
2.8	.997 44	.997 52	.997 60	.997 67	.997 74	2.8
2.9	.998 13	.998 19	.998 25	.998 31	.998 36	2.9
3.0	.998 65	.998 69	.998 74	.998 78	.998 82	3.0
3.1	.999 03	.999 06	.999 10	.999 13	.999 16	3.1
3.2	.999 31	.999 34	.999 36	.999 38	.999 40	3.2
3.3	.999 52	.999 53	.999 55	.999 57	.999 58	3.3
3.4	.999 66	.999 68	.999 69	.999 70	.999 71	3.4
3.5	.999 77	.999 78	.999 78	.999 79	.999 80	3.5
3.6	.999 84	.999 85	.999 85	.999 86	.999 86	3.6
3.7	.999 89	.999 90	.999 90	.999 90	.999 91	3.7
3.8	.999 93	.999 93	.999 93	.999 94	.999 94	3.8
3.9	.999 95	.999 95	.999 96	.999 96	.999 96	3.9

[a] Reproduced with permission from *Probability and Statistics in Engineering and Management Science*, 3rd edition, by W. W. Hines and D. C. Montgomery, Wiley, New York, 1990.

I. Cumulative Standard Normal Distribution (*continued*)

$$\Phi(z) = \int_{-\infty}^{z} \frac{1}{\sqrt{2\pi}} e^{-u^2/2} \, du$$

z	.05	.06	.07	.08	.09	z
.0	.519 94	.523 92	.527 90	.531 88	.535 86	.0
.1	.559 62	.563 56	.567 49	.571 42	.575 34	.1
.2	.598 71	.602 57	.606 42	.610 26	.614 09	.2
.3	.636 83	.640 58	.644 31	.648 03	.651 73	.3
.4	.673 64	.677 24	.680 82	.684 38	.687 93	.4
.5	.708 84	.712 26	.715 66	.719 04	.722 40	.5
.6	.742 15	.745 37	.748 57	.751 75	.754 90	.6
.7	.773 37	.776 37	.779 35	.782 30	.785 23	.7
.8	.802 34	.805 10	.807 85	.810 57	.813 27	.8
.9	.828 94	.831 47	.833 97	.836 46	.838 91	.9
1.0	.853 14	.855 43	.857 69	.859 93	.862 14	1.0
1.1	.874 93	.876 97	.879 00	.881 00	.882 97	1.1
1.2	.894 35	.896 16	.897 96	.899 73	.901 47	1.2
1.3	.911 49	.913 08	.914 65	.916 21	.917 73	1.3
1.4	.926 47	.927 85	.929 22	.930 56	.931 89	1.4
1.5	.939 43	.940 62	.941 79	.942 95	.944 08	1.5
1.6	.950 53	.951 54	.952 54	.953 52	.954 48	1.6
1.7	.959 94	.960 80	.961 64	.962 46	.963 27	1.7
1.8	.967 84	.968 56	.969 26	.969 95	.970 62	1.8
1.9	.974 41	.975 00	.975 58	.976 15	.976 70	1.9
2.0	.979 82	.980 30	.980 77	.981 24	.981 69	2.0
2.1	.984 22	.984 61	.985 00	.985 37	.985 74	2.1
2.2	.987 78	.988 09	.988 40	.988 70	.988 99	2.2
2.3	.990 61	.990 86	.991 11	.991 34	.991 58	2.3
2.4	.992 86	.993 05	.993 24	.993 43	.993 61	2.4
2.5	.994 61	.994 77	.994 92	.995 06	.995 20	2.5
2.6	.995 98	.996 09	.996 21	.996 32	.996 43	2.6
2.7	.997 02	.997 11	.997 20	.997 28	.997 36	2.7
2.8	.997 81	.997 88	.997 95	.998 01	.998 07	2.8
2.9	.998 41	.998 46	.998 51	.998 56	.998 61	2.9
3.0	.998 86	.998 89	.998 93	.998 97	.999 00	3.0
3.1	.999 18	.999 21	.999 24	.999 26	.999 29	3.1
3.2	.999 42	.999 44	.999 46	.999 48	.999 50	3.2
3.3	.999 60	.999 61	.999 62	.999 64	.999 65	3.3
3.4	.999 72	.999 73	.999 74	.999 75	.999 76	3.4
3.5	.999 81	.999 81	.999 82	.999 83	.999 83	3.5
3.6	.999 87	.999 87	.999 88	.999 88	.999 89	3.6
3.7	.999 91	.999 92	.999 92	.999 92	.999 92	3.7
3.8	.999 94	.999 94	.999 95	.999 95	.999 95	3.8
3.9	.999 96	.999 96	.999 96	.999 97	.999 97	3.9

II. Percentage Points of the t Distribution[a]

ν \ α	.40	.25	.10	.05	.025	.01	.005	.0025	.001	.0005
1	.325	1.000	3.078	6.314	12.706	31.821	63.657	127.32	318.31	636.62
2	.289	.816	1.886	2.920	4.303	6.965	9.925	14.089	23.326	31.598
3	.277	.765	1.638	2.353	3.182	4.541	5.841	7.453	10.213	12.924
4	.271	.741	1.533	2.132	2.776	3.747	4.604	5.598	7.173	8.610
5	.267	.727	1.476	2.015	2.571	3.365	4.032	4.773	5.893	6.869
6	.265	.727	1.440	1.943	2.447	3.143	3.707	4.317	5.208	5.959
7	.263	.711	1.415	1.895	2.365	2.998	3.499	4.019	4.785	5.408
8	.262	.706	1.397	1.860	2.306	2.896	3.355	3.833	4.501	5.041
9	.261	.703	1.383	1.833	2.262	2.821	3.250	3.690	4.297	4.781
10	.260	.700	1.372	1.812	2.228	2.764	3.169	3.581	4.144	4.587
11	.260	.697	1.363	1.796	2.201	2.718	3.106	3.497	4.025	4.437
12	.259	.695	1.356	1.782	2.179	2.681	3.055	3.428	3.930	4.318
13	.259	.694	1.350	1.771	2.160	2.650	3.012	3.372	3.852	4.221
14	.258	.692	1.345	1.761	2.145	2.624	2.977	3.326	3.787	4.140
15	.258	.691	1.341	1.753	2.131	2.602	2.947	3.286	3.733	4.073
16	.258	.690	1.337	1.746	2.120	2.583	2.921	3.252	3.686	4.015
17	.257	.689	1.333	1.740	2.110	2.567	2.898	3.222	3.646	3.965
18	.257	.688	1.330	1.734	2.101	2.552	2.878	3.197	3.610	3.922
19	.257	.688	1.328	1.729	2.093	2.539	2.861	3.174	3.579	3.883
20	.257	.687	1.325	1.725	2.086	2.528	2.845	3.153	3.552	3.850
21	.257	.686	1.323	1.721	2.080	2.518	2.831	3.135	3.527	3.819
22	.256	.686	1.321	1.717	2.074	2.508	2.819	3.119	3.505	3.792
23	.256	.685	1.319	1.714	2.069	2.500	2.807	3.104	3.485	3.767
24	.256	.685	1.318	1.711	2.064	2.492	2.797	3.091	3.467	3.745
25	.256	.684	1.316	1.708	2.060	2.485	2.787	3.078	3.450	3.725
26	.256	.684	1.315	1.706	2.056	2.479	2.779	3.067	3.435	3.707
27	.256	.684	1.314	1.703	2.052	2.473	2.771	3.057	3.421	3.690
28	.256	.683	1.313	1.701	2.048	2.467	2.763	3.047	3.408	3.674
29	.256	.683	1.311	1.699	2.045	2.462	2.756	3.038	3.396	3.659
30	.256	.683	1.310	1.697	2.042	2.457	2.750	3.030	3.385	3.646
40	.255	.681	1.303	1.684	2.021	2.423	2.704	2.971	3.307	3.551
60	.254	.679	1.296	1.671	2.000	2.390	2.660	2.915	3.232	3.460
120	.254	.677	1.289	1.658	1.980	2.358	2.617	2.860	3.160	3.373
∞	.253	.674	1.282	1.645	1.960	2.326	2.576	2.807	3.090	3.291

ν = degrees of freedom.

[a] Adapted with permission from *Biometrika Tables for Statisticians*, Vol. 1, 3rd edition, by E. S. Pearson and H. O. Hartley, Cambridge University Press, Cambridge, 1966.

III.   Percentage Points of the χ^2 Distribution[a]

ν	.995	.990	.975	.950	.500	.050	.025	.010	.005
1	0.00 +	0.00 +	0.00 +	0.00 +	0.45	3.84	5.02	6.63	7.88
2	0.01	0.02	0.05	0.10	1.39	5.99	7.38	9.21	10.60
3	0.07	0.11	0.22	0.35	2.37	7.81	9.35	11.34	12.84
4	0.21	0.30	0.48	0.71	3.36	9.49	11.14	13.28	14.86
5	0.41	0.55	0.83	1.15	4.35	11.07	12.38	15.09	16.75
6	0.68	0.87	1.24	1.64	5.35	12.59	14.45	16.81	18.55
7	0.99	1.24	1.69	2.17	6.35	14.07	16.01	18.48	20.28
8	1.34	1.65	2.18	2.73	7.34	15.51	17.53	20.09	21.96
9	1.73	2.09	2.70	3.33	8.34	16.92	19.02	21.67	23.59
10	2.16	2.56	3.25	3.94	9.34	18.31	20.48	23.21	25.19
11	2.60	3.05	3.82	4.57	10.34	19.68	21.92	24.72	26.76
12	3.07	3.57	4.40	5.23	11.34	21.03	23.34	26.22	28.30
13	3.57	4.11	5.01	5.89	12.34	22.36	24.74	27.69	29.82
14	4.07	4.66	5.63	6.57	13.34	23.68	26.12	29.14	31.32
15	4.60	5.23	6.27	7.26	14.34	25.00	27.49	30.58	32.80
16	5.14	5.81	6.91	7.96	15.34	26.30	28.85	32.00	34.27
17	5.70	6.41	7.56	8.67	16.34	27.59	30.19	33.41	35.72
18	6.26	7.01	8.23	9.39	17.34	28.87	31.53	34.81	37.16
19	6.84	7.63	8.91	10.12	18.34	30.14	32.85	36.19	38.58
20	7.43	8.26	9.59	10.85	19.34	31.41	34.17	37.57	40.00
25	10.52	11.52	13.12	14.61	24.34	37.65	40.65	44.31	46.93
30	13.79	14.95	16.79	18.49	29.34	43.77	46.98	50.89	53.67
40	20.71	22.16	24.43	26.51	39.34	55.76	59.34	63.69	66.77
50	27.99	29.71	32.36	34.76	49.33	67.50	71.42	76.15	79.49
60	35.53	37.48	40.48	43.19	59.33	79.08	83.30	88.38	91.95
70	43.28	45.44	48.76	51.74	69.33	90.53	95.02	100.42	104.22
80	51.17	53.54	57.15	60.39	79.33	101.88	106.63	112.33	116.32
90	59.20	61.75	65.65	69.13	89.33	113.14	118.14	124.12	128.30
100	67.33	70.06	74.22	77.93	99.33	124.34	129.56	135.81	140.17

ν = degrees of freedom

[a] Adapted with permission from *Biometrika Tables for Statisticians*, Vol. 1, 3rd edition by E. S. Pearson and H. O. Hartley, Cambridge University Press, Cambridge, 1966.

IV. Percentage Points of the F Distribution[a]

$F_{.25, \nu_1, \nu_2}$

ν_2	Degrees of Freedom for the Numerator (ν_1)																		
	1	2	3	4	5	6	7	8	9	10	12	15	20	24	30	40	60	120	∞
1	5.83	7.50	8.20	8.58	8.82	8.98	9.10	9.19	9.26	9.32	9.41	9.49	9.58	9.63	9.67	9.71	9.76	9.80	9.85
2	2.57	3.00	3.15	3.23	3.28	3.31	3.34	3.35	3.37	3.38	3.39	3.41	3.43	3.43	3.44	3.45	3.46	3.47	3.48
3	2.02	2.28	2.36	2.39	2.41	2.42	2.43	2.44	2.44	2.44	2.45	2.46	2.46	2.46	2.47	2.47	2.47	2.47	2.47
4	1.81	2.00	2.05	2.06	2.07	2.08	2.08	2.08	2.08	2.08	2.08	2.08	2.08	2.08	2.08	2.08	2.08	2.08	2.08
5	1.69	1.85	1.88	1.89	1.89	1.89	1.89	1.89	1.89	1.89	1.89	1.89	1.88	1.88	1.88	1.88	1.87	1.87	1.87
6	1.62	1.76	1.78	1.79	1.79	1.78	1.78	1.78	1.77	1.77	1.77	1.76	1.76	1.75	1.75	1.75	1.74	1.74	1.74
7	1.57	1.70	1.72	1.72	1.71	1.71	1.70	1.70	1.70	1.69	1.68	1.68	1.67	1.67	1.66	1.66	1.65	1.65	1.65
8	1.54	1.66	1.67	1.66	1.66	1.65	1.64	1.64	1.63	1.63	1.62	1.62	1.61	1.60	1.60	1.59	1.59	1.58	1.58
9	1.51	1.62	1.63	1.63	1.62	1.61	1.60	1.60	1.59	1.59	1.58	1.57	1.56	1.56	1.55	1.54	1.54	1.53	1.53
10	1.49	1.60	1.60	1.59	1.59	1.58	1.57	1.56	1.56	1.55	1.54	1.53	1.52	1.52	1.51	1.51	1.50	1.49	1.48
11	1.47	1.58	1.58	1.57	1.56	1.55	1.54	1.53	1.53	1.52	1.51	1.50	1.49	1.49	1.48	1.47	1.47	1.46	1.45
12	1.46	1.56	1.56	1.55	1.54	1.53	1.52	1.51	1.51	1.50	1.49	1.48	1.47	1.46	1.45	1.45	1.44	1.43	1.42
13	1.45	1.55	1.55	1.53	1.52	1.51	1.50	1.49	1.49	1.48	1.47	1.46	1.45	1.44	1.43	1.42	1.42	1.41	1.40
14	1.44	1.53	1.53	1.52	1.51	1.50	1.49	1.48	1.47	1.46	1.45	1.44	1.43	1.42	1.41	1.41	1.40	1.39	1.38
15	1.43	1.52	1.52	1.51	1.49	1.48	1.47	1.46	1.46	1.45	1.44	1.43	1.41	1.41	1.40	1.39	1.38	1.37	1.36
16	1.42	1.51	1.51	1.50	1.48	1.47	1.46	1.45	1.44	1.44	1.43	1.41	1.40	1.39	1.38	1.37	1.36	1.35	1.34
17	1.42	1.51	1.50	1.49	1.47	1.46	1.45	1.44	1.43	1.43	1.41	1.40	1.39	1.38	1.37	1.36	1.35	1.34	1.33
18	1.41	1.50	1.49	1.48	1.46	1.45	1.44	1.43	1.42	1.42	1.40	1.39	1.38	1.37	1.36	1.35	1.34	1.33	1.32
19	1.41	1.49	1.49	1.47	1.46	1.44	1.43	1.42	1.41	1.41	1.40	1.38	1.37	1.36	1.35	1.34	1.33	1.32	1.30
20	1.40	1.49	1.48	1.47	1.45	1.44	1.43	1.42	1.41	1.40	1.39	1.37	1.36	1.35	1.34	1.33	1.32	1.31	1.29
21	1.40	1.48	1.48	1.46	1.44	1.43	1.42	1.41	1.40	1.39	1.38	1.37	1.35	1.34	1.33	1.32	1.31	1.30	1.28
22	1.40	1.48	1.47	1.45	1.44	1.42	1.41	1.40	1.39	1.39	1.37	1.36	1.34	1.33	1.32	1.31	1.30	1.29	1.28
23	1.39	1.47	1.47	1.45	1.43	1.42	1.41	1.40	1.39	1.38	1.37	1.35	1.34	1.33	1.32	1.31	1.30	1.28	1.27
24	1.39	1.47	1.46	1.44	1.43	1.41	1.40	1.39	1.38	1.38	1.36	1.35	1.33	1.32	1.31	1.30	1.29	1.28	1.26
25	1.39	1.47	1.46	1.44	1.42	1.41	1.40	1.39	1.38	1.37	1.36	1.34	1.33	1.32	1.31	1.29	1.28	1.27	1.25
26	1.38	1.46	1.45	1.44	1.42	1.41	1.39	1.38	1.37	1.37	1.35	1.34	1.32	1.31	1.30	1.29	1.28	1.26	1.25
27	1.38	1.46	1.45	1.43	1.42	1.40	1.39	1.38	1.37	1.36	1.35	1.33	1.32	1.31	1.30	1.28	1.27	1.26	1.24
28	1.38	1.46	1.45	1.43	1.41	1.40	1.39	1.38	1.37	1.36	1.34	1.33	1.31	1.30	1.29	1.28	1.27	1.25	1.24
29	1.38	1.45	1.45	1.43	1.41	1.40	1.38	1.37	1.36	1.35	1.34	1.32	1.31	1.30	1.29	1.27	1.26	1.25	1.23
30	1.38	1.45	1.44	1.42	1.41	1.39	1.38	1.37	1.36	1.35	1.34	1.32	1.30	1.29	1.28	1.27	1.26	1.24	1.23
40	1.36	1.44	1.42	1.40	1.39	1.37	1.36	1.35	1.34	1.33	1.31	1.30	1.28	1.26	1.25	1.24	1.22	1.21	1.19
60	1.35	1.42	1.41	1.38	1.37	1.35	1.33	1.32	1.31	1.30	1.29	1.27	1.25	1.24	1.22	1.21	1.19	1.17	1.15
120	1.34	1.40	1.39	1.37	1.35	1.33	1.31	1.30	1.29	1.28	1.26	1.24	1.22	1.21	1.19	1.18	1.16	1.13	1.10
∞	1.32	1.39	1.37	1.35	1.33	1.31	1.29	1.28	1.27	1.25	1.24	1.22	1.19	1.18	1.16	1.14	1.12	1.08	1.00

Degrees of Freedom for the Denominator (ν_2)

ν = degrees of freedom

Degrees of Freedom for the Numerator (v_1)

v_2	1	2	3	4	5	6	7	8	9	10	12	15	20	24	30	40	60	120	∞
1	39.86	49.50	53.59	55.83	57.24	58.20	58.91	59.44	59.86	60.19	60.71	61.22	61.74	62.00	62.26	62.53	62.79	63.06	63.33
2	8.53	9.00	9.16	9.24	9.29	9.33	9.35	9.37	9.38	9.39	9.41	9.42	9.44	9.45	9.46	9.47	9.47	9.48	9.49
3	5.54	5.46	5.39	5.34	5.31	5.28	5.27	5.25	5.24	5.23	5.22	5.20	5.18	5.18	5.17	5.16	5.15	5.14	5.13
4	4.54	4.32	4.19	4.11	4.05	4.01	3.98	3.95	3.94	3.92	3.90	3.87	3.84	3.83	3.82	3.80	3.79	3.78	3.76
5	4.06	3.78	3.62	3.52	3.45	3.40	3.37	3.34	3.32	3.30	3.27	3.24	3.21	3.19	3.17	3.16	3.14	3.12	3.10
6	3.78	3.46	3.29	3.18	3.11	3.05	3.01	2.98	2.96	2.94	2.90	2.87	2.84	2.82	2.80	2.78	2.76	2.74	2.72
7	3.59	3.26	3.07	2.96	2.88	2.83	2.78	2.75	2.72	2.70	2.67	2.63	2.59	2.58	2.56	2.54	2.51	2.49	2.47
8	3.46	3.11	2.92	2.81	2.73	2.67	2.62	2.59	2.56	2.54	2.50	2.46	2.42	2.40	2.38	2.36	2.34	2.32	2.29
9	3.36	3.01	2.81	2.69	2.61	2.55	2.51	2.47	2.44	2.42	2.38	2.34	2.30	2.28	2.25	2.23	2.21	2.18	2.16
10	3.29	2.92	2.73	2.61	2.52	2.46	2.41	2.38	2.35	2.32	2.28	2.24	2.20	2.18	2.16	2.13	2.11	2.08	2.06
11	3.23	2.86	2.66	2.54	2.45	2.39	2.34	2.30	2.27	2.25	2.21	2.17	2.12	2.10	2.08	2.05	2.03	2.00	1.97
12	3.18	2.81	2.61	2.48	2.39	2.33	2.28	2.24	2.21	2.19	2.15	2.10	2.06	2.04	2.01	1.99	1.96	1.93	1.90
13	3.14	2.76	2.56	2.43	2.35	2.28	2.23	2.20	2.16	2.14	2.10	2.05	2.01	1.98	1.96	1.93	1.90	1.88	1.85
14	3.10	2.73	2.52	2.39	2.31	2.24	2.19	2.15	2.12	2.10	2.05	2.01	1.96	1.94	1.91	1.89	1.86	1.83	1.80
15	3.07	2.70	2.49	2.36	2.27	2.21	2.16	2.12	2.09	2.06	2.02	1.97	1.92	1.90	1.87	1.85	1.82	1.79	1.76
16	3.05	2.67	2.46	2.33	2.24	2.18	2.13	2.09	2.06	2.03	1.99	1.94	1.89	1.87	1.84	1.81	1.78	1.75	1.72
17	3.03	2.64	2.44	2.31	2.22	2.15	2.10	2.06	2.03	2.00	1.96	1.91	1.86	1.84	1.81	1.78	1.75	1.72	1.69
18	3.01	2.62	2.42	2.29	2.20	2.13	2.08	2.04	2.00	1.98	1.93	1.89	1.84	1.81	1.78	1.75	1.72	1.69	1.66
19	2.99	2.61	2.40	2.27	2.18	2.11	2.06	2.02	1.98	1.96	1.91	1.86	1.81	1.79	1.76	1.73	1.70	1.67	1.63
20	2.97	2.59	2.38	2.25	2.16	2.09	2.04	2.00	1.96	1.94	1.89	1.84	1.79	1.77	1.74	1.71	1.68	1.64	1.61
21	2.96	2.57	2.36	2.23	2.14	2.08	2.02	1.98	1.95	1.92	1.87	1.83	1.78	1.75	1.72	1.69	1.66	1.62	1.59
22	2.95	2.56	2.35	2.22	2.13	2.06	2.01	1.97	1.93	1.90	1.86	1.81	1.76	1.73	1.70	1.67	1.64	1.60	1.57
23	2.94	2.55	2.34	2.21	2.11	2.05	1.99	1.95	1.92	1.89	1.84	1.80	1.74	1.72	1.69	1.66	1.62	1.59	1.55
24	2.93	2.54	2.33	2.19	2.10	2.04	1.98	1.94	1.91	1.88	1.83	1.78	1.73	1.70	1.67	1.64	1.61	1.57	1.53
25	2.92	2.53	2.32	2.18	2.09	2.02	1.97	1.93	1.89	1.87	1.82	1.77	1.72	1.69	1.66	1.63	1.59	1.56	1.52
26	2.91	2.52	2.31	2.17	2.08	2.01	1.96	1.92	1.88	1.86	1.81	1.76	1.71	1.68	1.65	1.61	1.58	1.54	1.50
27	2.90	2.51	2.30	2.17	2.07	2.00	1.95	1.91	1.87	1.85	1.80	1.75	1.70	1.67	1.64	1.60	1.57	1.53	1.49
28	2.89	2.50	2.29	2.16	2.06	2.00	1.94	1.90	1.87	1.84	1.79	1.74	1.69	1.66	1.63	1.59	1.56	1.52	1.48
29	2.89	2.50	2.28	2.15	2.06	1.99	1.93	1.89	1.86	1.83	1.78	1.73	1.68	1.65	1.62	1.58	1.55	1.51	1.47
30	2.88	2.49	2.28	2.14	2.03	1.98	1.93	1.88	1.85	1.82	1.77	1.72	1.67	1.64	1.61	1.57	1.54	1.50	1.46
40	2.84	2.44	2.23	2.09	2.00	1.93	1.87	1.83	1.79	1.76	1.71	1.66	1.61	1.57	1.54	1.51	1.47	1.42	1.38
60	2.79	2.39	2.18	2.04	1.95	1.87	1.82	1.77	1.74	1.71	1.66	1.60	1.54	1.51	1.48	1.44	1.40	1.35	1.29
120	2.75	2.35	2.13	1.99	1.90	1.82	1.77	1.72	1.68	1.65	1.60	1.55	1.48	1.45	1.41	1.37	1.32	1.26	1.19
∞	2.71	2.30	2.08	1.94	1.85	1.77	1.72	1.67	1.63	1.60	1.55	1.49	1.42	1.38	1.34	1.30	1.24	1.17	1.00

Degrees of Freedom for the Denominator (v_2)

IV. Percentage Points of the F Distribution (*continued*)

$$F_{0.05,\ \nu_1,\ \nu_2}$$

ν_2	\multicolumn{19}{c}{Degrees of Freedom for the Numerator (ν_1)}
	1	2	3	4	5	6	7	8	9	10	12	15	20	24	30	40	60	120	∞
1	161.4	199.5	215.7	224.6	230.2	234.0	236.8	238.9	240.5	241.9	243.9	245.9	248.0	249.1	250.1	251.1	252.2	253.3	254.3
2	18.51	19.00	19.16	19.25	19.30	19.33	19.35	19.37	19.38	19.40	19.41	19.43	19.45	19.45	19.46	19.47	19.48	19.49	19.50
3	10.13	9.55	9.28	9.12	9.01	8.94	8.89	8.85	8.81	8.79	8.74	8.70	8.66	8.64	8.62	8.59	8.57	8.55	8.53
4	7.71	6.94	6.59	6.39	6.26	6.16	6.09	6.04	6.00	5.96	5.91	5.86	5.80	5.77	5.75	5.72	5.69	5.66	5.63
5	6.61	5.79	5.41	5.19	5.05	4.95	4.88	4.82	4.77	4.74	4.68	4.62	4.56	4.53	4.50	4.46	4.43	4.40	4.36
6	5.99	5.14	4.76	4.53	4.39	4.28	4.21	4.15	4.10	4.06	4.00	3.94	3.87	3.84	3.81	3.77	3.74	3.70	3.67
7	5.59	4.74	4.35	4.12	3.97	3.87	3.79	3.73	3.68	3.64	3.57	3.51	3.44	3.41	3.38	3.34	3.30	3.27	3.23
8	5.32	4.46	4.07	3.84	3.69	3.58	3.50	3.44	3.39	3.35	3.28	3.22	3.15	3.12	3.08	3.04	3.01	2.97	2.93
9	5.12	4.26	3.86	3.63	3.48	3.37	3.29	3.23	3.18	3.14	3.07	3.01	2.94	2.90	2.86	2.83	2.79	2.75	2.71
10	4.96	4.10	3.71	3.48	3.33	3.22	3.14	3.07	3.02	2.98	2.91	2.85	2.77	2.74	2.70	2.66	2.62	2.58	2.54
11	4.84	3.98	3.59	3.36	3.20	3.09	3.01	2.95	2.90	2.85	2.79	2.72	2.65	2.61	2.57	2.53	2.49	2.45	2.40
12	4.75	3.89	3.49	3.26	3.11	3.00	2.91	2.85	2.80	2.75	2.69	2.62	2.54	2.51	2.47	2.43	2.38	2.34	2.30
13	4.67	3.81	3.41	3.18	3.03	2.92	2.83	2.77	2.71	2.67	2.60	2.53	2.46	2.42	2.38	2.34	2.30	2.25	2.21
14	4.60	3.74	3.34	3.11	2.96	2.85	2.76	2.70	2.65	2.60	2.53	2.46	2.39	2.35	2.31	2.27	2.22	2.18	2.13
15	4.54	3.68	3.29	3.06	2.90	2.79	2.71	2.64	2.59	2.54	2.48	2.40	2.33	2.29	2.25	2.20	2.16	2.11	2.07
16	4.49	3.63	3.24	3.01	2.85	2.74	2.66	2.59	2.54	2.49	2.42	2.35	2.28	2.24	2.19	2.15	2.11	2.06	2.01
17	4.45	3.59	3.20	2.96	2.81	2.70	2.61	2.55	2.49	2.45	2.38	2.31	2.23	2.19	2.15	2.10	2.06	2.01	1.96
18	4.41	3.55	3.16	2.93	2.77	2.66	2.58	2.51	2.46	2.41	2.34	2.27	2.19	2.15	2.11	2.06	2.02	1.97	1.92
19	4.38	3.52	3.13	2.90	2.74	2.63	2.54	2.48	2.42	2.38	2.31	2.23	2.16	2.11	2.07	2.03	1.98	1.93	1.88
20	4.35	3.49	3.10	2.87	2.71	2.60	2.51	2.45	2.39	2.35	2.28	2.20	2.12	2.08	2.04	1.99	1.95	1.90	1.84
21	4.32	3.47	3.07	2.84	2.68	2.57	2.49	2.42	2.37	2.32	2.25	2.18	2.10	2.05	2.01	1.96	1.92	1.87	1.81
22	4.30	3.44	3.05	2.82	2.66	2.55	2.46	2.40	2.34	2.30	2.23	2.15	2.07	2.03	1.98	1.94	1.89	1.84	1.78
23	4.28	3.42	3.03	2.80	2.64	2.53	2.44	2.37	2.32	2.27	2.20	2.13	2.05	2.01	1.96	1.91	1.86	1.81	1.76
24	4.26	3.40	3.01	2.78	2.62	2.51	2.42	2.36	2.30	2.25	2.18	2.11	2.03	1.98	1.94	1.89	1.84	1.79	1.73*
25	4.24	3.39	2.99	2.76	2.60	2.49	2.40	2.34	2.28	2.24	2.16	2.09	2.01	1.96	1.92	1.87	1.82	1.77	1.71
26	4.23	3.37	2.98	2.74	2.59	2.47	2.39	2.32	2.27	2.22	2.15	2.07	1.99	1.95	1.90	1.85	1.80	1.75	1.69
27	4.21	3.35	2.96	2.73	2.57	2.46	2.37	2.31	2.25	2.20	2.13	2.06	1.97	1.93	1.88	1.84	1.79	1.73	1.67
28	4.20	3.34	2.95	2.71	2.56	2.45	2.36	2.29	2.24	2.19	2.12	2.04	1.96	1.91	1.87	1.82	1.77	1.71	1.65
29	4.18	3.33	2.93	2.70	2.55	2.43	2.35	2.28	2.22	2.18	2.10	2.03	1.94	1.90	1.85	1.81	1.75	1.70	1.64
30	4.17	3.32	2.92	2.69	2.53	2.42	2.33	2.27	2.21	2.16	2.09	2.01	1.93	1.89	1.84	1.79	1.74	1.68	1.62
40	4.08	3.23	2.84	2.61	2.45	2.34	2.25	2.18	2.12	2.08	2.00	1.92	1.84	1.79	1.74	1.69	1.64	1.58	1.51
60	4.00	3.15	2.76	2.53	2.37	2.25	2.17	2.10	2.04	1.99	1.92	1.84	1.75	1.70	1.65	1.59	1.53	1.47	1.39
120	3.92	3.07	2.68	2.45	2.29	2.17	2.09	2.02	1.96	1.91	1.83	1.75	1.66	1.61	1.55	1.55	1.43	1.35	1.25
∞	3.84	3.00	2.60	2.37	2.21	2.10	2.01	1.94	1.88	1.83	1.75	1.67							

IV. Percentage Points of the F Distribution (continued)

Degrees of Freedom for the Numerator (ν_1)

ν_2	1	2	3	4	5	6	7	8	9	10	12	15	20	24	30	40	60	120	∞
1	647.8	799.5	864.2	899.6	921.8	937.1	948.2	956.7	963.3	968.6	976.7	984.9	993.1	997.2	1001	1006	1010	1014	1018
2	38.51	39.00	39.17	39.25	39.30	39.33	39.36	39.37	39.39	39.40	39.41	39.43	39.45	39.46	39.46	39.47	39.48	39.49	39.50
3	17.44	16.04	15.44	15.10	14.88	14.73	14.62	14.54	14.47	14.42	14.34	14.25	14.17	14.12	14.08	14.04	13.99	13.95	13.90
4	12.22	10.65	9.98	9.60	9.36	9.20	9.07	8.98	8.90	8.84	8.75	8.66	8.56	8.51	8.46	8.41	8.36	8.31	8.26
5	10.01	8.43	7.76	7.39	7.15	6.98	6.85	6.76	6.68	6.62	6.52	6.43	6.33	6.28	6.23	6.18	6.12	6.07	6.02
6	8.81	7.26	6.60	6.23	5.99	5.82	5.70	5.60	5.52	5.46	5.37	5.27	5.17	5.12	5.07	5.01	4.96	4.90	4.85
7	8.07	6.54	5.89	5.52	5.29	5.12	4.99	4.90	4.82	4.76	4.67	4.57	4.47	4.42	4.36	4.31	4.25	4.20	4.14
8	7.57	6.06	5.42	5.05	4.82	4.65	4.53	4.43	4.36	4.30	4.20	4.10	4.00	3.95	3.89	3.84	3.78	3.73	3.67
9	7.21	5.71	5.08	4.72	4.48	4.32	4.20	4.10	4.03	3.96	3.87	3.77	3.67	3.61	3.56	3.51	3.45	3.39	3.33
10	6.94	5.46	4.83	4.47	4.24	4.07	3.95	3.85	3.78	3.72	3.62	3.52	3.42	3.37	3.31	3.26	3.20	3.14	3.08
11	6.72	5.26	4.63	4.28	4.04	3.88	3.76	3.66	3.59	3.53	3.43	3.33	3.23	3.17	3.12	3.06	3.00	2.94	2.88
12	6.55	5.10	4.47	4.12	3.89	3.73	3.61	3.51	3.44	3.37	3.28	3.18	3.07	3.02	2.96	2.91	2.85	2.79	2.72
13	6.41	4.97	4.35	4.00	3.77	3.60	3.48	3.39	3.31	3.25	3.15	3.05	2.95	2.89	2.84	2.78	2.72	2.66	2.60
14	6.30	4.86	4.24	3.89	3.66	3.50	3.38	3.29	3.21	3.15	3.05	2.95	2.84	2.79	2.73	2.67	2.61	2.55	2.49
15	6.20	4.77	4.15	3.80	3.58	3.41	3.29	3.20	3.12	3.06	2.96	2.86	2.76	2.70	2.64	2.59	2.52	2.46	2.40
16	6.12	4.69	4.08	3.73	3.50	3.34	3.22	3.12	3.05	2.99	2.89	2.79	2.68	2.63	2.57	2.51	2.45	2.38	2.32
17	6.04	4.62	4.01	3.66	3.44	3.28	3.16	3.06	2.98	2.92	2.82	2.72	2.62	2.56	2.50	2.44	2.38	2.32	2.25
18	5.98	4.56	3.95	3.61	3.38	3.22	3.10	3.01	2.93	2.87	2.77	2.67	2.56	2.50	2.44	2.38	2.32	2.26	2.19
19	5.92	4.51	3.90	3.56	3.33	3.17	3.05	2.96	2.88	2.82	2.72	2.62	2.51	2.45	2.39	2.33	2.27	2.20	2.13
20	5.87	4.46	3.86	3.51	3.29	3.13	3.01	2.91	2.84	2.77	2.68	2.57	2.46	2.41	2.35	2.29	2.22	2.16	2.09
21	5.83	4.42	3.82	3.48	3.25	3.09	2.97	2.87	2.80	2.73	2.64	2.53	2.42	2.37	2.31	2.25	2.18	2.11	2.04
22	5.79	4.38	3.78	3.44	3.22	3.05	2.93	2.84	2.76	2.70	2.60	2.50	2.39	2.33	2.27	2.21	2.14	2.08	2.00
23	5.75	4.35	3.75	3.41	3.18	3.02	2.90	2.81	2.73	2.67	2.57	2.47	2.36	2.30	2.24	2.18	2.11	2.04	1.97
24	5.72	4.32	3.72	3.38	3.15	2.99	2.87	2.78	2.70	2.64	2.54	2.44	2.33	2.27	2.21	2.15	2.08	2.01	1.94
25	5.69	4.29	3.69	3.35	3.13	2.97	2.85	2.75	2.68	2.61	2.51	2.41	2.30	2.24	2.18	2.12	2.05	1.98	1.91
26	5.66	4.27	3.67	3.33	3.10	2.94	2.82	2.73	2.65	2.59	2.49	2.39	2.28	2.22	2.16	2.09	2.03	1.95	1.88
27	5.63	4.24	3.65	3.31	3.08	2.92	2.80	2.71	2.63	2.57	2.47	2.36	2.25	2.19	2.13	2.07	2.00	1.93	1.85
28	5.61	4.22	3.63	3.29	3.06	2.90	2.78	2.69	2.61	2.55	2.45	2.34	2.23	2.17	2.11	2.05	1.98	1.91	1.83
29	5.59	4.20	3.61	3.27	3.04	2.88	2.76	2.67	2.59	2.53	2.43	2.32	2.21	2.15	2.09	2.03	1.96	1.89	1.81
30	5.57	4.18	3.59	3.25	3.03	2.87	2.75	2.65	2.57	2.51	2.41	2.31	2.20	2.14	2.07	2.01	1.94	1.87	1.79
40	5.42	4.05	3.46	3.13	2.90	2.74	2.62	2.53	2.45	2.39	2.29	2.18	2.07	2.01	1.94	1.88	1.80	1.72	1.64
60	5.29	3.93	3.34	3.01	2.79	2.63	2.51	2.41	2.33	2.27	2.17	2.06	1.94	1.88	1.82	1.74	1.67	1.58	1.48
120	5.15	3.80	3.23	2.89	2.67	2.52	2.39	2.30	2.22	2.16	2.05	1.94	1.82	1.76	1.69	1.61	1.53	1.43	1.31
∞	5.02	3.69	3.12	2.79	2.57	2.41	2.29	2.19	2.11	2.05	1.94	1.83	1.71	1.64	1.57	1.48	1.39	1.27	1.00

IV. Percentage Points of the F Distribution (continued)

$F_{0.01, v_1, v_2}$

Degrees of Freedom for the Numerator (v_1)

v_2	1	2	3	4	5	6	7	8	9	10	12	15	20	24	30	40	60	120	∞
1	4052	4999.5	5403	5625	5764	5859	5928	5982	6022	6056	6106	6157	6209	6235	6261	6287	6313	6339	6366
2	98.50	99.00	99.17	99.25	99.30	99.33	99.36	99.37	99.39	99.40	99.42	99.43	99.45	99.46	99.47	99.47	99.48	99.49	99.50
3	34.12	30.82	29.46	28.71	28.24	27.91	27.67	27.49	27.35	27.23	27.05	26.87	26.69	26.60	26.50	26.41	26.32	26.22	26.13
4	21.20	18.00	16.69	15.98	15.52	15.21	14.98	14.80	14.66	14.55	14.37	14.20	14.02	13.93	13.84	13.75	13.65	13.56	13.46
5	16.26	13.27	12.06	11.39	10.97	10.67	10.46	10.29	10.16	10.05	9.89	9.72	9.55	9.47	9.38	9.29	9.20	9.11	9.02
6	13.75	10.92	9.78	9.15	8.75	8.47	8.26	8.10	7.98	7.87	7.72	7.56	7.40	7.31	7.23	7.14	7.06	6.97	6.88
7	12.25	9.55	8.45	7.85	7.46	7.19	6.99	6.84	6.72	6.62	6.47	6.31	6.16	6.07	5.99	5.91	5.82	5.74	5.65
8	11.26	8.65	7.59	7.01	6.63	6.37	6.18	6.03	5.91	5.81	5.67	5.52	5.36	5.28	5.20	5.12	5.03	4.95	4.86
9	10.56	8.02	6.99	6.42	6.06	5.80	5.61	5.47	5.35	5.26	5.11	4.96	4.81	4.73	4.65	4.57	4.48	4.40	4.31
10	10.04	7.56	6.55	5.99	5.64	5.39	5.20	5.06	4.94	4.85	4.71	4.56	4.41	4.33	4.25	4.17	4.08	4.00	3.91
11	9.65	7.21	6.22	5.67	5.32	5.07	4.89	4.74	4.63	4.54	4.40	4.25	4.10	4.02	3.94	3.86	3.78	3.69	3.60
12	9.33	6.93	5.95	5.41	5.06	4.82	4.64	4.50	4.39	4.30	4.16	4.01	3.86	3.78	3.70	3.62	3.54	3.45	3.36
13	9.07	6.70	5.74	5.21	4.86	4.62	4.44	4.30	4.19	4.10	3.96	3.82	3.66	3.59	3.51	3.43	3.34	3.25	3.17
14	8.86	6.51	5.56	5.04	4.69	4.46	4.28	4.14	4.03	3.94	3.80	3.66	3.51	3.43	3.35	3.27	3.18	3.09	3.00
15	8.68	6.36	5.42	4.89	4.56	4.32	4.14	4.00	3.89	3.80	3.67	3.52	3.37	3.29	3.21	3.13	3.05	2.96	2.87
16	8.53	6.23	5.29	4.77	4.44	4.20	4.03	3.89	3.78	3.69	3.55	3.41	3.26	3.18	3.10	3.02	2.93	2.84	2.75
17	8.40	6.11	5.18	4.67	4.34	4.10	3.93	3.79	3.68	3.59	3.46	3.31	3.16	3.08	3.00	2.92	2.83	2.75	2.65
18	8.29	6.01	5.09	4.58	4.25	4.01	3.84	3.71	3.60	3.51	3.37	3.23	3.08	3.00	2.92	2.84	2.75	2.66	2.57
19	8.18	5.93	5.01	4.50	4.17	3.94	3.77	3.63	3.52	3.43	3.30	3.15	3.00	2.92	2.84	2.76	2.67	2.58	2.49
20	8.10	5.85	4.94	4.43	4.10	3.87	3.70	3.56	3.46	3.37	3.23	3.09	2.94	2.86	2.78	2.69	2.61	2.52	2.42
21	8.02	5.78	4.87	4.37	4.04	3.81	3.64	3.51	3.40	3.31	3.17	3.03	2.88	2.80	2.72	2.64	2.55	2.46	2.36
22	7.95	5.72	4.82	4.31	3.99	3.76	3.59	3.45	3.35	3.26	3.12	2.98	2.83	2.75	2.67	2.58	2.50	2.40	2.31
23	7.88	5.66	4.76	4.26	3.94	3.71	3.54	3.41	3.30	3.21	3.07	2.93	2.78	2.70	2.62	2.54	2.45	2.35	2.26
24	7.82	5.61	4.72	4.22	3.90	3.67	3.50	3.36	3.26	3.17	3.03	2.89	2.74	2.66	2.58	2.49	2.40	2.31	2.21
25	7.77	5.57	4.68	4.18	3.85	3.63	3.46	3.32	3.22	3.13	2.99	2.85	2.70	2.62	2.54	2.45	2.36	2.27	2.17
26	7.72	5.53	4.64	4.14	3.82	3.59	3.42	3.29	3.18	3.09	2.96	2.81	2.66	2.58	2.50	2.42	2.33	2.23	2.13
27	7.68	5.49	4.60	4.11	3.78	3.56	3.39	3.26	3.15	3.06	2.93	2.78	2.63	2.55	2.47	2.38	2.29	2.20	2.10
28	7.64	5.45	4.57	4.07	3.75	3.53	3.36	3.23	3.12	3.03	2.90	2.75	2.60	2.52	2.44	2.35	2.26	2.17	2.06
29	7.60	5.42	4.54	4.04	3.73	3.50	3.33	3.20	3.09	3.00	2.87	2.73	2.57	2.49	2.41	2.33	2.23	2.14	2.03
30	7.56	5.39	4.51	4.02	3.70	3.47	3.30	3.17	3.07	2.98	2.84	2.70	2.55	2.47	2.39	2.30	2.21	2.11	2.01
40	7.31	5.18	4.31	3.83	3.51	3.29	3.12	2.99	2.89	2.80	2.66	2.52	2.37	2.29	2.20	2.11	2.02	1.92	1.80
60	7.08	4.98	4.13	3.65	3.34	3.12	2.95	2.82	2.72	2.63	2.50	2.35	2.20	2.12	2.03	1.94	1.84	1.73	1.60
120	6.85	4.79	3.95	3.48	3.17	2.96	2.79	2.66	2.56	2.47	2.34	2.19	2.03	1.95	1.86	1.76	1.66	1.53	1.38
∞	6.63	4.61	3.78	3.32	3.02	2.80	2.64	2.51	2.41	2.32	2.18	2.04	1.88	1.79	1.70	1.59	1.47	1.32	1.00

IV. Percentage Points of the F Distribution (*continued*)

V.   Operating Characteristic Curves for the Fixed Effects Model Analysis of Variance[a]

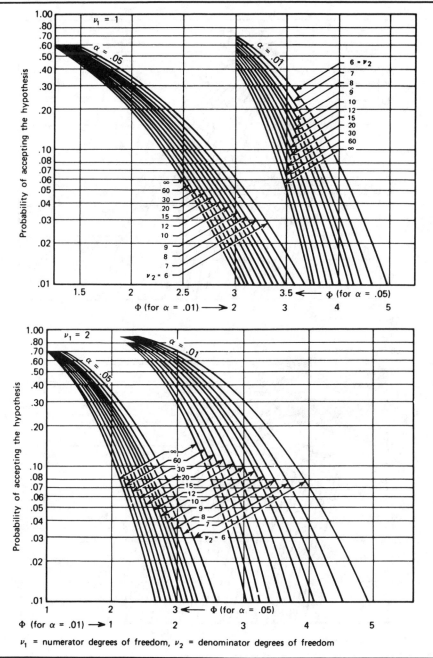

ν_1 = numerator degrees of freedom, ν_2 = denominator degrees of freedom

[a] Adapted with permission from *Biometrika Tables for Statisticians*, Vol. 2, by E. S. Pearson and H. O. Hartley, Cambridge University Press, Cambridge, 1972.

V. Operating Characteristic Curves for the Fixed Effects Model Analysis of Variance
(*continued*)

V.   Operating Characteristic Curves for the Fixed Effects Model Analysis of Variance
(*continued*)

V.   Operating Characteristic Curves for the Fixed Effects Model Analysis of Variance (*continued*)

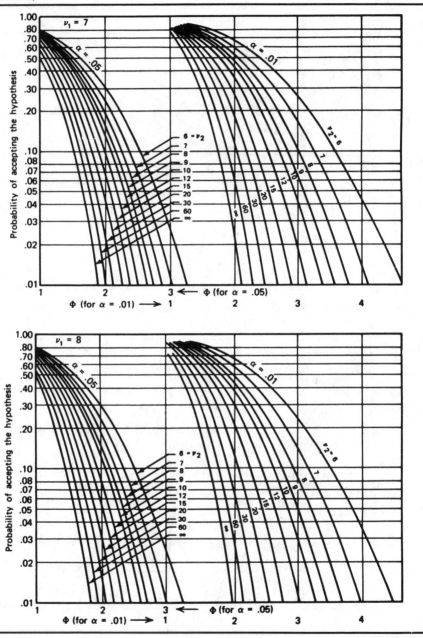

VI.   Operating Characteristic Curves for the Random Effects Model Analysis of Variance[a]

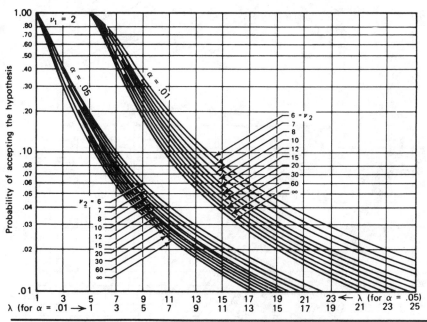

[a] Reproduced with permission from *Engineering Statistics*, 2nd edition, by A. H. Bowker and G. J. Lieberman, Prentice-Hall, Inc., Englewood Cliffs, N.J., 1972.

VI. Operating Characteristic Curves for the Random Effects Model Analysis of Variance (*continued*)

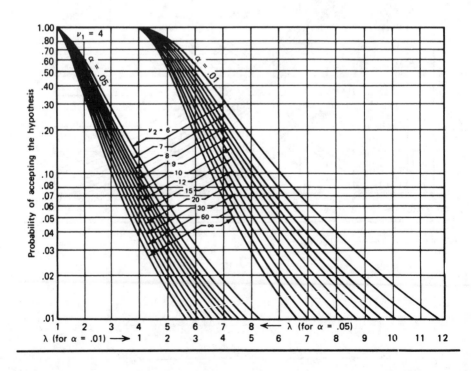

VI.   Operating Characteristic Curves for the Random Effects Model Analysis of Variance
(*continued*)

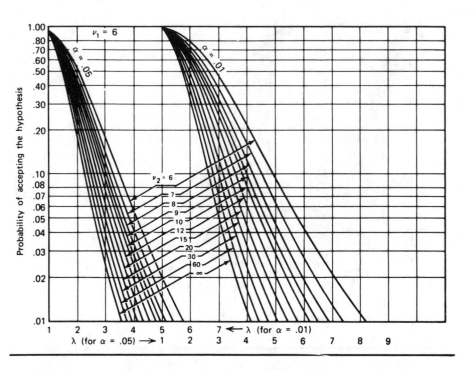

VI. Operating Characteristic Curves for the Random Effects Model Analysis of Variance (*continued*)

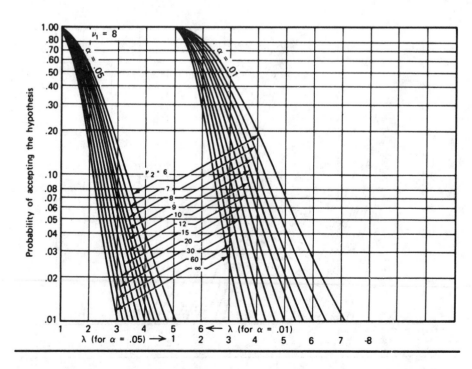

VII. Significant Ranges for Duncan's Multiple Range Test[a]

$$r_{.01}(p, f)$$

	p											
f	2	3	4	5	6	7	8	9	10	20	50	100
1	90.0	90.0	90.0	90.0	90.0	90.0	90.0	90.0	90.0	90.0	90.0	90.0
2	14.0	14.0	14.0	14.0	14.0	14.0	14.0	14.0	14.0	14.0	14.0	14.0
3	8.26	8.5	8.6	8.7	8.8	8.9	8.9	9.0	9.0	9.3	9.3	9.3
4	6.51	6.8	6.9	7.0	7.1	7.1	7.2	7.2	7.3	7.5	7.5	7.5
5	5.70	5.96	6.11	6.18	6.26	6.33	6.40	6.44	6.5	6.8	6.8	6.8
6	5.24	5.51	5.65	5.73	5.81	5.88	5.95	6.00	6.0	6.3	6.3	6.3
7	4.95	5.22	5.37	5.45	5.53	5.61	5.69	5.73	5.8	6.0	6.0	6.0
8	4.74	5.00	5.14	5.23	5.32	5.40	5.47	5.51	5.5	5.8	5.8	5.8
9	4.60	4.86	4.99	5.08	5.17	5.25	5.32	5.36	5.4	5.7	5.7	5.7
10	4.48	4.73	4.88	4.96	5.06	5.13	5.20	5.24	5.28	5.55	5.55	5.55
11	4.39	4.63	4.77	4.86	4.94	5.01	5.06	5.12	5.15	5.39	5.39	5.39
12	4.32	4.55	4.68	4.76	4.84	4.92	4.96	5.02	5.07	5.26	5.26	5.26
13	4.26	4.48	4.62	4.69	4.74	4.84	4.88	4.94	4.98	5.15	5.15	5.15
14	4.21	4.42	4.55	4.63	4.70	4.78	4.83	4.87	4.91	5.07	5.07	5.07
15	4.17	4.37	4.50	4.58	4.64	4.72	4.77	4.81	4.84	5.00	5.00	5.00
16	4.13	4.34	4.45	4.54	4.60	4.67	4.72	4.76	4.79	4.94	4.94	4.94
17	4.10	4.30	4.41	4.50	4.56	4.63	4.68	4.73	4.75	4.89	4.89	4.89
18	4.07	4.27	4.38	4.46	4.53	4.59	4.64	4.68	4.71	4.85	4.85	4.85
19	4.05	4.24	4.35	4.43	4.50	4.56	4.61	4.64	4.67	4.82	4.82	4.82
20	4.02	4.22	4.33	4.40	4.47	4.53	4.58	4.61	4.65	4.79	4.79	4.79
30	3.89	4.06	4.16	4.22	4.32	4.36	4.41	4.45	4.48	4.65	4.71	4.71
40	3.82	3.99	4.10	4.17	4.24	4.30	4.34	4.37	4.41	4.59	4.69	4.69
60	3.76	3.92	4.03	4.12	4.17	4.23	4.27	4.31	4.34	4.53	4.66	4.66
100	3.71	3.86	3.98	4.06	4.11	4.17	4.21	4.25	4.29	4.48	4.64	4.65
∞	3.64	3.80	3.90	3.98	4.04	4.09	4.14	4.17	4.20	4.41	4.60	4.68

f = degrees of freedom.

[a] Reproduced with permission from "Multiple Range and Multiple F Tests," by D. B. Duncan, *Biometrics*, Vol. 1, No. 1, pp. 1–42, 1955.

VII. Significant Ranges for Duncan's Multiple Range Test (*continued*)

$$r_{.05}(p, f)$$

f	2	3	4	5	6	7	8	9	10	20	50	100
						p						
1	18.0	18.0	18.0	18.0	18.0	18.0	18.0	18.0	18.0	18.0	18.0	18.0
2	6.09	6.09	6.09	6.09	6.09	6.09	6.09	6.09	6.09	6.09	6.09	6.0
3	4.50	4.50	4.50	4.50	4.50	4.50	4.50	4.50	4.50	4.50	4.50	4.5
4	3.93	4.01	4.02	4.02	4.02	4.02	4.02	4.02	4.02	4.02	4.02	4.0
5	3.64	3.74	3.79	3.83	3.83	3.83	3.83	3.83	3.83	3.83	3.83	3.8
6	3.46	3.58	3.64	3.68	3.68	3.68	3.68	3.68	3.68	3.68	3.68	3.6
7	3.35	3.47	3.54	3.58	3.60	3.61	3.61	3.61	3.61	3.61	3.61	3.6
8	3.26	3.39	3.47	3.52	3.55	3.56	3.56	3.56	3.56	3.56	3.56	3.5
9	3.20	3.34	3.41	3.47	3.50	3.52	3.52	3.52	3.52	3.52	3.52	3.5
10	3.15	3.30	3.37	3.43	3.46	3.47	3.47	3.47	3.47	3.48	3.48	3.4
11	3.11	3.27	3.35	3.39	3.43	3.44	3.45	3.46	3.46	3.48	3.48	3.4
12	3.08	3.23	3.33	3.36	3.40	3.42	3.44	3.44	3.46	3.48	3.48	3.4
13	3.06	3.21	3.30	3.35	3.38	3.41	3.42	3.44	3.45	3.47	3.47	3.4
14	3.03	3.18	3.27	3.33	3.37	3.39	3.41	3.42	3.44	3.47	3.47	3.4
15	3.01	3.16	3.25	3.31	3.36	3.38	3.40	3.42	3.43	3.47	3.47	3.4
16	3.00	3.15	3.23	3.30	3.34	3.37	3.39	3.41	3.43	3.47	3.47	3.47
17	2.98	3.13	3.22	3.28	3.33	3.36	3.38	3.40	3.42	3.47	3.47	3.47
18	2.97	3.12	3.21	3.27	3.32	3.35	3.37	3.39	3.41	3.47	3.47	3.47
19	2.96	3.11	3.19	3.26	3.31	3.35	3.37	3.39	3.41	3.47	3.47	3.47
20	2.95	3.10	3.18	3.25	3.30	3.34	3.36	3.38	3.40	3.47	3.47	3.47
30	2.89	3.04	3.12	3.20	3.25	3.29	3.32	3.35	3.37	3.47	3.47	3.47
40	2.86	3.01	3.10	3.17	3.22	3.27	3.30	3.33	3.35	3.47	3.47	3.47
60	2.83	2.98	3.08	3.14	3.20	3.24	3.28	3.31	3.33	3.47	3.48	3.48
100	2.80	2.95	3.05	3.12	3.18	3.22	3.26	3.29	3.32	3.47	3.53	3.53
∞	2.77	2.92	3.02	3.09	3.15	3.19	3.23	3.26	3.29	3.47	3.61	3.67

f = degrees of freedom.

VIII. Percentage Points of the Studentized Range Statistic[a]

$q_{.01}(p, f)$

f	2	3	4	5	6	7	8	9	10	11	12	13	14	15	16	17	18	19	20
1	90.0	135	164	186	202	216	227	237	246	253	260	266	272	277	282	286	290	294	198
2	14.0	19.0	22.3	24.7	26.6	28.2	29.5	30.7	31.7	32.6	33.4	34.1	34.8	35.4	36.0	36.5	37.0	37.5	37.9
3	8.26	10.6	12.2	13.3	14.2	15.0	15.6	16.2	16.7	17.1	17.5	17.9	18.2	18.5	18.8	19.1	19.3	19.5	19.8
4	6.51	8.12	9.17	9.96	10.6	11.1	11.5	11.9	12.3	12.6	12.8	13.1	13.3	13.5	13.7	13.9	14.1	14.2	14.4
5	5.70	6.97	7.80	8.42	8.91	9.32	9.67	9.97	10.24	10.48	10.70	10.89	11.08	11.24	11.40	11.55	11.68	11.81	11.93
6	5.24	6.33	7.03	7.56	7.97	8.32	8.61	8.87	9.10	9.30	9.49	9.65	9.81	9.95	10.08	10.21	10.32	10.43	10.54
7	4.95	5.92	6.54	7.01	7.37	7.68	7.94	8.17	8.37	8.55	8.71	8.86	9.00	9.12	9.24	9.35	9.46	9.55	9.65
8	4.74	5.63	6.20	6.63	6.96	7.24	7.47	7.68	7.87	8.03	8.18	8.31	8.44	8.55	8.66	8.76	8.85	8.94	9.03
9	4.60	5.43	5.96	6.35	6.66	6.91	7.13	7.32	7.49	7.65	7.78	7.91	8.03	8.13	8.23	8.32	8.41	8.49	8.57
10	4.48	5.27	5.77	6.14	6.43	6.67	6.87	7.05	7.21	7.36	7.48	7.60	7.71	7.81	7.91	7.99	8.07	8.15	8.22
11	4.39	5.14	5.62	5.97	6.25	6.48	6.67	6.84	6.99	7.13	7.25	7.36	7.46	7.56	7.65	7.73	7.81	7.88	7.95
12	4.32	5.04	5.50	5.84	6.10	6.32	6.51	6.67	6.81	6.94	7.06	7.17	7.26	7.36	7.44	7.52	7.59	7.66	7.73
13	4.26	4.96	5.40	5.73	5.98	6.19	6.37	6.53	6.67	6.79	6.90	7.01	7.10	7.19	7.27	7.34	7.42	7.48	7.55
14	4.21	4.89	5.32	5.63	5.88	6.08	6.26	6.41	6.54	6.66	6.77	6.87	6.96	7.05	7.12	7.20	7.27	7.33	7.39
15	4.17	4.83	5.25	5.56	5.80	5.99	6.16	6.31	6.44	6.55	6.66	6.76	6.84	6.93	7.00	7.07	7.14	7.20	7.26
16	4.13	4.78	5.19	5.49	5.72	5.92	6.08	6.22	6.35	6.46	6.56	6.66	6.74	6.82	6.90	6.97	7.03	7.09	7.15
17	4.10	4.74	5.14	5.43	5.66	5.85	6.01	6.15	6.27	6.38	6.48	6.57	6.66	6.73	6.80	6.87	6.94	7.00	7.05
18	4.07	4.70	5.09	5.38	5.60	5.79	5.94	6.08	6.20	6.31	6.41	6.50	6.58	6.65	6.72	6.79	6.85	6.91	6.96
19	4.05	4.67	5.05	5.33	5.55	5.73	5.89	6.02	6.14	6.25	6.34	6.43	6.51	6.58	6.65	6.72	6.78	6.84	6.89
20	4.02	4.64	5.02	5.29	5.51	5.69	5.84	5.97	6.09	6.19	6.29	6.37	6.45	6.52	6.59	6.65	6.71	6.76	6.82
24	3.96	4.54	4.91	5.17	5.37	5.54	5.69	5.81	5.92	6.02	6.11	6.19	6.26	6.33	6.39	6.45	6.51	6.56	6.61
30	3.89	4.45	4.80	5.05	5.24	5.40	5.54	5.65	5.76	5.85	5.93	6.01	6.08	6.14	6.20	6.26	6.31	6.36	6.41
40	3.82	4.37	4.70	4.93	5.11	5.27	5.39	5.50	5.60	5.69	5.77	5.84	5.90	5.96	6.02	6.07	6.12	6.17	6.21
60	3.76	4.28	4.60	4.82	4.99	5.13	5.25	5.36	5.45	5.53	5.60	5.67	5.73	5.79	5.84	5.89	5.93	5.98	6.02
120	3.70	4.20	4.50	4.71	4.87	5.01	5.12	5.21	5.30	5.38	5.44	5.51	5.56	5.61	5.66	5.71	5.75	5.79	5.83
∞	3.64	4.12	4.40	4.60	4.76	4.88	4.99	5.08	5.16	5.23	5.29	5.35	5.40	5.45	5.49	5.54	5.57	5.61	5.65

f = degrees of freedom.

[a] From J. M. May, "Extended and Corrected Tables of the Upper Percentage Points of the Studentized Range," Biometrika, Vol. 39, pp. 192–193, 1952. Reproduced by permission of the trustees, of Biometrika.

VIII. Percentage Points of the Studentized Range Statistic (continued)

$$q_{.05}(p, f)$$

f	2	3	4	5	6	7	8	9	10	11	12	13	14	15	16	17	18	19	20
1	18.1	26.7	32.8	37.2	40.5	43.1	45.4	47.3	49.1	50.6	51.9	53.2	54.3	55.4	56.3	57.2	58.0	58.8	59.6
2	6.09	8.28	9.80	10.89	11.73	12.43	13.03	13.54	13.99	14.39	14.75	15.08	15.38	15.65	15.91	16.14	16.36	16.57	16.77
3	4.50	5.88	6.83	7.51	8.04	8.47	8.85	9.18	9.46	9.72	9.95	10.16	10.35	10.52	10.69	10.84	10.98	11.12	11.24
4	3.93	5.00	5.76	6.31	6.73	7.06	7.35	7.60	7.83	8.03	8.21	8.37	8.52	8.67	8.80	8.92	9.03	9.14	9.24
5	3.61	4.54	5.18	5.64	5.99	6.28	6.52	6.74	6.93	7.10	7.25	7.39	7.52	7.64	7.75	7.86	7.95	8.04	8.13
6	3.46	4.34	4.90	5.31	5.63	5.89	6.12	6.32	6.49	6.65	6.79	6.92	7.04	7.14	7.24	7.34	7.43	7.51	7.59
7	3.34	4.16	4.68	5.06	5.35	5.59	5.80	5.99	6.15	6.29	6.42	6.54	6.65	6.75	6.84	6.93	7.01	7.08	7.16
8	3.26	4.04	4.53	4.89	5.17	5.40	5.60	5.77	5.92	6.05	6.18	6.29	6.39	6.48	6.57	6.65	6.73	6.80	6.87
9	3.20	3.95	4.42	4.76	5.02	5.24	5.43	5.60	5.74	5.87	5.98	6.09	6.19	6.28	6.36	6.44	6.51	6.58	6.65
10	3.15	3.88	4.33	4.66	4.91	5.12	5.30	5.46	5.60	5.72	5.83	5.93	6.03	6.12	6.20	6.27	6.34	6.41	6.47
11	3.11	3.82	4.26	4.58	4.82	5.03	5.20	5.35	5.49	5.61	5.71	5.81	5.90	5.98	6.06	6.14	6.20	6.27	6.33
12	3.08	3.77	4.20	4.51	4.75	4.95	5.12	5.27	5.40	5.51	5.61	5.71	5.80	5.88	5.95	6.02	6.09	6.15	6.21
13	3.06	3.73	4.15	4.46	4.69	4.88	5.05	5.19	5.32	5.43	5.53	5.63	5.71	5.79	5.86	5.93	6.00	6.06	6.11
14	3.03	3.70	4.11	4.41	4.64	4.83	4.99	5.13	5.25	5.36	5.46	5.56	5.64	5.72	5.79	5.86	5.92	5.98	6.03
15	3.01	3.67	4.08	4.37	4.59	4.78	4.94	5.08	5.20	5.31	5.40	5.49	5.57	5.65	5.72	5.79	5.85	5.91	5.96
16	3.00	3.65	4.05	4.34	4.56	4.74	4.90	5.03	5.15	5.26	5.35	5.44	5.52	5.59	5.66	5.73	5.79	5.84	5.90
17	2.98	3.62	4.02	4.31	4.52	4.70	4.86	4.99	5.11	5.21	5.31	5.39	5.47	5.55	5.61	5.68	5.74	5.79	5.84
18	2.97	3.61	4.00	4.28	4.49	4.67	4.83	4.96	5.07	5.17	5.27	5.35	5.43	5.50	5.57	5.63	5.69	5.74	5.79
19	2.96	3.59	3.98	4.26	4.47	4.64	4.79	4.92	5.04	5.14	5.23	5.32	5.39	5.46	5.53	5.59	5.65	5.70	5.75
20	2.95	3.58	3.96	4.24	4.45	4.62	4.77	4.90	5.01	5.11	5.20	5.28	5.36	5.43	5.50	5.56	5.61	5.66	5.71
24	2.92	3.53	3.90	4.17	4.37	4.54	4.68	4.81	4.92	5.01	5.10	5.18	5.25	5.32	5.38	5.44	5.50	5.55	5.59
30	2.89	3.48	3.84	4.11	4.30	4.46	4.60	4.72	4.83	4.92	5.00	5.08	5.15	5.21	5.27	5.33	5.38	5.43	5.48
40	2.86	3.44	3.79	4.04	4.23	4.39	4.52	4.63	4.74	4.82	4.90	4.98	5.05	5.11	5.17	5.22	5.27	5.32	5.36
60	2.83	3.40	3.74	3.98	4.16	4.31	4.44	4.55	4.65	4.73	4.81	4.88	4.94	5.00	5.06	5.11	5.15	5.20	5.24
120	2.80	3.36	3.69	3.92	4.10	4.24	4.36	4.47	4.56	4.64	4.71	4.78	4.84	4.90	4.95	5.00	5.04	5.09	5.13
∞	2.77	3.32	3.63	3.86	4.03	4.17	4.29	4.39	4.47	4.55	4.62	4.68	4.74	4.80					

IX. Critical Values for Dunnett's Test for Comparing Treatments with a Control[a]
$d_{.05}(a - 1, f)$
Two-Sided Comparisons

f	\multicolumn{9}{c}{a − 1 = Number of treatment means (excluding control)}
	1	2	3	4	5	6	7	8	9
5	2.57	3.03	3.29	3.48	3.62	3.73	3.82	3.90	3.97
6	2.45	2.86	3.10	3.26	3.39	3.49	3.57	3.64	3.71
7	2.36	2.75	2.97	3.12	3.24	3.33	3.41	3.47	3.53
8	2.31	2.67	2.88	3.02	3.13	3.22	3.29	3.35	3.41
9	2.26	2.61	2.81	2.95	3.05	3.14	3.20	3.26	3.32
10	2.23	2.57	2.76	2.89	2.99	3.07	3.14	3.19	3.24
11	2.20	2.53	2.72	2.84	2.94	3.02	3.08	3.14	3.19
12	2.18	2.50	2.68	2.81	2.90	2.98	3.04	3.09	3.14
13	2.16	2.48	2.65	2.78	2.87	2.94	3.00	3.06	3.10
14	2.14	2.46	2.63	2.75	2.84	2.91	2.97	3.02	3.07
15	2.13	2.44	2.61	2.73	2.82	2.89	2.95	3.00	3.04
16	2.12	2.42	2.59	2.71	2.80	2.87	2.92	2.97	3.02
17	2.11	2.41	2.58	2.69	2.78	2.85	2.90	2.95	3.00
18	2.10	2.40	2.56	2.68	2.76	2.83	2.89	2.94	2.98
19	2.09	2.39	2.55	2.66	2.75	2.81	2.87	2.92	2.96
20	2.09	2.38	2.54	2.65	2.73	2.80	2.86	2.90	2.95
24	2.06	2.35	2.51	2.61	2.70	2.76	2.81	2.86	2.90
30	2.04	2.32	2.47	2.58	2.66	2.72	2.77	2.82	2.86
40	2.02	2.29	2.44	2.54	2.62	2.68	2.73	2.77	2.81
60	2.00	2.27	2.41	2.51	2.58	2.64	2.69	2.73	2.77
120	1.98	2.24	2.38	2.47	2.55	2.60	2.65	2.69	2.73
∞	1.96	2.21	2.35	2.44	2.51	2.57	2.61	2.65	2.69

f = degrees of freedom.

[a] Reproduced with permission from C. W. Dunnett, "New Tables for Multiple Comparison with a Control," *Biometrics*, Vol. 20, No. 3, 1964, and from C. W. Dunnett, "A Multiple Comparison Procedure for Comparing Several Treatments with a Control," *Journal of the American Statistical Association*, Vol. 50, 1955.

IX.   Critical Values for Dunnett's Test for Comparing Treatments with a Control
$$d_{.01}(a - 1, f)$$
Two-Sided Comparisons

(continued)

f	\multicolumn{9}{c}{$a - 1 =$ number of treatment means (excluding control)}
	1	2	3	4	5	6	7	8	9
5	4.03	4.63	4.98	5.22	5.41	5.56	5.69	5.80	5.89
6	3.71	4.21	4.51	4.71	4.87	5.00	5.10	5.20	5.28
7	3.50	3.95	4.21	4.39	4.53	4.64	4.74	4.82	4.89
8	3.36	3.77	4.00	4.17	4.29	4.40	4.48	4.56	4.62
9	3.25	3.63	3.85	4.01	4.12	4.22	4.30	4.37	4.43
10	3.17	3.53	3.74	3.88	3.99	4.08	4.16	4.22	4.28
11	3.11	3.45	3.65	3.79	3.89	3.98	4.05	4.11	4.16
12	3.05	3.39	3.58	3.71	3.81	3.89	3.96	4.02	4.07
13	3.01	3.33	3.52	3.65	3.74	3.82	3.89	3.94	3.99
14	2.98	3.29	3.47	3.59	3.69	3.76	3.83	3.88	3.93
15	2.95	3.25	3.43	3.55	3.64	3.71	3.78	3.83	3.88
16	2.92	3.22	3.39	3.51	3.60	3.67	3.73	3.78	3.83
17	2.90	3.19	3.36	3.47	3.56	3.63	3.69	3.74	3.79
18	2.88	3.17	3.33	3.44	3.53	3.60	3.66	3.71	3.75
19	2.86	3.15	3.31	3.42	3.50	3.57	3.63	3.68	3.72
20	2.85	3.13	3.29	3.40	3.48	3.55	3.60	3.65	3.69
24	2.80	3.07	3.22	3.32	3.40	3.47	3.52	3.57	3.61
30	2.75	3.01	3.15	3.25	3.33	3.39	3.44	3.49	3.52
40	2.70	2.95	3.09	3.19	3.26	3.32	3.37	3.41	3.44
60	2.66	2.90	3.03	3.12	3.19	3.25	3.29	3.33	3.37
120	2.62	2.85	2.97	3.06	3.12	3.18	3.22	3.26	3.29
∞	2.58	2.79	2.92	3.00	3.06	3.11	3.15	3.19	3.22

IX.   Critical Values for Dunnett's Test for Comparing Treatments with a Control
$$d_{.05}(a - 1, f)$$
One-Sided Comparisons
(continued)

f	a − 1 = number of treatment means (excluding control)								
	1	2	3	4	5	6	7	8	9
5	2.02	2.44	2.68	2.85	2.98	3.08	3.16	3.24	3.30
6	1.94	2.34	2.56	2.71	2.83	2.92	3.00	3.07	3.12
7	1.89	2.27	2.48	2.62	2.73	2.82	2.89	2.95	3.01
8	1.86	2.22	2.42	2.55	2.66	2.74	2.81	2.87	2.92
9	1.83	2.18	2.37	2.50	2.60	2.68	2.75	2.81	2.86
10	1.81	2.15	2.34	2.47	2.56	2.64	2.70	2.76	2.81
11	1.80	2.13	2.31	2.44	2.53	2.60	2.67	2.72	2.77
12	1.78	2.11	2.29	2.41	2.50	2.58	2.64	2.69	2.74
13	1.77	2.09	2.27	2.39	2.48	2.55	2.61	2.66	2.71
14	1.76	2.08	2.25	2.37	2.46	2.53	2.59	2.64	2.69
15	1.75	2.07	2.24	2.36	2.44	2.51	2.57	2.62	2.67
16	1.75	2.06	2.23	2.34	2.43	2.50	2.56	2.61	2.65
17	1.74	2.05	2.22	2.33	2.42	2.49	2.54	2.59	2.64
18	1.73	2.04	2.21	2.32	2.41	2.48	2.53	2.58	2.62
19	1.73	2.03	2.20	2.31	2.40	2.47	2.52	2.57	2.61
20	1.72	2.03	2.19	2.30	2.39	2.46	2.51	2.56	2.60
24	1.71	2.01	2.17	2.28	2.36	2.43	2.48	2.53	2.57
30	1.70	1.99	2.15	2.25	2.33	2.40	2.45	2.50	2.54
40	1.68	1.97	2.13	2.23	2.31	2.37	2.42	2.47	2.51
60	1.67	1.95	2.10	2.21	2.28	2.35	2.39	2.44	2.48
120	1.66	1.93	2.08	2.18	2.26	2.32	2.37	2.41	2.45
∞	1.64	1.92	2.06	2.16	2.23	2.29	2.34	2.38	2.42

IX. Critical Values for Dunnett's Test for Comparing Treatments with a Control
$$d_{.01}(a - 1, f)$$
One-Sided Comparisons
(continued)

f	$a - 1$ = number of treatment means (excluding control)								
	1	2	3	4	5	6	7	8	9
5	3.37	3.90	4.21	4.43	4.60	4.73	4.85	4.94	5.03
6	3.14	3.61	3.88	4.07	4.21	4.33	4.43	4.51	4.59
7	3.00	3.42	3.66	3.83	3.96	4.07	4.15	4.23	4.30
8	2.90	3.29	3.51	3.67	3.79	3.88	3.96	4.03	4.09
9	2.82	3.19	3.40	3.55	3.66	3.75	3.82	3.89	3.94
10	2.76	3.11	3.31	3.45	3.56	3.64	3.71	3.78	3.83
11	2.72	3.06	3.25	3.38	3.48	3.56	3.63	3.69	3.74
12	2.68	3.01	3.19	3.32	3.42	3.50	3.56	3.62	3.67
13	2.65	2.97	3.15	3.27	3.37	3.44	3.51	3.56	3.61
14	2.62	2.94	3.11	3.23	3.32	3.40	3.46	3.51	3.56
15	2.60	2.91	3.08	3.20	3.29	3.36	3.42	3.47	3.52
16	2.58	2.88	3.05	3.17	3.26	3.33	3.39	3.44	3.48
17	2.57	2.86	3.03	3.14	3.23	3.30	3.36	3.41	3.45
18	2.55	2.84	3.01	3.12	3.21	3.27	3.33	3.38	3.42
19	2.54	2.83	2.99	3.10	3.18	3.25	3.31	3.36	3.40
20	2.53	2.81	2.97	3.08	3.17	3.23	3.29	3.34	3.38
24	2.49	2.77	2.92	3.03	3.11	3.17	3.22	3.27	3.31
30	2.46	2.72	2.87	2.97	3.05	3.11	3.16	3.21	3.24
40	2.42	2.68	2.82	2.92	2.99	3.05	3.10	3.14	3.18
60	2.39	2.64	2.78	2.87	2.94	3.00	3.04	3.08	3.12
120	2.36	2.60	2.73	2.82	2.89	2.94	2.99	3.03	3.06
∞	2.33	2.56	2.68	2.77	2.84	2.89	2.93	2.97	3.00

X. Coefficients of Orthogonal Polynomials[a]

x_i	n = 3		n = 4			n = 5				n = 6					n = 7					
	P_1	P_2	P_1	P_2	P_3	P_1	P_2	P_3	P_4	P_1	P_2	P_3	P_4	P_5	P_1	P_2	P_3	P_4	P_5	P_6
1	-1	1	-3	1	-1	-2	2	-1	1	-5	5	-5	1	-1	-3	5	-1	3	-1	1
2	0	-2	-1	-1	3	-1	-1	2	-4	-3	-1	7	-3	5	-2	0	1	-7	4	-6
3	1	1	1	-1	-3	0	-2	0	6	-1	-4	4	2	-10	-1	-3	1	1	-5	15
4			3	1	1	1	-1	-2	-4	1	-4	-4	2	10	0	-4	0	6	0	-20
5						2	2	1	1	3	-1	-7	-3	-5	1	-3	-1	1	5	15
6										5	5	5	1	1	2	0	-1	-7	-4	-6
7															3	5	1	3	1	1
$\sum_{i=1}^{n}\{P_i(x_i)\}^2$	2	6	20	4	20	10	14	10	70	70	84	180	28	252	28	84	6	154	84	924
λ	1	3	2	1	$\frac{10}{3}$	1	1	$\frac{5}{6}$	$\frac{35}{12}$	2	$\frac{3}{2}$	$\frac{5}{3}$	$\frac{7}{12}$	$\frac{21}{10}$	1	1	$\frac{1}{6}$	$\frac{7}{12}$	$\frac{7}{20}$	$\frac{77}{60}$

x_i	n = 8						n = 9						n = 10					
	P_1	P_2	P_3	P_4	P_5	P_6	P_1	P_2	P_3	P_4	P_5	P_6	P_1	P_2	P_3	P_4	P_5	P_6
1	-7	7	-7	7	-7	1	-4	28	-14	14	-4	4	-9	6	-42	18	-6	3
2	-5	1	5	-13	23	-5	-3	7	7	-21	11	-17	-7	2	14	-22	14	-11
3	-3	-3	7	-3	-17	9	-2	-8	13	-11	-4	22	-5	-1	35	-17	-1	10
4	-1	-5	3	9	-15	-5	-1	-17	9	9	-9	1	-3	-3	31	3	-11	6
5	1	-5	-3	9	15	-5	0	-20	0	18	0	-20	-1	-4	12	18	-6	-8
6	3	-3	-7	-3	17	9	1	-17	-9	9	9	1	1	-4	-12	18	6	-8
7	5	1	-5	-13	-23	-5	2	-8	-13	-11	4	22	3	-3	-31	3	11	6
8	7	7	7	7	7	1	3	7	-7	-21	-11	-17	5	-1	-35	-17	1	10
9							4	28	14	14	4	4	7	2	-14	-22	-14	-11
10													9	6	42	18	6	3
$\sum_{i=1}^{n}\{P_i(x_i)\}^2$	168	168	264	616	2184	264	60	2772	990	2002	468	1980	330	132	8580	2860	780	660
λ	2	1	$\frac{2}{3}$	$\frac{7}{12}$	$\frac{7}{10}$	$\frac{11}{60}$	2	3	$\frac{5}{6}$	$\frac{7}{12}$	$\frac{3}{20}$	$\frac{11}{60}$	2	$\frac{1}{2}$	$\frac{5}{3}$	$\frac{5}{12}$	$\frac{1}{10}$	$\frac{11}{240}$

[a] Adapted with permission from *Biometrika Tables for Statisticians*, Vol. 1, 3rd edition by E. S. Pearson and H. O. Hartley, Cambridge University Press, Cambridge, 1966.

XI. Random Numbers[a]

10480	15011	01536	02011	87647	91646	69179	14194	62590
22368	46573	25595	85393	30995	89198	27982	53402	93965
24130	48360	22527	97265	76393	64809	15179	24830	49340
42167	93093	06243	61680	07856	16376	39440	53537	71341
37570	39975	81837	16656	06121	91782	60468	81305	49684
77921	06907	11008	42751	27756	53498	18602	70659	90655
99562	72905	56420	69994	98872	31016	71194	18738	44013
96301	91977	05463	07972	18876	20922	94595	56869	69014
89579	14342	63661	10281	17453	18103	57740	84378	25331
85475	36857	53342	53988	53060	59533	38867	62300	08158
28918	69578	88231	33276	70997	79936	56865	05859	90106
63553	40961	48235	03427	49626	69445	18663	72695	52180
09429	93969	52636	92737	88974	33488	36320	17617	30015
10365	61129	87529	85689	48237	52267	67689	93394	01511
07119	97336	71048	08178	77233	13976	47564	81056	97735
51085	12765	51821	51259	77452	16308	60756	92144	49442
02368	21382	52404	60268	89368	19885	55322	44819	01188
01011	54092	33362	94904	31273	04146	18594	29852	71585
52162	53916	46369	58586	23216	14513	83149	98736	23495
07056	97628	33787	09998	42698	06691	76988	13602	51851
48663	91245	85828	14346	09172	30168	90229	04734	59193
54164	58492	22421	74103	47070	25306	76468	26384	58151
32639	32363	05597	24200	13363	38005	94342	28728	35806
29334	27001	87637	87308	58731	00256	45834	15398	46557
02488	33062	28834	07351	19731	92420	60952	61280	50001
81525	72295	04839	96423	24878	82651	66566	14778	76797
29676	20591	68086	26432	46901	20849	89768	81536	86645
00742	57392	39064	66432	84673	40027	32832	61362	98947
05366	04213	25669	26422	44407	44048	37937	63904	45766
91921	26418	64117	94305	26766	25940	39972	22209	71500
00582	04711	87917	77341	42206	35126	74087	99547	81817
00725	69884	62797	56170	86324	88072	76222	36086	84637
69011	65795	95876	55293	18988	27354	26575	08625	40801
25976	57948	29888	88604	67917	48708	18912	82271	65424
09763	83473	73577	12908	30883	18317	28290	35797	05998
91567	42595	27958	30134	04024	86385	29880	99730	55536
17955	56349	90999	49127	20044	59931	06115	20542	18059
46503	18584	18845	49618	02304	51038	20655	58727	28168
92157	89634	94824	78171	84610	82834	09922	25417	44137
14577	62765	35605	81263	39667	47358	56873	56307	67607

[a] Reproduced with permission from *Probability and Statistics in Engineering and Management Science*, 3rd edition, by W. W. Hines and D. C. Montgomery, Wiley, New York, 1990.

XI. Random Numbers (*continued*)

98427	07523	33362	64270	01638	92477	66969	98420	04880
34914	63976	88720	82765	34476	17032	87589	40836	32427
70060	28277	39475	46473	23219	53416	94970	25832	69975
53976	54914	06990	67245	68350	82948	11398	42878	80287
76072	29515	40980	07391	58745	25774	22987	80059	39911
90725	52210	83974	29992	65831	38857	50490	83765	55657
64364	67412	33339	31926	14883	24413	59744	92351	97473
08962	00358	31662	25388	61642	34072	81249	35648	56891
95012	68379	93526	70765	10592	04542	76463	54328	02349
15664	10493	20492	38391	91132	21999	59516	81652	27195

[a] Reproduced with permission from *Probability and Statistics in Engineering and Management Science*, 3rd edition, by W. W. Hines and D. C. Montgomery, Wiley, New York, 1990.

XII. Alias Relationships for 2^{k-p} Fractional Factorial Designs with $k \leq 11$ and $n \leq 64$

Designs with 3 Factors

(a) 2^{3-1}; 1/2 replicate of 3 factors in 8 runs   **Resolution III**

$$1 = ABC$$

<u>Aliases</u>

$A = BC$
$B = AC$
$C = AB$

Designs with 4 Factors

(b) 2^{4-1}; 1/2 replicate of 4 factors in 8 runs   **Resolution IV**

$$I = ABCD$$

<u>Aliases</u>

$A = BCD$
$B = ACD$
$C = ABD$
$D = ABC$
$AB = CD$
$AC = BD$
$AD = BC$

XII. Atlas Relationships for 2^{k-p} Fractional Factorial Designs with $k \leq 11$ and $n \leq 64$ (Continued)

Designs with 5 Factors

(c) 2^{5-2}; 1/4 replicate of 5 factors in 8 runs   **Resolution III**

$$I = ABD = ACE = BCDE$$

Aliases

$A = BD = CE$
$B = AD = CDE$
$C = AE = BDE$
$D = AB = BCE$
$E = AC = BCD$
$BC = DE = ACD = ABE$
$CD = BE = ABC = ADE$

(d) 2^{5-1}; 1/2 replicate of 5 factors in 16 runs   **Resolution V**

$$I = ABCDE$$

Aliases

Each main effect is aliased with a single 4-factor interaction.

$AB = CDE$
$AC = BDE$
$AD = BCE$
$AE = BCD$
$BC = ADE$
$BD = ACE$
$BE = ACD$
$CD = ABE$
$CE = ABD$
$DE = ABC$

2 blocks of 8: $AB = CDE$

Designs with 6 Factors

(e) 2^{6-3}; 1/8 replicate of 6 factors in 8 runs   Resolution III

$$I = ABD = ACE = BCDE = BCF = ACDF = ABEF = DEF$$

Aliases

$A = BD = CE = CDF = BEF$

$B = AD = CF = CDE = AEF$

$C = AE = BF = BDE = ADF$

$D = AB = EF = BCE = ACF$

$E = AC = DF = BCD = ABF$

$F = BC = DE = ACD = ABE$

$CD = BE = AF = ABC = ADE = BDF = CEF$

(f) 2^{6-2}; 1/4 replicate of 6 factors in 16 runs   Resolution IV

$$I = ABCE = BCDF = ADEF$$

Aliases

$A = BCE = DEF$	$AB = CE$	
$B = ACE = CDF$	$AC = BE$	
$C = ABE = BDF$	$AD = EF$	
$D = BCF = AEF$	$AE = BC = DF$	
$E = ABC = ADF$	$AF = DE$	
$F = BCD = ADE$	$BD = CF$	
$ABD = CDE = ACF = BEF$	$BF = CD$	
$ACD = BDE = ABF = CEF$		

2 blocks of 8: $ABD = CDE = ACF = BEF$

XII. Alias Relationships for 2^{k-p} Fractional Factorial Designs with $k \leq 11$ and $n \leq 64$ *(continued)*

(g) 2^{6-1}; 1/2 replicate of 6 factors in 32 runs   **Resolution VI**

$$I = ABCDEF$$

<u>Aliases</u>

Each main effect is aliased with a single 5-factor interaction.
Each 2-factor interaction is aliased with a single 4-factor interaction.

$ABC = DEF$	$ACE = BDF$
$ABD = CEF$	$ACF = BDE$
$ABE = CDF$	$ADE = BCF$
$ABF = CDE$	$ADF = BCE$
$ACD = BEF$	$AEF = BCD$

2 blocks of 16: $ABF = CDE$ 4 blocks of 8: BC

$$ABF = CDE$$
$$ACF = BDE$$

Designs with 7 Factors

(h) 2^{7-4}; 1/16 replicate of 7 factors in 8 runs   **Resolution III**

$$I = ABD = ACE = BCDE = BCF = ACDF = ABEF = DEF = ABCG$$
$$= CDG = BEG = ADEG = AFG = BDFG = CEFG = ABCDEFG$$

Aliases

$A = BD = CE = FG = CDF = BEF = BCG = DEG$
$B = AD = CF = EG = CDE = AEF = ACG = DFG$
$C = AE = BF = DG = BDE = ADF = ABG = EFG$
$D = AB = EF = CG = BCE = ACF = AEG = BFG$
$E = AC = DF = BG = BCD = ABF = ADG = CFG$
$F = BC = DE = AG = ACD = ABE = BDG = CEG$
$G = CD = BE = AF = ABC = ADE = BDF = CEF$

(i) 2^{7-3}; 1/8 replicate of 7 factors in 16 runs   **Resolution IV**

$I = ABCE = BCDF = ADEF = ACDG = BDEG = ABFG = CEFG$

Aliases

$A = BCE = DEF = CDG = BFG$	$AB = CE = FG$
$B = ACE = CDF = DEG = AFG$	$AC = BE = DG$
$C = ABE = BDF = ADG = EFG$	$AD = EF = CG$
$D = BCF = AEF = ACG = BEG$	$AE = BC = DF$
$E = ABC = ADF = BDG = CFG$	$AF = DE = BG$
$F = BCD = ADE = ABG = CEG$	$AG = CD = BF$
$G = ACD = BDE = ABF = CEF$	$BD = CF = EG$

$ABD = CDE = ACF = BEF = BCG = AEG = DFG$

2 blocks of 8: $ABD = ACF = CDE = ACF = BEF = BCG = AEG = DFG$

(j) 2^{7-3}; 1/4 replicate of 7 factors in 32 runs   **Resolution IV**

$I = ABCDF = ABDEG = CEFG$

Aliases

$A =$	$AF = BCD$	$CG = EF$
$B =$	$AG = BDE$	$DE = ABG$
$C = EFG$	$BC = ADF$	$DF = ABC$
$D =$	$BD = ACF = AEG$	$DG = ABE$
$E = CFG$	$BE = ADG$	$ACE = AFG$
$F = CEG$	$BF = ACD$	$ACG = AEF$
$G = CEF$	$BG = ADE$	$BCE = BFG$
$AB = CDF = DEG$	$CD = ABF$	$BCG = BEF$
$AC = BDF$	$CE = FG$	$CDE = DFG$
$AD = BCF = BEG$	$CF = ABD = EG$	$CDG = DEF$
$AE = BDG$		

2 blocks of 16: $ACE = AFG$ 4 blocks of 8: $ACE = AFG$
                                                    $BCE = BFG$

XII.   Alias Relationships for 2^{k-p} Fractional Factorial Designs with $k \leq 11$ and $n \leq 64$ (continued)

(k) 2^{7-2}; 1/2 replicate of 7 factors in 64 runs   **Resolution VII**

$$I = ABCDEFG$$

Each main effect is aliased with a single 6-factor interaction.
Each 2-factor interaction is aliased with a single 5-factor interaction.
Each 3-factor interaction is aliased with a single 4-factor interaction.

2 blocks of 32: ABC   4 blocks of 16: ABC
                                          CEF
                                          CDG

Designs with 8 Factors

(l) 2^{8-4}; 1/16 replicate of 8 factors in 16 runs   **Resolution IV**

Aliases

$I = BCDE = ACDF = ABEF = ABCG = ADEG = BDFG = CEFG = ABDH$
$\ = ACEH = BCFH = DEFH = CDGH = BEGH = AFGH = ABCDEFGH$

$A = CDF = BEF = BCG = DEG = BDH = CEH = FGH$	$AB = EF = CG = DH$
$B = CDE = AEF = ACG = DFG = ADH = CFH = EGH$	$AC = DF = BG = EH$
$C = BDE = ADF = ABG = EFG = AEH = BFH = DGH$	$AD = CF = EG = BH$
$D = BCE = ACF = AEG = BFG = ABH = EFH = CGH$	$AE = BF = DG = CH$
$E = BCD = ABF = ADG = CFG = ACH = DFH = BGH$	$AF = CD = BE = GH$
$F = ACD = ABE = BDG = CEG = BCH = DEH = AGH$	$AG = BC = DE = FH$
$G = ABC = ABE = BDF = CEF = CDH = BEH = AFH$	$AH = BD = CE = FG$
$H = ABD = ACE = BCF = DEF = CDG = BEG = AFG$	

2 blocks of 8: $AB = EF = CG = DH$

Resolution IV

(m) 2^{8-3}; 1/8 replicate of 8 factors in 32 runs

$I = ABCF = ABDG = CDFG = BCDEH = ADEFH = ACEGH = BEFGH$

Aliases

$A = BCF = BDG$	$AE = DFH = CGH$	$DE = BCH = AFH$
$B = ACF = ADG$	$AF = BC = DEH$	$DH = BCE = AEF$
$C = ABF = DFG$	$AG = BD = CEH$	$EF = ADH = BGH$
$D = ABG = CFG$	$AH = DEF = CEG$	$EG = ACH = BFH$
$E =$	$BE = CDH = FGH$	$EH = BCD = ADF = ACG = BFG$
$F = ABC = CDG$	$BH = CDE = EFG$	$FH = ADE = BEG$
$G = ABD = CDF$	$CD = FG = BEH$	$GH = ACE = BEF$
$H =$	$CE = BDH = AGH$	$ABE = CEF = DEG$
$AB = CF = DG$	$CG = DF = AEH$	$ABH = CFH = DGH$
$AC = BF = EGH$	$CH = BDE = AEG$	$ACD = BDF = BCG = AFG$
$AD = BG = EFH$		

2 blocks of 16: $ABE = CEF = DEG$ 4 blocks of 8: $ABE = CEF = DEG$

$ABH = CFH = DGH$

$EH = BCD = ADF = ACG = BFG$

XII. Alias Relationships for 2^{k-p} Fractional Factorial Designs with $k \leq 11$ and $n \leq 64$ (continued)

Resolution V

(n) 2^{8-2}; 1/4 replicate of 8 factors in 64 runs

$$I = ABCDG = ABEFH = CDEFGH$$

Aliases

$AB = CDG = EFH$	$BG = ACD$	$EF = ABH$	$ADH =$	$BFG =$
$AC = BDG$	$BH = AEF$	$EG =$	$AEG =$	$BGH =$
$AD = BCG$	$CD = ABG$	$EH = ABF$	$AFG =$	$CDE = FGH$
$AE = BFH$	$CE =$	$FG =$	$AGH =$	$CDF = EGH$
$AF = BEH$	$CF =$	$FH = ABE$	$BCE =$	$CDH = EFG$
$AG = BCD$	$CG = ABD$	$GH =$	$BCF =$	$CEF = DGH$
$AH = BEF$	$CH =$	$ACE =$	$BCH =$	$CEG = DFH$
$BC = ADG$	$DE =$	$ACF =$	$BDE =$	$CEH = DFG$
$BD = ACG$	$DF =$	$ACH =$	$BDF =$	$CFG = DEH$
$BE = AFH$	$DG = ABC$	$ADE =$	$BDH =$	$CFH = DEG$
$BF = AEH$	$DH =$	$ADF =$	$BEG =$	$CGH = DEF$

2 blocks of 32: $CDE = FGH$ 4 blocks of 16: $CDE = FGH$
                                              ACF
                                              BDH

Designs with 9 Factors

Resolution III

(o) 2^{9-5}; 1/32 replicate of 9 factors in 16 runs

I = ABCE = BCDF = ADEF = ACDG = BDEG = ABFG = CEFG = ABDH
= CDEH = ACFH = BEFH = BCGH = AEGH = DFGH = ABCDEFGH = ABCDJ
= DEJ = AFJ = BCEFJ = BGJ = ACEGJ = CDFGJ = ABDEFGJ = CHJ
= ABEHJ = BDFHJ = ABCDEFHJ = ADGHJ = BCDEGHJ = ABCFGHJ = EFGHJ

Aliases

A = FJ = BCE = DEF = CDG = BFG = BDH = CFH = EGH
B = GJ = ACE = CDF = DEG = AFG = ADH = EFH = CGH
C = HJ = ABE = BDF = ADG = EFG = DEH = AFH = BGH
D = EJ = BCF = AEF = ACG = BEG = ABH = CEH = FGH
E = DJ = ABC = ADF = BDG = CFG = CDH = BFH = AGH
F = AJ = BCD = ADE = ABG = CEG = ACH = BEH = DGH
G = BJ = ACD = BDE = ABF = CEF = BCH = AEH = DFH
H = CJ = ABD = CDE = ACF = BEF = BCG = AEG = DFG
J = DE = AF = BG = CH

AB = CE = FG = DH = CDJ = BFJ = AGJ = EHJ
AC = BE = DG = FH = BDJ = CFJ = EGJ = AHJ
AD = EF = CG = BH = BCJ = AEJ = DFJ = GHJ
AE = BC = DF = GH = ADJ = EFJ = CGJ = BHJ
AG = CD = BF = EH = ABJ = CEJ = FGJ = DHJ
AH = BD = CF = EG = ACJ = BEJ = DGJ = FHJ

2 blocks of 8: AB = CE = FG = DH = CDJ = BFJ = AGJ = EHJ

XII. Alias Relationships for 2^{k-p} Fractional Factorial Designs with $k \leq 11$ and $n \leq 64$ (continued)

(p) 2^{9-4}; 1/16 replicate of 9 factors in 32 runs   **Resolution IV**

I = BCDEF = ACDEG = ABFG = ABDEH = ACFH = BCGH = DEFGH = ABCEJ
  = ADFJ = BDGJ = CEFGJ = CDHJ = BEFHJ = AEGHJ = ABCDFGHJ

Aliases

A = BFG = CFH = DFJ	AD = CEG = BEH = FJ	BJ = ACE = DG = EFH
B = AFG = CGH = DGJ	AE = CDG = BDH = BCJ = GHJ	CD = BEF = AEG = HJ
C = AFH = BGH = DHJ	AF = BG = CH = DJ	CE = BDF = ADG = ABJ = FGJ
D = AFJ = BGJ = CHJ	AG = CDE = BF = EHJ	CJ = ABE = EFG = DH
E =	AH = BDE = CF = EGJ	DE = BCF = ACG = ABH = FGH
F = ABG = ACH = ADJ	AJ = BCE = DF = EGH	EF = BCD = DGH = CGJ = BHJ
G = ABF = BCH = BDJ	BC = DEF = GH = AEJ	EG = ACD = DFH = CFJ = AHJ
H = ACF = BCG = CDJ	BD = CEF = AEH = GJ	EH = ABD = DFG = BFJ = AGJ
J = ADF = BDG = CDH	BE = CDF = ADH = ACJ = FHJ	EJ = ABC = BFH = AGH
AB = FG = DEH = CEJ	BH = ADE = CG = EFJ	AEF = BEG = CEH = DEJ
AC = DEG = FH = BEJ		

2 blocks of 16: AEF = BEG = CEH = DEJ   4 blocks of 8: AEF = BEG = CEH = DEJ
                                                              AB = FG = DEH = CEJ
                                                              CD = BEF = AEG = HJ

Resolution IV

(q) 2^{9-3}; 1/8 replicate of 9 factors in 64 runs

$I = ABCDG = ACEFH = BDEFGH = CDEFJ = ABEFGJ = ADHJ = BCGHJ$

<u>Aliases</u>

$A = DHJ$	$AC = BDG = EFH$	$BF =$	
$B =$	$AD = BCG = HJ$	$BG = ACD = CHJ$	
$C =$	$AE = CFH$	$BH = CGJ$	
$D = AHJ$	$AF = CEH$	$BJ = CGH$	
$E =$	$AG = BCD$	$CD = ABG = EFJ$	
$F =$	$AH = CEF = DJ$	$CE = AFH = DFJ$	
$G =$	$AJ = DH$	$CF = AEH = DEJ$	
$H = ADJ$	$BC = ADG = GHJ$	$CG = ABD = BHJ$	
$J = ADH$	$BD = ACG$	$CH = AEF = BGJ$	
$AB = CDG$	$BE =$	$CJ = DEF = BGH$	
$DE = CFJ$	$GJ = BCH$	$AFJ = BEG = DFH$	
$DF = CEJ$	$ABE = FGJ$	$AGH = DGJ$	
$DG = ABC$	$ABF = EGJ$	$AGJ = BEF = DGH$	
$EF = ACH = CDJ$	$ABH = BDJ$	$BCE =$	
$EG =$	$ABJ = EFG = BDH$	$BCF =$	
$EH = ACF$	$ACJ = CDH$	$BDE = FGH$	
$EJ = CDF$	$ADE = EHJ$	$BDF = EGH$	
$FG =$	$ADF = FHJ$	$BEH = DFG$	
$FH = ACE$	$AEG = BFJ$	$BFH = DEG$	
$FJ = CDE$	$AEJ = BFG = DEH$	$CEG =$	
$GH = BCJ$	$AFG = BEJ$	$CFG =$	

2 blocks of 32: CFG   4 blocks of 16: CFG

$AGJ = BEF = DGH$

$ADE = EHJ$

Designs with 10 Factors   Resolution III

(r) 2^{10-6}; 1/64 replicate of 10 factors in 16 runs

I = ABCE = BCDF = ADEF = ACDG = BDEG = ABFG = CEFG = ABDH
  = CDEH = ACFH = BEFH = BCGH = AEGH = DFGH = ABCDEFGH = ABCDJ
  = DEJ = AFJ = BCEFJ = BGJ = ACEGJ = CDFGJ = ABDEFGJ = CHJ
  = ABEHJ = BDFHJ = ACDEFHJ = ADGHJ = BCDEGHJ = ABCFGHJ = EFGHJ = ABK
  = CEK = ACDFK = BDEFK = BCDGK = ADEGK = FGK = ABCEFGK = DHK
  = ABCDEHK = BCFHK = AEFHK = ACGHK = BEGHK = ABDFGHK = CDEFGHK = CDJK
  = ABDEJK = BFJK = ACEFJK = AGJK = BCEGJK = ABCDFGJK = DEFGJK = ABCHJK
  = EHJK = ADFHJK = BCDEFHJK = BDGHJK = ACDEGHJK = CFGHJK = ABEFGHJK

<u>Aliases</u>

A = FJ = BK = BCE = DEF = CDG = BFG = BDH = CFH = EGH = CJK
B = GJ = AK = ACE = CDF = DEG = AFG = ADH = EFH = CGH = FJK
C = HJ = EK = ABE = BDF = ADG = EFG = DEH = AFH = BGH = DJK
D = EJ = HK = BCF = AEF = ACG = BEG = ABH = CEH = FGH = CJK
E = DJ = CK = ABC = ADF = BDG = CFG = CDH = BFH = AGH = HJK
F = AJ = GK = BCD = ADE = ABG = CEG = ACH = BEH = DGH = BJK
G = BJ = FK = ACD = BDE = ABF = CEF = BCH = AEH = DFH = AJK
H = CJ = DK = ABD = CDE = ACF = BEF = BCG = AEG = DFG = EJK
J = DE = AF = BG = CH = CDK = BFK = AGK = EHK
K = AB = CE = FG = DH = CDJ = BFJ = AGJ = EHJ

AC = BE = DG = FH = BDJ = CFJ = EGJ = AHJ = BCK = AEK = DFK = GHK
AD = EF = CG = BH = BCJ = AEJ = DFJ = GHJ = BDK = CFK = EGK = AHK
AE = BC = DF = GH = ADJ = EFJ = CGJ = BHJ = ACK = BEK = DGK = FHK
AG = CD = BF = EH = JK = ABJ = CEJ = FGJ = DHJ = DEK = AFK = BGK = CHK
AH = BD = CF = EG = ACJ = BEJ = DGJ = FHJ = ADK = EFK = CGK = BHK

2 blocks of 8: AG = CD = BF = EH = JK = ABJ = CEJ
                                = FGJ = DHJ = DEK = AFK = BGK = CHK

Resolution IV

(s) 2^{10-5}; 1/32 replicate of 10 factors in 32 runs

I = ABCDF = ABCEG = DEFG = ABDEH = CEFH = CDGH = ABFGH = ACDEJ
= BEFJ = BDGJ = ACFGJ = BCHJ = ADFHJ = AEGHJ = BCDEFGHJ = BCDEK
= AEFK = ADGK = BCFGK = BDFHK = BEGHK = ACDEFGHK = ABJK
= CDFJK = CEGJK = ABDEFGJK = DEHJK = ABCEFHJK = ABCDGHJK = FGHJK

Aliases

A = EFK = DGK = CHK = BJK	AB = CDF = CEG = DEH = FGH = JK
B = EFJ = DGJ = CHJ = AJK	AC = BDF = BEG = DEJ = FGJ = HK
C = EFH = DGH = BHJ = AHK	AD = BCF = BEH = CEJ = FHJ = GK
D = EFG = CGH = BGJ = AGK	AE = BCG = BDH = CDJ = GHJ = FK
E = DFG = CFH = BFJ = AFK	AF = BCD = BGH = CGJ = DHJ = EK
F = DEG = CEH = BEJ = AEK	AG = BCE = BFH = CFJ = EHJ = DK
G = DEF = CDH = BDJ = ADK	AH = BDE = BFG = DFJ = EGJ = CK
H = CEF = CDG = BCJ = ACK	AJ = CDE = CFG = DFH = EGH = BK
J = BEF = BDG = BCH = ABK	AK = EF = DG = CH = BJ
K = AEF = ADG = ACH = ABJ	BC = ADF = AEG = HJ = DEK = FGK

BD = ACF = AEH = GJ = CEK = FHK
BE = ACG = ADH = FJ = CDK = GHK
BF = ACD = AGH = EJ = CGK = DHK
BG = ACE = AFH = DJ = CFK = EHK
BH = ADE = AFG = CJ = DFK = EGK
CD = ABF = GH = AEJ = BEK = FJK
CE = ABG = FH = ADJ = BDK = GJK
CF = ABD = EH = AGJ = BGK = DJK
CG = ABE = DH = AFJ = BFK = EJK
DE = FG = ABH = ACJ = BCK = HJK
DF = ABC = EG = AHJ = BHK = CJK

2 blocks of 16: AK = EF = DG = CH = BJ
4 blocks of 8: AK = EF = DG = CH = BJ

XII. Alias Relationships for 2^{k-p} Fractional Factorial Designs with $k \leq 11$ and $n \leq 64$ (continued)

Resolution IV

(t) 2^{10-4}; 1/16 replicate of 10 factors in 64 runs

$I = BCDFG = ACDFH = ABGH = ABDEJ = ACEFGJ = BCEFHJ = DEGHJ = ABCEK$
$= ADEFGK = BDEFHK = CEGHK = CDJK = BFGJK = AFHJK = ABCDGHJK$

Aliases

$A = BGH$	$AD = CFH = BEJ$	$BK = ACE = FGJ$
$B = AGH$	$AE = BDJ = BCK$	$CD = BFG = AFH = JK$
$C = DJK$	$AF = CDH = HJK$	$CE = ABK = GHK$
$D = CJK$	$AG = BH$	$CF = BDG = ADH$
$E =$	$AH = CDF = BG = FJK$	$CG = BDF = EHK$
$F =$	$AJ = BDE = FHK$	$CH = ADF = EGK$
$G = ABH$	$AK = BCE = FHJ$	$CJ = DK$
$H = ABG$	$BC = DFG = AEK$	$CK = ABE = EGH = DJ$
$J = CDK$	$BD = CFG = AEJ$	$DE = ABJ = GHJ$
$K = CDJ$	$BE = ADJ = ACK$	$DF = BCG = ACH$
$AB = GH = DEJ = CEK$	$BF = CDG = GJK$	$DG = BCF = EHJ$
$AC = DFH = BEK$	$BJ = ADE = FGK$	$DH = ACF = EGJ$
$EF =$	$GJ = DEH = BFK$	$AEG = BEH = CFJ = DFK$
$EG = DHJ = CHK$	$GK = CEH = BFJ$	$AEH = BEG$
$EH = DGJ = CGK$	$HJ = DEG = AFK$	$AFC = BFH = CEJ = DEK$
$EJ = ABD = DGH$	$HK = CEG = AFJ$	$AGJ = CEF = BHJ$
$EK = ABC = CGH$	$ABF = FGH$	$AGK = DEF = BHK$
$FG = BCD = BJK$	$ACG = BCH = EFJ$	$BCJ = EFH = BDK$
$FH = ACD = AJK$	$ACJ = EFG = ADK$	$BEF = CHJ = DHK$
$FJ = BGK = AHK$	$ADG = BDH = EFK$	$CDE = EJK$
$FK = BGJ = AHJ$	$AEF = CGJ = DGK$	$CFK = DFJ$

2 blocks of 32: $AGJ = CEF = BHJ$ 4 blocks of 16: $AGJ = CEF = BHJ$
$AGK = DEF = BHK$
$CD = BFG = AFH = JK$

Designs with 11 Factors

Resolution III

(u) 2^{11-7}; 1/128 replicate of 11 factors in 16 runs

I = ABCE = BCDF = ADEF = ACDG = BDEG = ABFG = CEFG = ABDH
= CDEH = ACFH = BEFH = BCGH = AEGH = DFGH = ABCDEFGH = ABCDJ
= DEJ = AFJ = BCEFJ = BGJ = ACEGJ = CDFGJ = ABDEFGJ = CHJ
= ABEHJ = BDFHJ = ACDEFHJ = ADGHJ = BCDEGHJ = ABCFGHJ = EFGHJ = ABK
= CEK = ACDFK = BDEFK = BCDGK = ADEGK = FGK = ABCEFGK = DHK
= ABCDEHK = BCFHK = AEFHK = ACGHK = BEGHK = ABDFGHK = CDEFGHK = CDJK
= ABDEJK = BFJK = ACEFJK = AGJK = BCEGJK = ABCDFGJK = DEFGJK = ABCHJK
= EHJK = ADFHJK = BCDEFHJK = BDGHJK = ACDEGHJK = CFGHJK = ABEFGHJK = ACL
= BEL = ABDFL = CDEFL = DGL = ABCDEGL = BCFGL = AEFGL = BCDHL
= ADEHL = FHL = ABCEFHL = ABGHL = CEGHL = ACDFGHL = BDEFGHL = BDJL
= ACDEJL = CFJL = ABEFJL = ABCGJL = EGJL = ADFGJL = BCDEFGJL = AHJL
= BCEHJL = ABCDFHJL = DEFHJL = CDGHJL = ABDEGHJL = BFGHJL = ACEFGHJL = BCKL
= AEKL = DFKL = ABCDEFKL = ABDGKL = CDEGKL = ACFGKL = BEFGKL = ACDHKL
= BDEHKL = ABFHKL = CEFHKL = GHKL = ABCEGHKL = BCDFGHKL = ADEFGHKL = ADJKL
= BCDEJKL = ABCFJKL = EFJKL = CGJKL = ABEGJKL = BDFGJKL = ACDEFGJKL = BHJKL
= ACEHJKL = CDFHJKL = ABDEFHJKL = ABCDGHJKL = DEGHJKL = AFGHJKL = BCEFGHJKL

Aliases

A = FJ = BK = CL = BCE = DEF = CDG = BFG = BDH = CFH = EGH = GJK = HJL = EKL
B = GJ = AK = EL = ACE = CDF = DEG = AFG = ADH = EFH = CGH = FJK = DJL = CKL
C = HJ = EK = AL = ABE = BDF = ADG = EFG = DEH = AFH = BGH = DJK = FJL = BKL
D = EJ = HK = GL = BCF = AEF = ACG = BEG = ABH = CEH = FGH = CJK = BJL = FKL
E = DJ = CK = BL = ABC = ADF = BDG = CFG = CDH = BFH = AGH = HJK = GJL = AKL
F = AJ = GK = HL = BCD = ADE = ABG = CEG = ACH = BEH = DGH = BJK = CJL = DKL
G = BJ = FK = DL = ACD = BDE = ABF = CEF = BCH = AEH = DFH = AJK = EJL = HKL
H = CJ = DK = FL = ABD = CDE = ACF = BEF = BCG = AEG = DFG = EJK = AJL = GKL
J = DE = AF = BG = CH = CDK = BFK = AGK = EHK = BDL = CFL = EGL = AHL
K = AB = CE = FG = DH = CDJ = BFJ = AGJ = EHJ = BCL = AEL = DFL = GHL

XII.   Alias Relationships for 2^{k-p} Fractional Factorial Designs with $k \leq 11$ and $n \leq 64$ (continued)

AD = EF = CG = BH = BCJ = AEJ = DFJ = GHJ = BDK = CFK = EGK = AHK = CDL = BFL = AGL = EHL = JKL
AE = BC = DF = GH = KL = ADJ = EFJ = CGJ = BHJ = ACK = BEK = DGK = FHK = ABL = CEL = FGL = DHL
AG = CD = BF = EH = JK = ABJ = CEJ = FGJ = DHJ = DEK = AFK = BGK = CHK = ADL = EFL = CGL = BHL
AH = BD = CF = EG = JL = ACJ = BEJ = DGJ = FHJ = ADK = EFK = CGK = BHK = DEL = AFL = BGL = CHL

2 blocks of 8: AE = BC = DF = GH = KL = ADJ = EFJ = CGJ = BHJ
= ACK = BEK = DGK = FHK = ABL = CEL = FGL = DHL

(v) 2^{11-6}; 1/64 replicate of 11 factors in 32 runs   **Resolution IV**

I = ABCF = BCDG = ADFG = CDEH = ABDEFH = BEGH = ACEFGH = ACDJ
= BDFJ = ABGJ = CFGJ = AEHJ = BCEFHJ = ABCDEGHJ = DEFGHJ = ADEK
= BCDEFK = ABCEGK = EFGK = ACHK = BFHK = ABDGHK = CDFGHK = CEJK
= ABEFJK = BDEGJK = ACDEFGJK = DHJK = ABCDFHJK = BCGHJK = AFGHJK = BDEL
= ACDEFL = CEGL = ABEFGL = BCHL = AFHL = DGHL = ABCDFGHL = ABCEJL
= EFJL = ADEGJL = BCDEFGJL = ABDHJL = CDFHJL = ACGHJL = BFGHJL = ABKL
= CFKL = ACDGKL = BDFGKL = ABCDEHKL = DEFHKL = AEGHKL = BCEFGHKL = BCDJKL
= ADFJKL = GJKL = ABCFGJKL = BEHJKL = ACEFHJKL = CDEGHJKL = ABDEFGHJKL

Aliases

A = BCF = DFG = CDJ = BGJ = EHJ = DEK = CHK = FHL = BKL
B = ACF = CDG = EGH = DFJ = AGJ = FHK = DEL = CHL = AKL
C = ABF = BDG = DEH = ADJ = FGJ = AHK = EJK = EGL = BHL = FKL
D = BCG = AFG = CEH = ACJ = BFJ = AEK = HJK = BEL = GHL
E = CDH = BGH = AHJ = ADK = FGK = CJK = BDL = CGL = FJL
F = ABC = ADG = BDJ = CGJ = EGK = BHK = AHL = EJL = CKL
G = BCD = ADF = BEH = ABJ = CFJ = EFK = CEL = DHL = JKL
H = CDE = BEG = AEJ = ACK = BFK = DJK = BCL = AFL = DGL
J = ACD = BDF = ABG = CFG = AEH = CEK = DHK = EFL = GKL
K = ADE = EFG = ACH = BFH = CEJ = DHJ = ABL = CFL = GJL
L = BDE = CEG = BCH = AFH = DGH = EFJ = ABK = CFK = GJK

AB = CF = GJ = KL   AG = DF = BJ   BD = CG = FJ = EL
AC = BF = DJ = HK   AH = EJ = CK = FL   BE = GH = DL
AD = FG = CJ = EK   AJ = CD = BG = EH   BH = EG = CL = FK
AE = HJ = DK   AK = DE = CH = BL   CE = DH = JK = GL
AF = BC = DG = HL   AL = FH = BK   EF = GK = JL

ABD = CDF = ACG = BFG = EFH = BCJ = AFJ = DGJ = BEK = GHK = AEL = HJL = DKL
ABE = CEF = DFH = AGH = EGJ = BHJ = BDK = CGK = FJK = ADL = FGL = CJL = EKL
ABH = DEF = AEG = CFH = BEJ = GHJ = CHJ = BCK = AFK = DGK = ACL = BFL = DJL = HKL
ACE = BEF = ADH = FGH = DEJ = CHJ = CDK = BGK = EHK = AJK = DFL = AGL = BJL
AEF = BCE = DEG = BDH = CGH = FHJ = DFK = AGK = BJK = CDL = BGL = EHL = AJL

2 blocks of 16: AB = CF = GJ = KL   4 blocks of 8: AB = CF = GJ = KL
                                                    AD = FG = CJ = EK
                                                    BD = CG = FJ = EL

XII. Alias Relationships for 2^{k-p} Fractional Factorial Designs with $k \leq 11$ and $n \leq 64$ (continued)

Resolution IV

(w) 2^{11-5}; 1/32 replicate of 11 factors in 64 runs

I = CDEG = ABCDH = ABEGH = ABFJ = ABCDEFGJ = CDFHJ = EFGHJ = BDEFK
= BCFGK = ACEFHK = ADFGHK = ADEJK = ACGJK = BCEHJK = BDGHJK = ADEFL
= ACFGL = BCEFHL = BDFGHL = BDEJL = BCGJL = ACEHJL = ADGHJL = ABKL
= ABCDEGKL = CDHKL = EGHKL = FJKL = CDEFGJKL = ABCDFHJKL = ABEFGHJKL

Aliases

A = BFJ = BKL

B = AFJ = AKL

C = DEG

D = CEG

E = CDG

F = ABJ = JKL

G = CDE

H =

J = ABF = FKL

K = ABL = FJL

L = ABK = FJK

CL = AFG = BGJ = DHK

DF = CHJ = BEK = AEL

DH = ABC = CFJ = CKL

DJ = CFH = AEK = BEL

DK = BEF = AEJ = CHL

DL = AEF = BEJ = CHK

EF = GHJ = BDK = ADL

EH = ABG = FGJ = GKL

EJ = FGH = ADK = BDL

EK = BDF = ADJ = GHL

AB = CDH = EGH = FJ = KL

AC = BDH = GJK = FGL

AD = BCH = EJK = EFL

AE = BGH = DJK = DFL

AF = BJ = DEL = CGL

AG = BEH = CJK = CFL

AH = BCD = BEG

AJ = BF = DEK = CGK

AK = DEJ = CGJ = BL

AL = DEF = CFG = BK

BC = ADH = FGK = GJL

EL = ADF = BDJ = GHK

FG = EHJ = BCK = ACL

FH = CDJ = EGJ

FK = BDE = BCG = JL

FL = ADE = ACG = JK

GH = ABE = EFJ = EKL

GJ = EFH = ACK = BCL

GK = BCF = ACJ = EHL

GL = ACF = BCJ = EHK

HJ = CDF = EFG

BD = ACH = EFK = EJL

BE = AGH = DFK = DJL

BG = AEH = CFK = CJL

BH = ACD = AEG

CD = EG = ABH = FHJ = HKL

CE = DG

CF = DHJ = BGK = AGL

CG = DE = BFK = AJK = AFL = BJL

CH = ABD = DFJ = DKL

CJ = DFH = AGK = BGL

CK = BFG = AGJ = DHL

HK = CDL = EGL

HL = CDK = EGK

ACE = ADG = FHK = HJL

AFH = BHJ = CEK = DGK

AFK = AJL = CEH = DGH = BJK = BFL

AHJ = BFH = CEL = DGL

AHK = CEF = DFG = BHL

AHL = CEJ = DGJ = BHK

BCE = BDG = HJK = FHL

CGH = DEH

2 blocks of 32: *AHJ* = *BFH* = *CEL* = *DGL*

4 blocks of 16: *AHJ* = *BFH* = *CEL* = *DGL*

AHK = *CEF* = *DFG* = *BHL*

FL = *ADE* = *ACG* = *JK*

Index